Novel Combustion Concepts for Sustainable Energy Development

Avinash K. Agarwal · Ashok Pandey
Ashwani K. Gupta · Suresh K. Aggarwal
Abhijit Kushari
Editors

Novel Combustion Concepts for Sustainable Energy Development

Springer

Editors
Avinash K. Agarwal
Department of Mechanical Engineering
Indian Institute of Technology Kanpur
Kanpur, Uttar Pradesh
India

Ashok Pandey
Centre for Biofuels and Biotechnology Division
NIIST
Thiruvananthapuram, Kerala
India

Ashwani K. Gupta
Department of Mechanical Engineering
University of Maryland
College Park, MD
USA

Suresh K. Aggarwal
Flow and Combustion Laboratory, Department of Mechanical Engineering
University of Illinois at Chicago
Chicago, IL
USA

Abhijit Kushari
Department of Aerospace Engineering
Indian Institute of Technology Kanpur
Kanpur, Uttar Pradesh
India

ISBN 978-81-322-2210-1 ISBN 978-81-322-2211-8 (eBook)
DOI 10.1007/978-81-322-2211-8

Library of Congress Control Number: 2014956215

Springer New Delhi Heidelberg New York Dordrecht London

Printed on acid-free paper

Springer (India) Pvt. Ltd. is part of Springer Science+Business Media (www.springer.com)

Preface

Combustion science and technology has made major strides in recent years with remarkable advances in analytical, computational, and diagnostic capabilities. High-speed computing along with improved fundamental understanding, and the development of modeling capabilities and detailed reaction mechanisms, have greatly enhanced our predictive capabilities. Concurrently, diagnostics has become significantly more accurate through the development of sophisticated instrumentation, particularly non-intrusive optical techniques, and is providing well-resolved spatial and temporal measurements of complex reacting flows in various combustion systems to support the modeling efforts. Consequently, CFD-based tools have now become ubiquitous in industry, and are increasingly being relied on for the improved design of existing systems and the validation of new design concepts and ideas. These developments have greatly helped in improving the performance of various combustion systems used in power generation, transportation, industrial, and residential applications. Advances in sensors and active control technology have further helped to improve system efficiencies, reduce emission levels, prolong equipment life, and mitigate catastrophic failures.

While the aforementioned advances in predictive and diagnostic capabilities have provided greatly improved combustion systems, energy and environmental issues are more important now than ever before due to rapidly increasing energy demand, diminishing fuel reserves, emission of toxins to the environment, and global warming. These issues continue to drive the fundamental and applied combustion research to achieve further gains in system performance and develop an integrated approach that incorporates low grade fuels and alternative energy sources for improving efficiency with cleaner environment. Moreover, global efforts must be made to develop common strategies and policies for a sustainable energy future.

An international workshop held at IIT Kanpur, India during January 2–5, 2014 provided a forum for discussions between eminent scientists and engineers from the USA, India, Japan, Thailand, Austria, who shared their views on various aspects of energy at this workshop. This research monograph is based on the topics covered at the workshop, and brings together a wealth of knowledge from world renowned experts on the latest development in selected technology areas with due focus on fundamentals, applications, and advanced education. A common theme of the monograph is the sustainable energy utilization of conventional and renewable fuels. Chapters on conventional fuels focus on their utilization for power generation and transportation, while renewable fuels provide insights into their production and utilization. Research on using blends of fossil and renewable fuels in current power and propulsion systems is also presented. In addition, novel technology developments pertaining to efficient fuel reforming, distributed combustion, and micro-combustion are discussed.

Another major outcome of the workshop was the formation of the International Society for Energy, Environment and Sustainability (ISEES). The main objective of the Society is to organize Workshops/Symposia/Conferences/Lectures/Courses for widening and dissemination of knowledge in areas related to energy, combustion, sustainability, and environment. We hope that the Society will bring many novel contributions in the future that will be of great benefit to the international community. We also hope that continued efforts in the future from the international scientific community at future workshops in the field will contribute to a greener planet through sustainable energy and environment.

We would like to express our gratitude to the authors for submitting their work in a timely manner and revising it at short notice. We acknowledge the support received from various agencies and organizations for the successful conduct of the workshop, which include the National Science Foundation, USA, Office of Naval Research Global, Singapore (with special thanks to Dr. R. Kolar), MHRD TEQIP center at IIT Kanpur, and TSI and TESSCORN, who were industry partners for the event.

We hope that the book would be of great interest to the postgraduate students and researchers involved broadly in the area of energy and environmental sustainability, and more specifically to those interested in the areas of applied energy, alternative fuels, combustion, power systems, and engines.

Kanpur, India — Avinash K. Agarwal
Kerala, India — Ashok Pandey
College Park, USA — Ashwani K. Gupta
Chicago, USA — Suresh K. Aggarwal
Kanpur, India — Abhijit Kushari

Contents

Part III Combustion of Fuels and Engine Performance

Part IV Emissions

Part V [illegible]

About the Editors

Prof. Avinash K. Agarwal is currently Poonam and Prabhu Goyal Endowed Chair Professor of Mechanical Engineering at Indian Institute of Technology Kanpur, India. Prof. Agarwal was at ERC, University of Wisconsin, Madison, as a Post-doctoral Fellow (1999–2001). His areas of interest include engine combustion, alternative fuels, biofuels, natural gas, hydrogen, conventional fuels, optical diagnostics, laser ignition, HCCI, emission and particulate control, and large bore engines. He has published more than 120 peer reviewed international journal papers and 70 peer reviewed international conference papers. Prof. Agarwal is an associate editor of ASME Journal of Energy Resources Technology, and International Journal of Vehicle Systems Modelling and Testing. He is editorial board member of 'IMechE Journal of Automobile Engineering', and 'International Journal of Oil, Gas, and Coal technologies' and has edited "Handbook of Combustion."

Prof. Agarwal is a Fellow of ASME in the class of 2013, and Fellow of SAE International, USA in the class of 2012. He is the recipient of several prestigious awards such as NASI-Reliance Industries Platinum Jubilee Award-2012, INAE Silver Jubilee Young Engineer Award-2012; Ralph R. Teetor Educational Award-2008; INSA Young Scientist Award-2007; UICT Young Scientist Award-2007; INAE Young Engineer Award-2005.

Prof. Ashok Pandey is Deputy Director at CSIR's National Institute for Interdisciplinary Science and Technology at Trivandrum and is heading the Centre for Biofuels and Biotechnology Division there. His research interests are on bio-based economy development (biomass-based biorefinery) for the production of fuels and chemicals. He has over 1,000 publications/communications (which include 14 patents and design copyrights, 37 books, 100 book chapters, 390 original, and review papers), with an *h* index of 64 and more than 17,700 citation (Goggle Scholar).

Professor Pandey is the recipient of many national/international awards and fellowships, which include Fellow of International Society for Energy, Environment and Sustainability; National Academy of Science (India); Biotech Research Society, India; International Organization of Biotechnology and Bioengineering and

Association of Microbiologists of India; Honorary Doctorate degree from Université Blaise Pascal, France; Thomson Scientific India Citation Laureate Award, USA; Visiting Professor in the University Blaise Pascal, France, Federal University of Parana, Brazil and EPFL, Switzerland, etc. He is Founder President of BRSI (www.brsi.in); General Secretary, IFIBiop (www.ifibiop.org) and Vice-President of ISEES (www.isees.org) and AIBA (www.aibaonline.com). Prof. Pandey is Editor-in-chief of *Bioresource Technology,* Honorary Executive Advisor of *Journal of Water Sustainability* and *Journal of Energy and Environmental Sustainability.*

Ashwani K. Gupta is Distinguished University Professor at the University of Maryland. He received his Ph.D. from Sheffield University, UK and higher doctorate D.Sc. from the University of Sheffield and also from the University of Southampton, UK. He was awarded an honorary doctorate from the University of Wisconsin Milwaukee USA and also from King Mungkut University of Technology North Bangkok, Thailand, which was bestowed by HRH the Queen of Thailand. He served as international consultant to the Japanese government METI for the NEDO project on novel high-temperature air-combustion technology (called HiTAC). He served as Director of Propulsion and Energy at AIAA. He is a Fellow of AIAA, ASME, and SAE. He is co-editor of an environmental energy book series and associate editor of four journals.

He has published over 650 papers, 3 books, 12 book chapters, and edited 11 books. He received the University of Maryland President Kirwan research award, College of Engineering research award, and several awards from ASME and AIAA for his research in combustion, propulsion, and air pollution. His research interests include high intensity distributed combustion, sulfur chemistry and recovery from acid gases, gasification, fuel reforming, biofuels, multi-micro sensors and combustion control, supersonic mixing and combustion, membranes for gas separation, and laser diagnostics.

Dr. Suresh K. Aggarwal received his Ph.D. in Aerospace Engineering from Georgia Institute of Technology in 1979. Since then, he has been at Princeton University as a member of the Professional Research Staff, and at Carnegie-Mellon University as a Senior Research Engineer, and since 1984 with The University of Illinois at Chicago, where he joined as Assistant Professor, and was promoted to the rank of Professor of Mechanical Engineering in 1995. His other appointments include being the Director of Graduate Studies, Visiting Scientist at Argonne National Laboratory, USA, Visiting Professor at Ecole Centrale-Paris, France, and Guest Professor at Jiangsu University, China.

Dr. Aggarwal's research and teaching interests include Combustion, Multiphase Reacting Flows, Renewable Fuels, Emissions, Clean Energy, Fire Suppression, and Microgravity Phenomena. He has authored over 300 journal and conference publications. His research has been funded by a number of federal agencies and companies, and has resulted in the graduation of 14 Ph.D. and 40 M.S. students. He is an ASME Fellow, AAAS Fellow, and AIAA Associate Fellow. He has served or currently serving as an Associate Editor of the AIAA Journal, the Founding Editor of the International Journal of Reacting Systems (now J of Combustion), and on the Editorial Boards of Journal of Green Energy, and Book Series on Sustainable Energy Developments.

Dr. Abhijit Kushari is a Professor in the Department of Aerospace Engineering at the Indian Institute of Technology Kanpur, India. Dr. Kushari earned his B.Tech. (Honours) in Aerospace Engineering from the Indian Institute of Technology Kharagpur in 1994 followed by an M.S. in Aerospace Engineering in 1996, and a Ph.D. in Mechanical Engineering in 2000 from Georgia Institute of Technology, Atlanta, Georgia. He joined IIT Kanpur as a visiting faculty in July 2001 and then as an Assistant Professor in the Department of Aerospace Engineering at IIT Kanpur in September 2001. Since then he has been a faculty at IIT Kanpur and was promoted to Professor in June 2014.

Dr. Kushari's research interests lie primarily in the area of Aerospace Propulsion. He is currently involved in various research projects in Turbo-machinery, Combustion, Liquid Atomization, Intake, and Nozzle flows with emphasis on the application of various flow control techniques in these processes and systems. He has authored two book chapters, 52 papers in archived journals, and has presented 54 papers in various national and international conferences on these topics. He is a member of various national and international societies in Combustion, Atomization, Aerospace Engineering, etc., and is on the editorial board of two international journals.

Part I
General

Introduction

Avinash K. Agarwal, Suresh K. Aggarwal, Ashwani K. Gupta, Abhijit Kushari and Ashok Pandey

Abstract The fuels and energy consumption worldwide has continued to increase due to increased standards of living and increased population. With the continuous depletion in hydrocarbon fuel reserves from growing demands on energy, it is important to develop new and novel technologies to reduce energy cost, improve efficiency, and reduce pollution from the existing energy utilization systems. The energy use, power generation, and propulsion pose direct impact on the environment with subsequent effects on local and global climate. Therefore, one must focus on efficient and sustainable "green" energy production from conventional and alternative energy sources and develop technologies for a range of applications from micro to mega scale. Although the air toxins from power plants have continued to decrease over the past few decades, CO_2 concentration in the atmosphere continues to grow, and CO_2 emissions remain a big challenge due to ever increasing demands on chemical energy.

Keywords Greenhouse gas emission (GHG) · Power generating plant efficiency · Computational fluid dynamics · Biofuels · Internal combustion engine combustion

The fuels and energy consumption worldwide has continued to increase due to increased standards of living and increased population. With the continuous

A.K. Agarwal (✉)
Department of Mechanical Engineering, Indian Institute of Technology Kanpur, Kanpur, Uttar Pradesh, India
e-mail: akag@iitk.ac.in

S.K. Aggarwal
Flow and Combustion Laboratory, Department of Mechanical Engineering, University of Illinois at Chicago, Chicago, IL 60607, USA

A.K. Gupta
Department of Mechanical Engineering, University of Maryland, College Park, MD, USA

A. Kushari
Department of Aerospace Engineering, Indian Institute of Technology Kanpur, Kanpur Uttar Pradesh, India

A. Pandey
Centre for Biofuels, Biotechnology Division, CSIR-National Institute for Interdisciplinary Science and Technology, Trivandrum, India

A.K. Agarwal et al. (eds.), *Novel Combustion Concepts for Sustainable Energy Development*, DOI 10.1007/978-81-322-2211-8_1

depletion in hydrocarbon fuel reserves from growing demands on energy, it is important to develop new and novel technologies to reduce energy cost, improve efficiency, and reduce pollution from the existing energy utilization systems. The energy use, power generation, and propulsion pose direct impact on the environment with subsequent effects on local and global climate. Therefore, one must focus on efficient and sustainable "green" energy production from conventional and alternative energy sources and develop technologies for a range of applications from micro to mega scale. Although the air toxins from power plants have continued to decrease over the past few decades, CO_2 concentration in the atmosphere continues to grow, and CO_2 emissions remain a big challenge due to ever increasing demands on chemical energy.

Combustion science has made remarkable strides over the last few decades on improved efficiency of all power, energy conversion, and propulsion systems, including power generation, gas turbine engines, automobile engines, rocket engines, industrial furnaces, and many other chemical to thermal to mechanical and electrical energy conversion systems. However, with rapid increase in energy demand and increased population, along with limited supply of fossil fuels and increased concern on global warming from global warming gases emission, significant efforts must be made to further improve the system efficiency and performance, reduce pollutants emission, and increase the life of combustion equipment with improved quality of product. The large reserves of coal in many countries make it attractive but coal as fuel emits twice the amount of CO_2 than natural gas on per unit energy basis (coal emits approximately 227 pounds of CO_2 while natural gas emits 117 pounds for equal BTU of energy). Recent announcement by the US Environmental Protection Agency on its proposed set of regulations to reduce annual greenhouse gas emissions from existing power plants by 30 % by 2030 makes it harder for coal to be used. We must enhance plant efficiency and performance with low carbon footprint fuels. The greenhouse gas emissions (mostly CO_2) in the USA alone were 6,525 million metric tons by the entire economy. Thermal generating stations contributed to more than 30 % of the CO_2. The global CO_2 concentration in 2014 has now exceeded over 400 ppm from about 315 ppm in 1960 and the current level continues to increase. We continue to blast at a rate that now exceeds over 2 ppm per year. Projected CO_2 concentration is expected to increase to over 850 ppm by some estimates that will lead to 4–8 °C global temperature rise and about 1 m sea level rise by year 2100. Today's contribution to the greenhouse effect on earth is from water vapor (36–70 %), carbon dioxide (9–26 %), methane (4–9 %), ozone (3–7 %), and other trace contributors. Major contributor is therefore CO_2 as water vapor is naturally present in the air with no real control technology for atmospheric water vapors.

Several developing economies face ever growing energy need in all sectors of power generation, transport, domestic, and industrial. The fuel used often depends on the local fuel availability. One key common issue is the growing environmental pollution for most of the Western countries as well the growing economies in India, China, and other countries. Western countries have made great strides to curtail most of the harmful emissions so that the air is much cleaner today than that in the immediate past. India and China must develop clean and efficient power and energy conversion devices to significantly reduce their emissions. Furthermore with the diminishing fossil fuel reserves worldwide, we must develop alternative energy sources to power the currently available engines at higher efficiency and increased performance. The USA consumes about 23 % of the annual world energy consumption to support only 5 % of the global population while in China and India, 20 % of the world energy consumption supports about 35 % of the global population. Industrial economy is much more dependent on the availability of clean fossil fuel sources worldwide so that enhanced efficiency of chemical to thermal energy systems with lower pollution potential will help mitigate high amounts of fossil fuel consumption. Therefore, developing economies of the world must continue research and development efforts to combat the common issues they face.

The identification of novel alternative energy sources as well as deployment of novel technologies for cleaner energy sources requires further research and development efforts. Sustained research efforts in the field of chemical to thermal energy conversion systems have motivated many researchers to seek novel combustion technologies for a wide range of applications that extend from power generation to industrial heat treatment processes and from automobiles to aerospace propulsion technologies. These applications encompass wide range of combustion intensities and time scales with different thermal and chemical behaviors.

Fundamental and applied combustion research has made significant advances in mathematical modeling, sensors, diagnostics, data acquisition, and control. The availability of high-speed computers has greatly enhanced the predictive capabilities to understand the complex fluid dynamics and chemistry associated with reacting flows in various flame and engine configurations. Fundamental research on combustion has made significant advances in both modeling and diagnostics. Validated computational fluid dynamics (CFD) models are rapidly emerging and playing a major role for the improved design of combustion systems through validated new design concepts. Sensors and diagnostics to monitor and control a plant and combustion unit are critical now than ever before. Their use in modern plants has grown in recent times and it is expected that future smart combustion systems and plants will use even more sensors and controls to closely monitor the local and global plant performance. Use of such techniques has provided improve system efficiency, significant pollution reduction, prolonged operation of combustion equipment, and enhanced safety. The development efforts from fundamental and applied research have resulted in the practical development and deployment with significant advancements in various sectors of chemical to thermal energy conversion technologies using all kinds of fossil and renewable fuels. International partnerships have grown and this volume brings together many international researchers and engineers, who provide their views on selected technologies and the path forward to meet the growing demands on energy and the associated issues on pollution curtailment.

This research monograph on Novel Combustion Concepts for Sustainable Energy Development provides both fundamental and applied innovations on several key emerging technologies on fossil fuel and alternative fuel utilization in various sectors of power, energy propulsion as well as biofuel production, and utilization in fossil fuel blends. High temperature air combustion technology, now used worldwide for industrial furnaces, is further developed and presented for stationary power applications. This green combustion turbine has shown several simultaneous benefits on energy savings, near zero emissions, uniform thermal field in the entire combustion chamber (improved pattern factor), low noise and longer combustor and turbine life to contribute to energy and environment sustainability.

The specific topics covered in this monograph include:

- Introduction to the monograph
- Biofuels from biomass
- Fluidized bed combustion
- Technical challenges for a sustainable energy efficient future
- High intensity distributed combustion/Green combustion turbine
- Emissions from internal combustion engines
- Chemical looping combustion for coals
- Future trends in commercial aviation
- Efficiency improvements of solid rocket motors
- Green electricity generation for refuse-derived fuel and land reclamation
- Enhancement of biofuel quality
- Combustion and emission behavior of fossil and biofuel blends
- Combustion of fossil and biofuel mixture in engines
- Biodiesel utilization effects on tribological properties of lubricating oils in engines
- Hydrogen rich syngas production from biomass steam gasification
- Spectroscopy methods and visualization methods for combustion diagnostics
- Droplet combustion for developing surrogates liquid transportation fuels
- Soot emission from partially premixed combustion
- Coherent structures and combustion instability
- Linear and nonlinear stability analysis of liquid sheet breakup and drop formation
- Lean blow-out detection in partially premixed dump combustor
- Polymer-derived ceramics for sustainable energy technologies

The above topics have been classified into five main groups of (i) General, (ii) Biofuels: production, properties, and application, (iii) Combustion of fuels and engine performance, (iv) Emissions, and (v) Sustainable energy systems and efficiency improvements. The spectroscopy and in-situ visualization topic is of common interest that spans from fundamental to applied technology for fossil and biofuels and their blends. The theoretical aspects presented for droplet formation and lean blow-out detection help to improve fuel distribution in the combustion system for improved combustion control. It is expected that knowledge gained from the selected efforts provided here will lead to improved and efficient use of fossil and biofuels for reduced pollution and enhance efficiency of the combustion system, therefore enhance the system performance. The state-of-the-art information provided here is aimed to serve as a valuable tool for all involved in every aspect of clean energy conversion while mitigating pollutants emitted to the environment.

The volume is expected to provide unique value to researchers from academia and industry, design and development engineers and to decision makers in the procurement of major plants, and air pollution control equipment as well as efficient and less polluting engines for near and long-term applications. The team of international experts invited herein shares their knowledge in their respective field of expertise. The list of extensive references on each topic will help provide additional reading materials. In many cases, more than one view is presented on the same topic to get a broader perspective from different experts on the subject area. This aspect will be of special interest to students, researchers, technology users, and technology developers in the respective fields.

Technical Challenges and Scientific Approach for a Sustainable Energy Efficient Future

Gabriel D. Roy

Abstract The energy landscape is rapidly changing and the availability of fossil fuels in abundance at affordable prices in the future cannot be assured. It is time now to focus attention to produce sustainable alternate fuels, further improve the energy conversion process, and develop new and more efficient energy conversion devices. Environmental constraints and geo physical changes restrict the selection and utilization of fuels for the future. Continuous availability of feed stock for alternate fuels at affordable prices and to convert those to fuels with required properties, as well as to control the combustion process, poses challenges. Alternate fuels for propulsion, in particular for aero-propulsion are required to meet varying and often contradicting performance parameters. These issues are elaborated and technical solutions are attempted in this chapter. Economics will play a major role in the successful use of alternate fuels. Future energy conversion devices (engines and turbines) will operate on higher pressures and temperatures and will offer challenges in the production of new materials and coatings.

Keywords High-energy-density fuels · Fuel injection strategy · Soot formation control · Alternative fuels · Noise foot print · Nano-size toxic particles

1 Introduction

The energy landscape is rapidly changing, in particular during the past few decades, and the energy demand from the developing and under developed countries is on the rise. Availability of fossil fuels in abundance at affordable prices in the future cannot be taken for granted. Along with the escalating depletion of natural resources, politics and international situations also play a role in causing uncertainties in the supply of fuel. Environmental regulations, eco awareness, and geophysical changes restrict the selection and utilization of fuels to meet full

G.D. Roy (✉)
CPnE Consultants, 9944 Great Oaks Way, Fairfax, VA 22030, USA
e-mail: roygd@aol.com

A.K. Agarwal et al. (eds.), *Novel Combustion Concepts for Sustainable Energy Development*, DOI 10.1007/978-81-322-2211-8_2

compliance. The challenge on the one hand is to make sure that the demand for the right type of fuels is met and on the other hand that the energy conversion process meets new standards. The technical challenges comprise of the production of the energy source (fuels), the energy conversion device (engines and turbines), and the energy conversion process (combustion/detonation) and vary with the energy application.

1.1 Energy Conversion Demands

Various criteria are to be met depending on the application such as stationary or mobile, on site, or remote site energy conversion.

1.1.1 Stationary Power

For stationary power conversion efficiency, maintenance, manpower, and environmental compliance are key factors. The goal is to have the lowest cost per kW of electricity. The energy conversion process should use efficient thermodynamic cycles, efficient components, and reliable control.

1.1.2 Road Transport

Cost of fuel, conversion efficiency, emission standards are key factors; and the aim is to get the best miles/gallon or km/liter from the fuels used. Emission regulations vary from country to country (and even state to state) requiring operational flexibility in order to be cost-effective.

1.1.3 Civil Aviation

Cost per passenger mile or km is the criterion. Lower cost of fuel, efficient engines, lower maintenance cost of aircrafts and infrastructure are the key factors including salaries of personnel (ground staff and flight crew), environmental compliance, and airport taxes.

1.1.4 Military Aviation

Thrust per unit of fuel is the major criterion. Increased energy content and density of the fuel and flexibility, wide operational envelope are key factors. Cost is an important issue now. The contradictory requirements pose additional technical challenges.

2 Scientific Approach

In order to meet the demands in energy conversion (of the afore-mentioned sectors), marginal evolutionary improvements will not suffice, rather revolutionary advances are in order. Development and production of alternate and high-energy fuels form an integral element, along with optimal conversion processes, for a sustainable energy future.

A carefully planned and well-executed international research and development program with participation from academia, government, and industry laboratories is essential with consumer interaction and participation. A "Fuels to Emissions" approach is needed in order to tackle the various aspects of the affordable, sustainable, eco-friendly energy scenario (Roy 1998). This approach comprises of the element shown in Fig. 1. In a nut shell, new fuels for specific applications need to be developed along with new thermodynamic cycles that will provide maximum conversion of the stored chemical energy of the fuels into thermal and/or mechanical energy. Basic studies targeted toward the reduced chemistry and reaction pathways of the fuels need to be done analytically, numerically and experimentally in order to burn the fuel most effectively in combustion or detonation modes. As the combustion process occurs at faster rate, pertinent diagnostics and measurement tools need to be developed. Current engines are designed for maximum efficiency, and minimum specific fuel consumption and emissions. However, even small improvements in these make a substantial impact in overall economy and ecology of the energy sector.

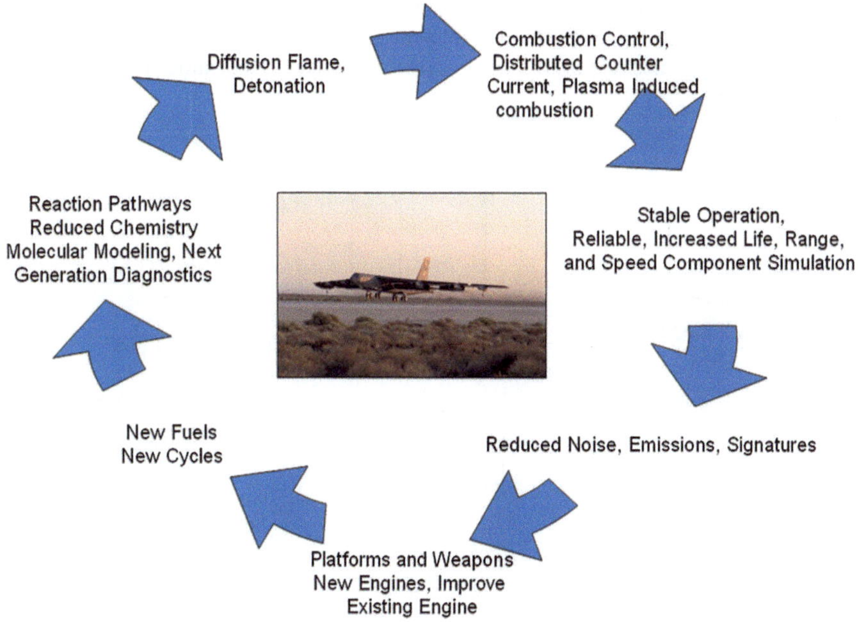

Fig. 1 Fuels to emissions approach to develop engines for stationary and propulsion applications

3 Fuels

Conventional fossil fuels are getting depleted rapidly due to the ever increasing demand. Political and economic instability causes difficulty not only in the production and acquisition of fuels, but also in transporting from place of origin to the demand sites. Nuclear energy is still controversial, and wind, geothermal, and other energy solutions are not going to be a full substitute. Under the circumstances, either new and more energetic hydrocarbon fuels or alternate fuels (such as biofuels) need to be developed. While the former provides an alternative to develop smaller engines with smaller fuel tanks to provide the same range as with a larger ones (for military applications), the latter provides a sustainable alternative for civilian and commercial applications.

3.1 High-Energy-Density (HED) Fuels

HED fuels are formed by manipulating the long-chain hydrocarbon molecules to form triangles, squares, and polygons and then further into tetrahedrons and cubes, etc. (Fig. 2). Fuels, such as, benzvalene, cubane, dihydrobenzvalene, and dimethyl cubane fall in this category.

Here, the increased density due to compacting the molecule, the strain energy added during the synthesis, and the higher carbon-to-hydrogen ratios account for the increased volumetric energy release rates.

These fuels will be of particular interest to military applications as these can increase the range and speed of a given system or provide the same speed and range

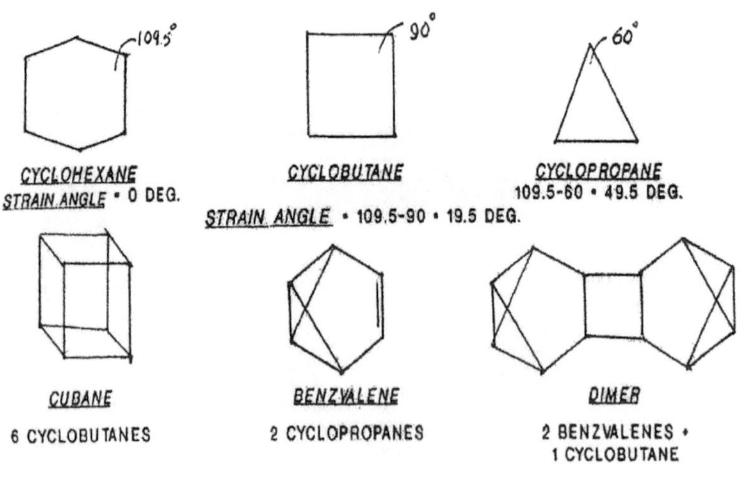

Fig. 2 High-energy-density (HED) fuels

with a smaller system. The high carbon-to-hydrogen ratio of these fuels makes higher soot emission in issue. In particular, the increased soot can increase the possibility of the source (platform) of the weapon being detected more easily. Research studies have shown that sequential timed fuel injection with respect to the incipient vortex can reduce the soot by orders of magnitude (Fig. 3). Also reactant stream velocity studies indicate that the soot production can be controlled by the relative velocity of fuel and oxidant streams (Fig. 4).

When the density of the fuels is increased, the gravimetric energy density usually decreases. But with this family of strained hydrocarbon fuels, energy density increases providing a double advantage (fuel density + energy density), thereby making it very advantageous for volume-limited propulsion systems.

However, there are several technical challenges. Fuels have been synthesized in small quantities only, and the number of steps involved in the synthesis should be reduced. Pilot plant to full-size production facility requires substantial investment. Development and facility buildup and maintenance could cost more than conventional plants. However, the advantages could justify the added cost for military applications.

Fig. 3 Effect of sequential fuel injection

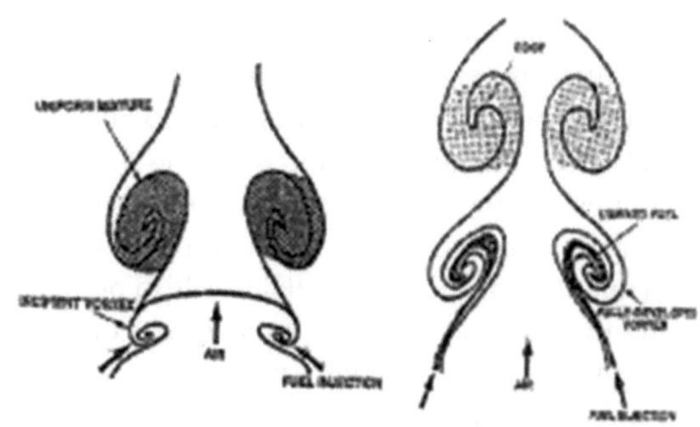

Fig. 4 Soot control by changing onsoot formation reactant stream velocity

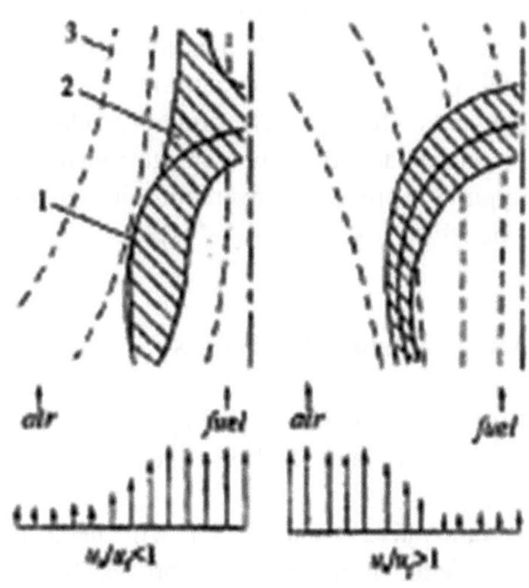

Combustion without soot emission is still a challenge and should be addressed in large continuous combustion test facilities. Though the physical and chemical properties are similar to conventional hydrocarbon fuels, engine modifications might be necessary. Further, properties of HED fuels can vary from batch to batch.

Pioneering research in the synthesis and combustion of this type of fuels has been done over two decades ago in the United States and currently pursued in India. However, due to the limited demand of such fuels, major oil manufactures have not played an active role in the development and production HED fuels.

3.2 Biofuels

The most plausible candidate now for alternate fuels is biofuels with the potential of sustainability and has drawn international attention in the past couple of decades. Government mandates and financial support, consumer advocates and green energy supporters together with the keen interest from academia, government, and industry—in particular small business—gave a boost to biofuel development. In the United States, organizations such as Commercial Alternate Aviation Fuels Initiative (CAAFI) with participation of government, industry, and academia leaders aid in promoting biojet fuel research and development. Biodiesel fuel development also gets equal attention, and gas stations selling biodiesel fuel are common sight in Asian countries. While biofuel development based on current technologies is an evolutionary approach, there are several limitations to those.

Transesterification of vegetable oils to produce biodiesel has the disadvantage of using food sources and only limited parts of the plant source. Further, it has higher viscosity and freeze point, poor cold flow properties, and limited shelf life. Conventional hydro-processing is low in efficiency, higher in hydrogen consumption and operating costs and results in undesirable byproducts. The entire plant material is not fully utilized. In conventional biochemical conversion, and ethanol produced from starch and grains divert valuable food sources. Cellulosic biomass to fuel is still under development with challenges in hemicellulose and lignin conversion to sugar. The process produces only gasoline additive and not a biojet or biodiesel fuel. Biomass/coal gasification produces pollutants (NO_x and SO_x) and requires expensive upstream air separation. Fischer–Tropsch process, though extensively used, involves expensive liquifaction.

3.2.1 Feedstock Challenges

Biofuel production is getting serious attention from the fundamental aspects to production plants. Some of the biofuel feedstocks with their oil content are shown in Fig. 5.

Innovative technologies will need novel catalysts and reactors. Sufficient knowledge of the chemistry, process flexibility, and pretreatment requirements must

Fig. 5 Typical biofuel feedstocks and oil content

be acquired before making large investments, and the long-term availability of the feedstock also should be determined. Rather than utilizing edible feedstock, attention should be paid in using various types of algae, marine sources, cellulosic and waste materials, and grass and weeds as feedstock options. Even with the same type of feedstock (due to the nonhomogenity), the fuel characteristics can vary from day to day and even from batch to batch. This requires not only the plant to be flexible, but also the engine or energy conversion device that uses the fuel.

As the production rate increases due to increasing demand of the fuel, fuel stock required should keep up the pace. "Irrespective of crop one acre of land, pond, or bioreactor can annually yield about enough biomass to field one motor vehicle (Walker 2009). If Virgin Atlantic is to power its entire fleet, half of UK's arable land will be required to get sufficient feedstock." This limits the production capability of several nations with limited free land available for this extra production of feedstock.

Oil content from plants and seeds could be season-dependent, which may require extra expenditure in growing the feedstock. In India, though substantial investment was made in growing Jatropha including the establishment of the Center for

Jatropha Promotion (CJP) for biodiesel production, the endeavor did not succeed as planned due the lower oil content and increased water demand. The increased cost made the private companies to drop out from further investment. Though edible feedstocks such as coconut, soybean, sugarcane have good oil content, these can no longer be used because they deprive of food stock and drive up food costs. Plants with good oil content used in cosmetic and pharmaceutical industry are not logical candidate for energy production.

3.2.2 Production Challenges

As steady world economic growth cannot be taken for granted, cost of production of biofuels will play a major role—in other words, these fuels need to be produced at competitive price with conventional fuels now available. Feedstock may be cultivated at remote places, and transportation costs to get the stock to production sites could be excessive. Transportation of cellulosic and biowaste materials needs extra care. Bacteria and other organism-based fuel production require extra protection from all possible harmful consequences. When cultivation of feedstock requires a certain seasonal cycle, storage could be a challenge.

3.2.3 Performance Challenges

Several engine and flight tests have demonstrated acceptable aircraft performance with biofuels. However, consistent energy density and specific thrust are key factors for military applications in particular. (If these fuels are used in air crafts that take off from flight decks, and in those demanding strict thrust performance.) Other challenges include the shelf life, long-term stability of the fuels, possible coating formation on components (by combustion products), and CO_2 emissions.

3.2.4 Possible Benefit from Biofuels

Noise produced from aircraft jet engine exhaust had been an issue for the past several decades, and engine and airplane manufacturers have invested substantially to reduce it. Cabins in modern jet aircrafts are quieter and together with noise canceling headphones provide a comfortable space. However, the noise produced from military fighter jets is extremely high, and a person exposed to these noise levels over a short period of time can become partially or fully deaf. Figure 6 shows a comparison of noise foot print from commercial and military aircrafts. As can be seen, the noise footprint of a small military aircraft is much larger than the jumbo Boeing 747 airplane.

The exhaust from the combustion of present jet fuels contains nano-size toxic particles that can enter the cochlea through the ear canal (Fig. 7) or by conduction through the skin increasing the risk of ear injury and hearing loss—as opposed to

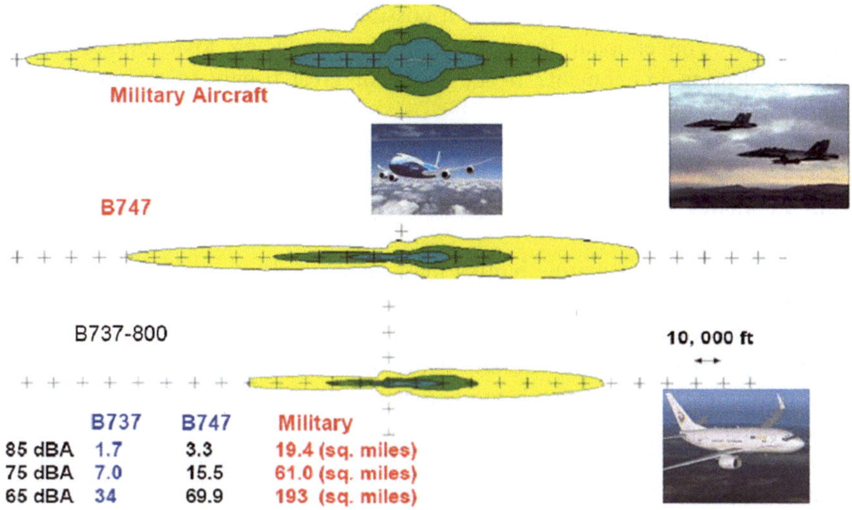

Fig. 6 Noise foot print comparison of commercial and military aircrafts (*Courtesy* Vishwanathan, Boeing Company)

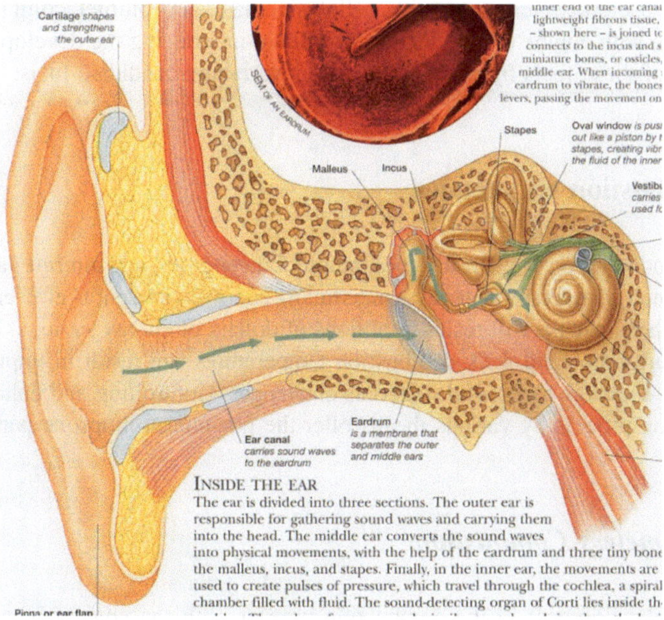

Fig. 7 Physiology of human ear

clean noise. The challenge here is, is it possible to develop less toxic biojet fuels compared to conventional fossil fuels, thereby reducing toxic emissions and reducing hearing impairment? If so, it will of great benefit to commercial and military aviation.

3.2.5 Economic Challenges

Ultimately, the land or marine biofuel must be available at competitive price (the drop in alternate fuels as envisioned by CAAFI) resulting in reduced cost/passenger mile (km) for commercial aviation, miles (km)/gallon (l) for transport, and specific thrust/gallon (l) for military aircrafts. Development of these fuels requires large investments, and the challenge is to get together governments and industries (both manufacturer and user), international collaboration, and exchange of scientific information over a long period of time. The Republic Acts, such as the Biofuel Act of 2006 and the Renewable Energy Act of 2008 executed in the Philippines, and consortia such as CAAFI, are vital for developing, coordinating, and sharing findings and promoting progress. Keeping up the momentum toward commercial production of a less toxic and sustainable biofuel is a challenge not only to the investors, but also to those who execute such programs. Continued basic research in genetically altering the algae growth and plant oil content, finding new tailor-made catalysts, bacteria interaction with cellulosic materials, biofuel combustion gas interaction with engine components, etc. will aid in reducing the development time and cost of these fuels by eliminating the unwarranted candidate fuels.

4 Combustion Challenges

Though modern power plant turbines, automobile engines, and aircraft gas turbines are designed to operate at today's maximum energy conversion efficiencies, any small further increase will result in billions of dollars of energy savings. Increasing the pressure ratios and peak combustion temperatures can result in improved performance, there are associated materials challenges. Controlling and enhancing the combustion process by various means offer the potential to improve performance.

4.1 Flameless Combustion

A recent development in the combustion process with the goal of increasing the energy release as well as reducing the heat losses is colorless distributed combustion or flameless combustion (Gupta 2003). This has been shown to produce a flat hat temperature profile, reduce NO_X formation, and improve combustion stability and

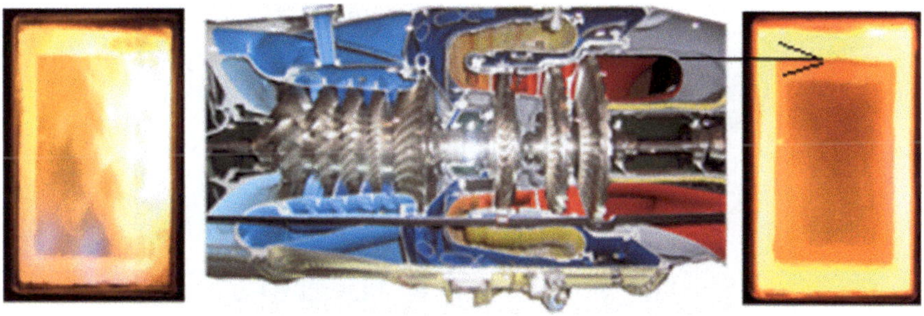

Fig. 8 Colorless distributed combustion, CDC (green combustion turbine)

efficiency. The physical color change from yellow flame to colorless flame (Fig. 8) and the reduction in exhaust temperatures have been observed at low pressures.

The challenge is to apply this to higher pressures involved in aircrafts. Though space is not a limitation in stationary applications (such as in electrical power plants), it is a challenge in aircraft engines as the add-on components will increase the weight and volume. Clever designs with minimal weight and volume penalty should be sought. Computational studies and system optimization consistent with new fuel characteristics are also challenges.

4.2 Porous Inserts

Studies have been made to evaluate performance of combustors with high-temperature foam material introduced at different locations within the combustion chamber (Ref. Agrawal). The idea is to stabilize the high-temperature combustion flow and to eliminate hot spots. High-temperature HfC/SiC-coated foam inserts at appropriate positions (as a ring at the inner periphery of the chamber) has been found to reduce combustion noise, mitigate combustion instability, and reduce CO emission (Fig. 9).

4.3 Supersonic Combustion

Supersonic combustion will play a major role in the combustion scenario in the near future due to the advent of hypersonic propulsion. For scramjet combustion, mixing of fuel and oxidant has been an issue and a number of researchers addressed this. Radial fuel injection, steps, and countercurrent combustion seem to improve the mixing and combustion. Recently, fin guided fuel injection (Ref. Yu) has shown to produce a twofold increase in fuel penetration and 45 % reduction in jet-induced shock strength (scramjets—Fig. 10).

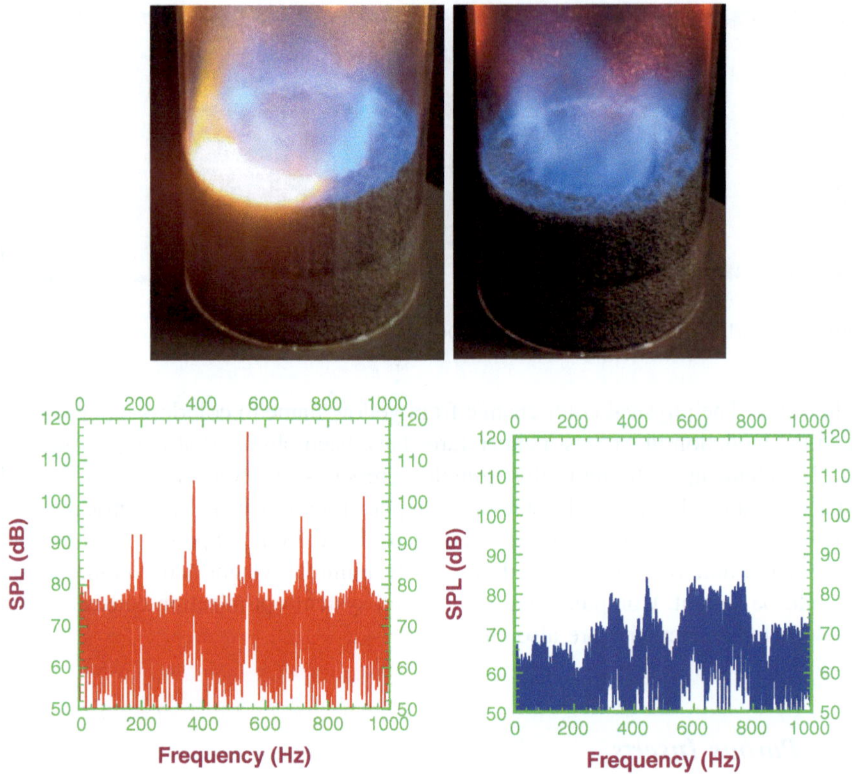

Fig. 9 Noise-level distribution in combustion chamber with foam inserts

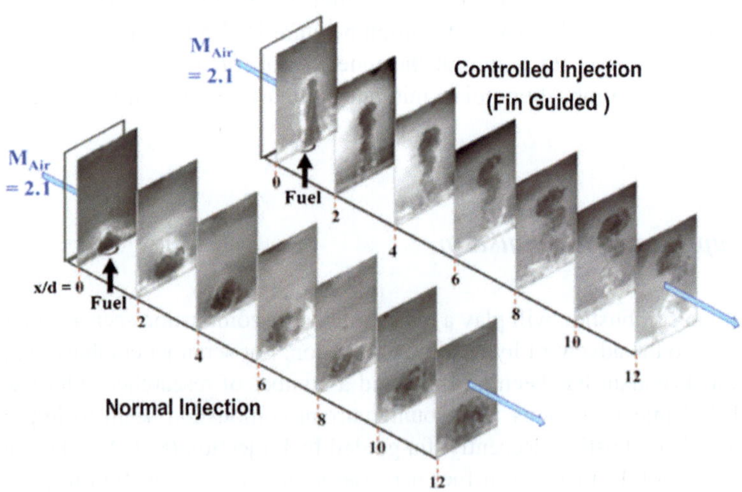

Fig. 10 Fin-guided combustion

5 Material Challenges

With the demise of the Concord aircraft, supersonic air transport came to a halt for commercial aviation. However, research and development of new engines for supersonic aircraft for military applications continues, and eco-friendly engines (low emission and noise) that operate at higher temperatures and speed pose additional challenges. Though commercial hypersonic aviation may not happen in the next few years, the challenges need to be solved as early as possible. As the speed, altitude, and operational temperatures increase, new materials need to be developed both for airframe and engine components. Hypersonic research has attracted worldwide interest, and a number of bilateral and multilateral collaborations have been signed off by several countries, and joint laboratories and development centers are already in operation. The major issues are wall cooling, environmental effects, high heat release rates, high speed flows and wall interactions, fatigue, crack propagation, abrasion, and drag. In order to solve these, research should focus on new light weight and stronger materials with fatigue and abrasion resistance. Further, self-healing coatings that heal themselves when a crack or skin damage occurs, is of interest. Other coatings include flexible coatings that will not crack when the base material is subjected to thermal and mechanical fatigue.

Research on materials and coatings for hypersonic applications will naturally benefit in manufacturing other products for longer life and eco-friendly operation.

6 Conclusions

As the energy landscape is changing, it is not prudent to depend on fossil fuels alone for the future. It is equally important to develop alternate sustainable fuels from land and marine sources, as finding new gas and oil resources. The selection of alternate fuels depends upon the natural resources and geography of the particular country. For example for countries such as Malaysia and Philippines with the extensive shore lines, algae based fuel is a natural choice. Nations with large available land areas such as the United States, India, and China, plant-based fuels present a reliable opportunity. The challenges in the production and utilization of alternate fuels, such as biofuels, must be thoroughly addressed before production to meet public and industry demand. Since economy plays a major role in the development of alternate fuels, attention should be paid to focus only on promising candidate fuels. Tailored fuels for specific applications are still in need, and clever means of converting the chemical energy contained in the fuels with maximum conversion efficiencies and minimum chemical and noise pollution need to be discovered. Functional materials, tailored coatings, and surface treatments to satisfy individual application requirements should be given utmost priority by the materials research community, as the future energy conversion systems will be operating in more demanding environments.

Interdisciplinary research, collaboration of government, producer, and consumer, and international cooperation are vital to meet these challenges and to look forward for a sustainable energy efficient future. The success of the effort will depend upon the passion and desire of those who invest and manage, and the dedication of those who make things happen.

Acknowledgments The author wishes to acknowledge the excellent research his Principal Investigators did, while he was managing the Propulsion Research Program at the US Office of Naval Research. Their work is used in this chapter.

References

Gupta AK (2003) High temperature air combustion: from energy conservation to pollution reduction. CRC Press, New York

Roy GD (1998) Propulsion combustion: fuels to emissions. Taylor and Francis, London

Walker DA (2009) Biofuels facts fantasy, and feasibility. J Appl Phycol 21:509

Part II
Biofuels: Production, Properties and Applications

Biofuels from Biomass

Anil Kumar Patel, Reeta Rani Singhania and Ashok Pandey

Abstract Biofuels are the only hope to decrease the dependency on fossil fuel so as to have carbon footprints for a healthier and more eco-friendly future. Technologies for various alternative fuels are sustainable needs to mature fast so as to replace petroleum-based fuels. Bioethanol production from lignocellulosic biomass would be unavoidable if liquid fossil fuels are to be replaced by renewable and sustainable alternatives, as it is ubiquitous in nature accounting for 50 % of all the biomass available in the world. Conversion of lignocellulosic biomass to fuel ethanol involves various process stages, and each stage contributes to the overall economy of the process. The utilization of lignocellulosic biomass for ethanol production for transportation sector necessitates the large-scale production technology to be cost-effective and environmentally sustainable. Biological conversion of lignocellulosics to fermentable sugars, employing hydrolyzing enzymes, is at present the most attractive alternative due to environmental concerns, though there are several hindrances on the way of developing economically feasible technology due to the presence of lignin. The present chapter discusses the various alternative fuels to petroleum-based fuels, Indian biofuel policy, status of biofuels specially bioethanol globally as well as in India and suitability of lignocellulosic biomass for bioethanol production and various technical steps which add major part of the cost to the overall process.

Keywords Greenhouse gas emission (GHG) · Bioalcohols · Biodiesel · Algal biofuels · Biohydrogen · Fermentation · Biofuels · Bioethanol · Lignocellulose · Biological conversion · Hydrolyzing enzymes

---

A.K. Patel (✉) · R.R. Singhania
DBT-IOC Advanced Bio-Energy Research Centre, Indian Oil Corporation, R&D Centre, Sector-13, Faridabad 121007, India
e-mail: anilkpatel22@gmail.com

A. Pandey
Centre for Biofuels and Biotechnology Division, NIIST, Thiruvananthapuram, Kerala, India

A.K. Agarwal et al. (eds.), *Novel Combustion Concepts for Sustainable Energy Development*, DOI 10.1007/978-81-322-2211-8_3

1 Introduction

Sustainable energy production is one of the most formidable global problems of the present century. The world is about to reach at the edge of the damage of global environment caused due to constant increase in greenhouse gas (GHG) emissions. An increased demand of energy globally for expanding population is a major cause of concern. Also, fossil fuels are depleting at an alarming rate, which necessitates the search for an alternative renewable source of energy. Thus, the efforts are continuing to decrease the dependency on fossil fuel so as to have carbon footprints for a healthier and more eco-friendly future. Biofuels are being looked up as a sustainable source of alternative renewable energy capable of solving major environmental and energy-related issues. In last two decades, biofuel research has been accelerated all over the world with several commercial plants coming up for bioethanol and biodiesel production for various sectors.

Global production of biofuels for the road transportation sector has grown at approximately 15 % per year since 2002, a rate over ten times that of oil, and has accelerated rapidly since 2005 when output totaled about 31 million tons (Mt), consisting of 27 Mt ethanol and 4 Mt biodiesel. This figure is expected to double by 2015 and would then represent slightly more than 3 % of world auto fuel consumption, compared with 1.3 % today (Fig. 1: Trend in world biofuel production).

Current biofuel output is mainly ethanol, most of which is processed from sugarcane and corn (maize). Ethanol production is approximately 10 times that of biodiesel, which is derived from vegetable oils including rapeseed, sunflower, palm, or soybean or from animal fats. Corn-based US ethanol and sugarcane-based Brazilian ethanol account for nearly 90 % of global production (http://www.dupontelastomers.com/autofocus/a4/af4.asp?article=biofuels).

Other than biodiesel and bioethanol, there are many alternative renewable sources of energy which could be promising for supplementing the conventional

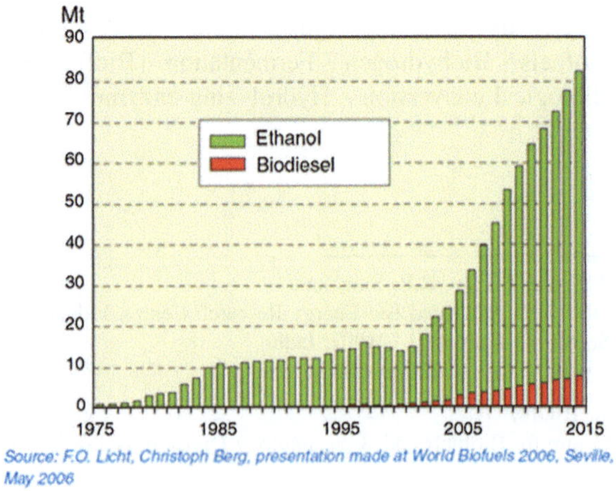

Fig. 1 Trend in world biofuel production

sources of energy to meet the growing demand of expanding population. Based on the raw materials available in a particular region, different alternative biofuel production should be encouraged.

2 Alternative Energy

Alternative energy is a broad term which includes all the energy that can replace the fossil fuels and could be served as a key to mitigating the environmental aspects to transport. Introduction of alternative fuels together with drastic energy efficiency could be a key to sustainable mobility globally. Biological sources such as plant and microbes could be potential sources for producing alternate fuels, termed as biofuels, including bioalcohols, biodiesel, algal biofuels, biohydrogen, and microbial fuel cells (Singhania and Pandey 2012).

2.1 Bioalcohols

Bioalcohol includes biologically produced alcohols from lignocellulosic (LC) biomass, chiefly bioethanol and biobutanol.

Bioethanol from LC biomass is among the most promising alternative energy sources. Due to abundant availability and ubiquitous nature of LC biomass, bioethanol produced from it offers potential benefits in terms of quality and quantity (Singhania et al. 2013). Ethanol has higher octane number than petroleum fuels and lesser CO_2 emissions which makes it energy-efficient fuel. It can be blended up to 15 % in gasoline without needing any change in the engine design. Its application as alternative transport fuel (as replacement to gasoline) has been well commercialized in Brazil since several decades (although that has been from sugarcane). Biobutanol could be a direct replacement of gasoline as it can be directly burned in gasoline engine without any engine modification. It is also claimed that it produces more energy and is less corrosive and less soluble in water than ethanol. Biotechnologists are trying to develop the microbial strains which could tolerate higher concentration of butanol and also give improved productivities using metabolic and genetic engineering approaches. Efforts are also being made to improve the extractive methods to remove it from the fermentation medium.

Multifaceted approaches are being adopted for developing economically feasible biomass conversion technologies. Over the period, the attention has shifted on biorefinery concept with aim of zero waste generation and value addition, either using the biochemical (sugar) platform, involving pretreatment, hydrolysis, and fermentation, or thermochemical platform, involving pyrolysis and or/gasification. Under biochemical platform, a suitable pretreatment process for the LC biomass is one of the most essential requirements, which could remove the lignin coating without removing sugar component from the plant cells, so as to overcome the recalcitrance of LC biomass and expose the cellulosic material to cellulase. Though globally huge efforts have been made to develop the pretreatment methods (physical, chemical, and biochemical) and tremendous successes have also been achieved, none of them have proved to be efficient and cost-effective, which could be considered as an ideal method. Moreover, each of them differs in their degree of efficiency with the type of biomass. Combinations of pretreatment methods have been also used for effective exposure of cellulosic material so as to be accessible to enzyme. The next step, the hydrolysis of biomass into sugars, employing cellulase is still a bottleneck in biochemical platform as major portion of the cost of this process is consumed by this step due to high cost of enzyme. Cellulases currently available commercially usually have a technical weakness—its beta glucosidase has weak tolerance for glucose, which interferes with the hydrolysis process. Researchers have applied recombinant DNA technology for developing microbial strains to produce cellulases with improved productivities and desirable properties, whose substrate specificity is yet another technical issue to be resolved for their efficient utilization. Designer cellulase could be a good possibility in the near future. For economic sustainability, it would be necessary to utilize the C-6 and C-5 hydrolysates efficiently, including production of value-added products. An example is the production if amino acids using C-5 hydrolysate by genetically modified *Corneybacterium* sp. Depending upon the pretreatment method used, there could be inhibitory chemicals such as furfural and acetic acid in the hydrolysate which interfere with the yeast fermentation. Thus, either these should be removed from the hydrolysate (additional process step causing cost increase) in such a way that sugars too are not removed, or yeast strains should be developed having tolerance to such chemicals, which requires biotechnological intervention. Two other approaches in this regard are simultaneous saccharification and fermentation and co-fermentation (C-6 and C-5 sugars) using genetically modified microbes. On larger perspectives, issues on land and water usage, carbon footprint, risks on venture capital for large-scale plants, etc., need to be considered together for LC bioalcohol program. *LC bioethanol will be discussed in detail further in this chapter.*

2.2 Biodiesel

Biodiesel is an alternative fuel similar to conventional or fossil diesel. It offers several environmental benefits compared to fossil diesel, the most important among which is, that it can be described as carbon neutral. That means the fuel produces no net carbon output in form of carbon dioxide. It is because the amount of CO_2 produced during its combustion is equal to the amount of CO_2 absorbed or utilized from the environment to be produced. It is safe, biodegradable and produces less air pollutant than petroleum-based diesel. It could be produced from vegetable oil, animal oil/fat, and waste cooking oil. These oils can be converted into biodiesel via transesterification. Oil crops such as rapeseed, soybean, and palm represent the most suitable source of oil for biodiesel. Though vegetable oil possesses the greatest potential for biodiesel

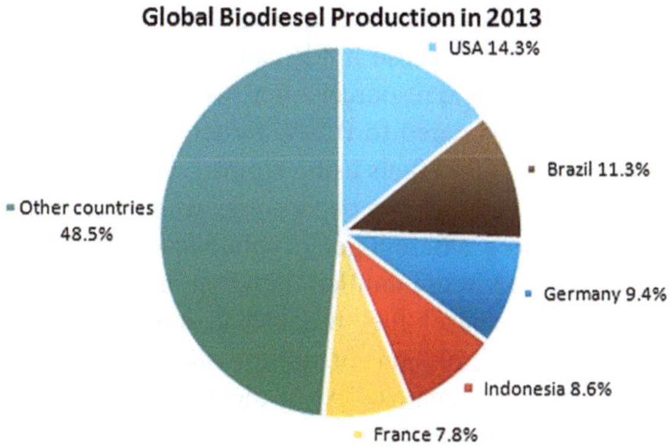

Fig. 2 Different countries involved in biodiesel production

production, it is not being done commercially because the raw oil is too expensive and after adding the cost of converting it to biodiesel, it becomes impossible to compete with fossil diesel. Also in country like India, food crops can never be selected for sustainable alternative bioenergy production where food security is a priority. In India, *Jatropa curcas* seeds were proposed for biodiesel production due to its high oil content, but the program was failed miserably due to unavailability of land for its production. It was earlier projected that barren lands would be used for its cultivation rather than agricultural fertile land, but it was a major failure.

Most of the industrial biodiesel is produced from waste vegetable oil sourced from restaurant, industrial food producers, chip shops, etc. Waste vegetable oil can be sourced for free which needs to be treated before conversion to biodiesel. Biodiesel produced from waste vegetable oil can compete with cost of petroleum-based diesel due to zero cost of the raw material.

During last decade, global biodiesel market has showed incredible growth. Production as well as demand has increased almost sevenfold since 2005. Today, the market still continues to grow mainly due to European Union countries. Brazil and Argentina with USA are lagging slightly behind as their main focus is bioethanol rather than biodiesel. In EU, Germany and France are leaders in regional biodiesel output; other main producers include Poland, the Netherlands, and Italy. EU is forecast to still dominate biodiesel market till the end of this decade; however, Latin American countries will also increase their shares in both biodiesel production and consumption in the near future. (http://mcgroup.co.uk/researches/biodiesel dated 06/21/2014). Figure 2 shows the contribution of different countries worldwide toward biodiesel production in 2013.

2.3 Algal Biofuels

Algal biofuels have attained high significance as a third-generation biofuel, due to their fast growth potential along with high lipid, carbohydrate, and nutrient content. Microalgae are known to produce and accumulate lipids in the form of triacylglycerols (TAG) within their cell mass, which could be as high as 30 % of their cell weight. These lipids can serve as feedstock for producing biodiesel and green diesel. Algal resources offer several advantages such as it can capture 10 % of sunlight as compared to 0.5 % by terrestrial plant, require less land area, can be converted to liquid fuels using simpler technology than those required for cellulose conversion, and have secondary uses which fossil fuel does not have. Additionally, it also reduces anthropogenic pollutants to the environment. Algal biodiesel can easily be used in unmodified diesel engines, and it has significant advantages over conventional diesel fuel because it is renewable and biodegradable and might produce lower emissions of sulfur oxides and particulates when burned. Though the life cycle assessment (LCA) of algal biofuels suggests that at present, it is not so attractive economically, they are environmentally better than the fossil fuels. Several biotechnological strategies have been adopted to increase the lipid content of the algal biomass, such as metabolic engineering and bioprocess modification and mixed culture cultivation. For large-scale cultivation, the pond system offers simplicity in operation, lower capital, and recurring expenditure but has issues of harvesting and control of contamination. Various photobioreactors have been developed which offer potential benefits to control the problems of pond system, but these require high-energy input for operation, which makes them economically unattractive presently.

2.4 Biohydrogen

Among other alternative energy sources, biohydrogen could be regarded as the most promising future energy carrier as is expected to contribute substantially for the future energy needs. Biohydrogen is recognized as clean energy carrier as it produces only water when combusted. It has a high calorific value (122 kJ/mol) of any known fuels which is 5 and 2.6 times greater than ethanol and gasoline, respectively, which offers a crosscut from electricity to transportation fuels and energy storage. In context of energy generation per unit mass, it has a great potential to be developed as an alternative fuel in the near future (Sabariswaran et al. 2013). Hydrogen has a high low heating value (LHV) and can be used in fuel cells and internal combustion engines. Worldwide, the compound annual growth rate of quantitative hydrogen production was estimated to grow by 5.6 % during the period 2011–2016 and the global market for hydrogen generation would mature from estimated $87.5 billion in 2011 to $118 billion by 2016 with a CAGR of 6.2 % (http://articles.pubarticles.com/hydrogen-generation-market-by-merchant-captive-type-distributed-centralized-1323076864,561523.html as on June 25, 2014).

Biohydrogen production is one among the most promising future energy sources which is expected to contribute substantially for the future energy needs for the mankind. It is unfortunate that only 5 % production from renewable sources has been achieved yet. Four mechanisms are known for biohydrogen production: direct biophotolysis, indirect biophotolysis, dark fermentation, and photo fermentation.

Among these, dark fermentations and photo fermentations are most widely studied for biohydrogen production because they accomplish waste reduction and energy generation and support sustainability and economic viability of the process. In dark fermentation, complex organic compounds are converted into biohydrogen, volatile fatty acids (VFAs), and CO_2 by fermentative bacteria at ambient temperature. However, for this, competent microbial cultures (*Clostridium, Ruminococcus, Fibrobacter, Enterobacter, Rhodobacter* sp.) are required to use substrate efficiently, which are usually complex in nature.

Development of technologies for biohydrogen production is still in early stages; however, some fermentation strategies have demonstrated with great efficiencies which are reasonable for implementation. To develop an efficient system to attain future hydrogen demands sustainably, it is mandatory to have a directed investment in a strategic biohydrogen research agenda. In such context, biohydrogen production via dark fermentation could be a promising method due to its higher hydrogen evolution rate in the absence of light and conversion of waste into environmentally benign material. Photo fermentation can be linked with dark fermentation with two stages of fermentation to increase process viability as the organic acid produced during dark fermentation can be converted into biohydrogen by second stage of photo fermentative bacteria. Apart from general approach of process optimization and metabolic engineering of biohydrogen pathway, thermophiles could be an option to get increased yield of biohydrogen. Main advantage of employing thermophiles is high temperature favors conversion toward hydrogen production due to favorable thermodynamic conditions, and many industrial organic wastewaters are discharged at elevated temperatures that can be directly used as feedstocks. Continuous fermentation would be another promising approach to increase hydrogen yield. To add economic viability of the process, it is desirable to minimize organic acid production and biomass generation for which significant fraction of energy was utilized from substrate. In continuous process, cell recycling and process optimization for maximum biohydrogen production would positively improve substrate to main product conversion with higher biohydrogen yield.

Some of the major bottlenecks for the development of biohydrogen production at commercial scale include low rate of H_2 production, low efficiency, incomplete substrate conversion, and its partial conversion into organic acids, viz. acetic, butyric, propionic acids, etc. Major challenges for biohydrogen production include developing bioreactor design, optimization of bioreactor conditions, and metabolic engineering of microbes for desired properties for higher production. Biohydrogen can be produced from a wide variety of primary energy sources. Abundant agro-residual biomass and waste effluents from various industries could be a source for biohydrogen production, where combination of waste treatment and energy production would be an advantage.

Hydrogen produces only water when combusted, generating a 2.75 times higher energy yield than hydrocarbon fuels and offers a crosscut from electricity to transportation fuels and energy storage, which is of special importance for a future dominated by renewable energies. It can be used for efficient power generation in fuel cells. In fuel cells, hydrogen offers high performance and can achieve 30–90 % efficiency.

However, it has a major disadvantage that it has low LHV per unit volume which means need of using high volume and heavy canisters. 'Well to wheel' analyses shows that biohydrogen is efficient than other alternative fuels in respect to efficiency and emissions.

2.5 Microbial Fuel Cells

Microbial fuel (biofuels) cells are being looked as new alternative class of energy source. There are several classes of biofuel cells: biofuel cells that convert chemical fuels to produce electrical power by use of catalytic enzymes, similar cells where microorganisms rather than enzymes are used to convert the fuel to energy, and hybrid biofuel cells that combine the utilization of photochemical chemistries and biological systems for the generation of electricity. The development of biofuel cells is still in its infancy, although there is immense potential for further improvement. Microbial fuel cells offer the advantage of using power supplies with high longevities that are capable of using complex biofuels formed from biological waste.

Thus, biofuels offer potential benefits as alternate energy, but their commercialization appears a distinct and far possibility at present unless technologies are developed offering chiefly economic but also social and environmental sustainability. Biotechnological interventions, in this regard, offer a silver ray of hope for these.

3 Biofuel in Indian Perspectives

Biofuel in Indian scenario has a distinct position from those of other parts of the world. India has limited land resources to be utilized for energy crop production rather than food crops. In India, most of the biofuel research focuses on utilization of agricultural residues that are available in surplus amount. These lignocellulosic residues could be utilized as raw material for cellulosic ethanol production. Exclusive information on availability of surplus amount of various feedstocks for exploitation in India has been available in NIIST-TIFAC report.

3.1 Feedstock Availability

The applicability of the technologies that relies on the feedstock widely available in India could reach to a commercialization stage. For example, companies such as DuPont and POET-DSM that use corn crops as a feedstock will not easily be able to transfer their technology to India as only a small amount of corn waste is available in India. Companies such as Abengoa, Beta Renewables, and Iogen which use agricultural waste, bagasse, and a variety of feedstocks will be much better placed to adapt the technology for the Indian situation. Hence, there was an urgent need to identify the feedstock available in surplus amount.

Table 1 gives an indication of the amount of waste biomass available in India from agricultural processes. These feedstocks are very different to the feedstocks being used in the first wave of cellulosic ethanol plants due to the makeup of the agricultural industries in which they are situated, so only give a vague indication of the amount of ethanol that could be produced. Some of these feedstocks may not be suitable for cellulosic ethanol production. The technologies that are licensed such as PROESA, from Beta Renewables and Abengoa Bioenergy; and those that gasify the biomass such as KiOR and INEOS Bio, look set to be able to use their technology within India at the present time.

However, Sukumaran et al. (2010) suggest that the vast majority of residues are used as fodder, manure, thatch, and direct fuel, sold to the paper industry, burned after harvest, or used in various other applications. As a result, perhaps, only 10 % of residues may be currently available for fuel use. Sugarcane tops are the residue produced in greatest surplus (79.4 MMT), but they are burnt in the field after harvest. Currently, the main residues available for exploitation are wheat straw, rice straw, corn stover, and sugarcane bagasse (Sukumaran et al. 2010), amounting to 25.5 MMT annually and potentially producing 5.7 MMT of ethanol. It would be expected some of these residues would be diverted to power generation, which has already achieved significant success in India through rural electrification plants (Singh and Gu 2010).

Table 1 Surplus agricultural residue availability in India

Residue	Surplus availability (MMTPA)	Ethanol potential (MMTPA)
Rice	8.9	2.0
Wheat	9.1	2.0
Sugarcane	79.4	17.5
Corn stover	1.1	0.2
Cotton	11.4	2.5
Chillies	0.5	0.1
Sorghum	1.6	0.4
Finger millet	0.5	0.1
Pearl millet	1.2	0.3
Pulses	5.7	1.3
Oilseeds	17.3	3.8
Bamboo	3.3	0.7
Pine	1.2	0.3
Water hyacinth	14.0	3.1
Total	155.2	34.1

Source NIIST-TIFAC report: 'Study on Availability of Indian Biomass Resources for Exploitation: A Report Based on a Nationwide Survey.' Reported through Sukumaran et al. (2010)

These figures do not take into account municipal solid waste or gas wastes from industry which could use technologies that are being rapidly developed globally to add significant capacity to India's ethanol production.

3.2 Current Ethanol Situation: India

India currently has a mandate for 5 % ethanol in fuel which equates to 1,050 million liters of ethanol. Domestic sugar mills can only produce 550 million liters of ethanol (52 % of demand) which leaves a 500 million liters shortfall which must be met by oil marketing companies by importing ethanol from abroad at a higher price. This pushes up the price of fuel and significantly affects the Indian economy. For this reason, cellulosic ethanol in India is a much needed technology which has great potential to address this shortfall problem.

4 National Biofuel Policy

The informations given under Sect. 4 have been derived from http://mnre.gov.in/file-manager/UserFiles/biofuel_policy.pdf.

Development objectives in Indian scenario are based on economic growth, equity, and human well-being. Energy being a critical input for socioeconomic development, the strategy of a country aims at efficiency, self-sufficiency, and security. Fossil fuels will continue to play a dominant role in the energy scenario in our country in the next few decades. However, conventional or fossil fuel resources are limited, non-renewable, and polluting and hence need to be used prudently. India is endowed with abundant renewable energy resources which are indigenous, non-polluting, and virtually inexhaustible. Therefore, every possibility for its effective utilization should be encouraged.

The world market price of crude oil has significantly increased in recent years and has reached higher than $ 140 per barrel. Petrol-based oil meets about 95 % of the requirement of the transportation fuel, and the demand is still increasing steadily. Only 23 % of oil requirement could be fulfilled by nation and the rest are being imported. India's energy security necessitates development of indigenous alternative source of energy. Biofuels derived from renewable biomass resources provide a strategic advantage to promote sustainable development and to supplement conventional energy sources. Biofuels can reduce the dependence on import of fossil fuels and provide a higher degree of national energy security and therefore are foreseen as a ray of hope in this direction being eco-friendly and sustainable.

The growth of biofuels around the globe is spurred largely by energy security and environmental concerns, and a wide range of market mechanisms, incentives, and subsidies have been put in place to facilitate their growth. The Indian approach to biofuels, in particular, is somewhat different from the international approaches.

It is based solely on non-food feedstocks to be raised on degraded or wastelands that are not suited to agriculture, thus avoiding a possible conflict of fuel versus food security. It is the endeavor of this policy to facilitate and bring about optimal development and utilization of indigenous biomass feedstocks for production of biofuels.

The goal of the policy is to ensure that a minimum level of biofuels become readily available in the market to meet the demand at any given time. An indicative target of 20 % blending of biofuels, both for biodiesel and bioethanol, by 2017 is proposed. Blending levels prescribed in regard to biodiesel are intended to be recommendatory in the near term.

The focus for development of biofuels in India is based on utilizing waste and degraded forest and non-forest lands only for cultivation of shrubs and trees bearing non-edible oilseeds for production of biodiesel. In India, bioethanol is produced mainly from molasses, a by-product of the sugar industry. In future too, it would be ensured that the next generation of technologies is based on non-food feedstocks so as not to compete with food and feed.

Plantations of trees bearing non-edible oilseeds will be taken up on government/ community wasteland, degraded or fallow land in forest and non-forest areas. Contract farming on private wasteland could also be taken up through the Minimum Support Price (MSP) mechanism proposed in the policy. Plantations on agricultural lands will be discouraged.

A major instrument of this policy is that a MSP for oilseeds should be announced and implemented with a provision for its periodic revision so as to ensure a fair price to the farmers. The details about implementation of the MSP mechanism will be worked out carefully after due consultations with concerned government agencies, states, and other stakeholders. It will then be considered by the 'Biofuel Steering Committee' and decided by the 'National Biofuels Coordination Committee' proposed to be set up under this policy. Different levels of MSP for oilseeds have already been declared by certain states. Employment provided in plantations of trees and shrub bearing non-edible oilseeds will be made eligible for coverage under the National Rural Employment Guarantee Programme (NREGP).

Ethanol is mainly being produced in the country at present from molasses, which is a by-product of the sugar industry. Blending up to 5 % of ethanol with gasoline has already been taken up by the Oil Marketing Companies (OMCs) in 20 states and 4 union territories. In order to augment availability of ethanol and reduce over supply of sugar, the sugar industry has been permitted to produce ethanol directly from sugarcane juice. The sugar and distillery industry will be further encouraged to expand production of ethanol to meet the blending requirements prescribed from time to time, while ensuring that this does not in any way create supply constraints in production of sugar or availability of ethanol for industrial use. Blending of 10 % ethanol with gasoline was decided to become effective from October 2008 in those states.

In order to take care of fluctuations in the availability of biofuels, OMCs will be permitted to bank the surplus quantities left after blending of biodiesel and bioethanol in a particular year and to carry it forward to the subsequent year when there may be a shortfall in their availability to meet the prescribed levels. The responsibility of storage, distribution, and marketing of biofuels would rest with OMCs.

In the determination of biodiesel purchase price, the entire value chain comprising production of oilseeds, extraction of bio-oil, its processing, blending, distribution, and marketing will have to be taken into account. The Minimum Purchase Price (MPP) for biodiesel by the OMCs will be linked to the prevailing retail diesel price. The MPP for bioethanol will be based on the actual cost of production and import price of bioethanol. The MPP, both for biodiesel and bioethanol, will be determined by the Biofuel Steering Committee and decided by the National Biofuel Coordination Committee. In the event of diesel or petrol price falling below the MPP for biodiesel and bioethanol, OMCs will be duly compensated by the government.

Financial incentives, including subsidies and grants, may be considered upon merit for new and second-generation feedstocks, advanced technologies and conversion processes, and production units based on new and second-generation feedstocks. If it becomes necessary, a National Biofuel Fund could be considered for providing such financial incentives. As biofuels are derived from renewable biomass resources, they will be eligible for various fiscal incentives and concessions available to the new and renewable energy sector from the Central and State governments. Bioethanol already enjoys concessional excise duty of 16 %, and biodiesel is exempted from excise duty. No other central taxes and duties are proposed to be levied on biodiesel and bioethanol. Custom and excise duty concessions would be provided on plant and machinery for production of biodiesel or bioethanol, as well as for engines run on biofuels for transport, stationary, and other applications, if these are not manufactured indigenously.

Intensive R&D work would be undertaken in the following areas

1. Biofuel feedstock production based on sustainable biomass with active involvement of local communities through non-edible oilseed bearing plantations on wastelands to include inter-alia production and development of quality planting materials and high sugar containing varieties of sugarcane, sweet sorghum, sugar beet, cassava, etc.
2. Advanced conversion technologies for first-generation biofuels and emerging technologies for second-generation biofuels including conversion of lignocellulosic materials to ethanol such as crop residues, forest wastes and algae, biomass-to-liquid (BTL) fuels, and biorefineries.
3. Technologies for end-use applications, including modification and development of engines for the transportation sector based on a large-scale centralized approach, and for stationary applications for motive power and electricity production based on a decentralized approach.
4. Utilization of by-products of biodiesel and bioethanol production processes such as oil cake, glycerin, and bagasse.

In regard to research and development in the area of biofuels, a subcommittee under the 'Biofuel Steering Committee' proposed in this policy comprising Department of Bio-Technology, Ministry of Agriculture, Ministry of New and

Renewable Energy, and Ministry of Rural Development would be constituted, led by Department of Bio-Technology, and coordinated by the Ministry of New and Renewable Energy.

The Bureau of Indian Standards (BIS) has already evolved a standard (IS-15607) for biodiesel (B 100), which is the Indian adaptation of the American Standard ASTM D-6751 and European Standard EN-14214. BIS has also published IS: 2796: 2008 which covers specification for motor gasoline blended with 5 % ethanol and motor gasoline blended with 10 % ethanol.

In a nutshell, the national policy on biofuels envisaged the biofuels to be the solution for India's containment of carbon emission control and energy vulnerability. Being derived from renewable biomass sources and less carbon emitting, they were trusted to be environmentally benign and cost-effective. It could lead to lesser dependence on fossil fuel, causing lesser expenditure on the import of the fossil fuel. Biofuel is also viewed as a catalyst for rural development and employment. The policy has also strategized and framed the technological, financial, and institutional interventions and its implementation mechanism. The goal of the policy is to achieve a biofuel blending of 20 % for biodiesel and bioethanol by 2017. Another goal of this policy was to ensure the availability of minimum level of biofuels in the market in accordance with the demands. The goals set so far are difficult to achieve. In India, lot of hope is lying with the second- and third-generation biofuels. But considering the importance of biofuels and the amount of dependability expected on the renewable energy sources, with time based on socio-economic conditions, it is needless to suggest that India needs to fulfill the biofuel targets. Taking into account the current unfulfilled targets and unevenness in the mandate and goals among the state policies in the country and considering the example of the other countries, there is a need for legislation and a strict mandate to be formulated (Sengupta and Poddar 2013).

All these were about the national policy for biofuels. Now let us have a look at scientific part of cellulosic ethanol production technology.

5 Bioethanol

5.1 Bioethanol Global Market Position

As shown in Fig. 3, global ethanol production and consumption are dominated by North America and Latin America. These two regions set the trend for global ethanol production and consumption. The plateau reached in North America in 2010 and 2011, combined with the decrease in Latin America between 2010 and 2011, resulted in global consumption remaining near its 2010 level in 2011 and lower production in 2011 than in 2010, and the first time global ethanol production has not

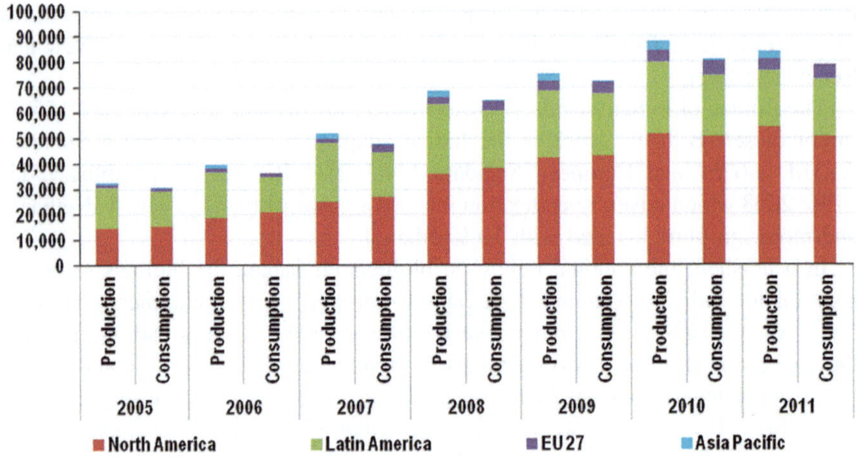

Fig. 3 Regional ethanol production and consumption, 2005–2011 (million liters)

grown for the period under review, 2005–2011. (http://www.globalbiofuelscenter.com/HomeSpecialReportDetails.aspx?CID=1778&Type=2 dated 21/06/2014).

A biofuel is a type of fuel which is produced from biomass and hence is a sustainable resource. Ethanol can be termed as bioethanol when it is produced from any waste plant material that is composed of cellulosic material rather than starch. Ethanol that is produced from starch- and sugar-based feedstocks such as corn and sugar cane is not termed as bioethanol. Starch-based feedstocks require little modification, and sugars are generated efficiently by hydrolysis of starch to be fermented by microorganisms to ethanol. Bioethanol is produced from waste biomass feedstocks of cellulosic nature in the agricultural and forestry process such as corn stover, rice straw, rice husks, sugarcane bagasse, and wheat straw. This is a much harder process as the cellulose is inaccessible due to the composition of the biomass.

Lignocellulosic feedstocks are composed mainly of cellulose (30–50 %), lignin (15–25 %), hemicellulose (20–40 %), extractives, and ash consisting of inorganic minerals (5–35 %) (Kumar and Murthy 2011). Cellulose is a glucose polymer linked by β-1,4-glycosidic bonds which is linear and packs well due to having few branches in the chain. Hence, it is a polysaccharide that can be hydrolyzed into glucose monomer units and the process is called as saccharification. Hemicellulose is a short highly branched polymer of C5 and C6 sugars composed mainly of xylose, arabinose (C5), galactose, glucose, and mannose (C6) (Brethauer and Wyman 2010). Due to its highly branched nature, it can be readily hydrolyzed into sugar monomers with a major product being xylose. Lignin is a phenolic polymer that cannot be utilized in the fermentation process and it binds cellulose and hemicellulose into a coherent structure which makes up the plant cell wall.

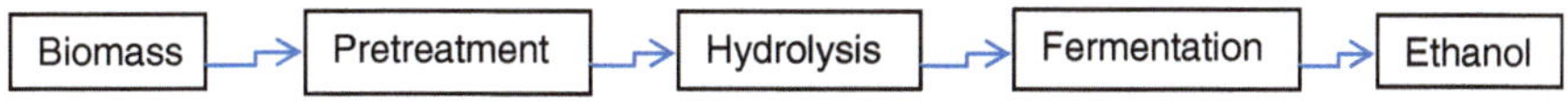

Fig. 4 General pathway for conversion of cellulosic feedstocks to bioethanol

For developing feasible technology for bioethanol production from lignocellulosic biomass, following challenges have been identified by various researchers worldwide:

1. Development of an efficient pretreatment method to open up the fibers, so as to overcome the recalcitrance of biomass.
2. Obtaining high titers of cellulase to hydrolyze cellulose into its monomers (could be by adopting cheaper fermentation technology, employing cheaper/ crude substrate for enzyme production, employing bioengineered organism, etc.).
3. Detoxification of inhibitors produced during pretreatment which are otherwise toxic for the yeast during ethanol fermentation.

The first two steps are critical for the production of ethanol from biomass. Thus, development of an efficient pretreatment method to open up the fibers and obtaining high titers of cellulase to hydrolyze cellulose into its monomers are among major challenges which adds cost to this technology. Although extensive studies have been carried out to meet the future challenges of bioenergy generation, there is no self-sufficient process or technology is available till date.

The process for bioethanol production from feedstocks consists of three key steps. Firstly, the biomass is pretreated to break up the macromolecular structure of the cells in order to make the cellulose more accessible for cellulolytic enzymes. The second stage is enzymatic hydrolysis/saccharification, where the cellulose is broken down into simple sugar monomers. In the third stage, these sugars are fermented by microorganisms to produce ethanol. Figure 4 shows general technical pathway for bioethanol production.

6 Current Cellulosic Ethanol Technology

6.1 Pretreatment

Pretreatment aims to decrease crystallinity of cellulose, increase biomass surface area, remove hemicellulose, and break lignin seal. Pretreatment makes cellulose more accessible to enzymes so that conversion of carbohydrate polymers into fermentable sugars can be achieved more rapidly and with more yields. Intensive research is going on globally to search for a robust and universal pretreatment method, but till date, all efforts have gone in vain as different biomass varies in its composition and the differences exist based on geographical region as well as seasonal difference is also significant.

Physical pretreatment is the preliminary step included in all pretreatment techniques to reduce the size of the biomass and break some of the cellulose crystallinity. Techniques used include milling and grinding which aim to get the particles as small as possible as this increases hydrolysis efficiency. However, the energy requirement also needs to be taken into consideration when analyzing the best pretreatment for a certain feedstock as the smaller the targeted particle size, the higher the energy requirement. After physical pretreatment, a second pretreatment step is employed. Dilute acid, alkali, hot water, and steam expansion are among the most commonly used pretreatments (Kumar and Murthy 2011).

Dilute acid pretreatment has been extensively studied and is a common stage in many pilot cellulosic ethanol projects. It is characterized by high temperature, and short residence time meaning the biomass can move on to the next stage in the process quickly which is suitable for a large-scale plant. In this stage, the hemicellulose is hydrolyzed so it becomes soluble, whereas the cellulose and lignin remain in the solid fraction. In this way, the hemicellulose can be separated from the cellulose allowing for a variety of ethanol production situations later on in the process. If the temperature or acid concentration is too high, then the hemicellulose could degrade into furans which inhibit fermentation. So it needs to be removed and the cellulose could start to hydrolyze into glucose. All of which are detrimental to the process making it more expensive.

Dilute alkali pretreatment also degrades the biomass structure but is less common. The alkali, at high temperature, reacts with the ester bonds of the lignin which are cross-linked with cellulose. In this way, lignin degradation is more favored and it is the lignin which becomes soluble and can be removed leaving the sugar containing compounds behind (Kumar and Murthy 2011). However, there will be some lignin left in the solid sugar component. Concentration of alkali significantly affects degradation of the biomass, and a balance must be struck to acquire the most effective pretreatment. At higher alkali concentration, hemicelluloses start to degrade and produce inhibitory compounds. Also, the high temperatures used and the cost of the alkali on a large scale make this unsuitable for industrial scale plants.

Hot water pretreatment is an auto-catalyzed process where high-temperature water is used as a solvent for the biomass which will degrade the hemicelluloses due to its branched structure. The acetic acid produced from degradation of hemicelluloses and self-ionization of water lower the pH hence causes the same structural change in the biomass as the dilute acid pretreatment. However, this process requires high pressure in order for the water to be kept liquid which is expensive and has not been demonstrated on a commercial scale.

In steam explosion, biomass is heated under pressure with water vapor. The pressure is then rapidly reduced causing a rapid expansion of the steam and a breakdown of the biomass structure. The process has been found to be most efficient for high-temperature short residence times or vice versa (Kumar et al. 2009). A major advantage of this technique is that it has been found to be effective on feedstock with large particle sizes thereby reducing the energy needed in physical pretreatment for size reduction (Sun and Chen 2002), and also, it does not release any acid or base as environmental pollutant.

Each pretreatment method has its own advantages and disadvantages. Though several other pretreatment methods exist such as aqueous ammonia, ionic liquid, and microwave, but all are confined to the research laboratories and none have been found suitable for industrial application other than dilute acid pretreatment. Once the biomass is pretreated, it is subjected to enzymatic hydrolysis.

6.2 Hydrolysis

The major bottleneck in the production of cellulosic ethanol through biological route is requirement of high dosage of cellulase. This step accounts for maximum part of the cost to this technology. Enzyme cost always remained a major hurdle to industrial second-generation ethanol production. However, enzyme costs have fallen 80 % over the last decade (Ethanol RFA 2013, http://ethanolrfa.3cdn.net/d9d44cd750f32071c6_h2m6vaik3.pdf) so more and more biofuel companies are partnering with enzyme producers such as Novozymes and Genecor as purchasing enzymes become competitive with producing enzymes on site from fungi such as *Trichoderma reesei* (Xu et al. 2009).

Cellulases are currently the third largest industrial enzyme worldwide, by dollar volume, because of their use in cotton processing, paper recycling, as detergent enzymes, in juice extraction, and as animal feed additives. However, cellulases will become the largest volume industrial enzyme, if ethanol from lignocellulosic biomass through enzymatic route becomes a major transportation fuel.

Cellulases are a complex enzyme mainly consisting of three enzymes, endo-cellulase, exo-cellulase and β-glucosidase, which work synergistically to hydrolyze cellulose completely into monomers, i.e., glucose (Singhania et al. 2010). Apart from these, several helper enzymes would also be present as xylanase and ligninase. The search for potential sources of cellulolytic enzymes is still continuing in the interest of successful bioconversion of lignocellulosic biomass. There are several microorganism especially filamentous fungi that produce cellulase with all the three enzyme at different proportion needed for the complete hydrolysis of cellulose. *T. reesei* is one among the most potent cellulase producers studied in detail. Currently, high-efficient cellulases are produced commercially using filamentous fungi mainly *Aspergillus niger* and *T. reesei. Penicillium* sp have come up as an alternative to *Trichoderma* as it also produces cellulase at high titers having all the other components required for synergistic action toward biomass hydrolysis (Gusakov 2013). On-site enzyme production utilizing low-value substrate could be a step toward developing an economically feasible process. Once fermentable sugars are available, it can be easily fermented into ethanol. Separate hydrolysis and fermentation (SHF) as well as simultaneous saccharification and fermentation (SSF) are highly adopted technologies for bioethanol production from biomass.

6.3 Fermentation

The cost of ethanol is highly dependent on the efficiency of conversion of biomass to sugars and then fermentation to ethanol. This depends on the substrate, bacteria used, concentration of enzyme and nutrients, pH, oxygen supply, and pressure.

A major hurdle to lowering the ethanol price is the lack of ability to ferment both C5 and C6 sugars. C6 sugar technology is well developed, but research is being undertaken to find a microorganism that will ferment both sugars and give a high yield of ethanol.

A variety of process methods have been suggested in the production of cellulosic ethanol. In separate hydrolysis and fermentation (SHF), the liquid and solid phases are separated after pretreatment. In the case of dilute acid pretreatment, the solid stream is washed and then subject to enzymatic hydrolysis. The resulting sugar solution is then fermented to ethanol with conventional yeast microbes. A suitable pentose fermenting strain can covert the liquid C5 sugar stream into ethanol after a detoxification step. These separate processes allow optimization for each step so you can get the highest yields possible. Alternatively, enzymes could be added to the whole pretreatment slurry without separation of the solid and liquid fractions, followed by fermentation of the C5 and C6 which is called separate hydrolysis and co-fermentation (SHcF). This is economically very attractive, but the fermentation step is much more challenging.

As cellulase enzymes are inhibited by their hydrolysis products, a favoured process is simultaneous saccharification and fermentation (SSF) which keeps sugar concentrations low. Despite the need to lower the temperature from optimum enzyme efficiency to accommodate the fermentation microorganisms, SSF was shown to achieve higher rates, yields, and concentrations than SHF by overcoming the effects of end product inhibition. In addition, SSF significantly reduces equipment demands and the presence of ethanol prevents contamination by unwanted microbes (Brethauer and Wyman 2010).

A new line of development has been consolidated bioprocessing of lignocellulose. This combines simultaneous saccharification with fermentation of the resulting sugars in a single process mediated by a single microbial mixture which produces the enzymes necessary for hydrolysis and has the pathways necessary to ferment the sugars produced under the same conditions. CBP has the potential to reduce the cost of conversion of biomass to ethanol significantly although a microorganism that produces ethanol at high yield has yet to be found (Xu et al. 2009).

Another area of development for fermentation is the concept of continuous fermentation. All commercial fermentation processes at the current time use a batch process whereby the sugar solution is put into the fermentation vessel and left to ferment until the ethanol concentration reaches a certain level. The vessel is drained and a new fermentation broth is created. In a continuous fermentation system, the substrate is constantly fed into the vessel at the same rate as the fermentation product is removed to keep the volume constant. The advantage of this technique is that it shows higher productivities and reduced vessel downtime, filling, and cleaning. However, there are other technological obstructions to overcome such as bacterial contamination and co-fermentation of C5 and C6 sugars (Brethauer and Wyman 2010).

Inspite of several challenges, we can see commercial plants coming up for biofuel production such as IneosBio, Iogen, Kior, Inbicon Coskata, Clariant, Beta-Renewable, and Dupont. Among these, Beta Renewables and Dupont are based on cellulosic ethanol production and can be suitably used in our Indian scenario as are based on multiple feedstocks.

7 Conclusion

Biofuels are no more a fantasy like in last decade but are real at present time. Lignocellulosic biomass is the only foreseeable source of raw materials for renewable energy source. After bioconversion of lignocellulosic biomass into its precursor molecules, glucose can be exploited for other value-added products also which is all possible theoretically, though some technological gaps persist. Available feedstock from agriculture and other sources are about 180 million tons per year, so literally no dearth of such biomass. Second-generation biofuels have the all potential to meet the need of transport sector. Commercial plants coming up globally for cellulosic ethanol production as well as other biofuel are an indication of successful research toward developing such technology; however, we need to wait for a while to see the economic feasibility and sustainability.

References

Brethauer S, Wyman CE (2010) Review: continuous hydrolysis and fermentation for cellulosic ethanol. Bioresour Technol 101:4862–4874

Gusakov AV (2013) Cellulases and hemicellulases in the 21st century race for cellulosic ethanol. Biofuels 4(6):567–569

Kumar D, Murthy GS (2011) Impact of pretreatment and downstream processing technologies on economics and energy in cellulosic ethanol production. Biotechnol Biofuels 4:27

Kumar P, Barrett DM, Delwiche MJ, Stroeve P (2009) Methods for pretreatment of lignocellulosic biomass for efficient hydrolysis and biofuel production. Ind Eng Chem Res 48:3713–3729

Sabariswaran K, Papitha P, Indumathi R, Selvakumar S, Raj SP (2013) Biohydrogen production from sago effluent. J Biodivers Environ Sci 3:17–23

Sengupta M, Poddar A (2013) National policy on biofuel under scanner. Int J Emerg Technol Adv Eng 3(3):521–526 (ICERTSD 2013)

Singh J, Gu S (2010) Biomass conversion to energy in India; A critique. Renew Sustain Energy Rev 14:1367–1378

Singhania RR, Pandey A (2012) Alternative energy—the role and scope of biotechnological interventions for developing biofuels, In: Oliver L (eds) Business guide Germany India 2012, 4th edn. Wegweiser, Germany, pp 146–147

Singhania RR, Sukumaran RK, Patel AK, Larroche C, Pandey A (2010) Advancement and comparative profiles in the production technologies using solid-state and submerged fermentation for microbial cellulases. Enzyme Microb Technol 46:541–549

Singhania RR, Parmeswaran B, Pandey A (2013) Advances in lignocellulosic bioethanol. In: Yang S-T, El-Enshasy HA, Nuttha T (eds) Bioprocessing technologies in biorefinery for sustainable production of fuels, chemicals and polymers. Wiley, Hoboken, pp 193–204. ISBN: 978-0-470-54195-1

Sukumaran RK, Surender VJ, Sindhu R, Binod P, Janu KU, Sajna KV et al (2010) Lignocellulosic ethanol in India: prospects, challenges and feedstock availability. Bioresour Technol 101:4826–4833

Sun Y, Chen J (2002) Hydrolysis of lignocellulosic materials for ethanol production: a review. Bioresour Technol 83:1–11

Xu Q, Singh A, Himmel E (2009) Perspectives and new directions for the production of bioethanol using consolidated bioprocessing of lignocellulose. Environ Biotechnol 20:364–371

Biomass Fuel Quality Enhancement and Respiratory Quotient (RQ) for Ranking Fossil and Biomass Fuels Based on CO_2 Emissions

Siva Sankar Thanapal, Dustin Eseltine, Kalyan Annamalai and Devesh Ranjan

Abstract Torrefaction is one of the thermo-chemical pretreatment techniques used to enhance the properties of biomass fuels. An inert environment is maintained during torrefaction process. Effect of using carbon dioxide (CO_2) as torrefaction medium for mesquite and juniper woody biomass was investigated in the current study. CO_2 torrefaction resulted in improved grindability and comparable energy retention in the case of mesquite and juniper. A three component model based on independent parallel reactions was developed to model the torrefaction process. Effect of torrefaction on behavior of the three components in the biomass: hemicellulose, cellulose and lignin was studied. Kinetic parameters for the three biomass components were obtained from literature. Results from the TCM were compared with the results obtained from TGA experiments carried out on mesquite and juniper samples. Further, concept of respiratory quotient (RQ) used in biological literature was applied to combustion applications. Global warming potential (GWP) of different fuels were estimated based on respective RQ. Higher the RQ of fuel, higher is its GWP. A term $RQ_{\text{PROCESS,BIOMASS}}$ was introduced to estimate the amount of CO_2 produced during processing a renewable fuel for combustion applications.

Keywords Torrefaction · Lignocellulosic biomass · Pyrolysis · Carbon dioxide emissions · Respiratory quotient · Thermo gravimetric analysis (TGA)

S.S. Thanapal · D. Eseltine · K. Annamalai (✉)
MS 3123 TAMU, Department of Mechanical Engineering, Texas A&M University, College Station, USA
e-mail: kannamalai@tamu.edu

D. Ranjan
Love 111, George W. Woodruff School of Mechanical Engineering, Georgia Institute of Technology, Atlanta, USA

A.K. Agarwal et al. (eds.), *Novel Combustion Concepts for Sustainable Energy Development*, DOI 10.1007/978-81-322-2211-8_4

1 Introduction

Coal accounts for 45 % of the total generated electricity in the United States (EIA 2013). Along with CO_2, harmful pollutants such as nitrogen oxides (NO_X), sulfur oxides (SO_X), and mercury are also released during the combustion of coal in power generating facilities. Usage of alternate renewable energy sources is investigated to reduce the impact of fossil fuels on the environment. Biofuels are one of the renewable energy sources which contribute about 10 % of the total energy in the world (EIA 2013). Desired characteristics of biomass crops which can be used for power generation were listed by Mckendry (2002). They include high yield, low energy input for production, low cost, low contaminants, and low nutrient requirements. Though wood was the dominant source of energy in the United States till 1850, its usage slowly decreased with the utilization of coal (Tillman et al. 2012). With current regulations on the emissions, the focus has again shifted toward the use of wood and other biomass fuels along with coal. The carbon dioxide emitted during the combustion of biomass fuels is supposed to be absorbed by the growing plants in the carbon cycle, and hence, biomass is considered to be carbon neutral. The term biomass includes wood, forest wood residues, agricultural wastes, energy crops, animal wastes, and municipal solid wastes.

Although biomass has been extensively studied as a renewable energy source, most research conducted used raw untreated biomass. While the use of raw biomass as a renewable energy source has potential, pretreatment of raw biomass could further enhance its capabilities as a fuel. Raw biomasses have relatively low energy contents and generally contain high amounts of moisture resulting in reduced conversion efficiency. The higher amount of oxygen contained within biomass also results in a lower calorific value. The torrefaction process directly affects the oxygen content via the liberation of volatiles that contain oxygen. This causes a decrease in the O/C ratio and results in a biomass with a larger higher heating value (HHV).

Torrefaction is a pretreatment process that aims to improve the quality of biomass by reducing the moisture content and increasing its calorific value via decreasing the oxygen percentage within the biomass. This is done by heating biomass at a constant temperature between 200 and 300 °C for a set period of time in an inert environment. Different gases which are used to maintain an inert environment includes nitrogen (Bridgeman et al. 2008; Arias et al. 2008; Almeida et al. 2010), argon (Prins et al. 2006b) and recently wet torrefaction using hot compressed water were studied to improve the energy density of biomass (Yan et al. 2010). Effect of using a small amount of oxygen on torrefaction was studied by Rousset et al. (2012) and Wang et al. (2012a). Effect of using CO_2 as the pretreatment medium was studied by Eseltine et al. (2013). Under certain torrefaction conditions, the resulting biomass can have properties comparable to low-rank coals (Chen and Kuo 2010a). Increasing the calorific content of biomass by lowering the oxygen–carbon ratio makes it more suitable for use as a fuel for gasification (Prins et al. 2006a). This coupled with the increased grindability makes torrefied biomass an attractive fuel for co-firing applications as well.

Hemicellulose, cellulose, and lignin are the main constituents that comprise the cell structure of lignocellulosic biomass. Changes in the percent of each constituent cause variation in the results obtained from the torrefaction of different biomasses. Recently, there has been a significant amount of research done on how the composition of lignocellulosic biomass influences the thermal degradation. Chen and Kuo (2010a, b) and Chen et al. (2011) have completed studies on the effect of torrefaction temperature on the lignocellulosic structure of several different biomasses. The effect torrefaction has upon the three main constituents is of particular interest due to the chemical makeup of hemicellulose, cellulose, and lignin. Understanding the chemical behavior of these components is integral in optimizing the torrefaction process. Demirbaş (2005) investigated the percent of each of the three main constituents of lignocellulosic biomass and their contribution to the HHV of the fuel. From the analysis of the structural composition, it was determined that, in general, an increase in the amount of lignin in a lignocellulosic biomass results in an increase in HHV. Elemental composition analysis of the lignin portion of each biomass indicated that lignin has an increased carbon content and decreased oxygen content when compared to hemicellulose and cellulose. This decrease in the O/C ratio between lignin and the other two constituents (hemicellulose and cellulose) is what causes lignin to have a higher HHV. Based upon the data obtained, it was determined that there is no direct relation between the hemicellulose and cellulose content of a biomass and its HHV. However, there is a good relation between the amount of lignin within a biomass and the HHV of the biomass. The importance of lignin becomes magnified in torrefied biomass due to most research (Arias et al. 2007; Chen and Kuo 2010a; Prins et al. 2006b), suggesting a mild torrefaction process (240–250 °C) for the pretreatment of biomass. Mild torrefaction, which causes a significant breakdown in hemicellulose, a moderate breakdown in cellulose, leaves lignin as the main contributor to the HHV of the resulting biomass.

The thermal breakdown of the lignocellulosic components of biomass was further studied by Wu et al. (2009) using a TGA. Three different compounds were used to model the three major components of biomass: xylan (a major component of hemicellulose), avicel (cellulose), and alkali lignin. Ultimate analysis of the three substances showed similar results compared to other literature (Demirbaş 2005), with lignin having a lower oxygen–carbon ratio than cellulose and hemicellulose. Weight trace curves and DTG analysis of the data clearly showed the thermal breakdown of each component. The majority of hemicellulose breakdown occurred between 210 and 320 °C with minimal weight loss occurring beyond 320 °C. Cellulose weight loss occurred predominantly between 310 and 390 °C with almost no weight loss occurring outside that temperature range. Lignin showed the widest range of thermal degradation with weight loss occurring between 200 and 550 °C. Related research done by Yang et al. (2007) on the pyrolytic behavior of hemicellulose, cellulose, and lignin had similar results, with the exception of lignin continuing to breakdown up to 900 °C. Since torrefaction occurs within the temperature range of 200–300 °C, thermal degradation of hemicellulose, cellulose, and lignin will occur during torrefaction, as previously shown (Chen and Kuo 2010b).

Most research on torrefaction of various biomasses has resulted in the same conclusions. Mild torrefaction temperatures (240–250 °C) produce biomass that is hydrophobic in nature (Arias et al. 2007), while the resulting heating value yield remains relatively high. Residence temperature has been shown to have a much greater effect upon the chemical changes than residence time. Previous work (Arias et al. 2007) has indicated that residence times beyond 1 h have no significant impact upon the fuel properties of biomass. The weight loss that occurs from mild torrefaction is highly dependent upon the hemicellulose fraction of the biomass being torrefied due to the thermal degradation of hemicellulose occurring below 280 °C. Limited studies were done on the torrefaction capability of CO_2. Considering the temperature limits for Boudouard reaction, the effect of using CO_2 under the pretreatment conditions (200–300 °C) should be studied further (Thanapal et al. 2014).

Thermogravimetric studies have been done to extract the kinetic constants: activation energy and pre-exponential factor from the biomass pyrolysis data. A number of studies have focused on determining the effect of heating rate on pyrolysis of biomass (Biagini et al. 2006; Vamvuka et al. 2003). Different methods have been used to determine the kinetic constants to predict the release of volatile matter from biomass. Some of the common methods which are used to determine the reaction kinetics includes Broido–Shafizadeh model for the pyrolysis of cellulose (Bradbury et al. 1979), Ozawa (Ozawa 1965), single reaction conventional Arrhenius (Chen 2012), independent parallel reactions (Hu et al. 2007; Wang et al. 2012b; Mészáros et al. 2004; Manya et al. 2003), successive reactions (Varhegyi et al. 1989), and distributed activation energy method (Anthony et al. 1976; Sonobe and Worasuwannarak 2008). The kinetic parameters have been determined for individual components of biomass materials: hemicellulose (Peng and Wu 2010), cellulose (Antal and Varhegyi 1995), lignin (Faix et al. 1988; Ferdous et al. 2002), and extractives using the above-mentioned methods. A comprehensive review by Di Blasi (2008) gives detailed information on studies done on the pyrolysis of biomass including different models for pyrolysis process.

Kinetics of pyrolysis of biomass and other fuels are useful for modeling the combustion reactions occurring within a burner. Also such kinetics can be used to determine the amount of mass loss which occurs during thermal pretreatment processes such as torrefaction. Limited studies have focused on utilizing the kinetics extracted from the pyrolysis of biomass constituents on the modeling of torrefaction. Prins et al. (2006c) used a two-step reaction mechanism to model the torrefaction of willow in the temperature range of 200–300 °C. Repellin et al. (2010) used three models to predict the mass loss during the torrefaction process. A simple model based on global weight loss kinetics, Di-blazi Lanzetta two-step reaction model, and Rousset model to study the torrefaction process.

In addition to reduced transportation costs, the improved quality results in lower CO_2 emission per unit amount of energy released. While biomass is considered as a renewable fuel and as such CO_2 emitted during combustion is not considered as emission, the production of such biomass fuel (e.g., ethanol) still involves use of fossil energy which releases CO_2. Thus, it is of interest to formulate a rating system for all fuels including renewable biomass fuels which will directly yield CO_2 emitted in tons per GJ of energy release. Biology literature defines RQ as a ratio of moles of CO_2 produced (or CO_2 eliminated) to stoichiometric oxygen (O_2) moles consumed typically during oxidation reaction, e.g., oxidation of nutrients in the body which results in release of energy for maintaining the normal warmth (37 °C) of the body. The RQ factor for fat, protein, and carbohydrates, the three basic nutrients of the body, is 0.7, 0.8, and 1, respectively (Hall 1905; Richardson 1929; Morinaka et al. 2012; McClave et al. 2003). However, senior citizens have problem of excreting CO_2, and hence, they are prescribed low RQ diet (e.g., fat with low RQ). Such a concept has been recently introduced in thermal systems in order to select low RQ fuel for power generation (Annamalai 2013). This method of RQ concept for estimating the CO_2 emission from different fuels was elaborated here to estimate the global warming potential (GWP) of fossil and as well as renewable fuels.

Current chapter presents the results obtained from TGA studies on CO_2 torrefaction of woody biomass which dominates the rangelands and grasslands of southwestern United States (mesquite and juniper), batch torrefaction of samples using a laboratory oven, modeling the weight loss during torrefaction using a three-component model and application of RQ concept to estimate GWP of different fuels.

2 Materials and Methods

2.1 Thermogravimetric Analyzer

All TGA tests were performed utilizing a TA Instruments Q600 thermal analyzer located at the Coal and Biomass Energy Laboratory (CABEL) at Texas A&M University. Ground biomass samples of size between 540 and 890 μm were used for the TGA tests. Around 10 mg sample was used for each TGA study. The instrument is capable of highly reproducible measurements with ±1 % accuracy. The purge gas flow was regulated to below 20 psi using a pressure regulating valve attached to the purge gas tank. During experiments, the flow rate of the purge gas was controlled via the TA Instrument Explorer software installed on the connected PC and regulated via the internal mass flow controller in the Q600. Purge air was used after each test run to cool the furnace back to an ambient temperature. The TA Instrument Explorer software allows for direct control of several test parameters and can also be utilized to create a step by step test procedure. Two separate test procedures were used for the torrefaction and pyrolysis tests completed with the Q600. A constant heating rate of 20 °C/min was used for the pyrolysis tests where the samples were heated from room temperature till 900 °C. For the torrefaction tests, the samples were heated at a constant heat rate till the torrefaction temperature and then maintained at the same temperature for a given time period. After the torrefaction process, the samples were again heated at a constant heating rate till 900 °C. All the tests were carried out in an inert environment by flowing nitrogen, argon, and carbon dioxide depending upon the environment which has to be used for the torrefaction and pyrolysis study.

2.2 Laboratory Oven and Grindability Tests

In order to obtain higher amount of samples for determining the fuel properties and study grindability, 5 g of samples were torrefied in a laboratory oven. The samples were prepared by heating at a constant rate of 20 °C/min from room temperature till the optimum torrefaction temperature. The samples were then torrefied at the given torrefaction temperature for a constant time period. Three torrefaction mediums were used for preparing the samples: nitrogen, carbon dioxide, and argon.

Both raw and torrefied samples were shipped to a third party commercial evaluation laboratory for fuel property analysis. After large-scale torrefaction using the laboratory oven, torrefied and raw biomass samples were tested for grindability. The grindability tests consisted of sieve analysis (size distribution analysis) and sample grinding using a Sweco model DM1 grinding mill. All samples were ground for 20 min so that the amount of energy used in grinding the samples was held constant.

2.3 Torrefaction Model

For a biomass material composed of hemicellulose, cellulose, and lignin represented in a carbon-normalized form, mole balance can be represented as

$$\mathrm{CH}_m\mathrm{O}_n \rightarrow n_{\mathrm{hemi}} \times \mathrm{CH}_{\mathrm{hh}}\mathrm{O}_{\mathrm{oh}} + n_{\mathrm{cell}} \times \mathrm{CH}_{\mathrm{hcell}}\mathrm{O}_{\mathrm{ocell}} + \mathrm{n}_{\mathrm{lig}} \times \mathrm{CH}_{\mathrm{hl}}\mathrm{O}_{\mathrm{ol}}$$

where m and n stand for the number of hydrogen and oxygen atoms, respectively, in the carbon-normalized fuel. From the atom balance; known percentage composition of hemicellulose, cellulose, and lignin; and H/C and O/C ratios for hemicellulose and cellulose, one can determine H/C and O/C for lignin. The heating value of the biomass can be represented in terms of the composition of the individual components.

$$\mathrm{HV}_{\mathrm{biomass}} = Y_{\mathrm{hemi}} \times \mathrm{HV}_{\mathrm{hemi}} + Y_{\mathrm{cell}} \times \mathrm{HV}_{\mathrm{cell}} + Y_{\mathrm{Lig}} \times \mathrm{HV}_{\mathrm{Lig}} \tag{1}$$

where Y represents the mass fraction of different biomass components. The conversion of each of the biomass component ($\mathrm{f_k}$) and the overall conversion (F) can be given as

$$f_k = \frac{m_k - m_{k,\mathrm{Char}}}{m_{k,o} - m_{k,\mathrm{Char}}} = \frac{Y_k - Y_{k,\mathrm{Char}}}{Y_{k,o} - Y_{k,\mathrm{Char}}} = \frac{\text{mass of } k \text{ that remains to be devolatilized}}{\text{Max mass of } k \text{ that could be devolatilized}}$$
$$k = \text{hemicellulose, cellulose and lignin.} \tag{2}$$

$$F = \frac{m}{m_0} = \frac{\sum m_k}{m_0} = \left\{ f_k \left(Y_{k,o} - Y_{k,Char} \right) + Y_{k,Char} \right\} = \frac{\text{mass left over}}{\text{Initial mass}} \tag{3}$$

where $Y_k = m_k/m_0$: m_0 is the initial dry ash-free mass of the component, and m_k is the dry ash-free mass of component after time t. $Y_{k,\text{char}}$ represents the mass fraction of k remaining in char at end of pyrolysis. The energy content ratio of the biomass after being heat treated for some time period can be determined from the remaining mass fraction of each component k of the treated sample and heating value of individual components (Eq. 1).

Assuming first-order pyrolysis for each component k, the decomposition of each of the biomass component can be determined using the following:

$$-\frac{\mathrm{d}m_k}{\mathrm{d}t} = B_k m_k \exp\left(\frac{-E_k}{\bar{\mathrm{R}}\mathrm{T}}\right), \quad k = \text{Hemicellulose, Cellulose, Lignin} \tag{4}$$

where m_k is the mass remaining in each of the components in sample (kg) which can be devolatilized. $m_k = (m_{k,0} - m_{k,\text{char}}) - m_{k,\text{lib}}$. Amount of char in each of the components was obtained from the pyrolysis data available elsewhere (Yang et al. 2007).

$$\frac{\mathrm{d}m_{k,lib}}{\mathrm{d}t} = B_k \left(m_{k,0} - m_{k,\text{char}} - m_{k,\text{lib}} \right) \exp\left(\frac{-E_k}{\bar{\mathrm{R}}\mathrm{T}}\right) \tag{5}$$

The above expression can be further simplified to

$$\frac{\mathrm{d}\alpha_k}{\mathrm{d}t} = B_k (1 - \alpha_k) \exp\left(\frac{-E_k}{\bar{\mathrm{R}}\mathrm{T}}\right) \tag{6a}$$

$$\frac{\mathrm{d}\alpha_k}{\mathrm{d}T} = \left(\frac{B_k}{\beta}\right) (1 - \alpha_k) \exp\left(\frac{-E_k}{\bar{\mathrm{R}}\mathrm{T}}\right) \tag{6b}$$

where α represents the liberated fraction,

$$\alpha_k = \frac{m_{k,\text{lib}}}{m_{k,0} - m_{k,\text{Char}}} = \frac{m_{k,0} - m_k}{m_{k,0} - m_{k,\text{Char}}} = \frac{Y_{k,0} - Y_k}{Y_{k,0} - Y_{k,\text{Char}}} \tag{7}$$

When the sample is heated at a constant heating rate β, Eqs. 6a and 6b reduce to

$$f_k = (1 - \alpha_k) = \exp\left[-\left(\frac{B_k}{\beta} \frac{E_k}{\bar{R}}\right) \cdot \left\{ \frac{\mathrm{E}_2(X_k)}{X_k} - \frac{\mathrm{E}_2\left(X_{k,0}\right)}{X_{k,0}} \right\} \right], \quad T < T_{\text{torr}} \tag{8}$$

where $\beta = \mathrm{d}T/\mathrm{d}t$ (Chen 2012). Once the torrefaction temperature T_{torr} is reached, the sample is maintained at T_{torr}, a constant value for a given residence time. From Eqs. (6a) and (6b), the mass loss during the constant temperature period for each of the component can be determined.

$$f_k = (1-\alpha_k) = \exp\left\{-B_k f_{k,T_{torr}} \exp\left(\frac{-E_k}{\bar{R}T_{torr}}\right)(t - t_{start,fixed\ T})\right\}, \quad T = T_{torr} \tag{9}$$

where B_k is the pre-exponential rate constant, E_k the activation energy for the kth component, and $\bar{R}$ is the universal gas constant. The overall conversion (F) during the torrefaction process can be determined from the conversion of each biomass component (f_k) using Eq. (3).

2.4 Respiratory Quotient (RQ)

Considering a power plant producing a fixed amount of power, heat input from the fuel and oxygen consumption for most fuels are fixed. Hence, a fuel with higher RQ factor will result in more CO_2 being released into the environment. Fuel chemical composition can be used to determine the fuel heating value using an empirical relation given by Boie (Annamalai and Puri 2007).

$$\begin{aligned}\mathrm{HHV(kJ/kmol)} = {} & 422{,}270 \times \mathrm{C} + 117{,}385 \times \mathrm{H} - 177{,}440 \times \mathrm{O} + 87{,}985 \times \mathrm{N} \\ & + 335{,}510 \times \mathrm{S}\end{aligned} \tag{10}$$

where C, H, O, N, and S are the number of carbon, hydrogen, oxygen, nitrogen, and sulfur atoms, respectively, in the fuel. Stoichiometric oxygen needed for complete combustion of a $C_C H_H N_N O_O S_S$ can be determined from the fuel chemical composition using the following relation.

$$\begin{aligned} v_{O_2}(\text{kg of oxygen/kg of fuel}) &= \frac{32 \times \left(\mathrm{C} + \frac{\mathrm{H}}{4} - \frac{\mathrm{O}}{2} + \mathrm{S}\right)}{M_{fuel}} \\ &= \frac{32 \times \mathrm{C} \times \left(1 + \frac{\mathrm{H}}{4\mathrm{C}} - \frac{\mathrm{O}}{2\mathrm{C}} + \frac{\mathrm{S}}{\mathrm{C}}\right)}{M_{fuel}} \end{aligned} \tag{11}$$

From Eqs. (10) and (11), Higher heating value of fuels per kg of stoichiometric oxygen (HHV_{O_2}) can be determined. It was estimated that HHV_{O_2} calculated for all the fuels was around 14,000 kJ/kg O_2 (Annamalai and Puri 2007; Thanapal 2014).

Based on the fuel composition, RQ of a particular fuel with C atoms of carbon which can produce C moles of carbon dioxide can be determined using the following equation.

$$RQ = \frac{1}{\left\{1 + \left(\frac{H}{4C}\right) - \left(\frac{O}{2C}\right) + \left(\frac{S}{C}\right)\right\}} \tag{12}$$

3 Results and Discussion

3.1 Fuel Preparation and Properties

Biomass fuels used for the current study, mesquite and juniper, were harvested from the rangelands in north central Texas near Vernon. Both mesquite and juniper are scattered over a large area which is estimated to be 200,000 ha. It should be noted that the species is not concentrated throughout the entire area but dispersed with a spacing of around 1 km. After the fuel is harvested using a chainsaw, a chipper is used to reduce the particle size from trunk and branches. It also includes the bark. It was observed that, when a freshly harvested biomass (moisture content around 45 %) was sent into the chipper for reducing the particle size, the chips produced after the chipping process had a lower moisture percentage of between 10 and 20 %. This might be because of the drying of the wood chips within the chipper using the heat produced as a result of the chipping process. Vermeer wood chippers were used for processing the biomass. Further details on the preparation of the samples are available elsewhere (Chen 2012). Properties of mesquite and juniper obtained from the commercial testing labs are shown in Table 1.

Table 1 Ultimate and Proximate analyses of mesquite and juniper

	Raw biomass	
	Mesquite	Juniper
Moisture (ar)	15.53	5.85
Volatile matter (ar)	66.09	77.99
Fixed carbon (ar)	16.71	14.25
Ash (ar)	1.67	1.91
Carbon (ar)	43.60	49.27
Oxygen (ar)	33.57	37.00
Hydrogen (ar)	4.98	5.68
Nitrogen (ar)	0.62	0.28
Sulfur (ar)	0.03	0.01
Dry ash-free basis		
VM (daf)	79.8	84.6
FC (daf)	20.2	15.4
HHV (kJ/kg)	16,666	18,987
HHV_{dry} (kJ/kg)	19,730	20,167
HHV_{DAF} (kJ/kg)	20,128	20,584
VM HHV_{DAF} (kJ/kg)	16,923	18,351
$HHV_{Boie,DAF}$ (kJ/kg)	21,059	21,509
HHV (kJ/kg st O_2)	13,652	13,632

3.2 *Thermogravimetric Analysis*

3.2.1 Mesquite

The thermograms for pyrolysis and torrefaction of mesquite using N_2, Ar, and CO_2 as the purge are shown in Figs. 1 and 2. The TGA thermograms for each purge gas have the same relative shape and can be broken down into three different zones: drying, active pyrolysis zone, and passive pyrolysis zone. For simplicity, the thermograms for each torrefaction temperature were overlaid on the same plot. For torrefaction, each thermogram follows the weight loss curve for pyrolysis until reaching the torrefaction temperature. At this point, the thermogram follows a vertical line resulting from weight loss due to the sample being held at a constant temperature for the duration of torrefaction. Overlaying the thermograms of samples that were pyrolyzed and torrified using different purge gasses gives more insight as to the effect of the purge gas upon the breakdown of the biomass. Overlays of both argon and carbon dioxide versus nitrogen (the reference gas) are shown in Figs. 1 and 2, respectively, (Eseltine 2011).

From initial inspection, it is clear that there is very little difference in the thermograms from the pyrolysis and torrefaction in nitrogen and argon. Both nitrogen and argon exhibit roughly the same amount of weight loss at a given torrefaction temperature. Slight differences in weight loss between these two purge gases can be attributed to differences in the biomass samples being tested. Comparison of the thermograms using nitrogen and carbon dioxide as the purge gas shows a noticeable difference between the two gases. For torrefaction temperatures above 240 °C, samples torrefied using CO_2 as the purge gas have a higher amount of weight loss

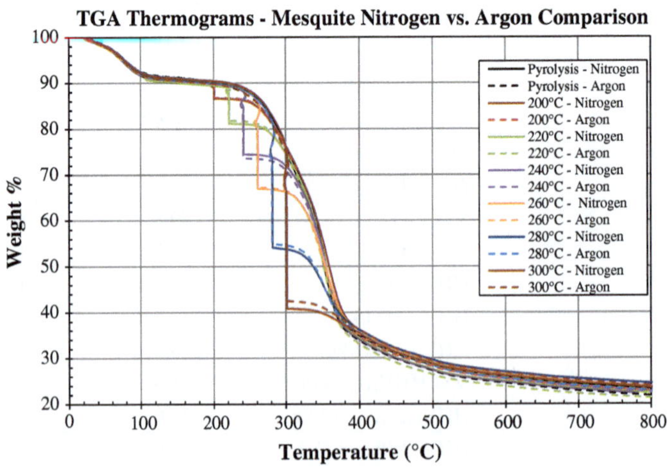

Fig. 1 Comparison of thermograms from pyrolysis and torrefaction of mesquite wood using N_2 and Ar as the purge gases, adapted from Eseltine (2011)

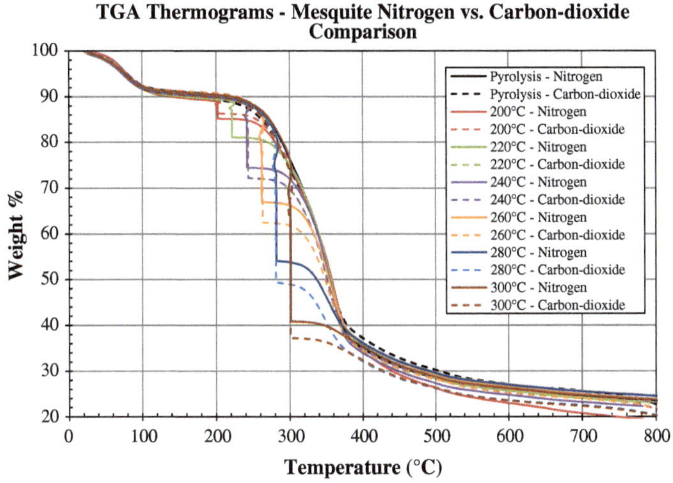

Fig. 2 Comparison of thermograms from pyrolysis and torrefaction of mesquite wood using N_2 and CO_2 as the purge gases, adapted from Eseltine (2011)

than samples torrefied using nitrogen. This is due mainly to chemical reactions occurring between CO_2 and the carbon within the fuel which will be discussed in further detail in a subsequent section.

3.2.2 Juniper

Data obtained from torrefaction and pyrolysis of juniper wood chips were analyzed in the same manner as the data from mesquite wood pyrolysis and torrefaction. The shape of the thermograms from juniper wood follows the same trends as the thermograms from mesquite wood. The only noticeable difference between the pyrolysis curves of mesquite and juniper is being the weight percent remaining after the active pyrolysis zone. Juniper wood loses more weight during the initial active pyrolysis zone when compared to mesquite due to differences in the chemical composition of the two biomasses.

Comparing the weight loss from the torrefaction of mesquite and juniper wood, it can be seen that mesquite wood has a higher amount of weight loss at a given torrefaction temperature. This is due mainly to differences in the chemical composition of both biomasses. Prins et al. (2006a) have stated that deciduous woods, such as mesquite, have a hemicellulose fraction comprised of mostly xylan. Xylan is being more reactive than mannan, the main constituent of the hemicellulose fraction found in coniferous woods (such as juniper). Due to the temperature range over which hemicellulose breaks down, a significant portion of the weight lost during torrefaction occurs from the decomposition of hemicellulose. Therefore, the hemicellulose fraction of mesquite being more reactive than that of juniper is the main cause for a higher weight loss during the torrefaction of mesquite.

For further analysis of the effect of the purge gas upon the torrefaction of juniper, thermograms from the torrefaction and pyrolysis using argon and carbon dioxide were overlaid with the thermograms using nitrogen as the purge gas. These overlays can be seen in Figs. 3 and 4, respectively. The trends observed from the comparison of argon and carbon dioxide to nitrogen appears similar to the ones observed from

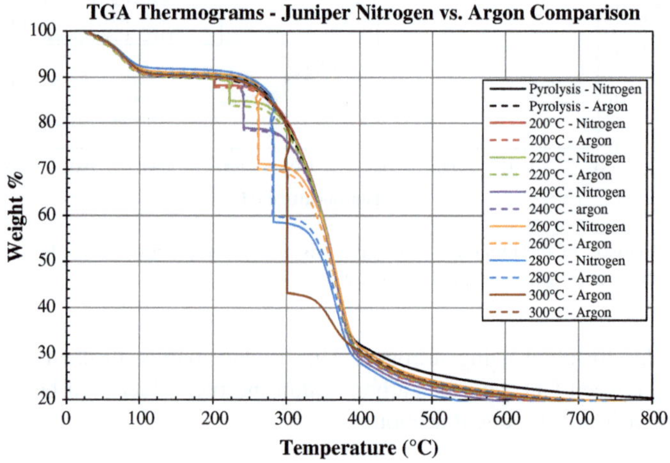

Fig. 3 Comparison of thermograms from pyrolysis and torrefaction of juniper wood using N_2 and Ar as the purge gasses, adapted from Eseltine (2011)

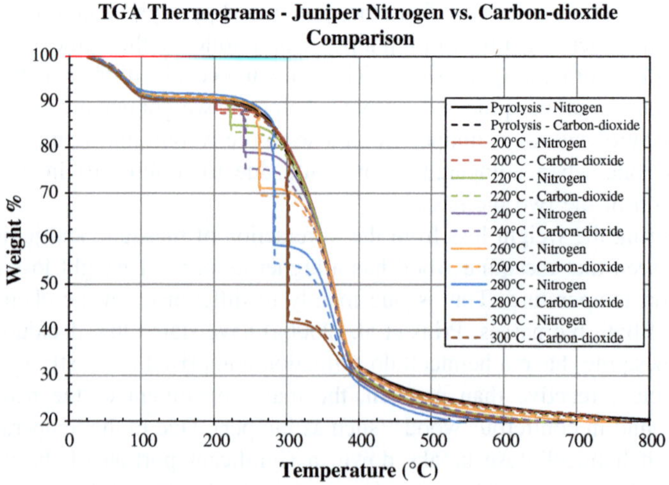

Fig. 4 Comparison of thermograms from pyrolysis and torrefaction of juniper wood using N_2 and CO_2 as the purge gases, adapted from Eseltine (2011)

previous experiments with mesquite. The slight differences are attributed to the variations in the samples tested. Comparison of the thermograms using nitrogen and carbon dioxide as the purge gas shows that using carbon dioxide results in a higher amount of weight loss.

3.2.3 Optimum Torrefaction Temperature

Based upon the TGA thermograms, an optimal torrefaction temperature can be determined. From the TGA thermograms presented for each biomass and purge gas at 240 °C, more than 70 % of the initial mass of the biomass remains after torrefaction. This is independent of both biomass and purge gas. Tests conducted at a torrefaction temperature of 260 °C resulted in roughly 58–65 % of the biomass remaining. Furthermore, an estimated percent of energy retained after torrefaction can be calculated based upon the raw biomass fuel properties and the TGA weight traces. The energy retention estimates were calculated based upon the HHV of the fuel on a dry ash-free (DAF) basis and the amount of fixed carbon (FC) and volatile matter (VM) in the biomass. If we use the DAF HHV of the biomass, the energy content comes entirely from the VM and FC and the following relation can be used.

$$\mathrm{HHV_{DAF}}\left(\frac{\mathrm{kJ}}{\mathrm{kg}}\right) = \mathrm{HHV_{VM}} \times \mathrm{VM} + \mathrm{HHV_{FC}} \times \mathrm{FC} \tag{13}$$

If we assume that the mass lost during torrefaction is entirely due to liberation of volatiles then the HHV of the torrefied biomass can be estimated as

$$\mathrm{HHV_{DAF,TB}}\left(\frac{\mathrm{kJ}}{\mathrm{kg}}\right) = (\mathrm{wt\%} \times \mathrm{VM} \times \mathrm{HHV_{VM}}) + (\mathrm{HHV_{FC}} \times \mathrm{FC}) \tag{14}$$

where wt% represents the weight remaining after torrefaction. Using the estimation proposed above to determine the HHV of the torrefied biomass on a DAF basis, it is possible to estimate the amount of energy retained after torrefaction using the following relation (Eseltine 2011).

$$\text{Energy Retained } (\%) = \left(\frac{M_{\mathrm{TB}} \times \mathrm{HHV_{TB}}}{M_{\mathrm{RB}} \times \mathrm{HHV_{RB}}}\right) \times 100 \tag{15}$$

Using data from the heat value analysis of the raw biomass, which gives the HHV of the VM, with the proximate analysis results an estimated HHV of the torrefied biomass can be calculated and therefore an estimated energy retained %. Based upon the mass loss observed from TGA thermograms as well as the estimated energy retention calculations, the optimal torrefaction temperature was determined to be 240 °C. For all of the biomass tested in each purge gas environment, torrefaction at 240 °C results in over 85 % energy retention as well as 70 % of the initial mass remaining.

3.2.4 Laboratory Oven Tests

After determining the optimal torrefaction temperature, large samples of each biomass were torrefied in a laboratory oven using nitrogen, argon, and carbon dioxide as the purge gas. Torrefied samples were then sent to commercial testing labs for analysis. Figure 5 shows the properties of the raw and torrefied samples. As expected, torrefaction of the raw biomass resulted in reduced moisture content in the biomass. Torrefaction of raw biomass also resulted in an increased ash and fixed carbon content for all biomass tested. The increase in ash and FC is mostly due to the decrease in moisture content and liberation of light weight volatiles during torrefaction. Biomass torrefied in an argon or nitrogen environment exhibits little difference in fuel properties. However, the fuel properties of biomass torrefied in carbon dioxide are slightly different.

Previous research (Lawrence et al. 2009) has shown that coal–biomass blends with a ratio of coal to biomass greater than 80:20 result in several combustion issues due to the fuel properties of the raw biomass. With the fuel properties of torrefied biomass being much better than raw biomass, it is entirely feasible that blended fuels with a ratio of coal to torrefied fuel greater than 80:20 could be used in co-firing applications.

Knowing the actual HHV of the torrefied biomass (HHV_{TB}), the actual amount of energy retained can be calculated (Sect. 3.2.3). Due to inherent mass loss from torrefaction, the energy retained provides the percentage of the raw biomass retained within the torrefied biomass. The resulting calculated values of energy retained for both biomasses in each purge gas environment can be seen in Table 2.

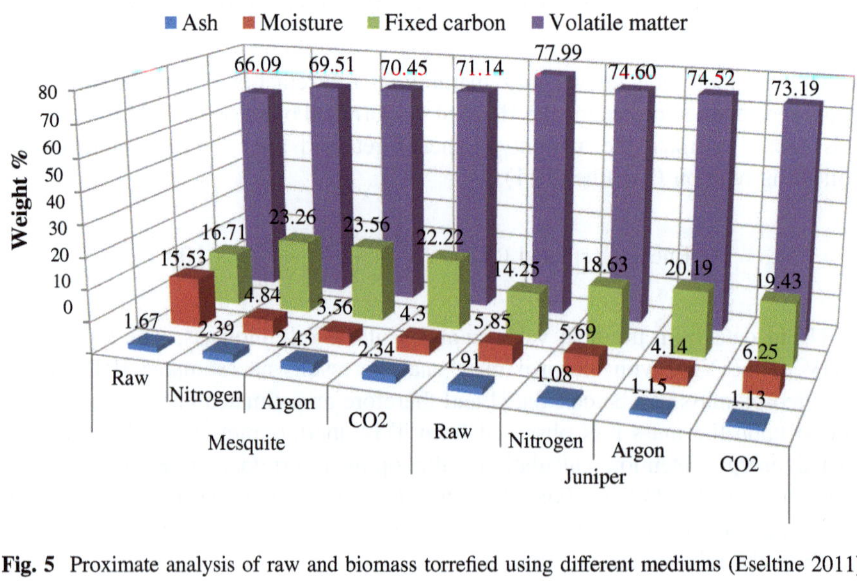

Fig. 5 Proximate analysis of raw and biomass torrefied using different mediums (Eseltine 2011)

Table 2 Percent energy retained in the torrefied biomass

Biomass	Purge gas		
	Nitrogen	Argon	Carbon dioxide
Mesquite	88.74	87.73	84.47
Juniper	83.68	83.16	78.86

The energy retention calculations show that torrefaction in a nitrogen environment resulted in the highest amount of energy being retained. Torrefaction using argon as the purge gas had a slightly lower amount of energy retention when compared to nitrogen. However, the energy retention of torrefaction in a carbon dioxide environment shows that using carbon dioxide as a purge gas results in the lowest energy retention for all three biomasses tested. The lower amount of energy being retained after torrefaction using CO_2 is a result of more volatiles being liberated via the Boudouard reaction.

3.2.5 Grindability Tests

The samples torrefied using the laboratory oven were then ground using an in-house vibratory grinding mill. From the grindability tests, torrefied samples exhibited improved grindability in comparison to raw biomass samples. For both mesquite and juniper, torrefaction using CO_2 resulted in a higher fraction of the ground biomass passing through an 840-μm mesh size. Further grindability tests conducted on samples torrefied in a bigger batch reactor by Thanapal et al. showed that using CO_2 as the torrefaction improved the grindability of torrefied samples by 30 %. More details on those tests are available elsewhere (Eseltine 2011).

3.2.6 Torrefaction Model

Properties of the Biomass Components

The ultimate and proximate analyses of mesquite and juniper are presented in Table 1. In addition, ultimate and proximate analyses of hemicellulose and cellulose were obtained from literature (Yoon et al. 2011; Van de Weerdhof 2010). The composition of lignin was then determined from the fuel composition and the data for hemicellulose and cellulose according to Eq. (1). Composition of cellulose and hemicellulose is presented in Table 3.

Based on the pyrolysis tests carried out on the three components, 32, 2.5, and 40 %, were left over as char from hemicellulose, cellulose, and lignin, respectively, on a dry ash-free basis (Raveendran et al. 1996). The estimated bulk char percentage on a DAF% based on the percentage of hemicellulose, cellulose, and lignin (Table 3) and the char content left after pyrolysis will be 15.64 and 17.44 % for hard wood mesquite and softwood juniper samples, respectively. The numbers

Table 3 Ultimate and proximate analyses of biomass components (Van de Weerdhof 2010; Yoon et al. 2011)

	Hemicellulose	Cellulose
Volatile matter (db)	74.11	91.64
Fixed carbon (db)	21.94	8.54
Ash (db)	3.95	0
C (daf)	43.77	43.58
H (daf)	5.91	6.09
O (daf)	50.26	50.27
N (daf)	0.05	0.05
S (daf)	0.02	0.01
Heating value (kJ/daf kg)	15,920	17,870

estimated for the bulk char percentage in the two samples were similar to the fixed carbon in mesquite and juniper on a dry ash-free basis (16.98 % for mesquite and 15.45 % for juniper from Table 2). The percentage of fixed carbon in the three components was used as a reference to determine the amount of volatile matter released from the components during the torrefaction process. Amount of these three components depends on the type of biomass. Hardwood was found to contain higher percentage of hemicellulose when compared to softwood. Percentage of the different components in hardwood and softwood was obtained from Liu et al. (2008). Table 4 shows the percentage of the three components along with the kinetic constants for the three components which were determined using error minimization technique.

Component Conversion and Mass Loss During Torrefaction: TCM

Volatilization of the three components at different torrefaction temperatures and residence times was studied. The results obtained for torrefaction of mesquite which is a hardwood at 240 °C for a residence time of 60 min is shown in Fig. 6. It was assumed that the samples are heated from room temperature to the desired torrefaction temperature at a heating rate of 20 °C/min.

Table 4 Composition of hardwood and softwood

	Hemicellulose (%)	Cellulose (%)	Lignin (%)
Softwood (Liu et al. 2008)	12.27	53.26	26.66
Hardwood (Liu et al. 2008)	28.97	53.95	9.43
B (1/min)	6.66E+08 (Cozzani et al. 1997)	6.83E+16 (Orfao et al. 1999)	1,000 (Varhegyi et al. 1997)
E (kJ/mol)	103.2 (Cozzani et al. 1997)	201 (Orfao et al. 1999)	65 (Varhegyi et al. 1997)

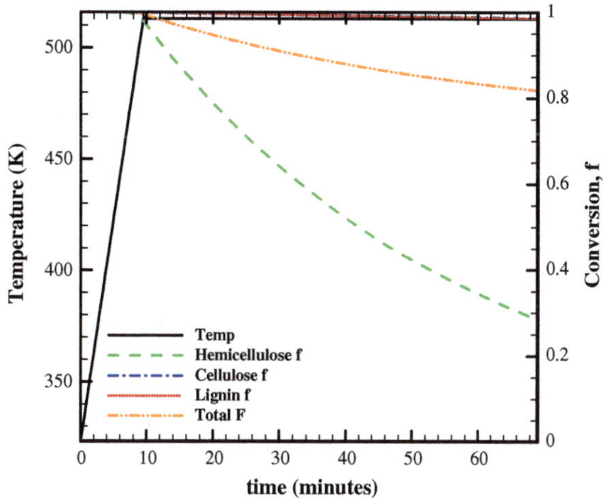

Fig. 6 Torrefaction of mesquite at 240 °C for 60 min: results predicted by TCM

The loss of volatiles from the three components of mesquite after 60-min time period at 240 °C was predicted to be 71.5, 1.39, and 1.46 % for hemicellulose, cellulose, and lignin, respectively, on a dry ash-free basis. The overall dry ash-free mass loss from the biomass sample was 18.1 %. The results obtained from previous torrefaction experiments carried out on smaller samples (10 mg) in a TGA on a dry ash-free basis were used as a reference for comparison with model results. It should also be noted that the juniper which is a softwood shows lower mass loss due to lower percentages of hemicellulose in the sample.

Effect of temperature on the torrefaction process was studied using TCM. Figure 7 compares the TCM results with the experimental results for a residence

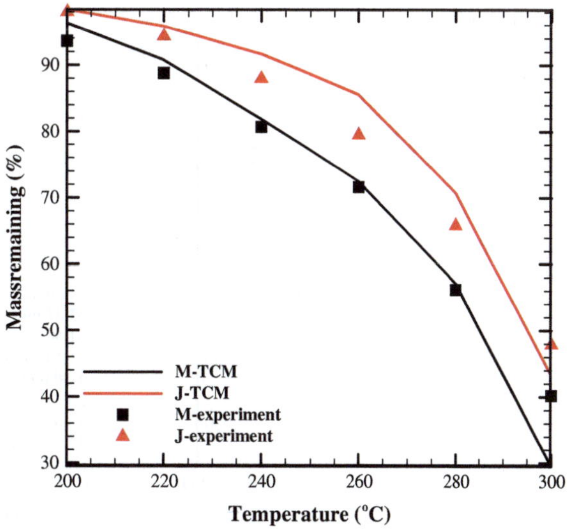

Fig. 7 Effect of temperature on the loss of volatile matter from the samples. Results from the model are compared with experimental results. *M* Mesquite; *J* Juniper; *TCM* Three-component model; t_{res} = 60 min

time of 60 min for both mesquite and juniper samples. The model results compare well with the experimental results except at the upper temperature limit. At temperature of 300 °C, TCM overpredicts the mass loss from the woody biomass. The kinetic constants of cellulose used to model the torrefaction process resulted in increased losses of cellulose at higher temperatures in the TCM. The variation in mass loss with increase in residence time for an initial heating rate of 20 °C/min is plotted in Fig. 8. Following residence times were studied: 30, 45, 60, and 120 min. Effect of higher residence time on mass loss was negligible at lower torrefaction temperatures, while an increase in temperature resulted in increased mass loss due to higher time and energy available for the volatilization of cellulose and lignin present in the biomass. Higher percentages of cellulose are released from the sample resulting in higher mass loss.

As the heating rate increased, the results showed an increased retention of mass at higher torrefaction temperatures when compared to the lower heating rates. Higher heating rates do not allow sufficient time for the volatiles to escape out from the biomass which causes lower mass loss during the initial heat up period. Hence, higher heating rates can be employed to reduce the loss of combustible volatile matter from the biomass at higher temperatures.

Energy Conversion Ratio (ECR) and Heating Value of the Torrefied Sample

The heating value and the chemical composition of the three components were used to monitor the change in heating value of the torrefied biomass with reference to the raw samples. Energy conversion ratio (ECR) which is defined as the ratio of the energy content of the torrefied sample to the energy content of the raw biomass and the ratio of the heating value of the torrefied biomass to raw biomass was also

Fig. 8 Variation in mass loss with increase in torrefaction temperature and residence times. *M* Mesquite; *J* Juniper

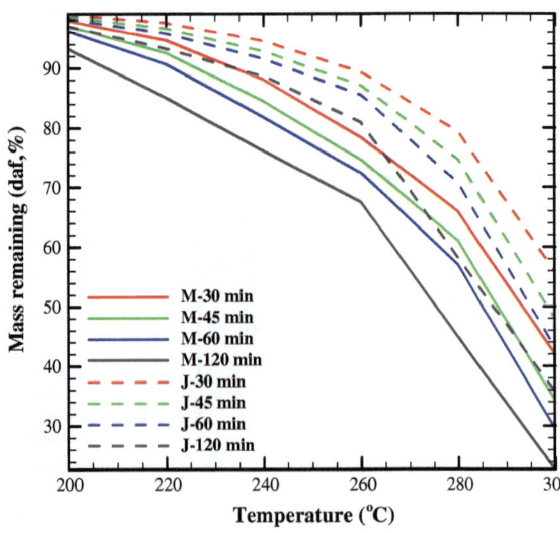

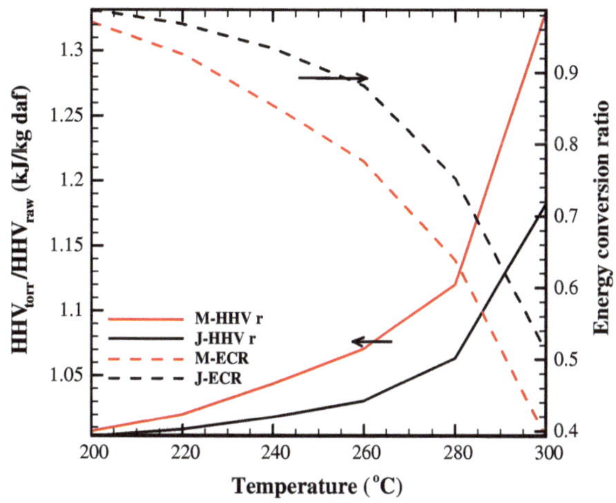

Fig. 9 Variation in heating value and ECR with increase in temperature. *M* Mesquite; *J* Juniper; *HHV r* Heating value ratio; *ECR* Energy conversion ratio

determined. Figure 9 shows the variation of heating value and ECR with respect to torrefaction temperatures.

As can be observed from Fig. 9, heating value of the torrefied sample increases with increase in torrefaction temperature as the hemicellulose and cellulose volatilizes and the sample becomes rich in char and lignin which have higher heating value. The ratio of the heating value of the torrefied samples to the raw biomass samples from the experiments at 240 °C for a residence time of 60 min was 1.06 and 1.05 for mesquite and juniper, respectively. The ratio of the heating values predicted by the model was lower. In the case of TCM, the composition of volatile matter released from the sample is assumed to remain constant and it also includes carbon and hydrogen along with oxygen. Hence, lower ratios were predicted by the model when compared to that of experiments. The ECR of the torrefied samples decreases due to loss of combustible volatile matter. As the energy content of the sample decreases with increase in temperature, ECR decreases. It is seen that ECR decreases rapidly after about 260 °C due to higher loss of cellulose. Thus, T_{torr} must be limited to $T_{torr} < 260$ °C. However, on a unit mass basis, heating value of the torrefied sample is higher than the raw biomass. Hence, the three-component model is a simple and powerful model to predict the mass loss from the biomass samples undergoing torrefaction. Such a model can also be applied to the fibrous biomass torrefaction. It should be noted here that the order of reaction for the volatilization of the three components is assumed to be one.

The kinetics used in the TCM also influences the amount of mass loss and change in heating value of the sample. In order to compare the effect of kinetic constants on the mass and energy yield, kinetics of pine wood components extracted using error minimization technique (Wang et al. 2012b) was used to predict the mass loss and ECR. The results obtained were compared to the heating value ratio determined using the kinetics presented in Table 4. Figure 10 plots the

Fig. 10 Variation in heating value with increase in torrefaction temperatures for two different kinetics. (i) Kinetics obtained from Wang et al. (2012b) and (ii) kinetics from Table 4

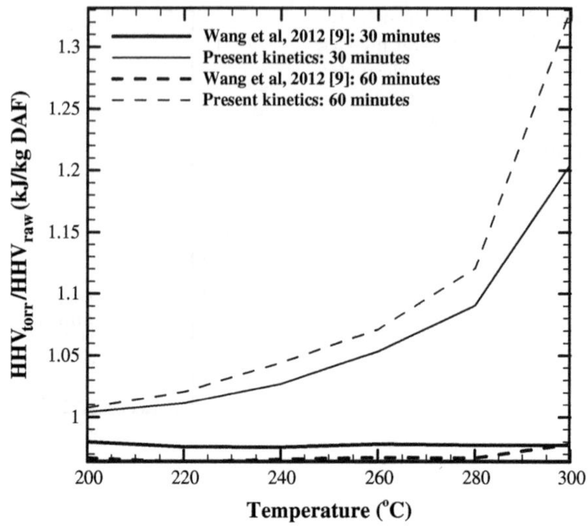

variation in heating value ratios obtained for two different residence times. Kinetics obtained from Wang et al. (2012b) resulted in higher losses in Lignin which decreased the heating value of the torrefied samples.

3.3 RQ Factor for Fuels

3.3.1 CO_2 Emissions

Boie equation can be used to derive an expression to estimate the CO_2 emitted in tons per GJ of energy input from the fuel chemical composition as given in the Eq. (16).

$$CO_2\text{in tons/GJ} = \frac{1 \times 44 \times 1{,}000}{\left\{422{,}270 \times \left(1 + \left(\frac{117{,}385}{422{,}270}\right) \times \left(\frac{H}{C}\right) - \left(\frac{177{,}440}{422{,}270}\right) \times \left(\frac{O}{C}\right)\right)\right\}} \quad (16)$$

The CO_2 in tons per GJ of energy input is given according to Eq. (17a) (Thanapal 2014).

$$CO_2 \text{ in tons per GJ energy input} = RQ \times \frac{\text{kg } O_2}{HHV_{O_2} \text{ in GJ}} \times \frac{\text{kmol } O_2}{32 \text{ kg}} \times \frac{44.01 \text{ kg}}{\text{kmol } CO_2} \times \frac{0.001 \text{ tons}}{\text{kg}} \quad (17a)$$

Assuming $HHV_{O_2} = 0.014$ GJ/kg O_2,

$$CO_2 \text{ in tons per GJ of energy input can be reduced to} \approx RQ \times 0.1. \quad (17b)$$

where the approximate sign is due to assumption of constant $HHV_{O_2} = 0.014$ GJ per kg of O_2 consumed. For RQ = 1 (pure carbon), CO_2 is about 0.1 tons per GJ or 100 g per MJ.

In order to validate approximate expression for CO_2, actual measured heating values of fuels for which compositions are well known (e.g., CH_4, C_8H_{18}, $C_{12}H_{23}$, C_2H_5OH, coal, and biomass) are used to estimate CO_2 in tons per GJ. Results are shown in Fig. 11. It is apparent from Fig. 11 that CO_2 in tons per GJ of energy input has a slope of 0.1 which confirms the approximation.

The RQ factor for fuels of known composition can be estimated using Eq. (12). Table 5 shows the variation of RQ factor for different fuels.

In general, the RQ factor increases with decrease in hydrogen-to-carbon ratio. Methane which has a H/C ratio of 4 has the lowest RQ factor for pure fuels with RQ = 0.5. It can be observed from Table 5 that the solid fuels have comparatively higher oxygen content. Higher oxygen content results in higher RQ factor for solid fuels. Gasoline and diesel fuels which are used in the automobiles have a comparatively lower RQ when compared to that of solid fuels. Lower O/C ratio of the liquid fuels is one of the factors behind the lower RQ values.

It is seen that most solid fuels (pure carbon RQ = 1, biomass fuels (mesquite and juniper) RQ = 0.94–0.97, most sweet sorghum sources = 0.98–1.0 (TAMU 2013),

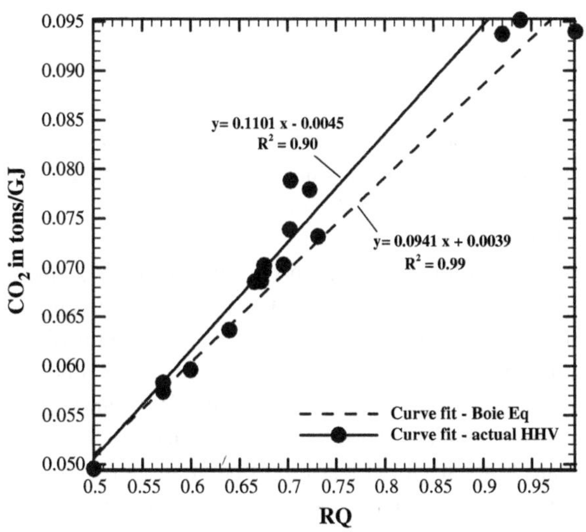

Fig. 11 CO_2 emitted in tons per GJ of energy input for fuels with different RQ factors. Fuel measured heating value and composition data was used in estimation of CO_2. Slope of both the trend lines (actual heating value and heating value estimated using Boie equation) was 0.1 which is same as the approximate value derived using Eq. (17b)

Table 5 RQ factor for different fuels along with their respective O/C and H/C ratios

Fuel	O/C	H/C	*RQ*
Gaseous fuels			
Methane	0.00	4.00	0.50
Ethane	0.00	3.00	0.57
Ethylene	0.00	3.00	0.57
Propane	0.00	2.67	0.60
Acetylene	0.00	1.00	0.80
Liquid fuels			
Methanol[a]	1.00	4.00	0.67
Ethanol[a]	0.50	2.99	0.67
Gasoline (C_8H_{18})	0.00	2.25	0.64
Diesel ($C_{12}H_{23}$)	0.00	1.92	0.68
Biodiesel[a]	0.10	1.90	0.70
Free fatty acid from peanut oil soap stock[a]	0.07	1.83	0.70
Canola oil[a]	0.08	1.63	0.73
Heavy oil	0.01	1.55	0.72
Bio oil (wood pyrolysis)[a]	0.50	1.34	0.92
Solid fuels			
Softwood (juniper)[a]	0.56	1.39	0.94
Hardwood (mesquite)[a]	0.58	1.37	0.95
Fibrous (rice straw)[a]	0.66	1.33	1.00
Animal based (LAPCDB)[a]	0.40	1.26	0.89
WYO coal	0.18	0.70	0.92
TXL coal	0.19	0.68	0.93

[a] Renewable fuels

coals RQ = 0.92–0.93, and animal wastes RQ = 0.92–0.95) have a RQ factor of around 0.95. Gaseous and liquid fuels have RQ between 0.50 and 0.80. It is noted that renewable biomass fuels have slightly higher RQ (e.g., coal with RQ of 0.92 and biomass with RQ of 0.97).

Figure 12 shows the plot for variation of "RQ" with H/C and O/C ratio of the fuels. Since HHV_{O_2} is constant for most fuels, then for given thermal input, the O_2 mol consumed will remain the same. Hence, a fuel with higher RQ produces more CO_2 for same thermal heat input, i.e., more tons of CO_2 per GJ. RQ scaling is applied only to oxidation processes, for example, RQ tends to ∞ for anaerobic digestion which produces CH_4 and releases CO_2 since no O_2 is consumed. It does not imply that it has highest global warming potential. Here, the production of CH_4 becomes important. Even in human body, senior people seem to have a higher RQ compared to young adults (Rizzo et al. 2005) due to anaerobic digestion in O_2 starved cells.

Using the Boie equation, one can estimate the amount of carbon emissions from a fuel. Carbon dioxide emitted on a mass basis (g/MJ or kg/GJ) determined for

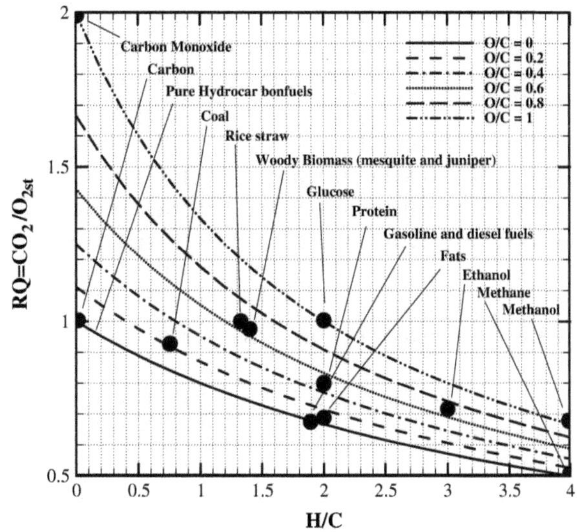

Fig. 12 Variation of RQ with H/C ratio and O/C ratio of the fuel. Gaseous and liquid fuels have lower RQ factor when compared to that of solid fossil and renewable fuels (Thanapal 2014)

different fuel compositions is shown in Fig. 13. Just as Environmental Protection Agency (EPA) sets limit on NO_x in lb per mmBTU or kg /GJ, the CO_2 amount must be estimated in kg per unit GJ or lb per mmBTU rather than kg of CO_2 per kg fuel since heat input must be maintained the same, when fuel is switched. Both Figs. 12 and 13 follow the same trend in terms of increased emissions with increase in oxygen content and C/H ratio in the fuel. From results on RQ factor and carbon dioxide emissions from fuels, it can be seen that the liquid fuels currently used in automobiles have the least RQ factor next only to natural gas. Biofuels produced from renewable energy sources are limited by the energy density and oxygen content.

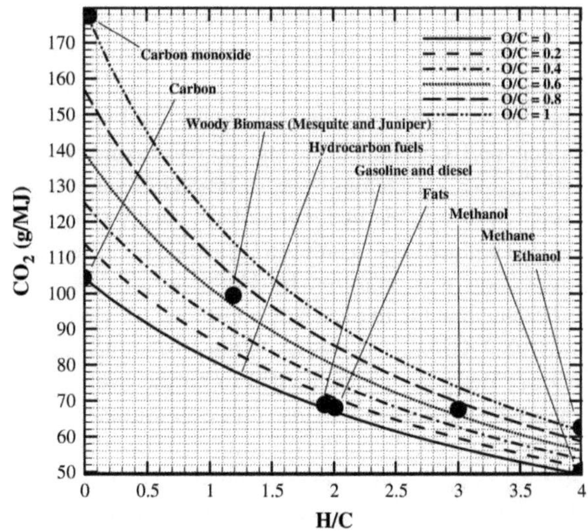

Fig. 13 Effect of H/C and O/C on the CO_2 emission from fuels. Fuels with higher RQ factor emit higher amounts of carbon dioxide

If the oxygen content can be reduced by using torrefaction of biomass, catalytic cracking, and hydrotreating (Jacobson et al. 2013) of bio-oils, the energy density of the biomass and bio-oils can be improved and in turn will also reduce the RQ factor of the fuels. But such a process also reduces the yield of bio oil. Though renewable fuels are argued to be carbon neutral, their potential to emit carbon is much higher (higher RQ when compared to coal). Coupled with its lower energy density, variability in production and process efficiencies and land usage pattern, studies should be done to effectively use fossil fuels with reduced RQ factor.

3.3.2 RQ Factor for Fuel Processing

Even though biomass is a renewable fuel where CO_2 produced by direct combustion is neutralized by CO_2 used in production of fuel, the processing of biomass (electrical power for pumping water, heat required for drying, fertilizer used for cultivation, fuels used for collection and transportation, etc.) releases CO_2 if fossil fuels are used to grow and process biomass. Then, one can define equivalent RQ value ($RQ_{process,biomass}$) for biomass processing by estimating tons of CO_2 released per GJ of heat content of biomass delivered to power plants. $RQ_{process,biomass}$ can be defined according to Eq. (18) (Thanapal 2014).

$$RQ_{process,biomass} = 219 \times \frac{RQ_{biomass}}{\left(\frac{C\%}{12.01}\right)} \left\{ \frac{RQ_1 \times p_{elec.}\left(\frac{GJ}{kg\ biomass}\right)}{\eta} + \frac{RQ_2 \times q_{heat}\left(\frac{GJ}{kg\ biomass}\right)}{\eta_{burner}} + RQ_3 + RQ_4 \right\} \tag{18}$$

where RQ_1 refers to fossil fuel used for electric power generation, RQ_2 refers to another fossil fuel used for direct heat production, η: power plant efficiency, η_{burner}: burner efficiency, RQ_3 refers to CO_2 released during collection and transportation, RQ_4 for CO_2 produced while producing fertilizers and C%, carbon mass% in biomass fuels. RQ (by direct oxidation) is relatively fixed for most biomass fuels. However, C% decreases when there is more oxygen, moisture, and ash in biomass fuel. Thus, higher amount of biomass would be needed for the same heat input and larger the CO_2 emitted during the processing of renewable biomass fuels. Hence, $RQ_{process,biomass}$ increases and if it reaches same as coal, then such a biomass is not useful as renewable fuel, i.e., criteria must be $RQ_{process,biomass} < RQ_{coal}$ if biomass is used to replace coal in order to reduce GWP. Lifecycle assessment of willow biomass production was analyzed by Heller et al. (2003). Based on the analysis, it was estimated that the total energy consumed for the production of willow based biomass was 98.3 GJ/ha over seven harvest rotations. Around 13.6 oven dry tonnes per hectare per year was estimated to be the average yield of the biomass. Assuming mesquite and juniper will also consume similar amount of energy when cultivated on a large scale solely for the utilization of these woody biomass for power production, $RQ_{process,biomass}$ can be determined based on the data provided by Heller et al. (2003). The $RQ_{process,biomass}$ for both mesquite and juniper was found to be 0.03 based on the lifecycle analysis data. Further, when the raw biomass is torrefied, moisture content in the torrefied samples will be low. In addition to lower moisture content, torrefied samples will exhibit improved grindability. Hence, $RQ_{process,biomass}$ for the torrefied samples will be lower than the raw biomass samples.

3.3.3 RQ Factor for Fuel Blends

The RQ factor for the blend of biomass with coal should be less than the RQ factor for pure coal so that using biomass will result in reduced CO_2 emissions. RQ for the fuel blend can be determined using the following formula (Thanapal 2014).

$$RQ_{fuel,blend} = RQ_{coal} \times HF_{coal} + \left(RQ_{biomass} + RQ_{process,biomass}\right) \times HF_{biomass} \quad (19)$$

where HF is the heat fraction contributed by the fuel present in the blend. Heat fraction of a fuel is defined as the ratio of heat content of a particular fuel to the total heat content of the blend (kJ of fuel/kJ of fuel blend). For the case of 90:10 blends of coal and woody biomass which was used in the current study, the $RQ_{fuel,blend}$ was determined to be 0.82 assuming $RQ_{biomass}$ to be zero since biomass is considered carbon neutral. RQ factor for the fuel blend is lower than the coal RQ value which is 0.92. This indicates lower amount of CO_2 will be released into the atmosphere when 10 % of biomass on mass basis is blended with coal. Table 6 shows the RQ_{blend} for the different biomass blends which were used in the current study.

Further, it can be seen from Table 6 that the RQ_{blend} for coal and torrefied biomass is lower (0.82) than the raw biomass blends (0.84). Higher heat values of torrefied biomass resulted in higher heat fraction for biomass which led to lower numbers for RQ_{blend}. Also it should be noted that energy consumed for processing torrefied biomass will be lower than the raw biomass if hot flue gas from boiler is used for torrefaction.

Table 6 RQ factor for the 90:10 blends of coal and biomass on a mass basis (Thanapal 2014)

Fuel	RQ_{coal}	HF_{coal}	$RQ_{biomass}$	$RQ_{biomass,processing}$	$HF_{biomass}$	RQ_{blend}
PRB + RM	0.92	0.91	0	0.03	0.09	0.84
PRB + RJ	0.92	0.90	0	0.03	0.10	0.83
PRB + TM	0.92	0.89	0	0.03	0.11	0.82
PRB + TJ	0.92	0.89	0	0.03	0.11	0.82

HF Heat fraction; *PRB* Powder river basin sub-bituminous coal; *RM* Raw mesquite; *RJ* Raw juniper; *TM* Torrefied mesquite; *TJ* Torrefied juniper

4 Summary and Conclusion

The use of nitrogen and argon as a purge gas during the torrefaction of the two biomass tested produced the same results. Torrefaction of mesquite and juniper using carbon dioxide as the purge gas resulted in a higher weight loss. The higher amount of weight loss when using CO_2 can be attributed to the reaction of the CO_2 purge gas with fixed carbon contained within the biomass.

Overall, the use of carbon dioxide as a torrefaction medium results in a biomass with a smaller HHV than the same biomass torrefied in nitrogen or argon. Therefore, given the higher amount of weight loss with CO_2 and the smaller HHV, the amount of energy retained in the torrefied biomass is lowest when torrefying in a CO_2 environment. However, grindability studies show that biomass that is torrefied in a CO_2 environment is easier to grind than biomass torrefied in N_2. That is biomass torrefied in a CO_2 environment will require less energy to pulverize.

Considering all of these factors, it is suggested that carbon dioxide is a better torrefaction medium than nitrogen and argon. Although, higher weight losses and more moisture absorption were observed when using CO_2, the increased grindability and possibility that less energy would be required to produce high-quality fuel make CO_2 the better torrefaction medium.

A simple TCM based on independent parallel reactions has been developed to model the torrefaction of woody and fibrous biomass in order to predict mass loss, heating value of biomass at any stage of torrefaction versus temperature (or time). Mass loss predicted by the TCM was compared with the experimental data obtained from TGA. It is shown that heat value of the torrefied biomass increased with increase in pretreatment temperatures, while the energy retention ratio decreased at high temperatures due to loss of higher amounts of combustible volatile matter at higher temperatures.

Chemical composition of the fuels and correlations developed to estimate the heating values can be used effectively to study the variation in heating values with respect to the fuel properties. The RQ term defined as the amount of CO_2 mol produced per unit mole O_2 consumed used extensively in the biological literature has been applied to combustion applications. RQ factor has been used to estimate the amount of CO_2 which is a measure of GWP. Lower the RQ, lower the amount of CO_2 produced for every mole of oxygen consumed for combustion process and lower the CO_2 in tons per GJ which is shown to be approximately equal to 0.1 times RQ.

Further, a term called $RQ_{process,biomass}$ is introduced to determine the effect of using a renewable fuel along with fossil fuels. $RQ_{process,biomass}$ for both mesquite and juniper (which does not use water or electricity for cultivation) was estimated to be 0.03 based on woody biomass lifecycle analysis. Method for estimating the RQ_{blend} for the blend of coal and biomass was presented, and the RQ_{blend} for 90:10 blends of coal and torrefied biomass was 0.82 (where RQ for biomass is zero) which is lower than the RQ factor of raw coal which is 0.92.

Acknowledgments A part of this work (3-component torrefaction model) was performed while Prof. Kalyan Annamalai spent his sabbatical at Energy laboratory, Mechanical Engineering, Mass. Inst of Technology, Cambridge, Mass. in Fall 2013 and KA gratefully acknowledges Prof. Ghoneim, MIT for providing facilities and many fruitful discussions.

References

Almeida G, Brito JO, Perré P (2010) Alterations in energy properties of eucalyptus wood and bark subjected to torrefaction: the potential of mass loss as a synthetic indicator. Bioresour Technol 101:9778–9784

Annamalai K (2013) Respiratory quotient (Rq), exhaust gas analyses, CO_2 emission and applications in automobile engineering. Adv Automobile Eng 2(2):e116

Annamalai K, Puri IK (2007) Combustion science and engineering. CRC Press, Boca Raton

Antal MJJ, Varhegyi G (1995) Cellulose pyrolysis kinetics: the current state of knowledge. Ind Eng Chem Res 34:703–717

Anthony DB, Howard JB, Hottel HC, Meissner HP (1976) Rapid devolatilization and hydrogasification of bituminous coal. Fuel 55:121–128

Arias B, Pevida C, Fermoso J, Plaza MG, Rubiera F, Pis JJ (2007) Influence of torrefaction on the grindability and reactivity of woody biomass. Fuel Process Technol 89(2):169–175

Arias B, Pevida C, Fermoso J, Plaza MG, Rubiera F, Pis JJ (2008) Influence of torrefaction on the grindability and reactivity of woody biomass. Fuel Process Technol 89:169–175

Biagini E, Barontini F, Tognotti L (2006) Devolatilization of biomass fuels and biomass components studied by TG/FTIR technique. Ind Eng Chem Res 45:4486–4493

Bradbury AG, Sakai Y, Shafizadeh F (1979) A kinetic model for pyrolysis of cellulose. J Appl Polym Sci 23:3271–3280

Bridgeman TG, Jones JM, Shield I, Williams PT (2008) Torrefaction of reed canary grass, wheat straw and willow to enhance solid fuel qualities and combustion properties. Fuel 87:844–856

Chen W (2012) Fixed bed counter current low temperature gasification of mesquite and juniper biomass using air-steam as oxidizer. PhD Dissertation, Texas A&M University

Chen WH, Kuo PC (2010a) A study on torrefaction of various biomass materials and its impact on lignocellulosic structure simulated by a thermogravimetry. Energy 35(6):2580–2586

Chen WH, Kuo PC (2010b) Torrefaction and co-torrefaction of hemicellulose, cellulose and lignin as well as torrefaction of the some basic constituents in biomass. Energy 36(2):803–811

Chen WH, Hsu HC, Lu KM, Lee WJ, Lin TC (2011) Thermal pretreatment of wood (Lauan) block by torrefaction and its influence on the properties of the biomass. Energy 36(5):3012–3021

Cozzani V, Lucchesi A, Stoppato G, Maschio G (1997) A new method to determine the composition of biomass by thermogravimetric analysis. Can J Chem Eng 75:127–133

Demirbaş A (2005) Estimating of structural composition of wood and non-wood biomass samples. Energy Sources 27(8):761–767

Di Blasi C (2008) Modeling chemical and physical processes of wood and biomass pyrolysis. Prog Energy Combust Sci 34:47–90

EIA (2013) EIA total energy data. http://www.eia.gov/totalenergy/data/annual/showtext.cfm?t=ptb0804a

Eseltine D (2011) Effect of using inert and non-inert gases on the thermal degradation and fuel properties of biomass in the torrefaction and pyrolysis region, MS thesis, Texas A&M University

Eseltine D, Thanapal SS, Annamalai K, Ranjan D (2013) Torrefaction of woody biomass (juniper and mesquite) using inert and non-inert gases. Fuel 113:379–388

Faix O, Jakab E, Till F, Székely T (1988) Study on low mass thermal degradation products of milled wood lignins by thermogravimetry-mass-spectrometry. Wood Sci Technol 22:323–334

Ferdous D, Dalai AK, Bej SK, Thring RW (2002) Pyrolysis of lignins: experimental and kinetics studies. Energy Fuels 16:1405–1412
Hall WS (1905) A Textbook of physiology, 2nd edn. Lea, New York
Heller MC, Keoleian GA, Volk TA (2003) Life cycle assessment of a willow bioenergy cropping system. Biomass Bioenergy 25:147–165
Hu S, Jess A, Xu M (2007) Kinetic study of chinese biomass slow pyrolysis: comparison of different kinetic models. Fuel 86:2778–2788
Jacobson K, Maheria KC, Kumar Dalai A (2013) Bio-oil valorization: a review. Renew Sustain Energy Rev 23:91–106
Lawrence B, Annamalai K, Sweeten JM, Heflin K (2009) Cofiring coal and dairy biomass in a 29 kWt furnace. Appl Energy 86(11):2359–2372
Liu Q, Wang S, Zheng Y, Luo Z, Cen K (2008) Mechanism study of wood lignin pyrolysis by using TG–FTIR analysis. J Anal Appl Pyrol 82:170–177
Manya JJ, Velo E, Puigjaner L (2003) Kinetics of biomass pyrolysis: a reformulated three-parallel-reactions model. Ind Eng Chem Res 42:434–441
McClave SA, Lowen CC, Kleber MJ, McConnell JW, Jung LY, Goldsmith LJ (2003) Clinical use of the respiratory quotient obtained from indirect calorimetry. J Parenter Enteral Nutr 27:21–26
McKendry P (2002) Energy production from biomass (part 1): overview of biomass. Bioresour Technol 83:37–46
Mészáros E, Várhegyi G, Jakab E, Marosvölgyi B (2004) Thermogravimetric and reaction kinetic analysis of biomass samples from an energy plantation. Energy Fuels 18:497–507
Morinaka T, Wozniewicz M, Jeszka J, Bajerska J, Limtrakul PN, Makonkawkeyoon L, Hirota N, Kumagai S, Sone Y (2012) Comparison of seasonal variation in the fasting respiratory quotient of young Japanese, Polish and Thai women in relation to seasonal change in their percent body fat. J Physiol Anthropol 31:10
Orfao JJM, Antunes FJA, Figueiredo JL (1999) Pyrolysis kinetics of lignocellulosic materials—three independent reactions model. Fuel 78:349–358
Ozawa T (1965) A new method of analyzing thermogravimetric data. Bull Chem Soc Japan 38:1881 1886
Peng Y, Wu S (2010) The structural and thermal characteristics of wheat straw hemicellulose. J Anal Appl Pyrol 88:134–139
Prins MJ, Kryzysztof JP, Janssen FJJG (2006a) More efficient gasification via torrefaction. Energy 31(15):3458–3470
Prins MJ, Kryzysztof JP, Janssen FJJG (2006b) Torrefaction of wood: part 2. Analysis of products. Energy 77(1):35–40
Prins MJ, Ptasinski KJ, Janssen FJ (2006c) Torrefaction of wood: part 1. Weight loss kinetics. J Anal Appl Pyrol 77:28–34
Raveendran K, Ganesh A, Khilar KC (1996) Pyrolysis characteristics of biomass and biomass components. Fuel 75:987–998
Repellin V, Govin A, Rolland M, Guyonnet R (2010) Modelling anhydrous weight loss of wood chips during torrefaction in a pilot kiln. Biomass Bioenergy 34:602–609
Richardson HB (1929) The respiratory quotient. Physiol Rev 9:61–125
Rizzo MR, Mari D, Barbieri M, Ragno E, Grella R, Provenzano R, Villa I, Esposito K, Giugliano D, Paolisso G (2005) Resting metabolic rate and respiratory quotient in human longevity. J Clin Endocrinol Metab 90:409–413
Rousset P, Macedo L, Commandré JM, Moreira A (2012) Biomass torrefaction under different oxygen concentrations and its effect on the composition of the solid by-product. J Anal Appl Pyrol 96:86–91
Sonobe T, Worasuwannarak N (2008) Kinetic analyses of biomass pyrolysis using the distributed activation energy model. Fuel 87:414–421
TAMU (2013) TAMU fuel data bank. http://cabel.tamu.edu/TAMU%20FDB.htm
Thanapal SS (2014) Effect of cofiring torrefied woody biomass with coal in a 30 kWt downfired burner. PhD dissertation, Texas A&M University

Thanapal SS, Chen W, Annamalai K, Carlin N, Ansley RJ, Ranjan D (2014) Carbon dioxide torrefaction of woody biomass. Energy Fuels 28:1147–1157

Tillman D, Duong D, Harding N (2012) Solid fuel blending; principles, practices, and problems, 1st edn. Heinemann, Butterworth

Vamvuka D, Kakaras E, Kastanaki E, Grammelis P (2003) Pyrolysis characteristics and kinetics of biomass residuals mixtures with lignite. Fuel 82:1949–1960

Van de Weerdhof MW (2010) Modeling the pyrolysis process of biomass particles. Masters Thesis, Eindhoven University of Technology

Varhegyi G, Antal MJ Jr, Szekely T, Szabo P (1989) Kinetics of the thermal decomposition of cellulose, hemicellulose, and sugarcane bagasse. Energy Fuels 3:329–335

Varhegyi G, Antal MJ Jr, Jakab E, Szabó P (1997) Kinetic modeling of biomass pyrolysis. J Anal Appl Pyrol 42:73–87

Wang C, Peng J, Li H, Bi XT, Legros R, Lim CJ, Sokhansanj S (2012a) Oxidative torrefaction of biomass residues and densification of torrefied sawdust to pellets. Bioresour Technol 127:318–325

Wang XY, Wan XJ, Chen MQ, Wang J (2012b) Kinetic model of biomass pyrolysis based on three-component independent parallel first-order reactions. Chin J Process Eng 12:1020–1024

Wu YM, Zhao ZL, Li HB, Fang HE (2009) Low temperature pyrolysis characteristics of major components of biomass. J Fuel Chem Technol 37(4):427–432

Yan W, Hastings JT, Acharjee TC, Coronella CJ, Vasquez VR (2010) Mass and energy balances of wet torrefaction of lignocellulosic biomass. Energy Fuels 24:4738–4742

Yang H, Yan R, Chen H, Lee DH, Zheng C (2007) Characteristics of hemicellulose, cellulose and lignin pyrolysis. Fuel 86:1781–1788

Yoon HC, Pozivil P, Steinfeld A (2011) Thermogravimetric pyrolysis and gasification of lignocellulosic biomass and kinetic summative law for parallel reactions with cellulose, xylan, and lignin. Energy Fuels 26:357–364

Effect of Biodiesel Utilization on Tribological Properties of Lubricating Oil in a Compression Ignition Engine

Avinash K. Agarwal and Jai Gopal Gupta

Abstract Biofuels are touted as a prime source of renewable and sustainable energy in near future. Biofuels have evolved from first to fourth generation. They mainly differ in feedstock and production technologies used. Biodiesel can be readily adopted as a partial substitute for mineral diesel in developing economies. Transesterification is an appropriate and well-established method for chemical modification of vegetable oils, which produces a suitable diesel fuel alternative "Biodiesel". Biodiesel has proved to be an engine friendly and environment friendly substitute for mineral diesel. However, increased use of biodiesel has led to need of investigating the effect of biodiesel on engine durability, lubricating oil as well as hardware compatibility. The effect of biofuels and of their blends with diesel on engine tribology is not completely understood. Use of biodiesel as replacement fuel may lead to substantial lubricating oil dilution, because biodiesel has higher boiling range than mineral diesel. Since fuel interacts with the lubrication oil in the sump via fuel dilution, it is necessary to understand the impact of biodiesel on the lubricant properties. Oil consumption in heavy-duty diesel engines lies in ranges of 0.1–0.2 % of the total fuel consumption.

Keywords Biofuels · Transesterification · Tribological properties · Engine wear and durability · Wear metal measurement · Atomic absorption spectroscopy (AAS)

1 Introduction

Biofuels are touted as a prime source of renewable and sustainable energy in near future. Biofuels have evolved from first to fourth generation. They mainly differ in feedstock and production technologies used. Biodiesel can be readily adopted as a

A.K. Agarwal (✉) · J.G. Gupta
Department of Mechanical Engineering, Indian Institute of Technology Kanpur, Kanpur, Uttar Pradesh 208016, India
e-mail: akag@iitk.ac.in

A.K. Agarwal et al. (eds.), *Novel Combustion Concepts for Sustainable Energy Development*, DOI 10.1007/978-81-322-2211-8_5

partial substitute for mineral diesel in developing economies. Transesterification is an appropriate and well-established method for chemical modification of vegetable oils, which produces a suitable diesel fuel alternative "Biodiesel". Biodiesel has proved to be an engine friendly and environment friendly substitute for mineral diesel. However, increased use of biodiesel has led to need of investigating the effect of biodiesel on engine durability, lubricating oil as well as hardware compatibility. The effect of biofuels and of their blends with diesel on engine tribology is not completely understood. Use of biodiesel as replacement fuel may lead to substantial lubricating oil dilution, because biodiesel has higher boiling range than mineral diesel. Since fuel interacts with the lubrication oil in the sump via fuel dilution, it is necessary to understand the impact of biodiesel on the lubricant properties. Oil consumption in heavy-duty diesel engines lies in ranges of 0.1–0.2 % of the total fuel consumption.

Several studies have been performed to evaluate the feasibility of biodiesel fueling in the vehicles. Materials compatibility issues observed with biodiesel showed are largely related to enhanced corrosion of engine components, primarily due to compositional differences among these fuels. On performance, emission, and engine durability aspect, there is a general higher consensus among researchers about reduced emissions as well as higher stickiness among components in addition to issues such as injector coking and fuel filter plugging (Fazal et al. 2011).

2 Tribological Behavior of Biodiesels

The tribological behavior of various types of biodiesels has been examined in bench tests, laboratory engine tests, and vehicle tests by several researchers (Arumugam and Sriram 2012; Uy et al. 2011; Celik and Aydin 2011; Sukjit and Dearn 2011; Haseeb et al. 2010; Wain et al. 2005). Few studies have investigated the effect of biodiesels on engine durability, but they focused mostly on the viscosity and neutralization number (total acid number [TAN] and total base number [TBN]) of the oils or on the amount of metallic debris in the tested oils (Munoz et al. 2011; Sinha and Agarwal 2010; Yuksek et al. 2009; Agarwal 2005).

Inspite of these studies looking at specific aspects, the tribological effects of biodiesel vis-a-vis mineral diesel are not well understood. During combustion in a diesel engine, biodiesel may be scrapped to the engine oil sump along with blow-by gases and can dilute the lubricating oil and severely affect its lubrication performance.

Agarwal (2005) reported that viscosity of lubricating oil decreases with engine run time, primarily due to increased dilution by biofuels. Fang et al. (2007) examined the wear effects of biodiesel (B100) using various methods such as electrical contact resistance measurement, high-frequency reciprocating rig, and four-ball testing. They consistently observed that while fresh biodiesel might actually decrease wear, aged biodiesel degradation products can interact with lubricating oil additives such as zinc-di-alkyl-di-thio-phosphate (ZDDP) and adversely affect lubricant's performance. They concluded that excessive fuel dilution in lubricating oil by biodiesel can lead to formation of complexes between oxidized biodiesel (i.e., aged biodiesel) components and ZDDP. They also reported that lubricating oil dilution with aged biodiesel may increase wear even with 5 % concentration (B05) or lower. Watson and Wong (2008) found in oxidation test that there is substantial increase in lubricant's acidity upon dilution by B100. Jech et al. (2008) used a model tribometer based on nanoscale wear volume coherence (nVCT) to investigate the lubricity (i.e., wear volume) effect of biodiesel dilution. They reported that biodiesel concentrations of 10 % in the mineral diesel (B10) increased the wear slightly, but higher concentrations (from 30 % upwards) induced lower progressive wear. Sulek et al. (2010) presented results on the effects of methyl esters derived from rapeseed bio-oils as additives of fuel, and they suggested that as little as 5 % of such additions may decrease wear by up to 50 % and coefficient of friction by up to 20 %.

A bus diesel engine with a mechanically controlled fuel injection system was tested with biodiesel (B100) produced from rapeseed. Biodiesel usage improved the pump plunger lubrication and reduced wear. Further, carbon deposits in the combustion chambers did not vary significantly, but they were noticeably redistributed. In a computational study, it was reported that variations in the nozzle discharge coefficient have to be taken into account, if high accuracy of numerical simulations is desired (Pehan et al. 2009).

Although the compatibility of biodiesel with key components of an engine such as cylinder, piston, piston rings, connecting rod, and bearings has posed a serious challenge to the tribologists, they are yet to come up with a solution to control/reduce tribological degradation of different metallic components. Some efforts have been made to understand the corrosion and wear of automotive materials in diesel and biodiesel. It was concluded beyond reasonable doubt that though biodiesel is more corrosive than mineral diesel, it provides better lubricity in terms of wear and friction (Fazal et al. 2014).

Continued legislative pressure to reduce exhaust emissions from CI engines has resulted in development of advanced fuel injection equipment. These advanced injection systems generate higher fuel pressures and temperatures at the injector tip, where deposit formation is initiated. In a research, an endurance test was carried out for 250 h on 2 fuel samples, namely mineral diesel as baseline fuel and JB20 (20 % Jatropha biodiesel and 80 % diesel) in a single-cylinder CI engine. Visual inspection of the in-cylinder components showed carbon deposits on injectors for both fuels. Scanning electron microscopy (SEM) and energy-dispersive X-ray spectroscopy (EDX) analysis showed greater carbon deposits on and around the injector tip for JB20-fueled engine compared to the mineral diesel-fueled engine. Similarly, lubricating oil analysis revealed excessive wear metal concentration and reduced viscosity for the engine fueled with JB20 (Liaquata et al. 2014).

In an experimental study done by Lin et al. (2013), petro-diesel (D100) and different concentrations of commercial biodiesel (B100, B50, B5) were blended with a commercial engine oil at a fixed volume ratio of 1:9 at two different temperatures (room temperature and 150 °C) to investigate the tribological effects of biodiesels on the anti-wear performance of the engine oil. The anti-wear performance of the engine oil blended with mineral diesel was reported to be worse than biodiesel blended fuels at both temperatures. At room temperature, the physical adsorption and local hydrodynamic effects of the blended oils dominate the tribological behavior, and at 150 °C, biodiesel seems to promote the growth of a reaction film with anti-wear additives present in lubricating oil and enhances the growth rate of chemical films. However, an excessive biodiesel concentration causes tribochemical wear and reduces anti-wear performance (Lin et al. 2013). Molina et al. (2014) concluded that any degree of mineral oil dilution by the tested biodiesels can reduce the wear protection performance and the viscosity of the engine oil even at small contaminating amounts, such as from 5 to 30 %.

3 Lubricating Oil Degradation

The lubricant transports protective chemicals to the sites where they are required and transports waste products away from the sites, where they are generated (Tung and McMillan 2004). Fuel chemistry plays an extremely important role in the performance and useful life of the lubricating oil (Fazal et al. 2011; Truhan et al. 1995). Properties of lubricating oil change with usage due to changes in chemical composition of oil and addition of external contaminants. These contaminants include (i) chemical constituents such as oil oxidation products and acids formed by combustion gases, (ii) particulates from dust and dirt from external sources, (iii) wear metal debris, and (iv) combustion soot (Truhan et al. 1995). Oxygenated chemical constituents of biodiesel make it more reactive than mineral diesel. Properties of lubricating oil change with its usage due to fuel dilution and addition of contaminants from the engine due to wear, carbonaceous materials and impurities entering along with intake air (Truhan et al. 1995). Biodiesel is comparatively more chemically reactive than mineral diesel due to presence of oxygen in the molecular structure of the biodiesel (Schumacher et al. 1996). Differences in physical properties of biodiesel and mineral diesel cause different levels of fuel dilution of lubricating oil. Hence, the effect of biodiesel on the lubricating oil degradation needs to be evaluated in long-duration engine tests before taking decision about their large-scale usage. There are very few studies reporting the long-term effects of biodiesel on the lubricating oil degradation in the open literature. Karanja and Jatropha oils are being considered as important feedstocks with a potential of producing biodiesel on large scale because they are available in surplus quantities throughout South Asia and other parts of the world (Agarwal and Dhar 2010, 2012).

4 Effect of Biodiesel on Engine Wear and Durability

In diesel engines, components such as piston ring, cylinder liner, piston, bearings, crankshaft, tappet, and valves are subjected to wear. The wear particles are removed by the lubricating oil and finally transported to the oil sump. Wear particles remain in suspension in the lubrication oil as minute metal debris. Therefore, sources of these elements and the engine condition (engine health) can be predicted and monitored by analyzing the concentration and variations in the metallic wear debris in the lubricating oil after a fixed operating interval.

Fazal et al. (2011) summarized comparison of wear of biodiesel- and mineral diesel-fueled engines in a review article. They concluded that either lower or similar wear takes place in biodiesel/biodiesel blend-fueled engines in comparison with mineral diesel-fueled engines, for both static engine tests as well as field trials (Fazal et al. 2011). Agarwal and Das (2001) also concluded from a 512-h-long endurance test, which they conducted on a single-cylinder engine using B20 of linseed biodiesel, that biodiesel does not have any significant adverse impact on wear of various vital moving components of the engine. Verhaeven et al. (2005) reported that there was no significant difference in the wear of the fuel injectors and fuel injection equipment, even after 100,000 km field trial with alternate fuels namely rapeseed oil methyl ester (RME) and used vegetable oil methyl ester (UVOME). Sinha and Agarwal (2010) reported physical measurements of various vital engine components and showed lower wear for B20-fueled engine in comparison with mineral diesel-fueled engine except for big end bearing, which showed slightly higher wear for B20-fueled compression ignition direct injection (CIDI) engine. Çetinkaya et al. (2005) observed almost identical level of carbon deposits on the fuel injectors of the two vehicles fueled with used cooking oil-based biodiesel and mineral diesel in a 7,500-km field test in winter conditions. Pehan et al. (2009) reported similar carbon deposits in the combustion chambers of biodiesel- and diesel-fueled engines. Agarwal and Dhar (2010, 2012) reported higher carbon deposits on the piston of straight vegetable oil- (Karanja oil) and blend-fueled engine in comparison with mineral diesel-fueled engine in a long-term endurance test.

Celik and Aydin (2011) investigated the effects of biodiesel on the fuel injector and fuel injection pump piston used in a diesel engine by using two identical engines with the same technical specifications, but operated on petroleum diesel and 100 % (B100) biodiesel for 200 h. SEM analyses showed significant structural changes on the surfaces of the injector nozzle used for B100. Less shrinkage was observed in the engine injector nozzle using petroleum diesel compared to the one using biodiesel (Fig. 1) (Celik and Aydin 2011).

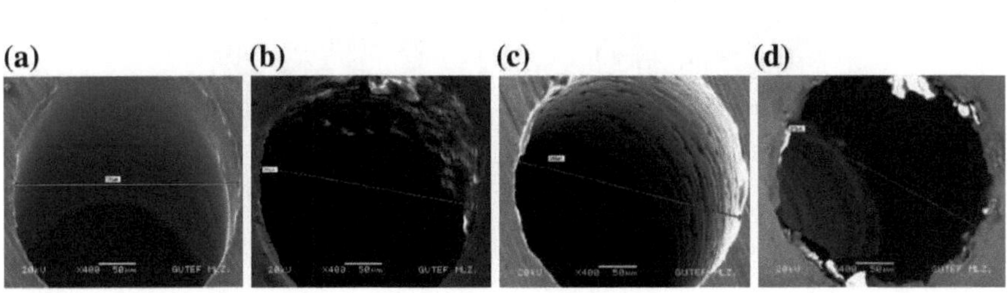

Fig. 1 SEM analyses of injector nozzles **a** before and **b** after diesel; **c** before and **d** after biodiesel usage (Celik and Aydin 2011)

Biodiesel reacts with Si and C in the combustion chamber and injector nozzle at higher temperatures and increases the density of Si and C on the surfaces. However, at relatively lower temperatures emanating from combustion, biodiesel reacts with C, V, Cr, and Mn on the surfaces of the injector pump piston and creates a passive layer on the metal surface and inhibits wear. Oxidation and corrosion processes due to chemical nature of biodiesel increased the level of wear (Celik and Aydin 2011). In another review paper, it was summarized that there is lower wear reported with biodiesel, and use of biodiesel would improve engine durability due to lower soot formation and inherent higher lubricity of biodiesel in comparison with diesel (Xue et al. 2011).

4.1 Carbon Deposits

Agarwal et al. (2003a, b) carried out experiments to investigate the effect of long-term engine operation for B20 of linseed oil biodiesel vis-à-vis mineral diesel by physical dimensioning and exhaustive lubricating oil analysis and characterization. For this test, two engines (diesel and biodiesel) were run for 512 h (32 cycles, each of 16 h continuous engine operation) at rated speed. Carbon deposits for biodiesel-fueled engines on cylinder head (Fig. 2), piston top, and injector tip were substantially lower than mineral diesel-fueled engine. It was reported that biodiesel-fueled engine deposits were 40 % lesser compared to the diesel-fueled engine. Also carbonization of biodiesel injector after 512 h of engine operation was significantly lower than diesel injector after 200 h of engine operation (Agarwal et al. 2003b). Sinha and Agarwal (2010) carried out an endurance test for 100 h on the CIDI engine with B20 of rice-bran biodiesel and observed lesser carbon deposits on the in-cylinder parts for B20 (Fig. 3).

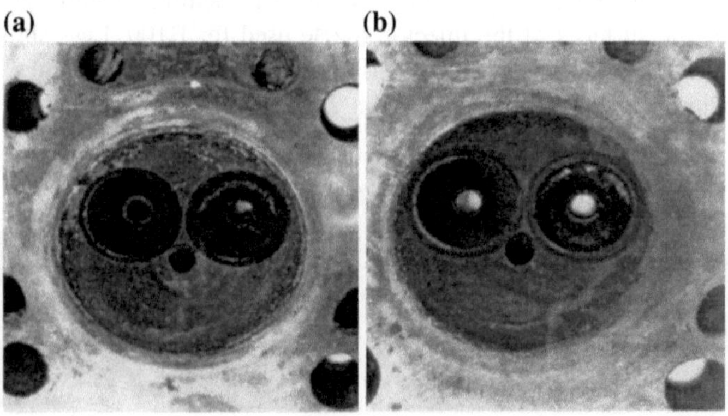

Fig. 2 Carbon deposits on the cylinder head for **a** diesel and **b** 20 % biodiesel-fueled engines after 512 h of single-cylinder DI engine operation (Agarwal et al. 2003b)

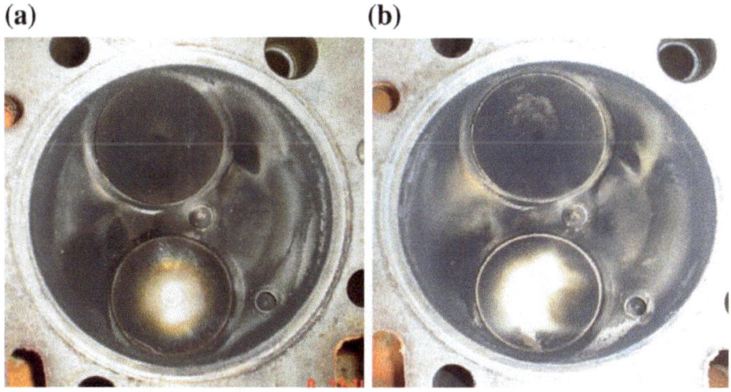

Fig. 3 Carbon deposits on the cylinder head of **a** diesel and **b** 20 % rice-bran biodiesel-fueled multi-cylinder DICI transportation engine after 100 h engine operation (Sinha and Agarwal 2010)

4.2 Physical Wear Measurement

Agarwal (1999) reported that the wear of vital moving parts (liner bore, piston rings, bearings, etc.) was about 30 % lower for B20-operated engine compared to mineral diesel-operated engine. Very low wear was observed for the cylinder liners. Sinha and Agarwal (2010) concluded that as per surface roughness profiles, various roughness parameters, and SEM micrographs (Fig. 4), wear was relatively higher on the anti-thrust side and at TDC location. Similar wear trend was reported in another study with B20-fueled engine, but overall wear was lower than diesel-fueled engine.

Shanta et al. (2011) employed a pin-on-disk tribometer for measuring specific wear and average friction force for five tested dilution lubricating oil–biodiesel mixtures and the non-contaminated lubricating oil reference (SAE15W40). Four different biodiesels (produced from peanut oil, canola oil, soybean oil, and chicken fat) and mineral diesel were employed. Engine oil dilution by each of these fuels

	Thrust side		Anti-thrust side	
	Mineral diesel	B20	Mineral diesel	B20
TDC	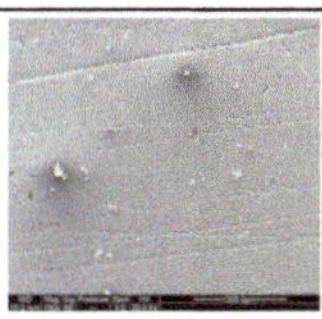		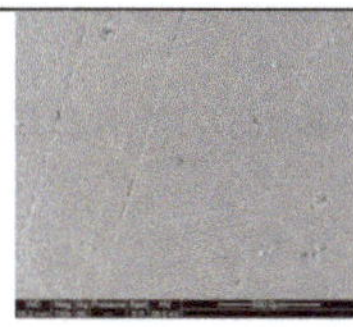	

Fig. 4 SEM of liner surfaces of mineral diesel- and biodiesel-fueled engine (Sinha and Agarwal 2010)

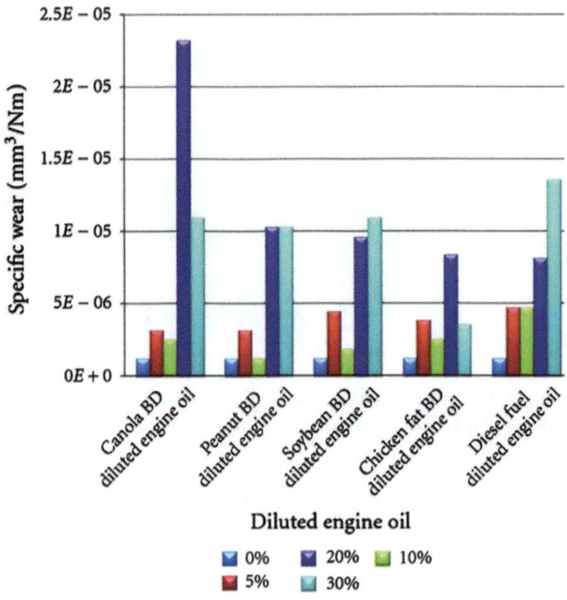

Fig. 5 Specific wear for mixtures of four biodiesels and mineral diesel diluting the lubricating oil (Shanta et al. 2011)

was in 5, 10, 20, and 30 % volume ratios. For all tested biodiesels, contaminated mixtures, the average friction force did not vary significantly as compared to that for mineral diesel contaminated and that for pure oil (Fig. 5). As per results for the 0–10 % dilution range, the observed wear for each biodiesel mixture was not substantially different and even slightly lower than that for same percentage of mineral diesel-fuel contamination. For the four tested biodiesel mixtures, the measured wear seemed to peak for around 20 % blends. However, the mineral diesel in oil mixture showed a monotonic increase of wear in the tested 0–30 % contamination interval (Shanta et al. 2011).

Arumugam and Sriram (2012) addressed the issue of friction and wear characteristics of diesel engine cylinder liner–piston ring combinations under different lubricating conditions using a pin-on-disk wear tribometer. They reported that the rapeseed oil-based bio-lubricants and biodiesel-contaminated synthetic lubricants exhibited better performance in terms of wear, friction, and frictional force under similar operating conditions.

4.3 Wear Metal Measurement Through Lubricating Oil Analysis

The quantitative evaluation of wear particles present in oil gives the magnitude of engine component deterioration, while qualitative analysis indicates its origin. Sinha and Agarwal (2010) analyzed wear metal in lubricating oil samples and observed lower wear for B20-fueled engine. Pb and Al were found slightly higher for B20-fueled engine, which may be due to the attack of biodiesel on paints and bearings (Sinha and Agarwal 2010). Agarwal et al. (2003b) demonstrated that wear metals' debris such as Fe, Cu, Al, Pb reduced with increasing biodiesel of palm oil into blends, which produced the lower level of wear concentration than that of mineral diesel, and the reason is the effect of the corrosion inhibitor in fuel and lubricating oil which controls corrosion as well as oxidation of the lubricating oil.

4.4 Viscosity

During engine operation at normal temperature, a minor quantity of fuel normally dilutes the lubricating oil, which escapes to the oil sump along with blow-by gases. Engine cold-starting, rich fuel–air mixtures, and low ambient temperatures promote fuel dilution of lubricating oil. Also dilution of engine oil occurs when the fuel is injected late in the combustion cycle, in order to regenerate the diesel particulate filter, typically used for trapping particulate emissions in modern engines (Uy et al. 2011). Fuel dilution reduces lubricating oil viscosity and the concentration of engine oil additives, potentially compromising lubricant's performance. Inadequate oil viscosity affects oil film formation and load-bearing capacity, which can potentially lead to excessive wear of bearings, journals, and other moving components, in addition to low oil pressure and poor oil economy. There are two main factors responsible for the viscosity changes affecting the oil in opposite direction. Formation of resinous products because of oil oxidation, evaporation of lighter fractions, depletion of anti-wear additives, and contamination by insolubles tend to increase the viscosity, while fuel dilution and shearing of viscosity index improvers tend to bring down the viscosity. Biodiesel usage may compound these issues due to its oxidative instability and its higher boiling point compared to mineral diesel, which increases its concentrate in the oil sump.

During the durability test of the engine for biodiesel, viscosity of lube oil samples was evaluated at 40 and 100 °C using kinematic viscometer (Agarwal 2003). The extent of lowering of viscosity of the lubricating oil was lesser in case of biodiesel-fueled engine compared to diesel-fueled engine. This may be because of lower fuel dilution. Fuel dilution is a direct consequence of clearance between piston rings and cylinder liner (Agarwal 2003).

4.5 Ash Content

In order to find the ash content, lubricating oil samples need to be kept in the furnace at 450 °C for 4 h and then 650 °C for 2 h to produce ash. The residual ash contains metallic wear debris. Ash content in the lubricating oils for biodiesel- and

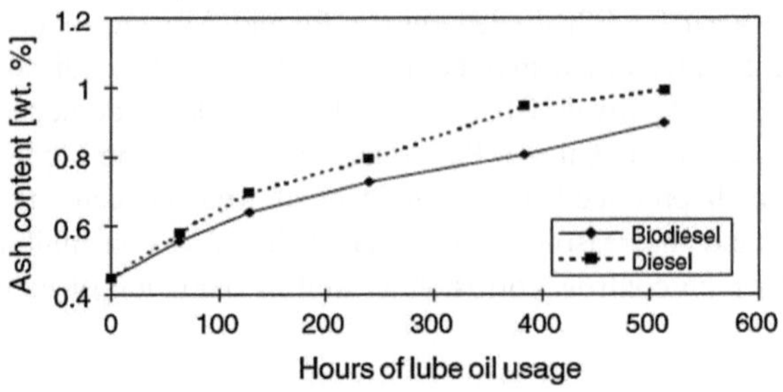

Fig. 6 Ash content of the lubricating oil samples from diesel- and biodiesel (B20)-fueled engines (Agarwal 2003)

diesel-fueled CI engine is shown in Fig. 6 (Agarwal 2003). Ash content for B20-fueled engine was lower than diesel-fueled engine, which shows that biodiesel-fueled engine produces lower wear debris (Agarwal 2003).

4.6 Atomic Absorption Spectroscopy

Agarwal et al. (2003b) and Agarwal (1999) carried out wear metal analysis using atomic absorption spectroscopy (AAS) of the lubricating oil samples drawn from biodiesel (B20)- and diesel-fueled engines (Figs. 7 and 8). The biodiesel-fueled engine produced lesser wear of engine moving parts as seen from lesser metallic debris, originating from different interfaces of moving parts and getting accumulated in the lubricating oil (Agarwal et al. 2003a, b).

Based on these exhaustive engine and tribological investigations of the lubricating oil, Agarwal (2005) concluded that biodiesel can overcome most of the operational durability concerns existing with vegetable oils such as fuel filter plugging, injector coking, carbon deposits in the engine combustion chamber, in addition to issues such as ring sticking and contamination of lubricating oils.

Fig. 7 Iron concentration as a function of lube oil usage in the lubricating oil samples from diesel- and biodiesel (B20)-fueled engines (Agarwal 2003)

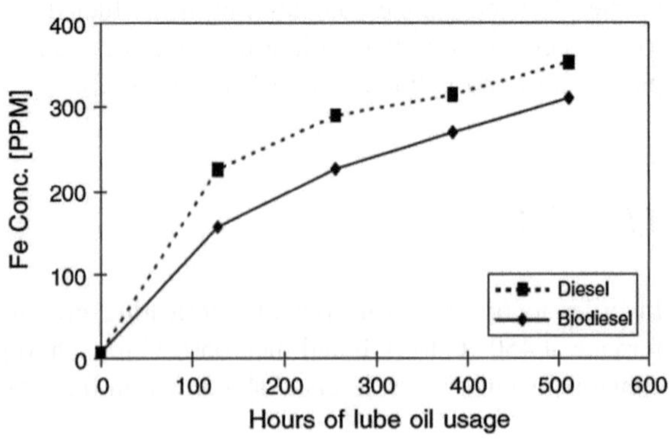

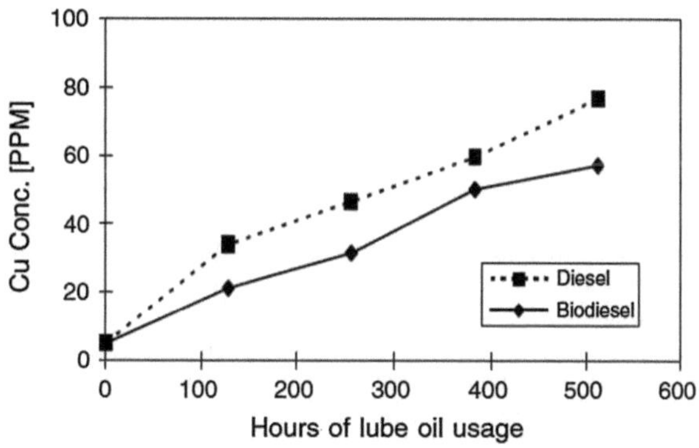

Fig. 8 Copper concentration as a function of lube oil usage in the lubricating oil samples from diesel- and biodiesel (B20)-fueled engines (Agarwal 2003)

A long-term endurance test proved that biodiesel can be successfully used for partially substituting mineral diesel. This increases the likelihood of biodiesel being adopted as an alternative fuel for the existing conventional diesel engines without any major modifications in the engine hardware.

5 Summary and Conclusions

Some results showed that any degree of mineral–oil dilution by the tested biodiesels can reduce the wear protection performance and the viscosity of the engine oil even at small contaminating amounts. It is expected that study would help understand interactions of methyl ester components with the engine oil anti-wear additives. Although there are the negative reports in wear, it is expectable that the use of biodiesel favors to improve durability of engine for biodiesel due to the lower soot formation and the inherent lubricity, compared with diesel. However, further detailed investigation of the effect of biodiesel properties on the composition and property degradation of the lubricating oil as well as engine combustion is required. These factors are primarily responsible for carbon deposits formation and wear of engine components and are therefore vital for large-scale implementation of biodiesel in transportation engines successfully.

References

Agarwal AK (1999) Performance evaluation and tribological studies on a biodiesel-fuelled compression ignition engine. PhD thesis, Center for Energy Studies, Indian Institute of Technology, Delhi, India

Agarwal AK (2003) Lubrication oil tribology of a biodiesel-fuelled compression ignition engine. In: ASME 2003, internal combustion engine division spring technical conference, American Society of Mechanical Engineers, pp 751–765

Agarwal AK (2005) Experimental investigations of the effect of biodiesel utilization on lubricating oil tribology in diesel engines. Proc Inst Mech Eng Part D: J Automobile Eng 219(5):703–713

Agarwal AK, Das LM (2001) Biodiesel development and characterization for use as a fuel in compression ignition engines. J Eng Gas Turbines Power 123:440–447

Agarwal AK, Dhar A (2010) Karanja oil utilization in direct-injection engine by preheating. Part 2: experimental investigations of engine durability and lubricating oil properties. Proc Inst Mech Eng Part D: J Automobile Eng 224:85–97

Agarwal AK, Dhar A (2012) Wear, durability, and lubricating oil performance of a straight vegetable oil (Karanja) blend fueled direct injection compression ignition engine. J Renew Sustain Energy 4(6):063138

Agarwal AK, Bijwe J, Das LM (2003a) Effect of biodiesel utilization on wear of vital parts in compression ignition engine. J Eng Gas Turbines Power 125(2):604–611

Agarwal AK, Bijwe J, Das LM (2003b) Wear assessment in biodiesel fuelled compression ignition engine. J Eng Gas Turbines Power 125(3):820–826

Arumugam S, Sriram G (2012) Effect of bio-lubricant and biodiesel-contaminated lubricant on tribological behavior of cylinder liner-piston ring combination. Tribol Trans 55(4):438–445

Celik I, Aydin O (2011) Effects of B100 biodiesel on injector and pump piston. Tribol Trans 54 (3):424–431

Çetinkaya M, Ulusoy Y, Tekìn Y, Karaosmanoğlu F (2005) Engine and winter road test performances of used cooking oil originated biodiesel. Energy Convers Manag 46(7):1279–1291

Fang HL, Whitacre SD, Yamaguchi ES, Boons M (2007) Biodiesel impact on wear protection of engine oils. SAE Technical Paper, 2007-01-4141

Fazal MA, Haseeb ASMA, Masjuki HH (2011) Biodiesel feasibility study: an evaluation of material compatibility; performance; emission and engine durability. Renew Sustain Energy Rev 15(2):1314–1324

Fazal MA, Haseeb ASMA, Masjuki HH (2014) A critical review on the tribological compatibility of automotive materials in palm biodiesel. Energy Convers Manag 79:180–186

Haseeb ASMA, Sia SY, Fazal MA, Masjuki HH (2010) Effect of temperature on tribological properties of palm biodiesel. Energy 35(3):1460–1464

Jech M, Wopelka T, Franek F (2008) Lubricating performance of mixtures of bio-diesel and mineral diesel investigated in a model-tribometer and with radioactive isotopes, IJTC2008-71161, Miami

Liaquata AM, Masjuki HH, Kalam MA, Fattah IMR (2014) Impact of biodiesel blend on injector deposit formation. Energy 72:813–823

Lin Y-C, Kan T-H, Chen J-N, Tsai C Jr, KuY Y, Lin K-W (2013) Tribological performance of engine oil blended with various diesel fuels. Tribol Trans 56(6):997–1010

Molina GJ, Soloiu V, Rahman M, Shanta SM, Powell W, Channing W, Aktaruzzaman FNU (2014) Wear effects of mineral-oil lubricant dilution with biofuels. 2014 STLE Annual Meeting & Exhibition

Munoz M, Moreno F, Monne C, Morea J, Terradillos J (2011) Biodiesel improves lubricity of new low sulphur diesel fuels. Renew Energy 36(11):2918–2924

Pehan S, Jerman MS, Kegl M, Kegl B (2009) Biodiesel influence on tribology characteristics of a diesel engine. Fuel 88(6):970–979

Schumacher L, Borgelt SC, Hires WG, Wetherell W, Nevils A (1996) 100,000 miles of fueling 5.9L cummins engines with 100 % biodiesel. SAE Paper, 1996:962233

Shanta SM, Molina GJ, Soloiu V (2011) Tribological effects of mineral-oil lubricant contamination with biofuels: a pin-on-disk tribometry and wear study. Adv Tribol 2011:820795

Sinha S, Agarwal AK (2010) Experimental investigation of the effect of biodiesel utilization on lubricating oil degradation and wear of a transportation CIDI engine. J Eng Gas Turbines Power 132(4):42801–42809

Sukjit E, Dearn KD (2011) Enhancing the lubricity of an environmentally friendly Swedish diesel fuel MK1. Wear 271(9–10):1772–1777

Sulek MW, Kulczycki A, Malysa A (2010) Assessment of lubricity of compositions of fuel oil with bio components derived from rape-seed. Wear 268(1–2):104–108

Truhan JJ, Covington CB, Wood L (1995) The classification of lubricating oil contaminants and their effect on wear in diesel engines as measured by surface layer activation. SAE Paper, 1995: 952558

Tung SC, McMillan ML (2004) Automotive tribology overview of current advances and challenges for the future. Tribol Int 37:517–536

Uy D, Zdrodowski RJ, O'Neil AE, Simko SJ, Gangopadhyay AK, Morcos M, Lauterwasser F, Parsons G (2011) Comparison of the effects of biodiesel and mineral diesel fuel dilution on aged engine oil properties. Tribol Trans 54(5):749–763

Verhaeven E, Pelkmans L, Govaerts L, Lamers R, Theunissen F (2005) Results of demonstration and evaluation projects of biodiesel from rapeseed and used frying oil on light and heavy duty vehicles. SAE Paper, 2005-01-2201

Wain KS, Perez JM, Chapman E, Boehman AL (2005) Alternative and low sulfur fuel options: boundary lubrication performance and potential problems. Tribol Int 38(3):313–319

Watson SA, Wong VW (2008) The effects of fuel dilution with biodiesel and low sulfur diesel on lubricant acidity, oxidation and corrosion—a bench scale study with CJ-4 and CI-4+ lubricants. ASME/STLE IJTC2008-71221, Miami

Xue J, Grift TE, Hansen AC (2011) Effect of biodiesel on engine performances and emissions. Renew Sustain Energy Rev 15:1098–1116

Yuksek L, Kaleli H, Ozener O, Ozoguz B (2009) The effect and comparison of biodiesel-diesel fuel on crankcase oil, diesel engine performance and emissions. FME Trans 37(2):91–97

Hydrogen-Enriched Syngas from Biomass Steam Gasification for Use in Land-Based Gas Turbine Engines

John Dascomb and Anjaneyulu Krothapalli

Abstract This paper presents the results from an experimental study on the production of hydrogen-enriched syngas through un-catalyzed steam biomass gasification. Wood pellets were gasified using a 100 kW_{th} fluidized bed gasifier at temperatures up to 850 °C. The syngas hydrogen concentration was found to increase with both bed temperature and steam–to-biomass weight ratio, reaching a maximum of 51 %. The overall energy conversion to syngas (based on heating value) also increased with bed temperature but was inversely proportional to the steam-to-biomass ratio. The maximum heating value measured was 16.1 MJ/m^3, occurring at a temperature of 697 °C, a S/B of 0.7, and a residence time of 1.9 s. The maximum energy conversion to syngas was found to be 68 %. Further analysis revealed that to produce syngas with high hydrogen concentration, inert-fluidized gasification was the most efficient method tested in this study. While fluidized bed gasification with CaO was found to be less efficient than inert-bed gasification, it does produce syngas with higher hydrogen concentration.

Keywords Biomass steam gasification · Syngas · Gas turbine engines · Energy conversion

1 Introduction

Worldwide, the use of energy from all sources will increase over the next few decades, primarily because of the demand driven by high GDP growth nations, such as China and India. Given the expectation that crude oil prices will remain relatively high, and petroleum and other liquid fuels are the world's slowest growing energy sources. Additionally, the concern about environmental consequences of green

---

J. Dascomb · A. Krothapalli (✉)
Department of Mechanical Engineering, Florida State University, Tallahassee, FL 32310, USA
e-mail: akrothapalli@fsu.edu

A.K. Agarwal et al. (eds.), *Novel Combustion Concepts for Sustainable Energy Development*, DOI 10.1007/978-81-322-2211-8_6

house gas emissions leads to a number of national governments to provide incentives in support of the development of alternative energy sources, making renewable sources of energy as the fastest growing in the outlook.

Gas turbines have emerged as the best means of transforming heat into mechanical energy and are now key components of the most efficient electrical generating systems. Hydraulic fracturing, or fracking, has increased the shale gas production, which accelerated the use of gas turbines for power production in lieu of coal power. However, many countries such as India and China do not have access to the shale gas and as a result will depend on alternative sources such as the syngas produced from either coal or biomass gasification. The more environmentally accepted solution would be syngas that is produced from biomass gasification. To be considered interchangeable with conventional fossil fuels (natural gas) and to ensure maximum flexibility, syngas heating value needs to be above 11 MJ/m^3. A high hydrocarbon content corresponds to a higher heating value for the syngas. Hence, a hydrogen-enriched synthesis gas with a higher heating value will help to mitigate the use of natural gas in land-based gas turbine applications. Hence, the subject of this paper is to present a cost effective means of biomass-derived synthesis gas production that is suitable for use in land-based gas turbines for power production.

The preferred feedstock for the current purpose is the cellulosic biomass that includes plant matters such as wood chips, switchgrass, wheat straw, corn stovers, sorghum, oilseed crop meals, and agriculture wastes. Cellulosic biomass is an advantageous feedstock because it is not an edible food crop feedstock. The cellulosic biomass consists primarily of cellulose, hemicellulose, and lignin, and these compounds pose significant challenges to conventional chemical and microbial processing methods. The process described in this paper can utilize different types of cellulosic biomass including lignocellulosic biomass and will ensure a continuous and renewable feedstock supply that is affordable. Because of the ability to convert cellulosic biomass, the proposed process has a significant advantage in full plant utilization. Other technologies have not addressed this important consideration.

1.1 Dual Fluidized Bed (DFB) Steam Gasification

Gasification is a well-proven technology that has been employed in various forms for almost 200 years (Klass 1998). It is essentially an oxygen limited thermochemical conversion of carbonaceous material to a useable gaseous fuel, synthesis gas or "syngas" consisting primarily of hydrogen (H_2) and carbon monoxide (CO), with lesser amounts of carbon dioxide (CO_2), methane (CH_4), higher hydrocarbons (C_2+), water (H_2O), and nitrogen (N_2). The oxidant used can be air, pure oxygen, or steam. Air-based gasifiers typically produce a product gas containing relatively high concentration of nitrogen with a low heating value (LHV) between 3 and 5 kJ/kg (4–7 MJ/Nm^3)[2]. Steam-based gasifiers, on the other hand, produced a product gas containing relatively high concentration of hydrogen and CO with heating values between 10 and 20 kJ/kg (10–20 MJ/Nm^3) (Paisley et al. 2001). Such a gas can be used in variety of applications, such as turbine-based power generation units, fuels, chemicals, and hydrogen production, with relatively simple downstream conversion technologies. A note that a high calorific value gas such as natural gas has a LHV of 30 MJ/Nm^3.

In a standard directly heated gasification process (Nguyen et al. 2010), the pyrolysis and gasification reactions occur in a single vessel. An oxidant, air, or oxygen combusts a portion of the biomass to provide the heat required for the endothermic reactions. Pyrolysis requires about 15 % of the heat of combustion of the feed to raise the reaction temperature and vaporize the products. In these systems, the reactor temperature is controlled by the oxidant feed rate. If the air is used as the oxidant, the product gas has a low heating value of 4–5 MJ/m^3 due to nitrogen dilution as shown in Table 1 (Roesch et al. 2011; Roesch 2011).

The results shown in the Table 1 were obtained in our laboratory, using an Ankur Scientific downdraft gasifier (Roesch 2011). As the gas composition results, shown in Table 1, indicate that a large percentage (~50 %) of the syngas contained high concentrations of N_2 for all the feedstocks considered in this study. The relatively low percentages of H_2, CO, and CH_4 are mainly responsible for the low lower heating values of the syngas.

By gasifying biomass with an N_2-free agent, the product gas will not be diluted with large concentrations of the inert gas. Using either O_2 or H_2O as the gasifying agent produces medium grade product gas (~11 MJ/Nm^3) (Ciferno and Marano 2002). In addition, catalysts in the gasifier and downstream can aid in the conversion of tars and char to usable gas. With these and other advanced techniques, a gas with a LHV of up to 20 MJ/Nm^3 is possible. The drawback to this higher quality gas is the absence of the heat produced by the burning of char. The reactions that break up the higher carbon chains in the biomass to produce H_2 and CO are highly endothermic. Heat must be provided to the process without polluting the produced gas with combustion products.

The best solution found to the problem of product stream dilution has been to burn the char produced in the reactor in a separate chamber and transfer the heat

Table 1 Syngas composition for different feedstocks

Pelletized feedstock	Component mol% dry basis						
	H_2	O_2	N_2	CH_4	CO	CO_2	HHC
Alfalfa	12.823	0.020	53.004	2.067	12.245	18.589	1.253
Algae	10.768	0.035	59.395	1.564	9.911	17.317	1.011
Field grass	12.454	0.023	51.621	2.860	13.630	17.795	1.617
Hemp	10.362	0.019	58.546	2.098	12.622	14.891	1.463
Miscanthus	9.147	0.039	53.152	2.685	17.509	16.194	1.274
Peanut shells	13.410	0.033	51.017	2.635	15.833	15.735	1.337
Pine	12.300	0.498	53.422	2.696	16.568	13.368	1.150
Municipal waste	11.323	0.029	59.493	1.899	9.615	16.257	1.384

back to the gasification process, commonly referred to as indirectly heated gasification technology. A successful scheme to achieve the indirectly heated gasification process is the dual fluidized bed (DFB) system as shown in Fig. 1. The unreacted char and bed material in the gasifier is carried to a combustion chamber where air is injected. The heat produced is then transferred back into the gasifier by the bed material. The gasifier utilizes a fluidized bed usually consisting of sand. When a gas is passed through the bed of fine solids, the bed takes on the characteristics of a liquid, allowing it to flow between the respective chambers. The continuously circulating bed greatly increases heat transfer and interaction time between the biomass and gasifying agent, while the sand breaks up the biomass quicker leading to higher gas production. The bed material easily flows when aerated, and the liquidity of the bed aids in the transfer of char to the different chambers of the system. This approach separates the combustion reaction from the remaining gasification reactions producing a product gas that has insignificant percentage of N_2 and has a heating value within a range of 15–18 MJ/Nm^3.

A list of dual fluidized bed gasification facilities that are used both in research and power plant scenarios are listed in Table 2 (Paisley et al. 2001; Kappotz et al. 2009, 2011; van der Meijden et al. 2007a, b; Corella et al. 2007; Herguido et al. 1992; Xu et al. 2006; Wei et al. 2006; Matsuoka et al. 2008). Most comprehensive studies of DFB gasification are more recently carried out at Vienna University of Technology in Austria (Koppotz et al. 2009, 2011) using both laboratory and pilot-scale facilities. The key findings from their work using the CO_2 absorption enhanced reforming process were a significantly higher H_2 content (50–60 %) in the producer gas as compared to the conventional dual fluidized bed steam gasification. The high hydrogen concentrations were obtained at gasification temperatures of about 700 °C and using repeated cycles of carbonation and calcination.

All the systems cited above use steam as the gasification agent in a typical DFB system. Because of the high heat of vaporization of steam and gasification being

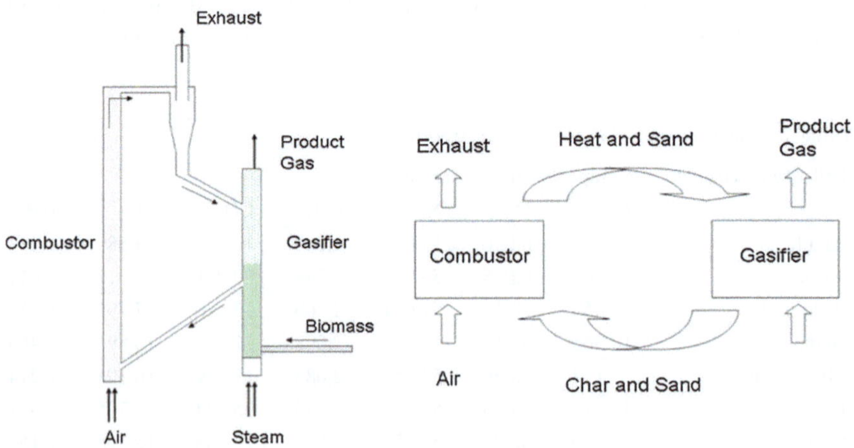

Fig. 1 Flow schematics for a dual fluidized bed gasification system

Table 2 Dual fluidized bed gasification facilities

Gasification facility	Size	Bed material	H_2 (vol%)	CO (vol%)	CH_4 (vol%)	HV MJ/m^3
McNeil plant, Vermont; DFB	50 MW		22	44.4	15.6	17.3 (HHV)
Gussing, Austria; demo DFB plant	8 MW	Limestone	50.6	16.5	12.9	
ECN (DEN) pilot-scale DFB	800 kW	Austrian olivine	21.2	37.2	12.1	
FSU, Florida; single bed	100 kW	Silica	28.7	43.5	14.6	16.5 (HHV)
University of Vienna; lab-scale DFB	100 kW	Calcite	65.1	9.3	8.8	13.1–16.5 (LHV)
University of Zaragoza, DFB	45 kW	Catalyst	30.1			15.8 (LHV)
ECN (DEN) lab-scale DFB	30 kW	Austrian olivine	27.3	27.5	9.5	
University of Zaragoza, single bed	20 kW	Silica	50	22	6	12.75 (LHV)
Yokohoma, Japan; DFB	25 kW	Silica	23.2	37.8	16.9	15.8 (HHV)
Dalin, China; dual moving beds	5 kW	Olivine	40	30	9	
Onogawa, Japan; circulating DBFB	0.22 kW	Alumina	35	27.5	7.5	

endothermic, most of the systems cited above require additional heating to maintain required reactor bed temperature. Since steam generation requires a significant fraction of the energy and the affordability of the process is questionable. Hence, only the 8-MW Gussing, Austria plant is currently in operation for power production (Koppatz et al. 2011). As a result, a careful investigation of process thermodynamics along with energy balance needs to be carried out for the design of an optimal DFB gasifier. With this in mind, a program of research was undertaken at FSU to produce affordable hydrogen-enriched synthesis gas using multiple biomass feedstock.

Figure 2 shows a sustainable DFB gasification facility schematic for a complete hydrogen-enriched syngas/H_2 production plant. The process begins with renewable biomass feedstock. The advantage of this facility is that it can process any and all parts of a biomass, even cellulosic wastes. This increases the total energy efficiency of the syngas production process. The biomass is chipped and dried with the combustion exhaust before being inserted into the gasification reactor. Inside the reactor, superheated steam is injected at up to 800 °C to fluidize and react with the bed material. Char formed in the gasification reaction is transferred to the combustor where it is burned by the addition of air. The heat generated by this combustion is then put back into the gasifier to continue the reactions. The superheated steam and product gas mixture leaves the gasification reactor where it is cooled in a heat exchanger exiting at a temperature of 350–400 °C. The gas then goes through the

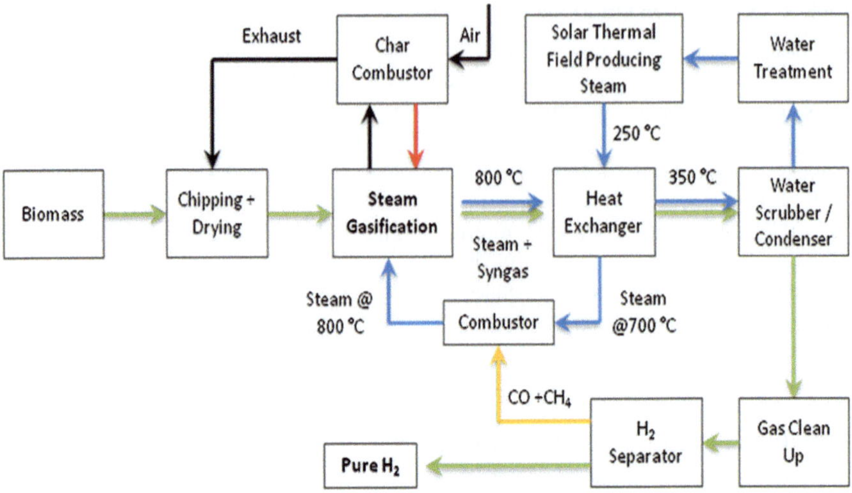

Fig. 2 Sustainable hydrogen-enriched syngas/hydrogen production process

scrubber/condenser that eliminates any tars and condenses out the fluidizing agent—the steam. The condensed water is processed before being reheated in a solar thermal field and the heat exchanger. The heated steam is brought up to its final temperature in a combustor before being once again injected into the gasifier. Back at the water scrubber, the cooled product (synthesis) gas is cleaned some more before the H_2 is separated out. All other combustibles are reused in the facility to superheat steam or heat the gasification reactor.

2 Experimental Facility and Procedures

A 100 kW-capacity dual fluidized bed thermal steam gasifier is proposed at FSU. As the first step, a single reactor bed was built to test many of the sub-systems of the facility and determine the thermal efficiency of the chemical process. A schematic of the single bed system is shown in Fig. 3. In the absence of the char combustor, heat is externally supplied to the reactor using series of electric heaters controlled with variable transformers. To precisely control the amount of steam used in the reactor, a standard electric boiler/superheater unit was installed. The advantage to both these changes is it gives the operator a highly controlled and measured process. The hot synthesis gas is then sent through a cyclone filter to remove any solid particles, followed by a heat exchanger, where the gas is cooled to about 400 °C. The gas is then passed through a venturi water scrubber/condenser and filters to further improve the gas quality. The resulting gas chemical characteristics are monitored in situ using a gas chromatograph before it is burned in a controlled Bunsen burner flame. The complete thermal characteristics of the system, as shown in Fig. 4, will be monitored and controlled.

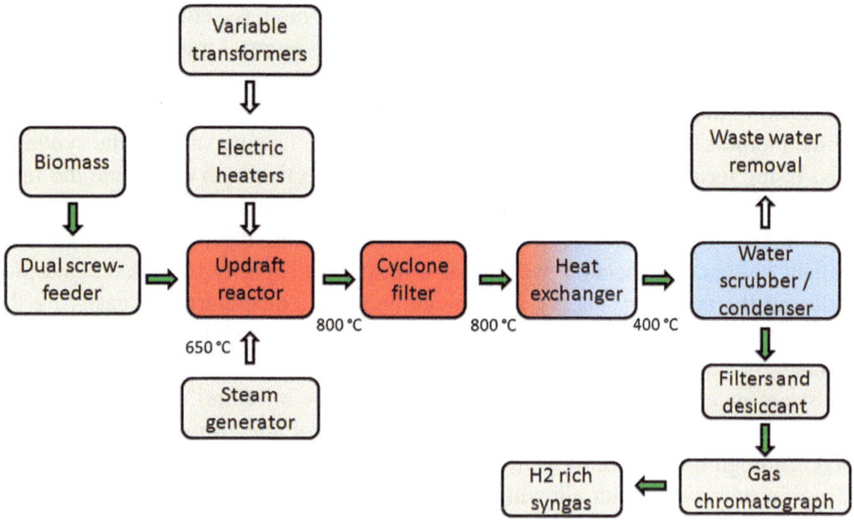

Fig. 3 Single bed hydrogen-enriched synthesis gas production process

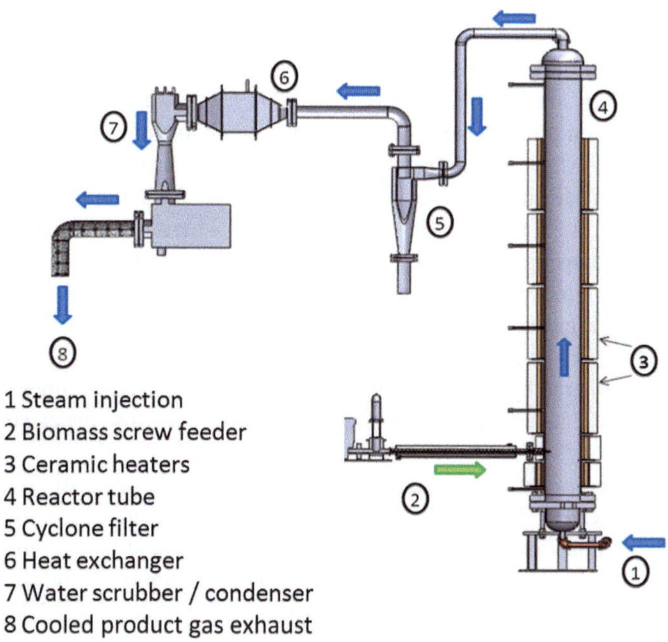

Fig. 4 Single bed hydrogen-enriched synthesis gas production plant

The main control parameters that govern the biomass conversion chemical process are as follows: bed temperature, steam-to-biomass ratio, biomass mass throughput, and catalytic properties. The energy balance of the system is determined from the measurement of the all the energy inputs such as the heating value of the biomass, steam energy, electric energy, and the heating value of the synthesis gas. Using measurement and analysis tools, the objective is to determine the most cost-efficient reactor conditions for desired gas composition and the heating value.

The settling chamber and reactor are heated using independently controlled radiant heaters. The heating system has a maximum capacity of 29 kWe and provides all the necessary energy for maintaining bed temperature and gasifying the biomass feedstock. The system is capable of gasifying up to 20 kg/h (115 kWth) of biomass pellets at 650 °C and 9 kg/h (50 kWth) at 850 °C.

Steam enters the reactor through 5 tuyeres (gas injectors). The tuyeres were constructed out of steel piping (24 mm OD) and protrude 76 mm into the gasifier bed. Through holes (1 mm) were drilled radially into the tuyeres to ensure uniform steam injection while not allowing material to flow back into the settling chamber. The gasifier reactor is filled with inert silica sand to a static height of 1.0 m. The pure SiO_2 sand has an average particle size of 0.28 mm. When fluidized, the bed height reaches 1.5–2.5 m depending on the steam flow rate. The reactor was constructed from 316 stainless steel with an inner diameter of 0.21 m and a height of 3.1 m.

The biomass used for gasification is stored in a sealed hopper before being injected into the reactor. The pellets have an average diameter of 8 mm and maximum length of 32 mm. The pellets are carried from the hopper by a metering auger. The pellets are then fed by a fast injection feeder into the reactor just above the steam injection point. A water-cooled jacket prevents premature pyrolysis in the feeding system. After the biomass and steam mixture pass through the reactor, a cyclone mechanical filter removes any solid bed materials and unreacted char from the gas stream. The gas then flows to a condenser where it is cooled to 30 °C. In addition to condensing out steam, many of the heavy tars and ash remaining in the syngas are also removed. Following the condenser, the product gas composition is analyzed using a gas chromatograph. A micro-pump carries a small amount of the gas to a filter system where it is cleaned and dried before analysis. The bulk syngas stream is then filtered by a bed of sawdust and fabric filter. The volumetric flow rate of the filtered stream is measured using a custom venturi flow meter before being disposed.

For each experiment, the reactor is first heated to the desired testing temperature by the radiant heaters. The heaters are connected to variable transformers, which are controlled by LabView process software. All facility control and measurement hardware utilized a NI compact DAQ chassis and input/output modules. The calibrated transformers tune the voltage sent to each of the heaters to maintain uniform temperature along the length of the reactor. To achieve this, seven K-type thermocouples are evenly spaced along the reactor. The temperatures are recorded using a 24-bit thermocouple module (NI-9213). Once the reactor becomes thermally stable, the electrical load of the heaters is measured to determine the thermal

Table 3 Biomass feedstock Analysis

Parameter	Value	Units
C	51.34	Moist. free wt%
O	42.50	Moist. free wt%
H	5.98	Moist. free wt%
N	0.17	Moist. free wt%
S	<0.01	Moist. free wt%
Ash	<0.6	Moist. free wt%
Moisture	5.82	wt%
LHV	20,242	kJ/kg

losses. The calibrated variable transformers have a relative error of 2.5 %. The current draw of the heaters is measured by ammeters that have a relative error of 5.0 %. Next, the steam generator and superheaters are started. The steam flow rate is determined by measuring the liquid feed rate (relative error of 4.3 %) which intermittently feeds the boiler. The flow rates used for calculations are averaged over an hour of testing. The electrical load of the boiler and superheaters is not directly measured, but calculated using the enthalpy gain from liquid water at standard temperature and pressure to its state entering the settling chamber.

The calibrated metering feeder controls the biomass feed rate. The feeder has a measured error of 6.9 %, which is caused by variations in pellet size, hopper fill height, and pellet structural integrity. The pellets contain no binders and tend to break up when agitated. The biomass feedstock used for all testing is north Florida pine. Table 3 shows the chemical analysis for the feedstock. Green Circle Bio Energy manufactures the pellets locally from native trees. The pellets have a low ash content of less than 0.6 %, which helps prevent conglomeration in the reactor. After gasification and cooling, the syngas mixture is analyzed by an Agilent Micro GC 3000, which has a relative error of 6.4 %. The chromatograph detects the dry relative concentrations of all compounds with atomic mass less than propylene (C_3H_6). During operation, samples are drawn from the produced stream every 4 min for analysis. Once the syngas is filtered and analyzed, the volume flow rate is then measured using a custom venturi type flow meter that has a relative error of 5.3 %. Another K-type thermocouple measures the gas density at the venturi flow meter.

3 Results and Discussion

To verify gasifier results, the syngas hydrogen concentration and cold gas efficiency are compared to published data. Figure 5 compares the hydrogen concentration found in this study to published results at similar steam-to-biomass (S/B) ratios and residence times. The agreement in the data confirms gasifier operation and the gas analysis techniques used in this study.

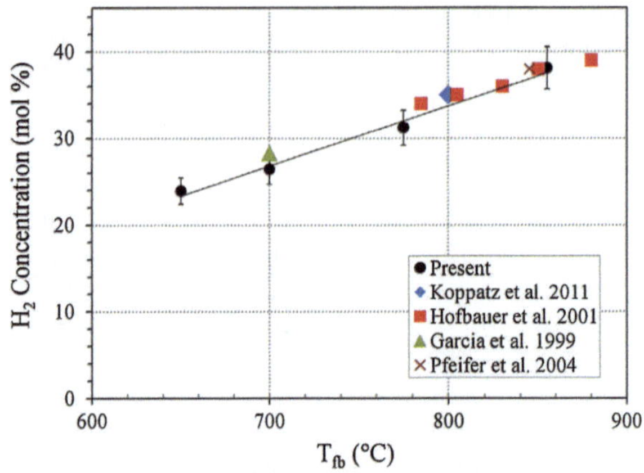

Fig. 5 Comparison of hydrogen concentration at similar run conditions (S/B = 0.8 ± 0.25, $\tau = 2.0 \pm 0.5$)

The cold gas efficiency is also compared to published data at similar temperatures and S/B ratios and is shown in Fig. 6 (Corella et al. 1991; Franco et al. 2003; Xu et al. 2004). The cold gas efficiency is reported less frequently than hydrogen concentration, as a result the residence times could not be perfectly matched. The cold gas efficiency is defined as follows:

$$\text{Cold Gas Efficiency (CGE)} = \frac{\dot{m}_{\text{syngas}} \cdot \text{HHV}_{\text{syngas}}}{\dot{m}_{\text{b}} \cdot \text{HHV}_{\text{b}}}$$

where subscript b denotes the biomass. Variation in residence time, gasifier shape, feedstock, feedstock entry points, and method of heating all effect CGE. Wei et al. (2006) used a dual fluidized bed gasifier system that burned a portion of the char in a separate chamber. This is a possible reason why their reported efficiency is lower. Additionally, Corella et al. (1991) injected biomass at the top of the fluidized bed, resulting in lowered gasification efficiency. Despite these differences, the trend of

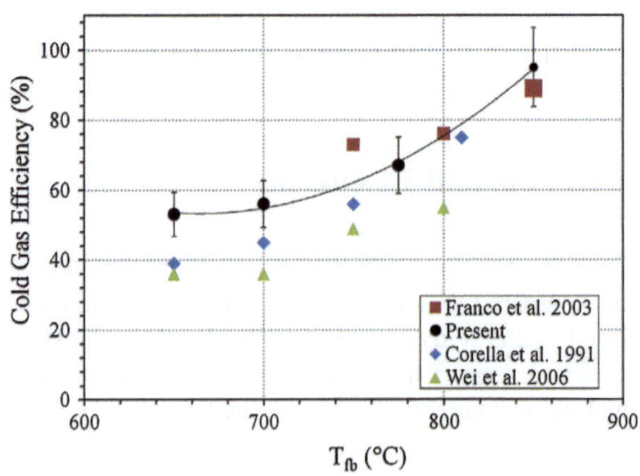

Fig. 6 Comparison of temperature effect on cold gas efficiency (S/B = 0.8, $\tau = 2.2$ s)

increased CGE with increased temperature holds for all of the facilities. The CGE found in the present study matches (within limits of error) the results obtained by Franco et al. at temperature above 750 °C. The agreement in the CGE values confirms proper operation of the gasifier energy measurement systems.

3.1 Syngas Composition

The effect of the reactor temperature on syngas composition was studied between 650 and 850 °C. Figure 7 shows the concentration of the major syngas constituents for moderate S/B ratios and residence times. At 650 °C, carbon monoxide was the most dominant syngas component at 32 mol%, followed closely by hydrogen at 30 mol%. When the reactor temperature was raised, the hydrogen concentration increased, while all of the other constituents gradually decreased. At 775 °C, hydrogen was the most dominant species at 38 mol% and carbon monoxide levels had dropped to 30 mol%. This trend continued when the reactor temperature was raised to 850 °C, where a hydrogen concentration of 44 mol% was recorded. Part of the reason for the increase in hydrogen concentration was the cracking of hydrocarbons. Methane levels dropped from 13 to 10 mol%, and higher hydrocarbon (HHC) levels dropped from 5 to 3 mol%.

The effect of temperature on HHC concentration is shown in Fig. 8. Ethylene (C_2H_4) levels increased as the reactor temperature was raised from 650 to 775 °C before falling to its lowest level at 850 °C. The increase in ethylene concentration could have been from the cracking of propylene (C_3H_6). Propylene contains two carbon atoms joined by a double bond (like ethylene) and with a methyl (methane) group attached to it. All of the other HHCs decreased in concentration with each subsequent increase in temperature. The resulting sum of the HHC's decreased with

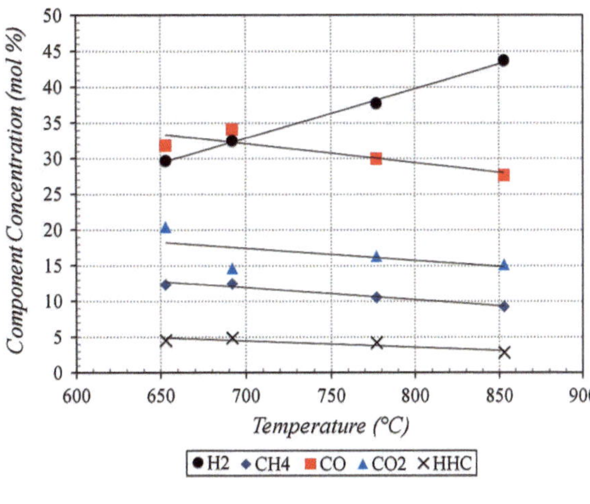

Fig. 7 Effect of temperature on gas composition (S/B = 2.0 ± 0.1, τ = 2.0 ± 0.4 s)

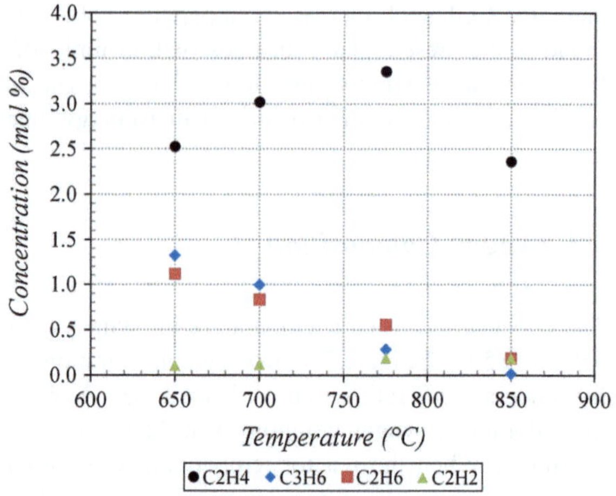

Fig. 8 Effect of temperature on higher hydrocarbon concentrations (S/B = 2.0 ± 0.1, τ = 2.0 ± 0.4 s)

temperature. The hydrocarbons crack at high temperatures to produce lighter molecular weight compounds.

The trend of increasing hydrogen concentration with increased temperature held for all S/B ratios and residence times tested. Figure 9 shows the similar rate of increase in hydrogen concentration for three different bed conditions. Hydrogen concentration increased by an average of 7 mol% per 100 °C increase in reactor temperature. The hydrogen concentration increased due to the increase in gas-phase steady-state concentration as well as increased char reactivity. The char–steam gasification reaction is endothermic and occurs at very low rates at low temperatures. The results presented are in agreement with published data (Franco et al. 2003; Koppatz et al. 2011). A hydrogen concentration maximum was not found in the present experimental study, and an increase in temperature would almost certainly increase hydrogen concentration further. In order to achieve an average bed temperature of 900 °C, the reactor exterior temperature in the main heating

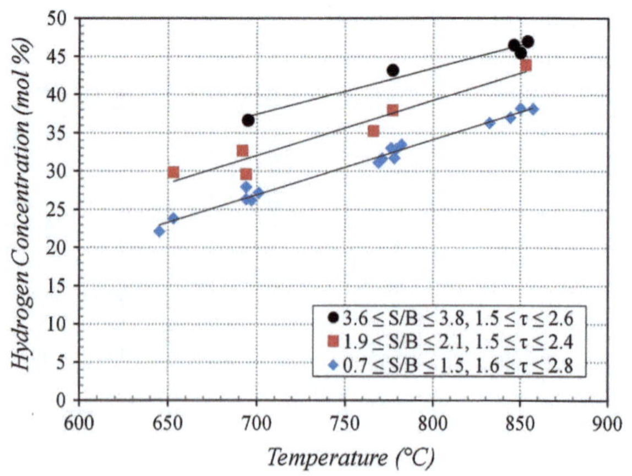

Fig. 9 Effect of temperature on hydrogen concentration for different S/B and τ

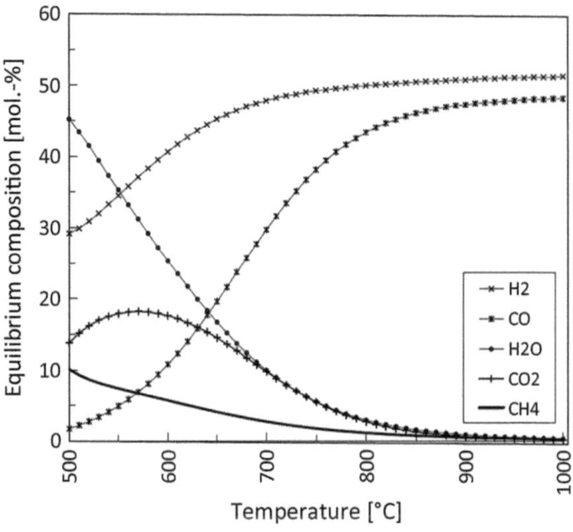

Fig. 10 Equilibrium syngas composition (Koppataz et al. 2011)

region was required to be in excess of 1,100 °C. At this temperature, corrosion of the reactor exterior surface occurred rapidly. As a result, the reactor surface would shed oxidized metal dust and flakes that accumulated in the gap between the heaters and reactor surface. This collection of solids created an electrical bridge between the heaters and the reactor, short circuiting the heaters. The resulting electrical discharge destroyed heaters and deformed the reactor surface. To test gasification at higher temperatures, a reactor material capable of withstanding such high temperatures is necessary, or oxygen must be eliminated from the environment.

Due to the difficulty of reaching gasification temperatures above 850 °C, most experimental facilities do not test above this temperature. Equilibrium models, such as the one by Koppatz et al. (2011) shown in Fig. 10, predict a slow increase in hydrogen concentration until 1,000 °C. Equilibrium models usually over predict hydrogen concentration at low temperatures due to slow reaction rates that resist changes in composition. When reactor temperature is increased to 800 °C and above, experimental results tend to more closely match equilibrium predictions. This explains the linear trend seen in the hydrogen concentration measured in this study. It is expected that at higher temperatures, the hydrogen concentration will taper off similar to the prediction of the equilibrium models.

3.2 Syngas Heating Value

The syngas heating value is an important parameter when determining the possible end uses for syngas. When used as a fuel to power an engine or turbine, the heating value must be within acceptable limits of the particular device. Hydrogen-enriched syngas can have a high heating value per unit mass, while having a low heating value

Fig. 11 Effect of temperature on syngas heating value

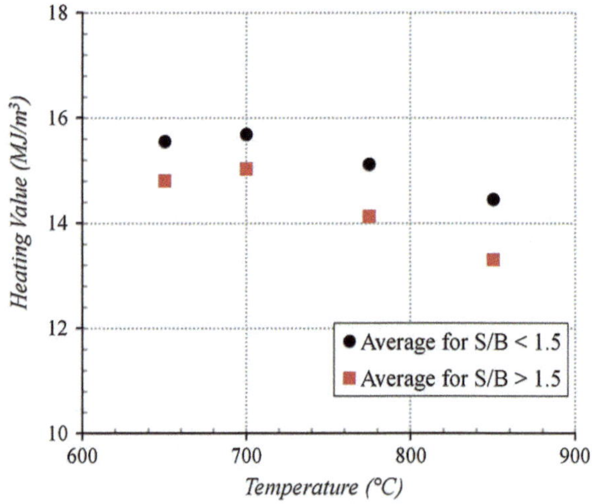

per unit volume compared to other gaseous fuels such as propane or natural gas. The relatively low density of syngas is due to the relatively high concentration of hydrogen, which has the lowest density of any gas. The heating value of gaseous fuels is usually reported on a volume basis. Hence, the following analysis is given as such. It is important to keep in mind that heating value presents qualitative information on the produced syngas and does not inform about the quantity of energy produced.

Figure 11 shows the effect of temperature on the syngas heating value. The heating value was highest at a reactor temperature of 700 °C. At this temperature, carbon monoxide was the most abundant gas constituent. The maximum heating value measured was 16.1 MJ/m^3, occurring at a temperature of 697 °C, S/B of 0.7, and a residence time of 1.9 s. At this condition, the methane concentration was 13.4 mol%, which its highest measured, and the carbon monoxide concentration was also near its maximum at 38 mol%. The main reason that the heating was highest because of the hydrogen concentration being relatively low at 26.3 mol%. The heating value dropped as the reactor temperature was increased. The minimum heating value measured was 12.7 MJ/m^3. This occurred at a reactor temperature of 853 °C, S/B of 2.9, and a gas residence time of 4.5 s. The hydrogen concentration during this particular experiment was the highest measured for un-catalyzed gasification at 51 mol%. The high hydrogen concentration increased the total combustible fraction and volume of the syngas, but decreased the heating value.

3.3 Energy Efficiency

One of the most important parameters determining the viability of steam gasification. is the thermodynamic conversion of energy, or energy efficiency (EE). This quantitative tool is a good first estimation of the cost of producing hydrogen-enriched syngas. The energy efficiency describes the conversion of all energy sources to syngas (on a heating value basis). It is observed that the energy flow rate of the syngas increases with reactor temperature, gas residence time, and S/B ratio (Dascomb 2013). The drawback to this increased power production is that increasing any of the three major parameters increases the auxiliary heating requirement. Increasing reactor temperature obviously increased conduction losses, but it also increased the sensible heat of the gas. For facilities without a heat exchanger, such as the one used presently, that sensible energy is not recovered. The non-heat recovery energy efficiency gives a conservative estimate of the impact of increasing reactor temperature. To analyze the efficiency with a heat exchanger, one can simply assume a recovery efficiency of a heat exchanger.

It takes a considerable amount of energy to generate superheated steam. Steam has a high heat of vaporization, and most of the energy requirement is needed simply to make saturated steam. Because the reactor thermal losses were fixed at each temperature, when less biomass was injected into the reactor, the relative amount of thermal losses increased. In this way, increasing gas residence time increases the relative amount of thermal losses. In other words, the amount of joules lost per kg of biomass is greater when less kilograms of biomass are inserted into the reactor. The gasifier conditions that maximize the thermal conversion of all energy sources to syngas heating rate was discussed in a recent publication (Dascomb et al. 2013). The energy conversion efficiency used is defined as

$$\frac{\dot{m}_{\text{syngas}} \cdot \text{HHV}_{\text{syngas}}}{\dot{m}_{\text{b}} \cdot \text{HHV}_{\text{b}} + \dot{Q}_{\text{h}} + \dot{m}_{\text{b}} \cdot \Delta H_{\text{s}}}$$

Figure 12 shows the effect of temperature on energy efficiency. The energy conversion efficiency increased with increasing reactor temperature despite the increased conduction losses of the reactor and increased sensible heat requirement

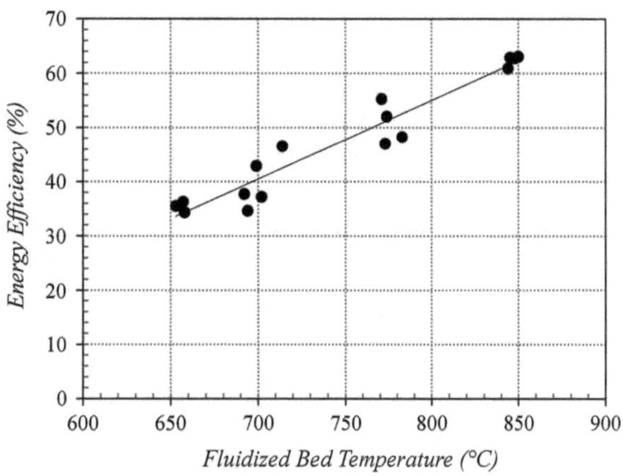

Fig. 12 Effect of temperature on energy efficiency. (S/B = 1.8 ± 0.4)

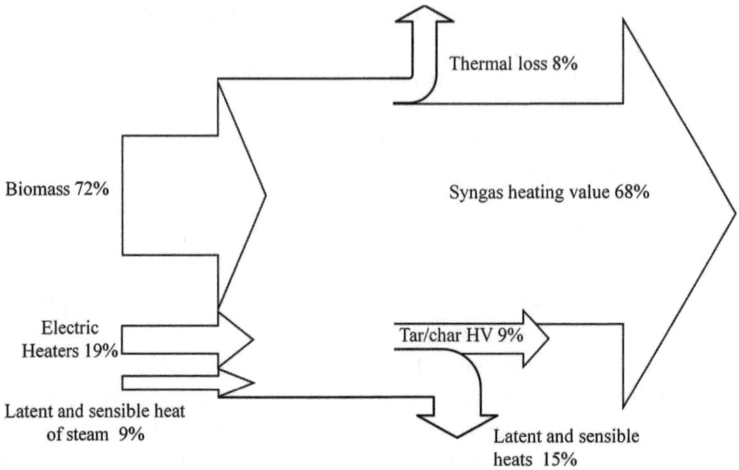

Fig. 13 Energy flow diagram for maximum energy efficiency case. (T = 857 °C, S/B = 0.75, τ = 2.4 s)

of the syngas. As shown in the previous section, the heating value of the syngas decreased as the reactor temperatures rose, but the total heating content (heating value x flow rate) increased. An increase in temperature would most likely increase the conversion efficiency further. Adding a heat exchanger could also increase the energy efficiency. Recovery of any of the sensible heats of the steam or syngas mixture would reduce the required energy input and increase energy efficiency.

A heat flow diagram for the maximum energy efficiency case is shown in Fig. 13. The biomass heating value accounted for 72 % of the energy input, while 19 % was required to heat the reactor bed and 9 % for the generation of steam. The total process energy converted to chemical potential in the syngas was 68 %. Reactor thermal losses consumed 8 % of the input energy. This portion can be reduced by better insulating the reactor. The thermal loss during this test was 7.8 kW. Lower thermal losses were accomplished after this particular test with better insulation wrapping. The results presented here are using the measured losses to accurately portray testing.

The Latent and sensible heat contained by the steam and syngas accounted for 15 % of the total energy. This energy is available to be recycled back into the system using heat exchangers. Because of the higher tar content produced at low S/B ratios, some form of tar treatment will be needed to prevent reduction in heat exchanger efficiency. The easiest way to prevent tar buildup in a heat exchanger is only cool the syngas mixture down to 400 °C. In this way, most of the tars are prevented from condensing. While effective, this method will limit the amount of recoverable heat. Another option is to use a catalytic tar reformer. This heated vessel will remove most of the tars and thus allow the gas to pass through a heat exchanger that can utilize the full energy potential of the heated gas. Although this adds infrastructure and energy cost to maintain another bed at high gasification temperatures.

A carbon balance was performed to determine the amount of carbon not converted to syngas. At the maximum energy efficiency case, 18 % of the carbon in the biomass did not form usable syngas. This corresponded to any total energy loss of 9 %. There are multiple ways to increase the conversion of carbon and increase energy efficiency; higher gasification temperatures, the addition of catalysts, and a dual bed gasification system. The tar and char energy content is the least accurate value given in the energy balance due to the indirect form of measurement. It is necessary to note that the heat content of the outputs was 16 % larger than the inputs. This was most likely due to the accumulation of char in the reactor bed. By the time the measurements were taken for this analysis, a significant amount of char had probably built up in the bed and was reacting with steam adding heating content to the syngas. To account for this in the energy flow, the difference was subtracted from the tar and char heat content value. This gave the most accurate reporting for steady-state gasification.

3.4 Gasification with Calcium Oxide

To further improve the energy conversion efficiency, a catalyst was used in the fluidized bed. This section presents the experimental findings of gasification performed with a CaO reactant. The temperatures tested were between 650 and 700 °C. At lower temperatures, low char reaction rates decrease biomass utilization and correspondingly the thermodynamic conversion efficiency. At higher temperatures, CO_2 reaction rates with limestone decrease. A full parametric study to determine the effect of S/B ratio or gas residence time was not performed. Test conditions were chosen based on the gasifier behavior observed in un-catalyzed testing. Moderate S/B ratios of between 2 and 3 were chosen to balance hydrogen production efficiency and energy efficiency. It was observed without a reagent, higher S/B ratios corresponded to increasing hydrogen production efficiency but lowered energy efficiency. The biomass feed rates were kept low to increase the duration of testing. Testing times were limited by the mass of limestone in the reactor. When the limestone becomes saturated with CO_2, the CO_2 concentration returns to levels seen in un-catalyzed testing. It was desired during this study to look at the behavior of gasification before this transition occurs, when the gasifier is at a quasi-steady state.

Once all of the CaO has reacted, it must be regenerated to continue testing. The ideal way to do this is to implement a second reactor. In this case, the spent quicklime is carried to the second reactor, regenerated, and returned to the fluidized gasification bed. With a single fluidized bed, as the case for this study, CaO gasification requires cycling the bed. This results in three distinct phases occurring as follows:

Biomass gasification phase: In this phase, biomass is inserted into the fluidized bed and gasified at temperatures between 650 and 700 °C. CO_2 produced during gasification reacts with CaO to produce $CaCO_3$, removing CO_2 from the reactor. The resulting shift in gas-phase composition creates more elemental hydrogen. At temperatures at and below 700 °C, the reactivity of char is low. As a result, char builds up in the reactor during gasification. The volume of char increases until its total surface area is large enough to react at a rate equal to what is being produced by the addition of biomass feedstock.

Char gasification phase: In this phase, no biomass was inserted into the fluidized bed but steam still flowed through it. The steam slowly gasified the solid char, producing H_2, CO, CO_2, and CH_4. The limestone reacted with CO_2, provided that there was still unreacted CaO, and the CO_2 partial pressure was higher than the equilibrium pressure. At 650 °C, the equilibrium partial pressure of CO_2 is 0.8 % of atmospheric pressure and at 700 °C, it is 2.6 %.

CaO regeneration phase: In this phase, the temperature of the bed was increased to promote the release of CO_2. The temperature required to raise the equilibrium pressure of CO_2 above atmospheric pressure is roughly 900 °C. However, if CO_2 is removed from the environment, lowering its partial pressure, CaO will regenerate at lower temperatures. The dry gas stream produced in this phase consists mainly of CO_2 and small amounts of hydrogen and carbon monoxide.

The resulting gas concentrations from gasification using CaO are shown in Table 4. The maximum average hydrogen concentration seen was 69.4 %. This was higher than any concentration seen without the use of CaO. The reason for the high H_2 levels was the elimination of CO_2. During the maximum H_2 concentration test, the CO_2 concentration was 12.0 mol%, which is lower than any concentration seen during un-catalyzed testing. The CO concentration at the maximum H_2 concentration case was 7.5 %. This is nearly three times lower than the lowest levels seen during un-catalyzed testing. The low carbon monoxide levels were a result of Le Chatelier's principle working on the CO-shift reaction. This confirms that the CO-shift reaction is one of the most important reactions determining the syngas composition.

The effect of CaO on hydrocarbon levels varied. All the hydrocarbon concentrations measured during catalytic testing were similar to levels seen at 850 °C in

Table 4 Results from gasification with CaO

Test #	1	2	3
Reactor temperature (°C)	657	690	701
S/B ratio	2.9	2.9	2.1
Gas residence time (s)	2.7	2.6	2.7
Syngas component	Average gas conc. (dry mol%)		
Hydrogen	65.5	69.4	68.3
Methane	11.1	8.8	8.7
Carbon monoxide	10.8	7.5	9.3
Carbon dioxide	9.4	12.0	11.3
Ethylene	1.6	1.3	1.4
Ethane	0.5	0.8	0.6
Acetylene	0.1	0.1	0.1
Propylene	0.4	0.8	0.4
HHV (dry MJ/m^3)	15.6	14.2	14.3

un-catalyzed testing except for propylene which was higher than levels seen at 850 °C. The increased cracking of tars into propylene may have caused this. The gas filters were noted during testing to be cleaner than during un-catalyzed testing. This suggests that tar levels were the lowest when using CaO, although this cannot be proven because tars were not directly measured in this study. The fact that hydrocarbon levels were not lower than the concentrations seen at 850 °C also tells us that CO_2 and CO are not a product of hydrocarbon cracking. Other catalysts types are needed to break down hydrocarbons, which will form additional elemental hydrogen.

The results shown in Table 4 were averaged over the entire test. Figure 14 shows the gas concentration for over 2 h of testing. The hydrogen concentration decreased slightly, and the CO concentration increased over the course of the test. This was due to the saturation of the CaO. As the fraction of reacted CaO increased, the CO_2 removal rate decreased. If the gasifier was to be continued to be run, the hydrogen concentration would eventually drop to the levels seen during un-catalyzed testing.

The thermodynamic efficiencies of gasification are shown in Table 5. The highest cold gas efficiency seen was 76 %. This CGE was higher than any seen during un-catalyzed testing at 700 °C, but much less than the maximum values seen at 775 °C or 850 °C. The maximum energy efficiency seen was 47 %. This was the same maximum energy efficiency seen during un-catalyzed testing at 700 °C but much lower than the maximum values seen at higher temperatures. The low temperatures required for proper CaO reactivity prevented the full utilization of the biomass feedstock in this phase of operation.

A complete breakdown of the thermodynamic inputs/outputs for the maximum energy efficiency case is shown in Fig. 15. The biomass heat content accounted for 61 % of the input energy. Heating the superheated steam before it entered the fluidized bed consumed 23 % and external reactor heating 6 %. Unlike previous testing, these were not the only energy inputs. The reaction between CaO and CO_2

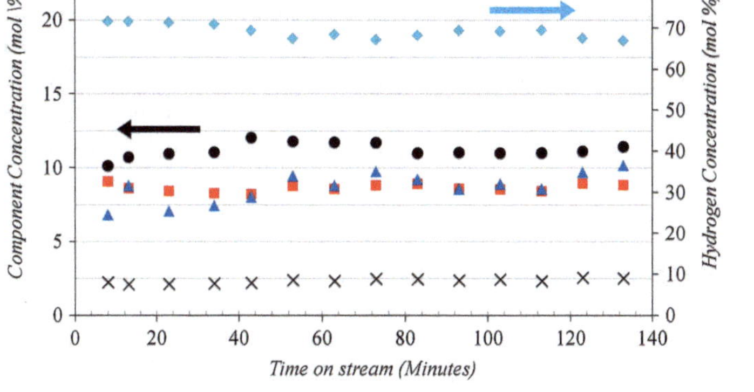

Fig. 14 Syngas composition for gasification using CaO (test number 3, T = 701 °C, S/B = 2.1, τ = 2.7 s)

Table 5 Efficiency of gasification with CaO

Test #	1	2	3
Reactor temperature (°C)	657	690	701
S/B ratio	2.9	2.9	2.1
Gas residence time (s)	2.7	2.6	2.7
Cold gas efficiency (%)	56	70	76
Energy efficiency (%)	32	39	47
Hydrogen production efficiency (%)	16	23	27

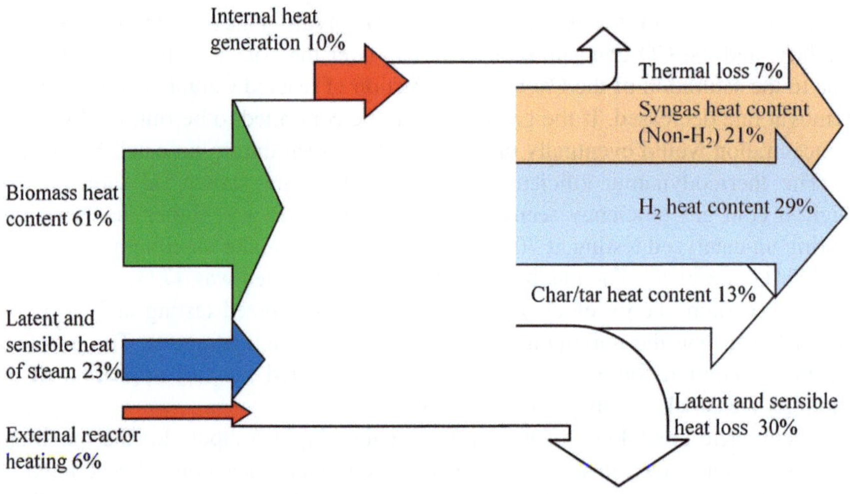

Fig. 15 Thermodynamic inputs/outputs for gasification with CaO for maximum efficiency case (T = 701 °C, S/B = 2.1, τ = 2.7 s)

releases 170.5 kJ/mol and added 10 % of the energy input to the system during this particular test. This heat content can be thought of as a battery, while energy was not inserted externally during testing, an equal amount of energy must be added during the regeneration phase. The energy addition from the CaO reaction was added to the energy efficiency and hydrogen production efficiency equations to give a more accurate measure of energy use during this phase. This is more useful when analyzing catalytic testing for use in a dual fluidized bed, which is its most likely use.

4 Summary and Conclusions

Currently, the most effective method for producing hydrogen-enriched synthesis gas from biomass is steam gasification. Steam gasification has been shown to produce a synthetic gas with high hydrogen concentration (>70 %) under certain conditions, but little information exists on the thermal conversion efficiency during these conditions. Researchers have expressed doubt in steam gasification citing its large energy demand.

The syngas produced during fluidized bed steam gasification had a maximum hydrogen concentration of 51 mol%. During this test, the conversion of energy to syngas heating content and hydrogen heat content was 58 and 26 %, respectively. These values were obtained at the highest temperature tested (850 °C), a high steam-to-biomass ratio (2.9 by weight), and a high gas residence time in the reactor (4.1 s). A significant fraction of energy was needed to generate steam during this test, which limited the thermal conversion of energy to syngas. At the lowest steam-to-biomass ratio tested (0.75), the conversion of energy to syngas heat content was 68 % but the conversion to hydrogen was only 20 %. This was due to the relatively low hydrogen concentration (38 mol%).

Steam gasification with CaO produced the highest H_2 concentrations tested in this study (69 mol%). The maximum thermal conversion of energy to syngas heat content and hydrogen heat content was 47 and 27 %, respectively. The conversion efficiencies were lower than those observed during inert-bed testing because at the optimal temperature for CaO gasification (700 °C), and char conversion was low. During CaO regeneration, the char was gasified, producing a syngas stream with over 80 mol% hydrogen. The energy produced during regeneration equaled the energy requirement during this phase.

References

Ciferno JP, Marano JJ (2002) Benchmarking biomass gasification technologies for fuels, chemicals and hydrogen production. Internal Report, National Energy Technology Laboratory

Corella J, Aznar MP, Delgado J, Aldea E (1991) Industrial and engineering chemistry research, pp 2252–2262

Corella J, Toledo JM, Molina G (2007) A review on dual fluidized-bed biomass gasifiers. Ind Eng Chem Res 46:6831–6839

Dascomb J (2013) Thermal conversion efficiency of producing hydrogen enriched synthesis gas from steam gasification. Ph.D. Dissertation, Florida State University

Dascomb J, Krothapalli A, Fakhrai R (2013) Thermal conversion efficiency of producing hydrogen enriched syngas from biomass steam gasification. Int J Hydrogen Energy 38:11790–11798

Franco C, Pinto F, Gulyurtlu I, Cabrita I (2003) The study of reactions in quenching the biomass steam gasification process. Fuel 82:835

Herguido J, Corella J, Gonzalez-Saiz J (1992) Steam gasification of lignocellulosic residues in a fluidized bed at a small pilot scale. Effect of the type of feedstock. Ind Eng Chem Res 31:1274–1282

Klass DL (1998) Biomass for renewable energy, fuels and chemicals. Academic Press, New York

Koppatz S, Pfeifer C, Rauch R, Hofbauer H, Marquard-Moellenstedt T, Specht M (2009) H_2 rich product gas by steam gasification of biomass with in situ CO_2 absorption in a dual fluidized bed system of 8 MW fuel input. Fuel Process Technol 90:914–921

Koppatz S, Pfeifer C, Hofbauer H (2011) Comparison of the performance behavior of silica sand and olivine in a dual fluidized bed reactor system for steam gasification of biomass at pilot plant scale. Chem Eng J 175:468–483

Matsuoka K, Kuramoto K, Murakami T, Suzuki Y (2008) Steam gasification of woody biomass in a circulating dual bubbling fluidized bed system. Energy Fuels 22:1980–1985

Nguyen DT,Parimi K, Krothapalli A, Greska B, Dascomb J (2010) Bio-jet fuels from nonedible bio-oils and cellulosic biomass. In: Proceedings of the advanced propulsion meeting, Office of Naval Research, Crystal City, VA, pp J9–J21

Paisley MA, Irving JM,Overend RP (2001) A promising power option—the fercosilvagas biomass gasification process—operating experience at the Burlington gasifier. In: Proceedings of ASME turbo expo, New Orleans, LA, USA, 4–7 June 2001

Roesch H (2011) Downdraft gasification of various biomass feedstocks for energy production. M.S. thesis, Florida State University

Roesch H, Dascomb J, Greska B, Krothapalli A (2011) Prediction of producer gas composition for small scale commercial downdraft gasifiers. In: Proceedings of 19th European biomass conference and exhibition, Berlin, Germany, pp 1594–1601

van der Meijden CM, Veringa HJ, van der Drift A, Vreugdenhil BJ (2007a) Bioenergy II: scale-up of the MILENA biomass gasification process. Int J Chem Reactor Eng 7:Article A53

van der Meijden CM, van der Drift A, Vreugdenhil BJ (2007b) Experimental results from the allothermal biomass gasifier MILENA. In: 15th European biomass conference, Berlin, Germany, pp 7–11

Wei L, Xu S, Liu J, Lu C, Liu S, Liu C (2006) A novel process of biomass gasification for hydrogen-rich gas with solid heat carrier: preliminary experimental results. Energy Fuels 20:2262–2273

Xu S, Liu J, Lu C, Liu S, Liu C (2004) Hydrogen production in steam gasification of biomass with CaO as CO_2 absorbent. Energy Fuels 2266–2273

Xu G, Murakami T, Suda T, Matsuzawa Y, Tani H (2006) Gasification of coffee grounds in dual fluidized bed: performance evaluation and parametric investigation. Energy Fuels 20:2695–2704

Part III
Combustion of Fuels and Engine Performance

Future Trends in Commercial Aviation Engines' Combustion

Hukam C. Mongia

Abstract This article gives an overview of the current rich-dome combustion system design, requirements and challenges; followed by the first alternative to rich-domes that have been successfully introduced as products; the lowest levels of achievable NO_x (so called entitlement) as determined from small scale rig testing; summary of recent engine emissions data with "green" alternative fuels; description of the 2nd alternative to rich-dome products that may be of interest to the OEM's for the N+2 and N+3 generation aviation engines; a brief discussion on the modeling and correlation accuracy expectations from future efforts in this area; the 3rd alternative to rich domes which was shown promising for autoignition times closer to 0.2 ms. The article concludes with a short section on operability and dynamics. Several large low-NO_x rich domes' takeoff NO_x emission index is reproduced well by a simple correlation $NO_{xRD-L} = 0.0303PR^{1.9722}w$ / $R^2 = 0.9906$ including Talon II and Trent1000. However, the LTO NO_x is correlated well by a similarly good quality curve only for the group of combustors without Trent1000, $DP / F_{00} = 0.6793PR^{1.2241}$ compared to lower value expression for Trent1000 alone given by $DP / F_{00} = 0.1292PR^{1.6327}$. Consistent with the NO_x stringency pattern set by CAEP4, CAEP6 and CAEP8 and longtime goal for achieving 85 % reduction in takeoff NO_x at 30OPR, we propose the long-term LTO regulatory standard of $CAEP / 18 = -37.763 + 2\pi$ effective December 31, 2033. The combustor inlet temperatures for desired overall pressure ratio at sea-level standard day static condition can be estimated by using $T_{3,SSS} = 317.544OPR^{0.2704}$; $320.1955OPR^{0.272}$, respectively for the N+1 and N+2 generation engines. This along with generally accepted requirements for combustor operability, we have to manage significantly increased range of P_3, T_3 and fuel/air ratio, viz. T_3: 216.67–1084 K; P_3: 0.33–60 atm; ΔP: ~ (0.1–1.2) ΔP_{design}; and FAR: FAR_{min} - FAR_{max}. The numerical values of the dome design pressure drop (ΔP_{design}), minimum and maximum fuel air ratios (FAR_{min} and FAR_{max}) depend upon the combustion system design and its potential applications. We will assume typical values of these variables, respectively, 3–5 %, 0.005–008 and 0.025–0.040 for the aviation engines. The 1st generation of lean dome products met

H.C. Mongia (✉)
School of Mechanical Engineering, Purdue University, West Lafayette, Indiana, USA
e-mail: hmongia43@hotmail.com

A.K. Agarwal et al. (eds.), *Novel Combustion Concepts for Sustainable Energy Development*, DOI 10.1007/978-81-322-2211-8_7

their original objectives of achieving lower NO_x within the specified design constraints including cooling technology. They were immediately followed by the 2nd generation lean dome products known popularly as TAPS in GEnx in addition to planned LEAP-X and GE9X. The takeoff NO_x of GEnx is given by $NO_{xGEnx} = 1.079 \times 10^{-5}\ OPR^{3.971}\ W\ /\ R^2 = 0.991$. These products will be able to meet the proposed long-term LTO NO_x regulatory standard within the generally accepted design modification and refinement process. TALON-X and recently introduced P&W TALON-Axially Staged Combustion concept will provide credible alternatives to TAPS; an interesting technologies competition. The 2nd generation lean domes produce an order of magnitude lower exhaust smoke number than the rich domes. For all other design requirements both the lean and rich domes have comparable characteristics. The effect of FT fuel blends on combustion efficiency and NO_x is insignificant; but its benefits in regard to particulate emissions are enormous in terms of both the number density and mass emissions due primarily to significantly lower aromatic and sulfur contents. Future CFD and semi-empirical model should be developed with the proposed long term accuracy goal expressed in term of the standard deviation σ goal: 3 % of takeoff NO_x, 7.5 % and 15 % respectively of idle CO and HC emission indices.

Keywords Rich-quench-lean (RQL) combustors · NO_x emission · Alternative fuels · Reynolds-averaged Navier–Stokes · Flame temperature

1 Introduction

Figure 1 provides an appropriate start for this article because of the following reasons. It introduces a new engine architecture involving fan-drive gear system [known popularly as Geared Turbofan (GTF)] with attendant propulsion system level impacts (Kurzke 2009). Compared to a ten fan bypass ratio (BPR), conventional two-spool 103 kN rated thrust turbofan engine, it reduces the number of the low-pressure (LP) compressor stages from 7 to 2, and that of the LP turbine (LPT) from 9 to 3 for an optimum 5.9 LPT expansion ratio. Simultaneously it reduces the LPT stage loading from 3.2 to 1.05 with attendant increase in its isentropic efficiency from 92.8 to 93.4 % which compares very well with 93.6 % of the present day six BPR turbofan engines. GTF reduces LPT spool torque from 77 to 27 kN-m with attendant smaller load for the fan-drive gear system. However, the LPT AN^2 increases from 6.7 to 41.9 m RPM $\times 10^{-6}$ resulting in significant increase in the mechanical design challenges; the corresponding value for the conventional six BPR engine is 18.5. Both engines produce comparable takeoff thrust specific fuel consumption (SFC) of 11.44 and 11.4 g/kN-s compared to 13.0 with the conventional engines resulting 12.3 % reduction in the takeoff fuel burn. GTF gives slightly smaller fan diameter (2.23 m) compared to 2.32 m but both are considerably larger than 1.78 m of the current engines. It will lead to higher installation losses without using advanced nacelle technology as implied in Fig. 1. Furthermore,

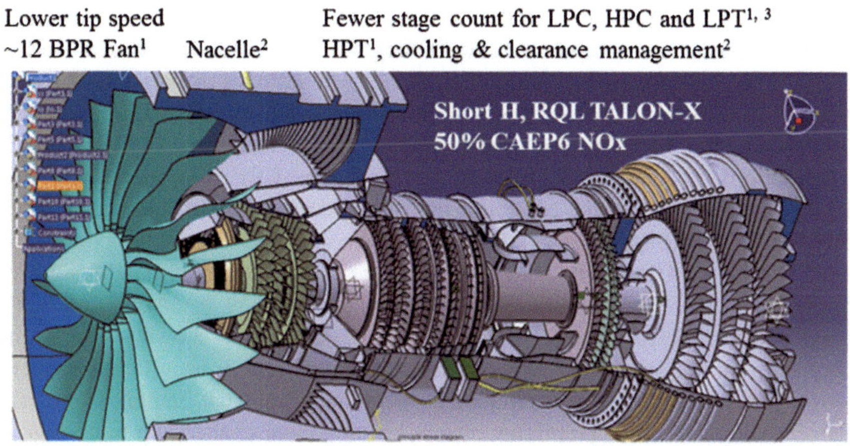

Fig. 1 The Pratt and Whitney PW1000G engine cross-section highlighting important technologies; Modified a picture grabbed from http://grabcad.com/library/pratt-and-whitney-turbofan-engine-pure-power-1000g on July 13, 2014. *1* Advanced 3-D aerodynamics for improved efficiency. *2* Advanced technology. *3* Higher speed

GTF provides reasonable engine architecture for future 14 BPR engines without increasing LPC and LPT stage count, with moderate increase in the LP spool torque (from 29.3 to 32.8kN-m), and LPT stage loading from 1.05 to 1.29 with additional fuel burn reduction of 6 %. In summary, GTF has invigorated competitive fights among the engine original equipment manufacturers (OEM) for the next generation of propulsion engines, so-called $N + 1$ and $N + 2$ generation engines with operating overall pressure ratios (OPR) approaching or exceeding 60 (Kumar et al. 2012a, b, c; Panda et al. 2012a, b). It is therefore interesting to note that the first GTF product, the PW1000G, will be used on the 130–180-seat A320neo family with 12 BPR, replacing the IAE V2500; the 70–90-seat Mitsubishi Regional Jet with 9 BPR; the 110–130-seat Bombardier CSeries with 12 BPR; and the 150–212-seat Irkut MS-21 family. The projected 12.2 BPR for the A320neo has forced the LEAP-X with conventional turbofan architecture to 11 BPR, both claiming 16 % reduction in mission fuel burn (Arvai 2011).

The medium-size GTFs and competing high BPR and OPR engines have to use smaller size high-pressure cores requiring a new set of fuel/air mixing devices for high combustor inlet temperature (T_3) and temperature rise (ΔT) (Mongia 2011a, b). Moreover, TALON-X as highlighted in Fig. 1 claims to take the rich-quench-lean (RQL) combustion technology developed progressively since the early 1980s (viz. Mongia 2013a, b, c, d) to yet another lower level (McKinney et al. 2007), leading to reassessing the combustion mixers technology path taken in the last 40 years (Mongia 2011c, d, e). If the statements made by AirInsight (2011) turn out true in regard to the PW1000G and LEAP-X operating pressure and HPT gas path temperature levels (viz. "The LEAP will need to run at hotter temperatures and have higher pressure ratios than the GTF—LEAP is expected to have gas path temperatures 200 °C hotter than the CFM56. GTF's gas temperatures are projected to be 60–130 °C lower than LEAP. etc."), then we must reassess the benefits of lean domes over rich domes for NO_x emissions. It is interesting to note that a second generation twin-annular premixing swirl-stabilized flames (TAPS), TAPS II will be used in the three LEAP engine models, namely LEAP-1Afor the A320neo, LEAP-1B for the B737 MAX, and LEAP-1C for the Comac C919 with more than 5,600 LEAP-X engine orders and commitments as of October 31, 2013 according to the Snecma website (Snecma 2014). If LEAP-X takeoff pressure ratios are comparable with that of the GEnx with TAPS which gave the characteristic landing–takeoff (LTO) NO_x of 35 % CAEP6 at 35 OPR but increased to 59 % CAEP6 at 47 OPR compared to TALON-X claim of achieving 50 % CAEP6 as shown in Fig. 1, this may set into motion another round of innovations in the GE lean dome products.

The combustion system durability including flashback, autoignition, and flameholding within the premixing mixers become more challenging to manage as OPR levels exceed 45. Therefore, alternative mixer technologies need to be explored for the $N + 2$ and $N + 3$ cycles (Lee et al. 2013; Mongia 2013e; Tacina et al. 2005a; Tacina et al. 2014).

This article therefore gives an overview of the current rich-dome combustion system design, requirements, and challenges (Sect. 2), followed by the first alternative to rich domes that have been successfully introduced as products (Sect. 3), the lowest levels of achievable NO_x (so-called entitlement) as determined from small-scale rig testing (Sect. 4), summary of recent engine emissions data with "green" alternative fuels (Sect. 5), description of the second alternative to rich-dome products (Sect. 6) that may be of interest to the OEM's for the $N + 2$ and $N + 3$ generation aviation engines; a brief discussion on the modeling and correlation accuracy expectations from future efforts in this area (Sect. 7), and the third alternative to rich domes which was shown promising for autoignition times closer to 0.2 ms in Sect. 8. Finally, a short section on operability and dynamics is followed by Summary, References, and Acknowledgments. Limited by the page count normally associated with any article, most of the material presented here is from the author's personal experience, and this should not be misinterpreted by the reader for lack of relevance and/or respect for the outstanding quality works conducted by numerous combustion research, technology, and product development teams.

2 Current Combustion System Designs and Challenges

2.1 Modern Rich-Quench-Lean Combustors RQL

Figure 2 shows a typical cross-section of a modern rich-dome aviation engine combustor along with its major subcomponents including diffusion system that reduces the compressor discharge Mach number by a factor of 1.3–2.0 before dumping the flow and dividing it into three annular streams, respectively, for the dome, outer, and inner annuli around the combustion chamber with the

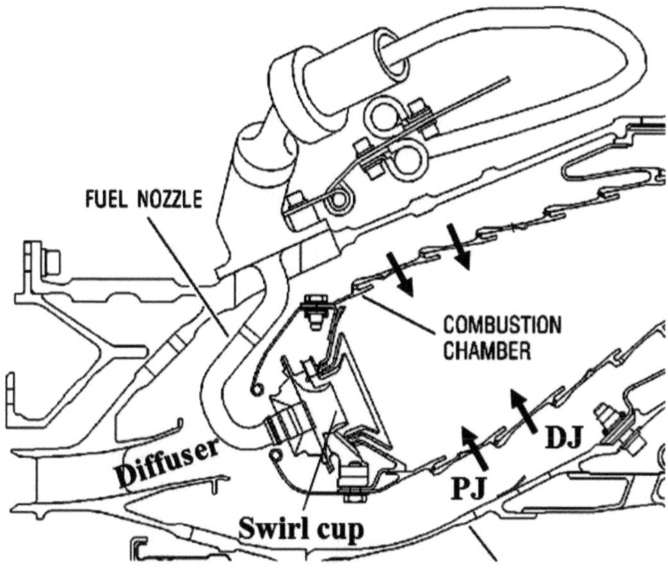

Fig. 2 A typical rich-dome single annular combustor (SAC); adopted from Mongia (1997)

corresponding nominal flow splits of 30, 40, and 30 % Wa_3 involving usually ±10 % Wa_3 variations between the different product engine designs (Mongia et al. 2004a, b, c, d, e, f). Here Wa_3 is the total compressor discharge air flow rate, and part of this bypasses the combustor (viz. 7–23 % Wa_3) as bleed, turbine cooling, and secondary flows with attendant difference between the combustor discharge temperature T_4 and the HP turbine rotor inlet temperature, T_{41}. Therefore, air flow rate through the combustor is given by $Wa_{3.1} \approx (0.93-0.77)\ Wa_3$. The combustor through flow air $Wa_{3.1}$ is used for combustion to achieve essentially 100 % combustion efficiency; dilution and trimming for providing the required turbine inlet temperature quality including pattern factor (PF) and radial profile; and dome and liner cooling in order to provide the combustor structure temperature levels and gradients consistent with the material design capabilities and durability requirements.

The combustion air is usually divided into air injected through a discrete set of primary/dilution jets, fuel nozzles, and dome-mounted swirlers which in many of the modern rich-dome engines are combined into one functional hardware, namely fuel/air mixing devices or simply called mixers, viz. swirl cup (Figs. 3, 4 and 5) and high-shear injector, Fig. 6; the former has been used by GE since the early 1970s and the latter in the P&W TALON-X. The swirl cup along with the conventionally designed primary zone (viz. the primary-zone air equivalence ratio ϕ closer to unity, providing comfortable margins for ignition and flameout) followed by optimally spaced dilution jets to allow soot burnout, and meet all the requirements of the first generation of GE engines. However, in order to further reduce high-power NO_x along with idle CO and unburned hydrocarbons HC, a new set of design rules emerged for the resulting GE low-emissions combustors LEC in the middle 1990s starting with

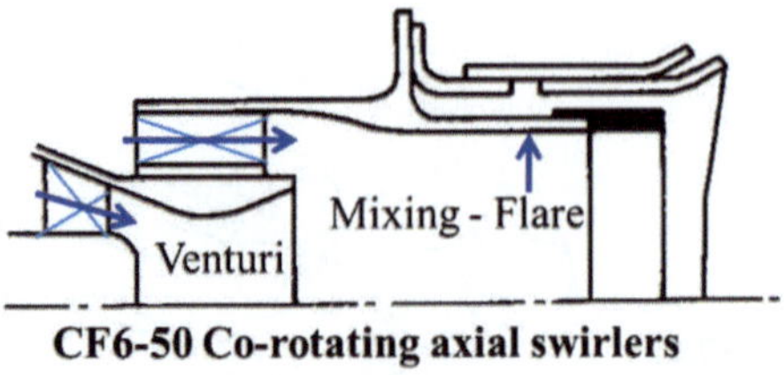

Fig. 3 Co-rotating swirl cup-based combustion products, viz. CF6-50 engine emission tested in 1979; adopted from Bahr and Gleason (1975)

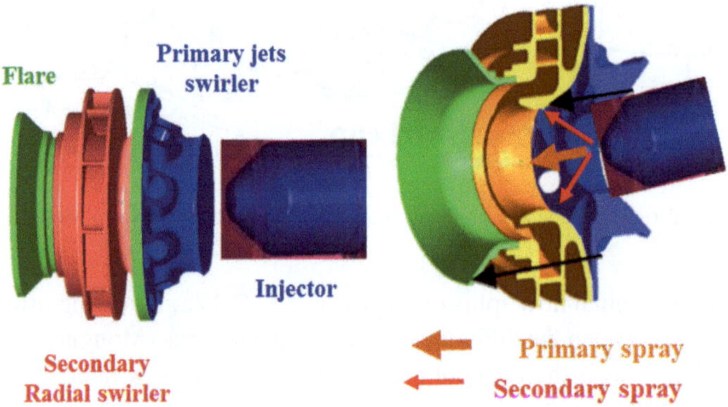

Fig. 4 Counter-rotating jet/radial swirlers used in swirl cups of the CFM56-2/3 ICAO engine emissions tested in 1983; adopted from Mongia (2011d)

Fig. 5 Counter-rotating twin radial swirlers used in several swirl cups; adopted from Mongia (2011d)

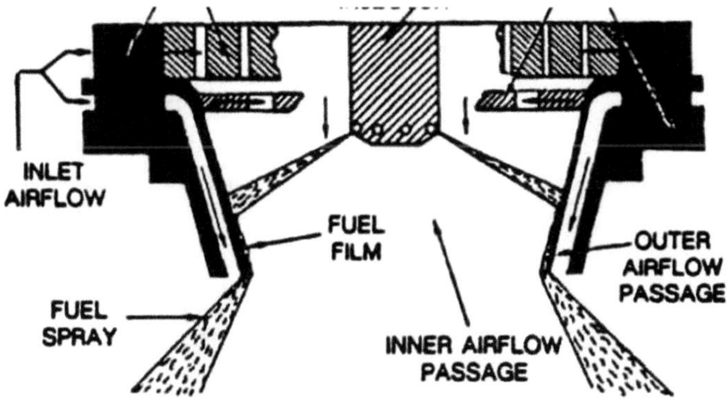

Fig. 6 Schematic of high-shear injector showing subcomponents for optimizing the design; reproduced from McKinney et al. (2007)

the CF6-80C2 LEC engine emission test in 1995 (Mongia 1997). Subsequently, all new GE rich domes have used LEC technology comprised of new mixers with variations in the airflow splits between the primary and secondary swirlers, carbon sweep air, and flare air, followed by closer spacing between primary and dilution jets along with optimized distribution of primary, dilution, and cooling air in order to meet all the design requirements as described in Sects. 2.2–2.5 (Mongia 2013a, b, c).

The mixer or injector and dome-mounted swirler designs have played a key role in producing low exhaust smoke as the residence time of the zone between the primary and dilution jets (intermediate zone) is minimized in order to get lower high-power NO_x. Some OEM's seem to indicate that it is possible to achieve lower PF with a single row of primary/dilution jets downstream of a rich primary zone burning at $\phi \geq 1.4$–1.5, a value normally considered reasonable for RQL concepts to produce competitive high-power NO_x and exhaust smoke levels. If it were true, then the OEM has developed a new technological capability than demonstrated previously in the 1980s for high ΔT combustors (Mongia 2011a).

The importance of idle CO tradeoff versus high high-power NO_x has been known since the early 1970s, and Fig. 7 represents a typical example from a low emission technology program as described by Fear (1976) and Bruce et al. (1977) which showed strong correlation between idle CO and HC. A critical evaluation of all low-emissions combustion technology and product development efforts should point to a key observation: idle CO is a function of takeoff NO_x whereas idle HC is a function of idle CO (Mongia 2013a, b, c). Therefore, advances in low emissions technology should be measured conceptually by the technology arrow direction shown in Fig. 7 thereby implying that none of the five modifications investigated for this particular combustor (identified as Concept 2) has made any improvement over the baseline configuration; the latter has certainly achieved improvement over the production combustor. In other words, one has to make substantial changes in the design in order to impact the CO-NO_x characteristics. Therefore, several swirl cup configurations have been developed by the GE combustion teams over the

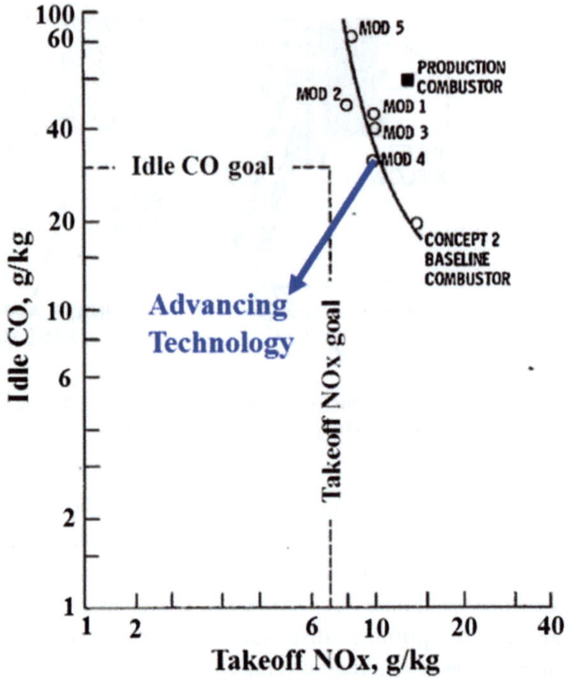

Fig. 7 A typical example of idle CO versus takeoff NO_x tradeoff in low-emissions combustion technology; adopted from Fear (1976)

years including the co-rotating axial swirlers for the CF6-50 reported by Bahr and Gleason (1975), Fig. 3. The five major subcomponents of swirl cups (viz. injector, venturi, flare, primary, and secondary swirlers) have been configured in several ways (Figs. 3, 4 and 5) in order to achieve the design objective; and the similar statement is perhaps applicable to another equally innovative mixer shown in Fig. 6.

Any initial successful demonstration of a particular mixer is followed by its own set of optimum aerothermal-mechanical design features for the rest of the combustor (viz. dome, combustor liner, airflow distribution, and attendant geometrical features) guided by experience, CFD, rig and engine testing leading to final certified products that must meet the combustion system requirements briefly described in Sect. 2.2.

2.2 Design Requirements

Table 1 gives a partial list of super-system level requirements from a typical aviation engine expected from the combustion system; which in practice runs into a design requirement document containing several pages with quantitative details which can vary with different applications. These requirements translate into combustion system level design requirements (Table 2) to be used as a guide across several interdependent disciplines including aero, heat transfer (thermal), mechanical, materials, manufacturing, controls and accessories, CA^2TM^3—or simply

Table 1 Engine level requirements from combustion system

1	Minimize mission fuel consumption
2	Meet ICAO and customer emissions requirements with margin to allow for future engine growth
3	Operate over wide range of thrust, rapid, acceleration/deceleration, steady operation in extreme rain and hail
4	Ground start, altitude relight
5	Hot-section (turbine) durability requirement
6	Combustor durability
7	Low weight and cost

Table 2 Combustion system level requirements

1	High combustion efficiency (>99.9 %) and low-pressure loss
2	Minimize emissions and smoke consistent with other requirements with attendant complex tradeoff process
3	Wide stability/operability limits. Sustain flame during water and hail ingestion
4	Reliable and smooth ignition; flame propagation; steady-state combustion with high combustion efficiency
5	Minimize combustor exit temperature nonuniformity consistent with specified radial temperature distribution, core as well as near-wall regions
6	Low structure temperature levels and gradients. Coke-free fuel insertion system
7	Affordable cost material/mechanical/manufacturing design and material along with low hot-streaks, reduced length and diameter

control your ATM machine otherwise you will run out of time and money in producing a new combustion system requested by the project. We will discuss some of these requirements in the following subsections relevant to this article.

2.3 Combustion Efficiency

The need for high combustion efficiency is quite obvious along the engine operating lines from sea-level to cruise-altitude operation as it directly affects the mission fuel burn. As shown in Figs. 8, 9 and 10, well-designed combustion systems have been giving very good combustion efficiency since the early 1970s. The data shown in these figures are for TFE731-2, the first engine for the author to learn from about the art of combustion design and technology development when he started with AiResearch Manufacturing Company, Phoenix in 1972. This dataset reported by Diehl and Biaglow (1974) was from an altitude tank that simulated engine operating conditions from minimum to maximum thrust at four different altitudes, namely 640, 4,570, 9,140, and 13,200 m. The data at 640 m included static (zero Mach flight number M) engine operation simulating idle, 30 % and maximum takeoff

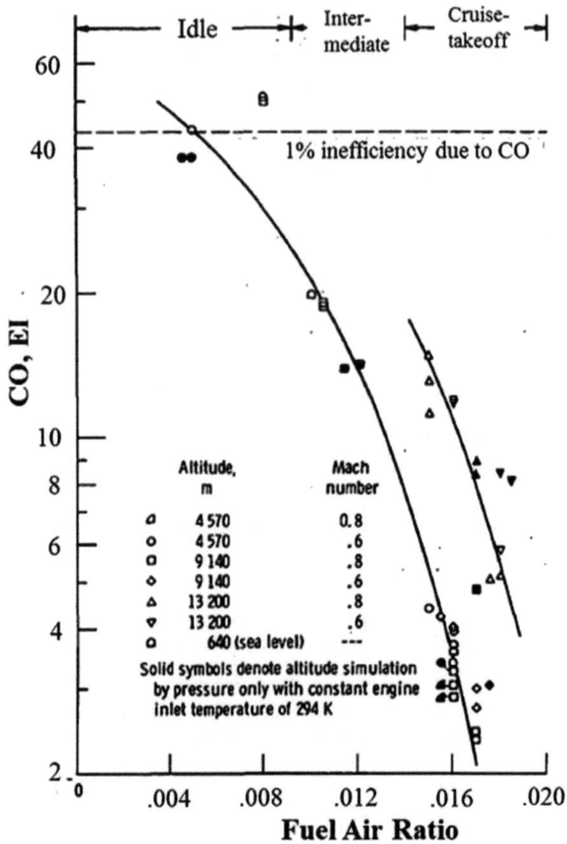

Fig. 8 Typical CO emission index versus fuel/air ratio of the TFE731-2 engine data from an altitude tank at simulated altitude operation from idle to maximum power; reported by Diehl and Biaglow (1974)

points; whereas for the other altitudes, the testing included two flight Mach numbers, namely 0.6 and 0.8. Therefore, the data covers a broad range of operation in terms of fuel/air ratio, combustor inlet pressure and temperature, namely $207 \leq P_3 \leq 1{,}150$ kPa, $416 \leq T_3 \leq 670$ K, and $0.0045 \leq \text{FAR} \leq 0.0185$. Combustion efficiency (%) is calculated from the measured gaseous emissions of CO and unburned hydrocarbons HC expressed as EI, g/kg of fuel, namely $\eta_C = 100 - (0.1 \times \text{EIHC} + 0.02334 \times \text{EICO})$. This means that approximately 43 g of CO/kg fuel accounts for one percent decrease in combustion efficiency, whereas 10 g of unburned fuel expressed as g CH_2/kg fuel has similar impact. Therefore, Mongia in pursuit of all of his technology programs has used 30EI for CO and 3EI for HC at sea-level idle operation conditions as his long-term goal since his first program (Bruce et al. 1977), giving an idle combustion efficiency of 99 %. Fortunately, most of the GE's modern rich-domes' idle HC and CO are represented well by $\text{EIHC}_{\text{idle}} = 0.1498\text{EICO}_{\text{idle}} - 1.4585$ (Mongia 2013a) which justifies Mongia's assumption for correlation between CO and HC at idle.

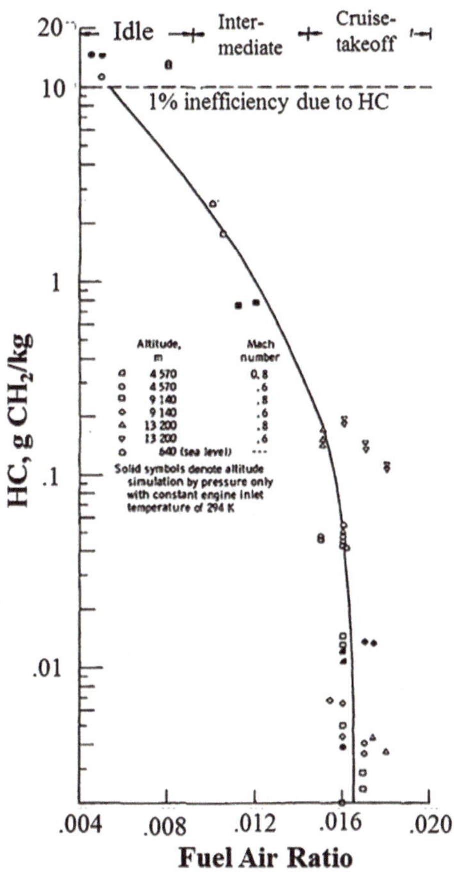

Fig. 9 Typical unburned hydrocarbon emission index (g CH_2/kg fuel) versus fuel/air ratio of the TFE731-2 engine data from an altitude tank at simulated altitude operation from idle to maximum power; reported by Diehl and Biaglow (1974)

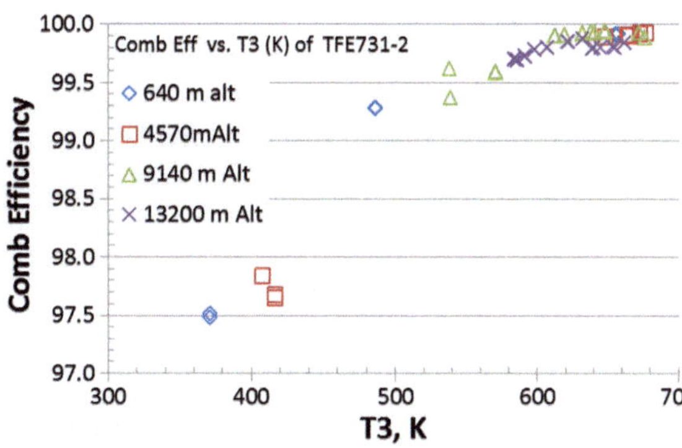

Fig. 10 Typical calculated combustion efficiency (η_C) versus T_3 of the TFE731-2 engine in a simulated altitude tank at 640, 4,570, 9,140, and 13,200 m

Several of the TFE731-2 engine operating points at cruising altitude (viz. 13,200 m with FAR $\sim$ 0.016) have CO and HC levels close to 15EI and 0.2EI giving combustion efficiency of 99.6 %, considered very acceptable level in the 1970s. However, for the leading $N + 1$ generation engines, this number has notched up to 99.9 %.The minimum acceptable combustion efficiency levels during altitude relight, and for avoiding flameout during inclement weather (viz. hail storm) vary from engine to engine models, but 80 % is considered a good guideline.

Potential air quality benefits and global climate impact are the primary reasons for interest in the alternative fuels in addition to other factors including price volatility and lack of diversity of the sources of conventional fuels. Carter et al. (2011) provides a good summary of the viability of some of the selective approaches for producing synthetic fuels including Fischer–Tropsch (FT) coal- or biomass to liquid and Hydroprocessed Renewable Jet (HRJ) processes. Research on quantifying combustion characteristics of synthetic fuels has been active since the middle 1970s (Longwell 1977). Recent emphasis (viz. Bester and Yates 2009; Bulzan et al. 2010, Chisty et al. 2011; Corporan et al. 2004; Davison and Chisty 2011; Moses and Roets 2008; Wey and Bulzan 2013, etc.) has been on investigating the impact of alternative fuels or their blends with conventional fuels on engines performance, emissions, operability, and durability in order to qualify drop in fuels.

We summarize combustion efficiency data of the CFM56-2C engine (which was ICAO emissions tested in 1983) reported by Bulzan et al. (2010) reproduced here as Fig. 11 showing that for its sea-level operating line, the five fuels tested (viz. JP 8,

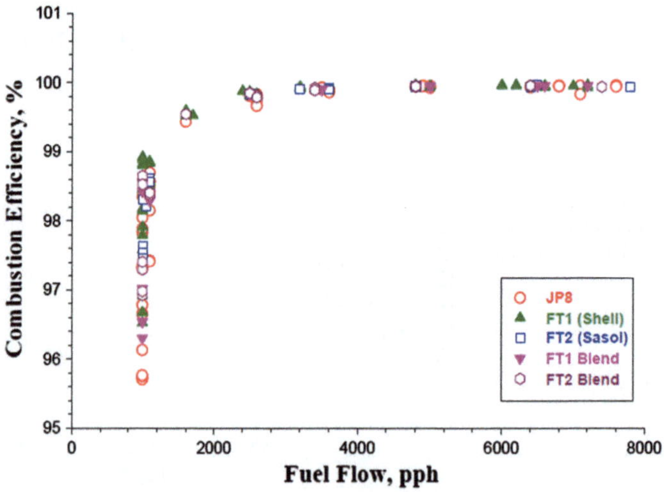

Fig. 11 Combustion efficiency (computed from gaseous emissions) versus fuel flow rate of the CFM56-2C1 engine sea-level operation with the base fuel (JP8), two FT fuels and their blends; reproduced from Bulzan et al. (2010)

alternative fuels FT1 and FT2 and their 50 % blends with JP 8) produce similar results. The minimum engine fuel flow operation shows a wide scatter in the combustion efficiency (viz. 95.5–99.0 %) that needs to be explained, for which we turn to the publicly available ICAO engine emission database, a source of certified emissions data of all the commercial propulsion engines subject to the ICAO regulatory NO_x standards described in Sect. 2.4. The CFM56-2/3 class of engine models cover the ICAO takeoff thrust ranging between 82 and 107 kN with the corresponding OPR between 21 and 26. Their idle emission indices of HC and CO range between 1.1 and 2.9, 24 and 38, respectively, with the corresponding idle efficiency ranging between 99.3 and 98.8 %. Their data for idle CO and HC is represented well by $EIHC_{idle} = 0.1178EICO_{idle} - 1.7351$ which is indistinguishable from the data of the most recent GE's rich domes. Therefore for a well-designed modern rich domes, combustion efficiency levels along the engine operating lines do not need to be improved, and one should not seek any guidance from the modeling community on CO and HC (and therefore combustion efficiency) as expressed in Table 3 except for the lean domes where we have a formidable challenge for demonstrating designs that meet simultaneously combustion efficiency and high-power NO_x requirements; the latter is discussed extensively in this article.

Table 3 Assess the adequacy of current turbulent combustion models to predict quantities of interest to industry

Quantity of interest	Status
Cold-flow aerodynamics/pressure drops	
Exit temperature profile/contours	
NOx Trends (op cond, geom., rich, lean burn)	
CO Trends (rich and lean burn)	
Unburned hydrocarbons trends	
Radiation/liner wall temperatures	
Soot trends	
Ignition, Lean-blowout	
Combustion dynamics	
Keys	
	Adequate/can get by
	Inadequate/ low to some confidence
	Hopeless/ very low confidence or methods not well developed

Can certain quantities be predicted significantly better than others? Which ones?

Prepared by M.S. Anand for the MACCCR workshop (2011)

2.4 Landing–Takeoff (LTO) NO_x

The first ICAO NO_x regulatory standard (viz. ICAO Committee on Aviation Engine Emissions, CAEE) was adopted in 1981 followed by the second (viz. ICAO Committee on Aviation Environment Protection, CAEP/2) with additional 20 % reduction for newly certified engines produced after December 31, 1999. The CAEE also finalized the landing–takeoff cycle for commercial subsonic aviation engines as depicted in Fig. 12. It comprises of four engine operating modes for the sea-level, static, standard-day (1) takeoff duration of 0.7 min at the rated thrust F_{00}, followed by (2) 2.5 min of climbout at constant thrust of 85 % F_{00}. On the other hand, the final approach of the aircraft descending from 3,000 ft. to the runway touchdown is simulated by assuming that it takes (3) 4 min at 30 % F_{00}. The aircraft taxi-in and taxi-out to the takeoff location is simulated by assuming engine (4) run time of 26 min at 7 % F_{00}.

Therefore, the gaseous pollutant during the LTO cycle and its normalized value by F_{00} (viz. DP/F_{00}) can be calculated by using the following two equations:

$$\text{LTO cycle pollutant, } P_i[\text{g of pollutant}] = \sum_j \text{EIP}_{ij} W_{Fj} tj; \text{ LTO Pollutant}(\text{NO}_x, \text{CO}, \text{HC})$$

$$= \frac{DP}{F_{00}}, \text{ g/kN} = P_i/F_{00}$$

Here the pollutant P_i is in EI units, namely gm of pollutant/kg fuel. Even though OEM's marketing offices campaign about their engines' achievements in terms of LTO NO_x regulatory levels which have become progressively more stringent through the years, CAEP2, CAEP4, CAEP6, and CAEP8 regulations need to be discussed later; combustion technologists perhaps get more insight by plotting rich domes' takeoff NO_x as a function of takeoff pressure ratio (viz. Mongia 2013a, b, c) or stoichiometric flame temperature (viz. Mongia 2010a, b); idle CO as a function of

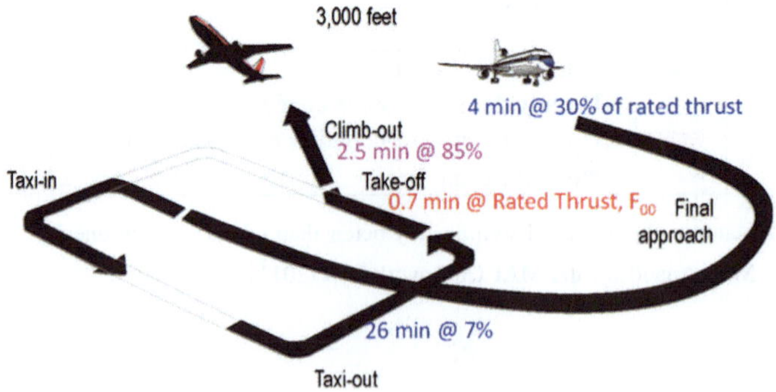

Fig. 12 ICAO Landing–Takeoff Cycle for commercial aviation engines comprised of idle, takeoff, climbout, and approach

takeoff NO_x (see Fig. 7), and idle HC as a function of idle CO (Mongia 2013a, b, c). Consequently, very interesting conclusions emerge as summarized in Fig. 13 which shows that several of the recent large rich-dome engines produced by GE, P&W, and RR have very similar characteristics in terms of the takeoff $EINO_x$ versus SLSS takeoff pressure ratio. The resulting empirical correlation identified by the legend RD-L in Fig. 13 $EINO_{xRD-L} = 0.0303PR^{1.9722}w/R^2 = 0.9906$ gives good agreement with the data; namely the correlation coefficient $R^2 = 0.9906$, root-mean-square errors (σ) of 0.87 g/kg and 2.7 % which is comparable with the measurement repeatability as determined by the average standard deviation of the LTO NO_x. Figure 13 also shows takeoff NO_x emissions of the CFM56-2/3 engine family to which belongs the engine tested by Bulzan et al. (2010); its combustion efficiency data was presented in Fig. 11 with more to follow in Figs. 38, 39 and 40 for NO_x and particulate emissions along its SLS engine operating line. It is interesting to see that takeoff NO_x of the CFM56-2/3 tested in 1983 is approximately 20 % higher than the best group of rich domes including Trent1000 tested in 2009, as shown by the dotted green curve which represents 1.2 time RD-L values. The author is hopeful that the $N + 1$ generation GTF TALON-X combustor will show significant lower takeoff NO_x than RD-L. Otherwise, we will have to conclude that rich-domes takeoff NO_x levels have settled down to the entitlement level given by the empirical correlation $EINO_{xRD-L} = 0.0303PR^{1.9722}w/R^2 = 0.9906$ and that it is only17 % lower than the CFM56-2/3 engine tested in 1983; albeit we have learned to scale up from 25–45 OPR engines during the last 30 years.

We should be interested in estimating mission NO_x emissionswhich depends on the knowledge of engine operating points (viz. P_3, T_3, and fuel flow rates) along the airplane flight path. We rely on engine performance models and weather conditions to make accurate predictions for the fuel burn; it will be equally desirable to have simple models for mission NO_x calculations as discussed next.

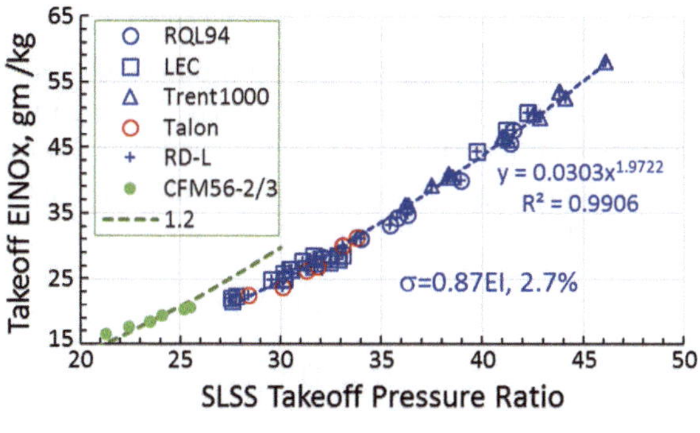

Fig. 13 Combined large rich-dome technology of GE, P&W, and RR identified as RD-L showing takeoff $EINO_x$ plotted as a function of SLSS takeoff pressure ratio; also shown the CFM56-2/3 and a dotted green curve showing 1.2 × RD-L trendline line which is given by $EINO_{xRD-L} = 0.0303\ OPR^{1.9722}$

NO_x formation in rich-dome combustors of propulsion engines can be described reasonably well by assuming that it is represented well by an extended Zeldovich mechanism (Gupta et al. 1991):

$$N_2 + O \Leftrightarrow NO + N; \quad (1)$$

$$N + O_2 \Leftrightarrow NO + O; \quad (2)$$

$$OH + N \Leftrightarrow NO + H \quad (3)$$

Assume $\frac{dN}{dt} = 0$, *then generalized rate of NO formation rate is given by*:

$$\frac{d[NO]}{dt} = 2K_1[O][N_2]\left\{\frac{1 - [NO]^2/K_C[O_2][N_2]}{1 + K_{-1}[NO]/K_2[O_2] + K_3[OH]}\right\} \quad (4')$$

Combining reactions 1 and 2 leads to an expression for K_C given in above equations, namely: $N_2 + O_2 \Leftrightarrow 2NO$ with equilibrium constant K_C as $K_C = \frac{K_1}{K_{-1}}\frac{K_2}{K_{-2}}$.

The next step of assumption (viz. O is in equilibrium with O_2) can generally be considered over simplification. But if it were true, it leads to the rate of NO_x formation being proportional to $p^{0.5}$ because the generalized equation reduces to:

$$\frac{d[NO]}{dt} = 2K_1K_{C5}[N_2][O_2]^{1/2}; \quad (4)$$

here, K_{C5} *is chemical equilibrium constant of*

$$\frac{1}{2}O_2 \Leftrightarrow O. \quad (5)$$

In the regions close to the flame front, better agreement with data has been achieved by making the super O equilibrium assumption which is achieved by assuming that the radicals can be calculated by regarding reactions (6, 7 and 8) as fast reactions in addition to CO reaction (9) being equilibrated. This process leads to the two expressions for [O] (viz. Equations 10 and 11) in terms of equilibrium constants of reactions 6–9, namely K_{C6}, K_{C7}, K_{C8}, and K_{C9}. Either of these equations can be used for plugging [O] in Eq. 4′ to calculate the rate of formation of [NO] under super-equilibrium O condition.

$$H + O_2 \Leftrightarrow OH + O; \quad (6)$$

$$H_2 + O \Leftrightarrow OH + H \quad (7)$$

$$H_2 + OH \Leftrightarrow H_2O + H; \tag{8}$$

$$CO + OH \Leftrightarrow CO_2 + H \tag{9}$$

Leading to the following two expression for O:

$$[O] = \frac{K_{C6}K_{C8}[H_2][O_2]}{[H_2O]}; \tag{10}$$

$$[O] = \frac{K_{C6}K_{C9}[H_2O][CO][O_2]}{[CO_2]} \tag{11}$$

Integration of either of the NO rate expression (viz. 4′ or 4) through time-temperature history leads to a simplistic model for NO emission provided it were a one-dimensional flame problem. The product combustors exhibit highly complex transient flow field characteristics which clearly cannot be duplicated by this simple model.

An alternative approach comprised of empirical correlation along with airflow distribution was successfully used by Danis et al. (1996) to correlate NO_x along the engine sea-level operating points. But the airflow distribution information is generally not available publicly. Moreover, it varies significantly between engine combustors. If one could use certified publicly available information on rated thrust, fan bypass ratio, overall pressure ratio, and fuel flow rate, in addition to LTO emission data and fuel flow rates, one can formulate a simple approach for evaluating emission characteristics of all the rich-dome combustors. The feasibility of this approach was demonstrated by Mongia 2010a for the sea-level operating lines of the 7 rich domes out of the 12 rich-lean dome combustors discussed by Mongia (2008). In order to get information about P_3 and T_3, one has to rely on the use of performance model calibrated with the ICAO database (Kumar et al. 2012b). The feasibility of this approach was also demonstrated for altitude operation, viz. the publicly available data of the TFE731-2. This methodology clearly shows that rich dome's NO_x emissions along the engine operating lines from sea-level to altitude operation can be represented as a function of adiabatic stoichiometric flame temperature, and there is no need to use NO_x severity parameter for correlating NO_x from different rich domes (Mongia 1997).

This approach led to very interesting results as shown in Fig. 14 which presents $EINO_x$ of one of the N-generation rich-dome engine (viz. combustor 7) and the TFE731-2. It should be pointed out that Combustor 7 data is for the sea-level operation only whereas the TFE731-2 takeoff OPR is only 14. These two datasets can be compared with Fig. 38 for the CFM56-2C, a 24 OPR engine. The TFE731-2 data presented in Fig. 14 encompasses NO_x data from sea-level to 13,200 m altitude operation with the corresponding pressure ratios ranging between 1.9 and 11.4. Since only one engine's NO_x data from sea-level to the altitude operation has been used to show feasibility of the recommended approach (viz. $EINO_x$ as a function of

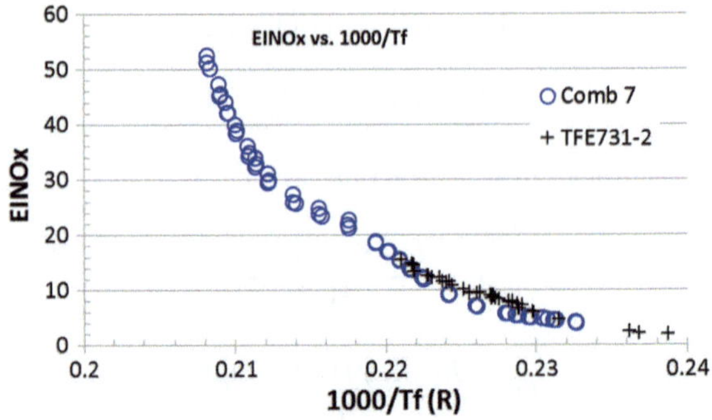

Fig. 14 $EINO_x$ versus $1000/T_f$ of Comb 7 (from Mongia 2010a) compared with the TFE731-2 data from Diehl and Biaglow (1974); the latter is identified as the first generation modern propulsion engine and the former as *N* generation. T_f is the adiabatic stoichiometric flame temperature computed from reported P_3 and T_3 data

stoichiometric flame temperature), it is premature to recommend this approach for all rich-dome engines to calculate the mission NO_x emission. The similarity of these two engines' NO_x versus stoichiometric curves and those in Fig. 13 discussed earlier forces one to wonder how much rich-dome NO_x reduction we can claim, and therefore providing motivation for alternative design approaches.

Recalling from Fig. 13, the takeoff NO_x emissions of the set of several low rich-dome combustors are represented well by a simple empirical correlation of their takeoff pressure ratio. However, when the LTO NO_x data of these engines expressed in terms of the average LTO NO_x, the correlation quality deteriorates significantly, namely $\left[\frac{DP}{F_{00}}\right]_{RD-L} = 1.7319PR^{0.9483} w/R^2 = 0.8459$. When data fit of these engines is sought in terms of the characteristic LTO NO_x (Fig. 15), the resulting expression is worst off, namely $\left[\frac{DP}{F_{00}}\right]_{RD-L} = 2.6692PR^{0.8552} w/R^2 = 0.6593$.

However the quality of the empirical fit improves considerably by redefining RD-L, namely when it does not include the Trent1000 average LTO NO_x. This is obvious by looking at Fig. 15 which shows characteristic LTO NO_x versus takeoff

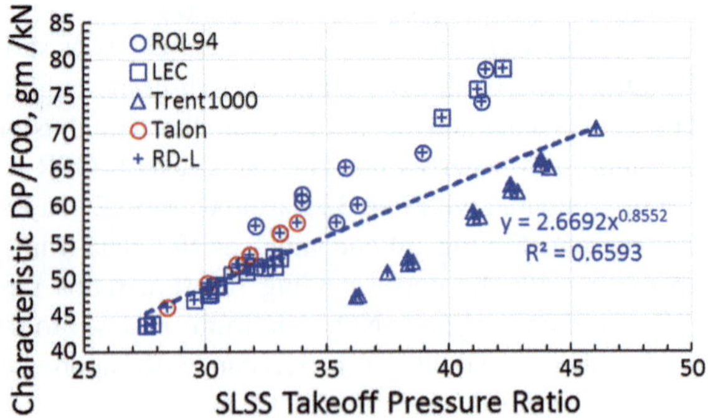

Fig. 15 Characteristic LTO NO_x of the lowest NO_x rich-dome large engines as defined in Fig. 13

pressure ratio for the entire set. Clearly, the Trent1000 appears to be a new brand of rich-dome concept whose details are not available publicly. Nevertheless, its takeoff and climbout $EINO_x$ value are not much different from their counterpart large rich dome. Therefore, by dividing the average LTO dataset of the RD-L combustors into two groups (viz. Trent1000 and the remainder identified as RDL-T), the average LTO NO_x expressions for the two groups of rich dome combustors are:

$$\left[\frac{DP}{F_{00}}\right]_{\text{Trent1000}} = 0.1292\text{PR}^{1.6327} w/R^2 = 0.9963;$$

$$\left[\frac{DP}{F_{00}}\right]_{\text{RDL}-\text{T1000}} = 0.6793PR^{1.2241} w\ R^2 = 0.9689 \text{ as 1st group;}$$

It will be interesting to see whether the future TALON-X-based GTF combustors will belong to the first group or establish a trend line lower than that of the Trent1000.

A few comments should be made concerning the slope of the NO_x regulatory line in addition to speculation about future ICAO NO_x standards because it impacts strongly the future combustion technology course. We will focus our discussion for medium- to large-size engines with OPR ≥ 30. The current rich-dome LTO NO_x data can be approximated by a slope of 2.0. Therefore, instead of across the board reduction in NO_x with attendant change in slope (viz. 20 % reduction of CAEP/2 and slope of 1.6 compared to slope of 2 for CAEE or CAEP/1), the future NO_x regulatory requirements were coined in terms of NO_x stringency, namely 16.25 % relative to CAEP/2 (viz. CAEP4) for engines certified after December 31, 2003, 12 % relative to CAEP/4 (viz. CAEP/6) for engines manufactured after December 2007, and the most recent being 15 % relative to CAEP/6, namely CAEP/8 with an effective date of December 31, 2013 with the attendant equation given by $CAEP/8 = -9.884 + 2\pi$.

In other words, according to the CAEP/8 regulation, aviation engines at 30 OPR cannot emit more than 50 gm/kN during the ICAO-specified landing–takeoff cycle compared to 100 allowed by the first regulation, namely CAEE.

One could therefore surmise that the potential future regulations, each with additional 15 % NO_x stringency spaced 4 years apart, are given by the following equations:

$$\text{CAEP}/10 = -17.401 + 2\pi;\ CAEP/12 = -23.791 + 2\pi;$$
$$\text{CAEP}/14 = -29.223 + 2\pi;\ CAEP/16 = -33.839 + 2\pi$$

Finally, we speculate that the long-term CAEP_LT is $\text{CAEP}/18 = -37.763 + 2\pi$ effective December 2033 giving a regulatory value of 22 gm of NO_2/kN at 30 OPR compared to 100, 78 % reduction. However, the author prefers to have 30 % margin on the regulatory standard so that when technology is transitioned to product, there is flexibility in the process of tradeoffs. This means that the long-term $N + 2$ and $N + 3$ products goal is 15 LTO value at 30 OPR with resulting reduction of 85 % from the base. But the slope of CAEP_LT will be 2.0 different from the process used in the technology community. The combustion technology community

Fig. 16 Characteristic LTO NO_x data of the Trent1000 and GEnx engines compared against the CAEP/8 standards, technology initiatives' (viz. ERA and N+ 3) goals

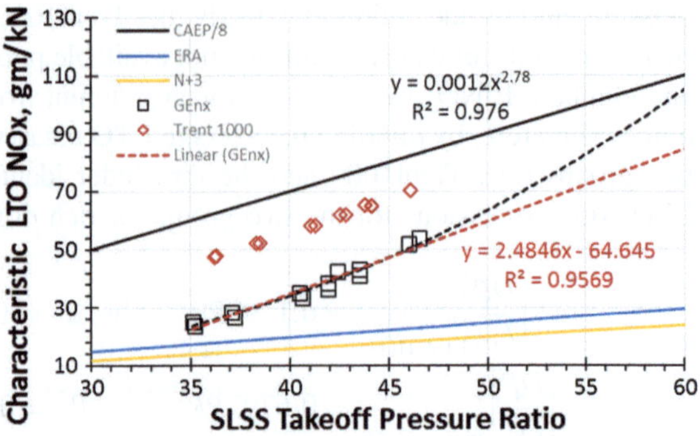

has expressed its goals in a different manner, namely 50 and 30 % CAEP2 for AST and UEET, 25 % and 20 % CAEP6 for ERA and N + 3 initiatives, respectively, with the corresponding 30 OPR LTO values of 40, 24, 15, and 12; the latter not much different from 15. However, these technology goals go with the corresponding reduction in the regulatory NO_x slope, viz. the slopes of the ERA and N + 3 lines in Fig. 16 are 0.5 and 0.4, respectively, unlike the slope of 2 for the regulatory standards after CAEP2, as shown in Fig. 16.

Figure 16 also shows characteristic LTO NO_x versus takeoff pressure ratio of the Tren1000 (the lowest NO_x rich dome) and GEnx, a second generation lean dome combustor to be discussed in Sect. 3. Also shown are the CAEP/8, regulatory standards for large engines with OPR ≥ 30. OEM's generally prefer to have 30 % margin from the standards for at least its lowest rated thrust engines so that the highest rated thrust engine does not have a margin lower than 10 %. The Trent1000 with OPR between 36 and 46 with the corresponding characteristic LTO NO_x of 48 and 71 gm of NO_2/kN thrust and 24 and 14 % margin from CAEP/8 barely meet the desired limits. The GEnx with 35–47 OPR and the corresponding NO_x of 24 and 55 g/KN has margins of 66 and 39 % from CAEP/8, a considerable margin from the future potential increased NO_x stringencies. However, it appears that GEnx if extrapolated to higher operating pressures may lose its advantage on margin rather rapidly compared to rich-dome combustors. Unfortunately, neither power nor linear extrapolations (viz.$NO_x = 0.0012\ OPR^{2.78} = 2.4846\ OPR - 64.645$) provide reasonable estimates as clear in Fig. 16. Therefore, one needs to find alternative lean dome technology with considerably lower impact of pressure ratio than GEnx. With OPR tending toward 60, OPR along with its higher T_3, should one come up with more robust lean dome technologies? Effect of P_3 and T_3 on ignition delay times therefore needs to be discussed.

2.5 N + 2 and N + 3 Generation Engines' Challenges

As engines' operating pressures during takeoff increase along with no relief on the minimum P_3, T_3, and FAR because of the operability requirements (Sect. 2.6), the ratios of peak to minimum P_3 and T_3 increase with attendant challenges on durability of the combustion structure and fuel injector. Moreover, for a given thrust class of engines, increase in takeoff pressure and bypass ratio to improve the overall system efficiency will lead to smaller high-pressure core size with several implications for the combustion system design options. Smaller engines generally will have fewer number of fuel nozzles, smaller combustor length and channel height with attendant higher surface to volume ratio and therefore requiring higher fraction of combustor air for cooling and more challenging profile and pattern factor issues. This will therefore lead to higher combustion temperatures which will make it difficult for lean domes to have lower NO_x. Therefore product versions of the basic advanced cooling technologies with potential cooling flow reductions up to 80 % (viz. Acosta 1987) as described in Anon (1982), Nealy et al. (1985), Paskin et al. (1990, 1993). Here we provide two figures in order to make summary statements.

Nealy et al. (1985) share their advanced cooling technology experience which is further illustrated by a simplified Fig. 17 from Acosta (1987). The conventional film-cooled liners relative to the baseline called splash film have large design bands in regard to the range of reduction in cooling flow as it can be attributed to several design factors including dome mixer/injector design features. Johansen et al. (1977) made convincing illustrations for managing impact of flame streaking without which almost all advanced schemes will not realize their full potential. Convective film cooling coupled with transpiration cooling schemes have the potential for reducing cooling flow requirement up to 70 %, again with a wide scatter band. Further reduction in cooling flow rate can be realized by minimally cooled ceramic structure, a technology still in TRL3 stage for combustor application. Technological

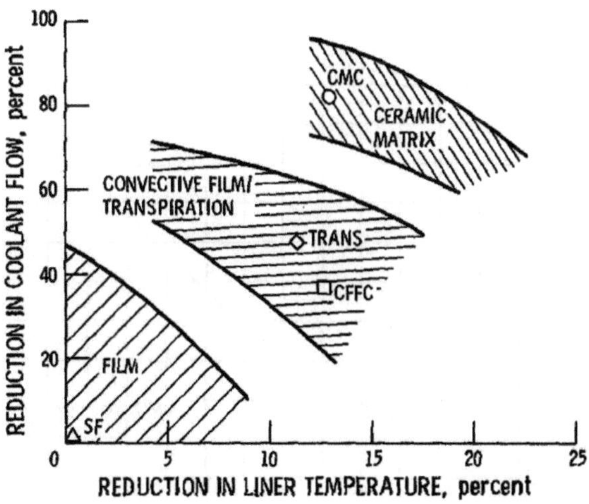

Fig. 17 Conventional film cooling including splash flame (*SF*) performance in terms of reduction in cooling flow rate versus liner temperature for different advanced cooling schemes including convective film/transpiration, transpiration (*TRANS*), counter flow film cooling (*CFFC*), compliant matrix composite (*CMC*), and ceramic matrix; reproduced from Acosta (1987)

feasibility of one of the most practical techniques, Compliant Matrix Ceramic (CMC) is reported by Paskin et al. (1990, 1993) which realized approximately 80 % reduction in the liner cooling.

One of the guidelines for assigning the amount of cooling air required during the preliminary design phase of a new combustor is illustrated in Fig. 18 which shows empirical correlation of small reverse flow combustor products as a function of P_3. The NASA CMC combustor for 275 psi inlet pressure was designed to use less than 0.004 lb/s-in^2 (2.8 kg/s-m^2) realizing approximately 80 % reduction from the current technology as presented in Paskin et al. (1990, 1993).

Ignition delay times, flashback, and flame holding in the premixing systems are strongly impacted by P_3, T_3, and equivalence ratio ϕ. OEMs have empirical know-how along with highly reliable tools to predict relationships between P_3 and T_3. Combustion research community interested in technology-relevant studies needs to have simpler tools to quickly calculate T_3 for a given P_3, which was developed as part of the activities summarized in Kumar et al. (2012a, b, c), and Panda et al. (2012a, b). These studies led to a very useful conclusion that simple isentropic calculations by assuming ideal air properties and isentropic efficiency η_{isen} approximately 87 % compare well with the publicly available ICAO cycle data on the $N + 1$ engines. On the other hand, η_{isen} value of 89 % is assumed for the $N + 2$ generation aviation engines. We also assume that a typical fully deteriorated compression system will lose its η_{isen} by approximately 1.0 %. These studies also led to the recommendation that we can assume η_{isen} approximately 86 and 84 %, respectively, for the N and the first generation of modern propulsion engines' compression systems; lower set of values should be used for smaller engines.

Further corroboration of these assumptions comes from typical sea-level engine operating conditions for the first generation modern engines as summarized in

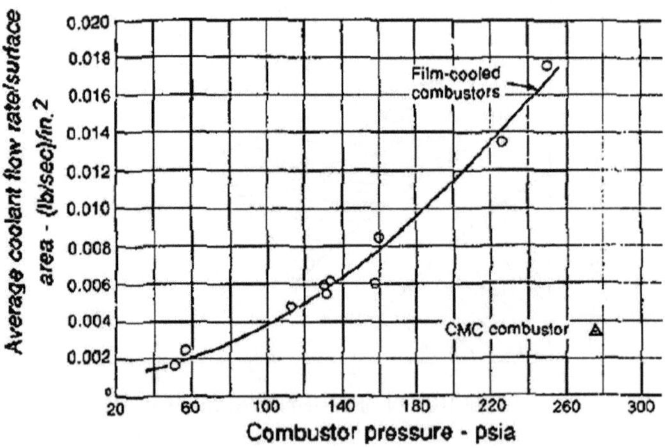

Fig. 18 Empirical curve of average liner cooling flow rate (lb./s-in^2) versus combustor pressure (psi) of conventional film-cooled liners compared to CMC combustor; reproduced from Paskin et al. (1990)

Table 4 Typical sea-level combustor relevant values of two small and large first generation modern aviation engines

	Thrust (kN)	P_3 (atm)	T_3 (K)	Isentropic efficiency	FAR	Wa (kg/s)	Wc (kg/s)	Vr (m/s)
Small engine combustor from Bruce et al. (1977)								
Idle	0.89	1.97	365.4		0.0100	2.31	1.32	10.88
Approach	4.67	5.10	499.8	80.0 %	0.0118	5.76	1.49	14.33
Climbout	14.01	12.45	652.0	81.8 %	0.0150	12.43	1.50	16.53
Takeoff	15.57	13.61	668.7	82.1 %	0.0156	13.38	1.50	16.70
SL ratio		7			1.6		1.14	
Opp ratio		41	3.1					
Large engine combustor from Bahr and Gleason (1975)								
GIDL	7.42	2.96	437.4	<75 %	0.0110	13.93	5.80	18.56
	11.2	3.69	463	<75 %	0.0103	17.3	5.94	19.6
	15.7	4.55	489	77.2 %	0.0100	21.3	6.10	20.7
FIDL	21.3	5.54	514	79.9 %	0.0099	25.3	6.10	21.4
	44.8	9.05	589.0	82.5 %	0.0116	38.6	6.10	22.3
Approach	67.27	11.81	631.9	84.0 %	0.0138	48.17	6.04	23.29
	100.9	15.85	691	83.5 %	0.0164	61.00	5.96	24.00
	190.5	20.89	745	<85 %	0.0190	76.7	5.90	24.7
Climbout	195.7	25.82	791.9	<85 %	0.0215	90.81	5.83	25.18
	206.2	27.49	807	<85 %	0.0224	95.3	5.80	25.3
Takeoff	224.2	29.44	826.3	<85 %	0.0236	100.6	5.79	25.51
SL ratio		10			2.4		1.05	
Opp Ratio		88	3.8					

Table 4. Listed are the sea-level, static, standard-day rated thrust; combustor inlet pressure and temperature (P_3, T_3); combustor exit fuel/air ratio and airflow rate (FAR, Wa); and corrected airflow rate ($\mathrm{Wc} = \mathrm{Wa}\sqrt{T3/288.15}/P_3$). Combustor reference velocity is also listed so that it can be related to Fig. 21 for discussion on ignition and lean blowout LBO FAR. The first entry is for a small engine, the TFE731-2 as reported by Bruce et al. (1977). NO_x emissions of this engine from sea-level to altitude cruise has been covered in Fig. 14. The sea-level operation of the small engine at idle, approach, climbout, and takeoff for P_3, T_3, and FAR values changing from (1.97, 365, and 0.01) to (13.6, 669, and 0.016) may be compared with those of the large engine entry [viz. the CF6-50 provided by Bahr and Gleason (1975)], namely (3, 437, and 0.011) and (29.4, 826, and 0.024). Calculated isentropic η_{isen} efficiency for the large engine varies between 75 % and <85 % as engine operates from several ground idle (GIDL), two flight idle (FIDL), three approach, two climbout, and rated takeoff operating conditions. However, its η_{isen} for the area of interest for the ICAO landing–takeoff points (viz. Fig. 12) varies between 80 % and <85 %.

In the 1970s, arbitrary classification of combustors into four categories (viz. large, medium, small, and very small sizes) was based on the corrected flow rates being >5, >2, between 1 and 2, and less than 1.0 kg/s, respectively. With the increasing takeoff pressure ratio since the 1990s, these numbers have been coming down but not fast enough for the medium- and small-size combustors to greatly influence the overall design strategy. However, if we jump to 55 OPR cruise for the single aisle airplanes (viz. Kumar et al. 2012c), Wc at takeoff goes down to 1.1 kg/s with attendant small-size design challenges. Similarly, the future energy-efficient large-size engines will progressively approach the limitations of the medium-size combustors suddenly subject to takeoff pressure ratio approaching 60. This will impact the durability characteristics of both the fuel injection system and the combustion structure requiring innovations in aerothermal, mechanical, materials, and manufacturing. Let us illustrate the impact of P_3 and T_3 as shown in Figs. 19 and 20; the former is for the sea-level, standard-day, static SLSS takeoff pressure ratio (also known as ICAO takeoff pressure ratio) whereas the latter is for a typical hot-day. The two generation of engines (viz. $N + 2$ and $N + 1$ generations) are represented, respectively, by isentropic efficiencies η_{isen} of 89 and 87 %, whereas 86 % η_{isen} represents a fully deteriorated $N + 1$ generation engine. Since the interest here is on ≥40 OPR engines, the x-axis value starts at 40 and goes up to 60, hoping that we do not have to deal with >60 OPR engines in the next 20 years. The empirical fits for SLSS as shown in Fig. 19 are $T_{3,\text{SSS}} =$ 317.544 $\text{OPR}^{0.2704}$; 320.1955 $\text{OPR}^{0.272}$ and 321.5627 $\text{OPR}^{0.2728}$, respectively, for the $N + 2$, $N + 1$, and its deteriorated engines. The corresponding fits for the hot-day, sea-level takeoff from Fig. 20 are $T_{3,\text{SShot}} = 356.430\ \text{OPR}^{0.2662}$; 359.2229 $\text{OPR}^{0.2679}$ and 360.6961 $\text{OPR}^{0.2688}$.

Model-based predictions for ignition delay, flashback, and/or flameholding F^3 are nonexistent at least for practical configurations because of the complexity and lack of information on the hardware details and attendant effects. Therefore, we will present ignition delay times τ_{ign} calculated by Prof. Chih-Jen Sung by using a most recent set of detailed kinetic schemes for Jet fuels. Table 5 gives a summary for the two values of OPR, namely 45 and 60, so that we can see the reduction in τ_{ign} relative to the most recently introduced lean dome engine (viz. GEnx) that has been in the field since 2011 without having encountered any incidence of hardware

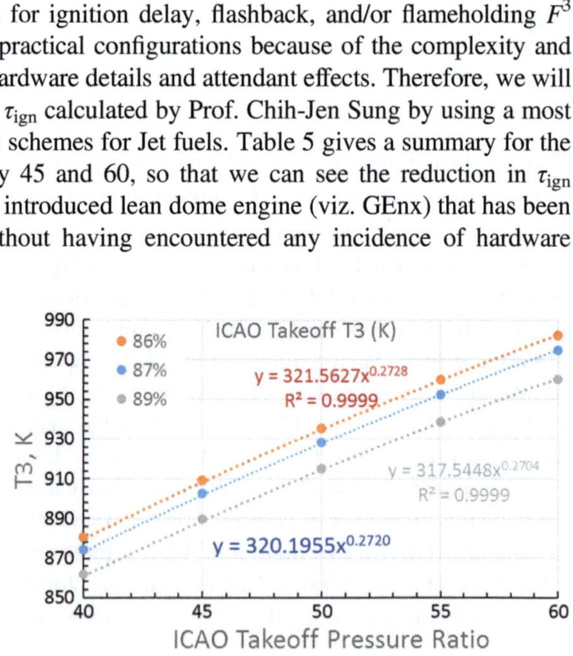

Fig. 19 Sea-level static (*SLS*) standard-day takeoff combustor inlet temperature versus ICAO takeoff pressure ratio for different isentropic efficiencies

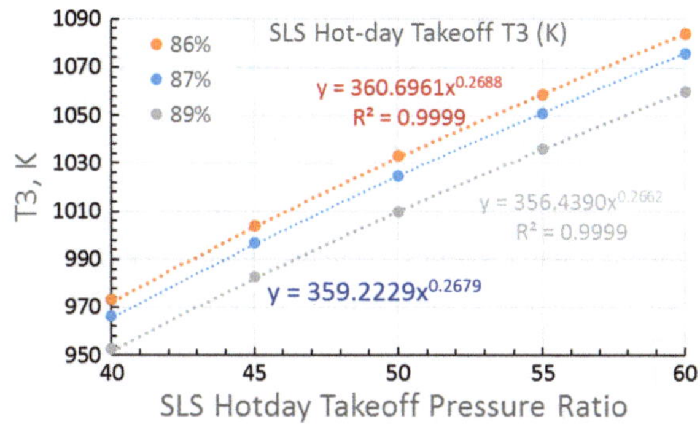

Fig. 20 SLS hot-day takeoff combustor inlet temperature versus ICAO takeoff pressure ratio for different isentropic efficiencies

Table 5 Academic ignition delay times (ms) for 45 and 60 takeoff pressure ratio from Sung (2014)

	SLS takeoff pressure ratio			
	45		60	
ICAO T3*, K	902.6	889.6	974.8	960.4
τ_{ign} (for 0.7ϕ), ms	1.82	1.83	1.02	0.82
τ_{ign} (for 1.0ϕ), ms	1.19	1.19	0.73	0.76
Det. Hot-day** T_3, K	1004.0	989.9	1084.0	1068
τ_{ign} (for 0.7ϕ), ms	1.08	1.27	0.34	0.41
τ_{ign} (for 1.0ϕ), ms	0.83	0.98	0.27	0.32
*η_{isen}, %	87	89	87	89
**η_{isen}, %	86	88	86	88

damage. The minimum τ_{ign} of 0.83 ms for the hot-day, fully deteriorated engine at stoichiometric condition can therefore be considered safe for this engine. If one moves forward with an $N + 1$ generation 60 OPR engine, then the calculated τ_{ign} goes down by a factor of 3–0.27 ms with potential for requiring changes in the combustor design in order to address F^3 concerns, if any. However, there will be some increase in τ_{ign} to 0.32 ms for the $N + 2$ generation engine.

2.6 Operability Challenges

Combustion system is a unique part of the gas turbines in that its design requirements have to be met from engine start to the most severe operating conditions including the highest engine pressure and temperature levels, as discussed in the previous section. There are several demanding requirements beyond low-emissions including durability, shop visits, and combustor exit temperature quality that rarely get the attention of the people beyond the combustion design community. An aviation engine must perform reliably over its flight envelope. The theoretical calculation of the flight envelopes involves expertise from several disciplines including gas turbine engine performance, aerodynamic, and structural analysis of the vehicle of interest. The database in these areas are known accurately only by OEMs followed by continuous refinement as the integration of the design proceeds for particular application. However for academic exercise, one can use commercially available software (viz. GasTurb™) and a vehicle with specified flight dynamic characteristics to create a flight envelope. The process of creating an actual flight envelope is beyond the scope of this article except to make the following observations.

Perhaps the weakest link in the process of finalizing the flight envelope is the combustion designer's limited analytical capability for "predicting" the ignition, flame propagation and acceleration, and lack of accurate information about the engine conditions during the conceptual and preliminary design phases of operation. Therefore, the initial phase of estimates based on empirical database is followed by extensive experimental evaluation and refinement of the flight envelope.

Engine start conditions are very demanding including the available combustor pressure drops (ΔP) an order of magnitude smaller than design ΔP and resulting corrected flow (Wc) less than one third of design Wc. This combined with lower combustor inlet pressure and temperature as the airplane climbs poses enormous challenge for achieving good spray atomization and combustion efficiency and thereby places limit on the maximum altitude ignition capability.

The stagnation pressure and temperature upstream of the engine inlet is straight forward as given by one-dimensional gas dynamics relations, namely the normalized ratios of stagnant temperature and pressure, respectively:

$$\frac{T_0}{T} = 1 + \frac{\gamma - 1}{2} M^2; \quad \frac{p_0}{p} = \left(1 + \frac{\gamma - 1}{2} M^2\right)^{\gamma/(\gamma-1)}$$

Here M is the flight Mach number and γ is the ratio of specific heats at constant pressure and volume. The process of determining windmill conditions even for a gas generator is very complicated and only the engine core test in a full-scale test facility provides a reliable data on P_3 and Wa_3 (Stearns et al. 1982). The engine (comprised of fan, low-pressure turbine, core, and accessories) and nacelle system has to be installed in a flying test bed to provide quantitatively accurate conditions for P_3, T_3, and Wa_3. This flight data is used to estimate fuel flow rate (Wf) for ignition and flame propagation from empirical curves based on full-scale rig data.

OEM's use proprietary empirical curves relating lean blowout and ignition fuel/air ratios to combustor operating conditions. Experimental values of ignition and lean blowout fuel flow rates acceptable at the engine system level are determined through an extensive tradeoff effort leading finally to establishing operating limits. Here we show the one reproduced from Bruce et al. (1977), namely Fig. 21 which will vary with different engines, their models, and vehicles. The maximum relight altitude is also determined from the flight test bed followed by some corrections later because of the differences in the nacelle and airplane interaction. For this

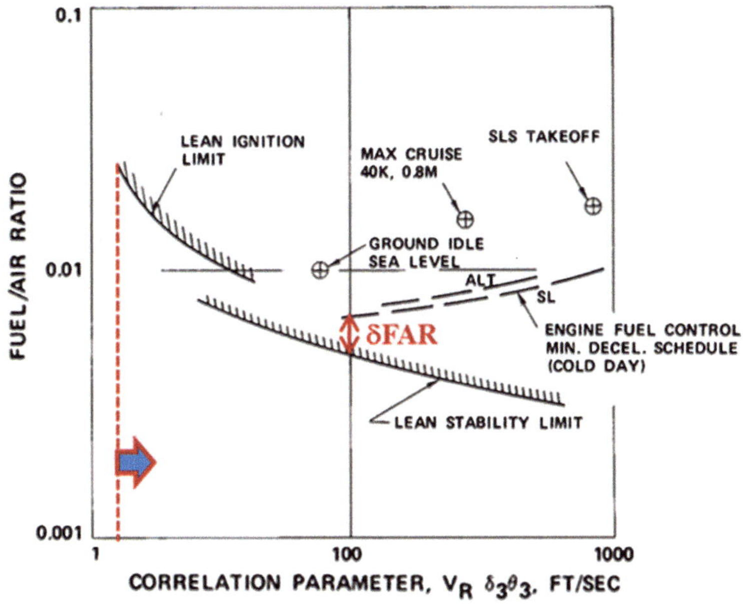

Fig. 21 Ignition and lean stability limits versus normalized reference velocity in addition to key engine operating points; reproduced from Bruce et al. (1977)

article, we will use a rule of thumb (viz. $\delta_{\text{isen}} = 0.33$) with attendant altitude relight capability of 35,000 ft. 0.85 cruise Mach number.

Another design requirement introduced with the twin-engine aircraft operation is in regard to restarting the engine at the cruising altitude after an extended exposure to the local thermal condition, namely $\theta = 0.7958$ or 0.7519 when flying at 35,000 ft. or 40,000 ft. unlike the older requirement for 3- or 4-engine airplanes where one could relight the hot engine with attendant higher T_3 levels. Clearly the cruise altitude relight for the latter case is much easier to accomplish than the former with a cold-soak engine.

When a twin-engine airplane is operating with one engine only and the pilot turns off the good engine by mistake, the engine must restart immediately while essentially running turbomachinery at the cruising speed. In other words, the relight capability at the cruising altitude must be met on both the low- and high-corrected flow limits with attendant big variation in the corrected reference velocity as shown in Fig. 21.

The range of the resulting full-scale annular rig data is very wide as listed in Fig. 21. Some of the key operating points (viz. sea-level ground idle, max cruise flying 0.8 Mach number at 40,000 ft. and sea-level takeoff, engine fuel schedule for cold day and altitude operation) are shown in order to identify the minimum value of $V_R\delta_3\theta_3$ below which ignition would not occur and the corresponding engine windmilling points, viz. pairs of altitude and Mach numbers along the left-side of the flight envelope. Also shown in the figure is the minimum margin on fuel air ratio, identified as δFAR which is known popularly as a pinch point.

Finally, engine must start reliably at ground on the all-time record coldest day at the airports of interest; viz. −42, −39 and −34 °C, respectively, in Moscow, Anchorage, and Montreal. The numerical values of the dome design pressure drop (ΔP_{design}) and the minimum and maximum fuel air ratios (FAR_{min} and FAR_{max}) depend upon the combustion system design and its potential applications. We will assume typical values of these variables, respectively, 3–5 %, 0.005–0.008, and 0.025–0.040 for the aviation engines.

In summary, combustor must meet its design requirements for the inlet pressures and temperature ranges listed for the $N + 2$ generation engines operating at takeoff 60 OPR; T_3: 216.67–1084 K; P_3: 0.33–60 atm; ΔP: ~(0.1–1.2) $\Delta P_{\text{design};}$ and FAR: $\text{FAR}_{\text{min}} - \text{FAR}_{\text{max}}$. These ranges can be compared with the first generation small and large engines presented previously in Table 4. The ratio of the operating pressures will go up from 41 to 180 and that of T_3 from 3.1 to 5.0. The operability challenge for the lean domes is obviously more than that of the rich dome.

3 First Alternative to Rich Domes

3.1 First Generation LDI in Product

Bahr and Gleason (1973) conceived and started the technology development effort on a dual-annular combustor DAC (Fig. 22) which may be compared with the rich-dome single annular combustor (SAC) presented earlier in Fig. 2. Its outer annulus (Pilot) is similar to SAC in operation from engine start to some intermediate power setting, generally approach, beyond which the inner annulus (Main) combustor gets fueled followed immediately by ignition and flame propagation with the hot

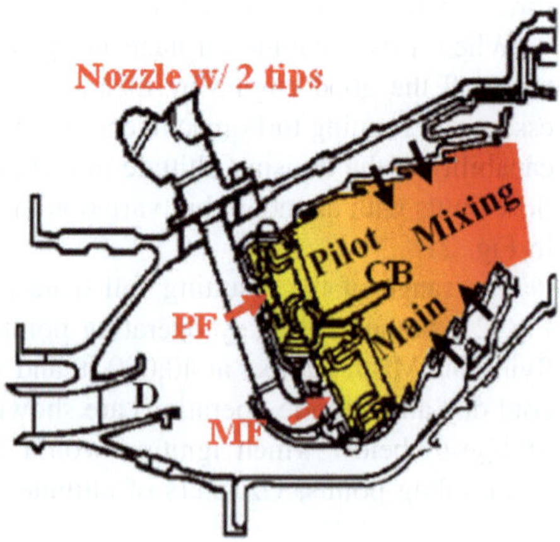

Fig. 22 A typical dual-annular combustor (*DAC*) cross-section; from Mongia (2003)

combustion products from the pilot zone which is separated by a center body (CB) so that its primary zone combustion process is shielded from the colder nonreacting air of the main combustor. The fuel nozzle has two tips, namely pilot and main tips identified as PF and MF in Fig. 22. The height of the pilot and main primary zone combined is obviously bigger than its counterpart SAC because the nozzle circumferential spacing is generally closer to the primary zone height. In order to distribute combustion air through the pilot and main, the preferred diffuser (*D*) should be a dual-passage which may not be easy to incorporate as retrofit in product engines with SAC. The fuel flow split between the pilot and main changes from 100 % with pilot only operation to lower values at higher power setting, normally decided by NO_x and/or combustor exit profile considerations. One has to develop a more complex control system in addition to requiring more hardware including the number of fuel manifolds compared to SAC. If the main fuel nozzles are not allowed to dribble during low-power setting in order to avoid tip coking, cooled main nozzle tip with pilot fuel and the associated plumbing will be required. The mixing of the pilot and main streams to provide combustor exit profile along with low pattern factor pose a big design challenge especially when the dilution and trimming air fractions go down significantly compared to SAC, because the main mixer is designed to operate lean in contrast to rich mixers in SAC's. Bahr and Gleason (1973) were fully aware of DAC's challenges; but its lower NO_x emission potentials relative to the first generation rich dome SAC provided the motivation to develop DAC technology in order to address its deficiencies as described in Gleason et al. (1975), Gleason and Bahr (1979), Stearns et al. (1982) and Burrus et al. (1984).

DAC technology successfully transitioned into product as retrofits for the CFM56-5 and CFM56-7 engine models (ICAO engine emissions tested during 1994 and 1997) and in a new centerline engine, the GE90 emissions tested in February 1995. The CFM56 engine models gave an average margin of 50 % from the CAEP2 regulatory standard with the average takeoff pressure ratio of 28. However, the corresponding margin for the GE90 was 25 % at the average takeoff pressure ratio of 38; attributed primarily to the use of less-effective liner cooling than the segmented structure used in the Energy Efficient Engine demonstration program (Stearns et al. 1982). It was possible to pursue the use of advanced liner cooling with attendant increase in the combustion and lower NO_x emissions in combination with enhancing fuel/air mixing by improving the mixer design (Mongia 2013e). However, seeing no support for advancing cooling technology, we pursued formulation and development of a new low-NO_x technology with the long-term plan for achieving premixed combustion NO_x entitlement in addition to addressing ALL combustion system design requirements including durability; profile and pattern factor; low-power CO and HC; ignition, relight, and flameout; combustion dynamics; and fewer number of lower cost nozzles fabricated by Direct Metal Laser Sintering (DMLS) spearheaded by Marie McMasters. The next section describes this approach.

3.2 TAPS Technology and Product Introduction

The formulation of the second generation lean dome technology benefitted from the lessons learned since the early 1970s including DAC discussed above and the references quoted; in addition to the two axially-staged combustion technologies (viz. Bruce et al. 1977, 1978, 1981; Roberts et al. 1977, Dubiel 1986 and Segalman et al. 1993); the recuperated automotive gas turbine combustion technologies (viz. Sanborn et al. 1983; Ross et al. 1983; Rizk and Mongia 1990); and the high-speed research combustion, US Patent 6530223, March 11, 2003. Almost all of these approaches except Sanborn et al. (1983) should have some degree of difficulty in meeting the profile requirement without which turbine stages cannot operate at their best with attendant fuel burn impact. Therefore, multiconcentric flames technology development (Fig. 23) mostly for applications in SACs (Fig. 24) became the focus of the new initiative sponsored under the NASA Advanced Subsonic Transport (AST) and the follow-on Ultra Energy Efficient Turbine (UEET) programs complimented by GE in-house technology funding.

Different types of pilot configurations could be envisioned but only a limited few have been shared through several US patents. Unlike swirl cups, fuel spray is not intended to impinge on the venturi. The fuel spray from pressure atomizer or other means including pure or piloted airblast nozzle interacts with the surrounding set of swirlers bounded by at least two annular confinements to produce fuel/air mixture fit for performing the low-power operation including ignition and flame propagation and idle to intermediate power setting with the combustion efficiency level matching that of the rich domes. Its NO_x emission contribution becomes more critical as one proceeds with mixers producing successively lower NO_x implying application of features that can control degree of mixedness. The pilot assembly can have co- or counter-rotating swirlers with the overall swirl strength being very weak, weak, or moderate. One could see the application of high-swirl inner swirler surrounded by the weak counter-rotating outer swirler. In summary, the pilot could be interpreted as a swirl cup; some could see it as an

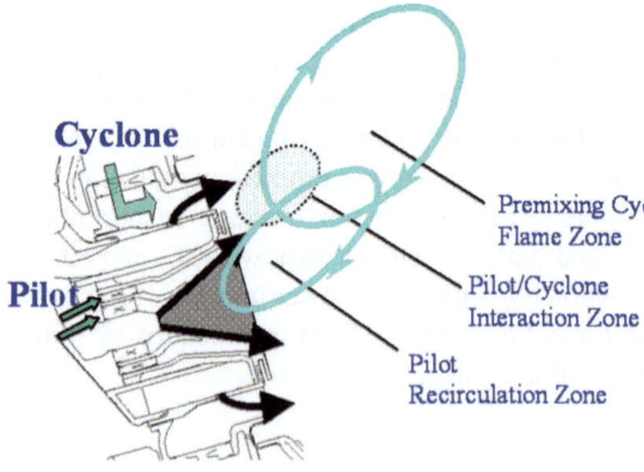

Fig. 23 Conceptual description of Twin-Annular partially Premixing Swirl-stabilized flames (TAPS); reproduced from Mongia (2003)

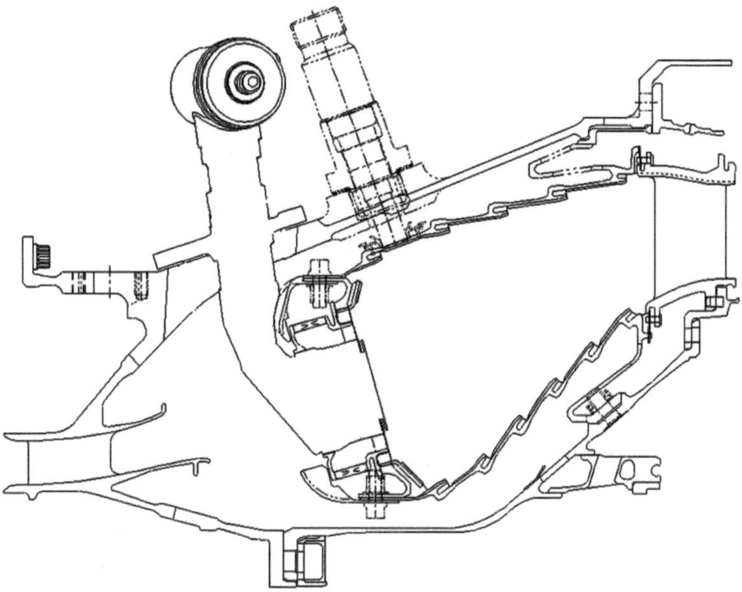

Fig. 24 A typical single annular TAPS combustor; from Mongia (2003)

extension of the swirl-venturi lean direct injection (SV-LDI), Sect. 6 or even low-swirl burner (Therkelsen et al. 2012).

The main fuel turned on at the intermediate to high power comprises of single or multiple swirlers in co- or counter-rotating arrangement along with discrete liquid jets with or without coaxial air injected into "cavity" producing partially premixed fuel/air mixture for interaction with three distinct external flow regions including the pilot, the annular gap between the pilot and main exit, and the outer expansion regions near the wall and circumferential between the cups. The "cavity premixing" main concept is intended to morph into robust design similar to Sanborn et al. (1983) as engine OPR increases beyond certain critical value in order to ensure absolutely no incidence of flashback, autoignition, and/or flame holding with attendant hardware integrity concerns.

In summary, the second generation lean domes have three primary functional features, namely twin-annular flames; degree of premixing of these flames are optimized for meeting all design requirements; and finally these flames are swirl-stabilized, TAPS. Only one of the three secondary flame regions (viz. pilot/cyclone interaction region in Fig. 23) has been shown in different publications (Mongia 2003; 2011e; Foust et al. 2012). The other two regions play an important role in regard to dynamics, durability, and flame propagation.

3.2.1 TAPS1 Through TAPS4 Technology

Development of the TAPS1 technology through single cup, sector, and full annular rig testing followed by extensive demonstrator engine testing for performance, emissions, and durability was described by Mongia (2003). A follow-on paper (viz. Mongia 2011e) provides insights on the engineering aspects of the various stages of TAPS technology mixers starting from TAPS1 through the most recent TAPS4 features. We therefore provide only two figures (viz. Figs. 25 and 26) to illustrate their connection to Figs. 23 and 24.

All the combustion air enters through the dome, Fig. 24; and therefore one can use a more robust single-passage diffuser instead of multipassage diffusers required for DAC and Dry Low Emissions (DLE) of aeroderivative power generation (Joshi et al. 1998). Even though more than 80 % of combustor air enters through the dome, it is still desirable to use domes with cowling as shown in Fig. 24. The hardware details of the pilot schematic of Fig. 23 are presented in Fig. 25 which

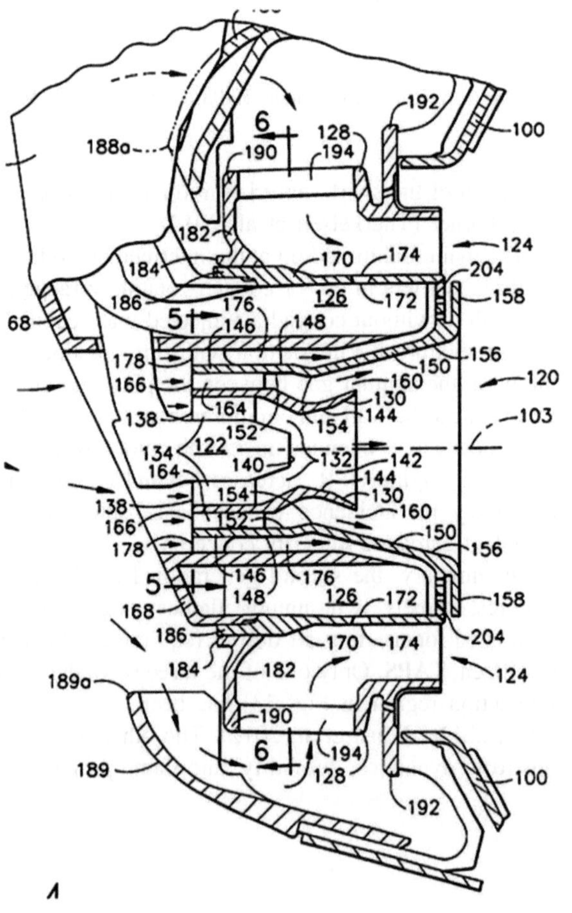

Fig. 25 Gen-1 TAPS premixing multi-swirler mixers with emphasis on improving geometry details around the pilot tips to influence pilot flame characteristics and the items impacting the pilot/cyclone interaction zone (viz. Fig. 23); reproduced from US Patent 6,389,815 May 21, 2002

shows a pressure atomizing injector surrounded by an axial swirler followed by a converging diverging venturi in order to further improve atomization and fuel/air mixing. The shape of the venturi is optimized to ensure no spray impingement on its wall unlike swirl cup venturi. The conditioned spray and air mixing stream is further improved by another concentric axial swirler followed by a diverging passage to help continue further fuel/air mixing with minimal spray impingement on the walls. The fraction of fuel impinging on the walls increases slightly with power because when the pilot fuel is cut back at approach and higher power points, we would like to have the equivalence ratio ϕ of the pilot recirculation zone only slightly higher than flameout. In other words, we want to have the spray concentrated on the outer peripheries of the pilot so that it helps continued flame sustaining in the pilot/cyclone interaction zone (viz. Fig. 23) which is also impacted by the fuel/air ratio, velocity, and turbulence profiles at the main exit plan, namely item 124 in Fig. 25. Several geometrical features including items 156, 158, and 204 have a direct bearing on the annular pilot/cyclone interaction zone's performance and stability. Several features of the cyclone (viz. 170, 172, 174, 182, 190, and 194) influence swirl number and mixing characteristic, flame holding, and flashback characteristics of the main mixer.

In order to further reduce NO_x and improve operability characteristics, several changes will be required for the TAPS4 technology; only limited details are illustrated in Fig. 26. For example, there is provision for having two fuel circuits for the pilot, one for the pressure atomizing tip and the other one to inject into crossflow; the latter is added to increase the operating envelope of the pilot and improve its operability including protection against hail storm and dynamics abatement. The main cavity is fed through a combination of axial and radial inflow swirlers generally counter-rotating in order to improve fuel/air mixing and lower overall swirl strength of the main. The plane fuel jets in crossflow (JIC) have been augmented by coaxial air streams which significantly impact the hardware details of the fuel circuit as shown. Moreover, the axial location of JIC is optimized relative to the shear layer formed between the counter-rotating main swirlers with attendant impact on mixedness and NO_x emission.

3.2.2 GEnx TAPS Engine Emissions

The TAPS1 technology substantiation was completed in 2003 (viz. Mongia 2003) and the first aviation engine product (viz. GEnx) emissions certified data was completed on October 2009 for the 13 engine models with a wide range in SLSS takeoff pressure ratio and thrust, namely BPR: 8–9.4; OPR: 35.1–46.5; F_{00}: 255–345 kN; and margin from CAEP8:60–35 %. The product team worked very hard and vigorously since 2004 in order to produce a reliable second generation lean dome engine that meets all the design requirements 6 years after technology substantiation. Its takeoff NO_x as a function of pressure ratio is shown in Fig. 27 which is presented well by an empirical correlation $EINO_{xGEnx} = 1.079 \times 10^{-5}\ OPR^{3.971}\ w/R^2 = 0.991$. This expression may be compared with the low rich-dome large engines' takeoff NO_x presented earlier

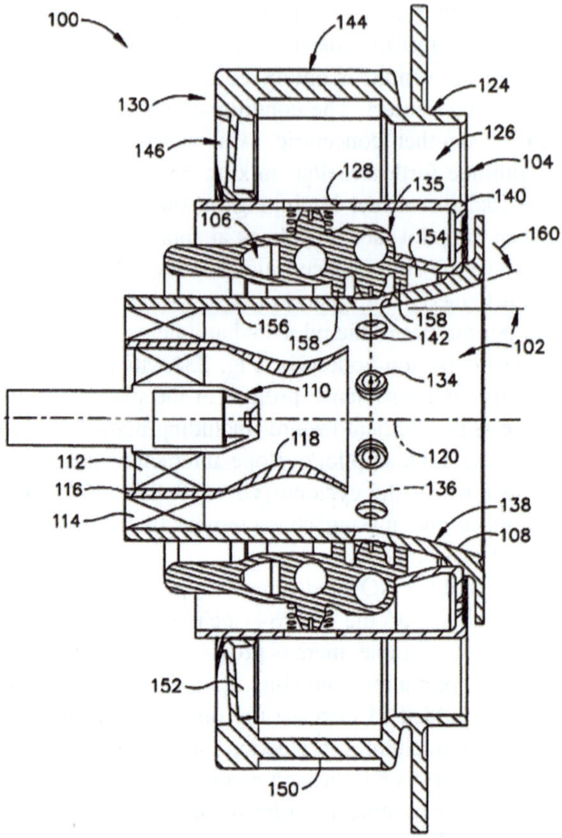

Fig. 26 Gen-4 TAPS focus is on continuing improvements including providing additional mixing and fuel circuit for the pilot to further improve NO_x and operability characteristics; reproduced from US Patent 7,762,073 July 27, 2010

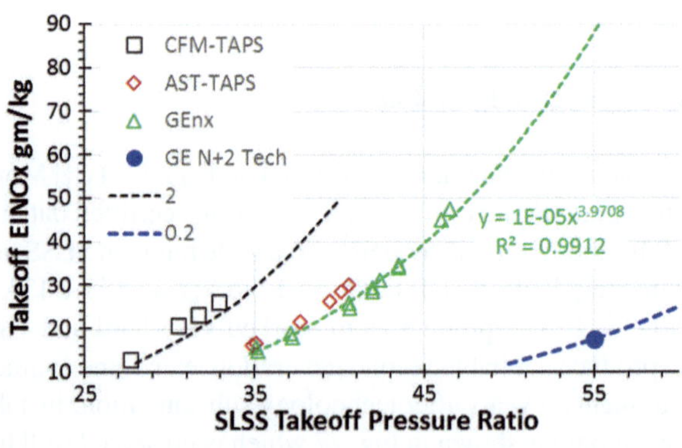

Fig. 27 Takeoff $EINO_x$ versus SLSS takeoff pressure ratio of the TAPS-based engines. The AST TAPS engine data scaled from a full-scale annular rig testing for the GE90-94 whereas the CFM-TAPS data is from a technology demonstrator engine. GE $N + 2$ Tech data from a sector test reported by Lee et al. (2013)

$EINO_{xRD-L} = 0.0303PR^{1.9722}$. Therefore, the ratio of lean to rich-dome takeoff NO_x is 3.56×10^{-4} OPR^2, which means that the ratio of NO_x at 35 OPR of 0.4 successively increases to 0.72 and 1.07 as OPR increases to 45 and 55. In other words, the relative advantage of lean domes over rich domes diminishes with increasing OPR unless one can introduce the next jump in lean technology, namely TAPS1 into TAPS2, TAPS3, and eventually TAPS4.

The TAPS technology team was able to demonstrate the feasibility of TAPS2 on a sector rig (viz. Lee et al. 2013) identified as $N + 2$ in Fig. 27. As clearly stated by the TAPS team in Lee et al. (2013), TAPS2 design required more than 70 % combustion air with enhanced fuel/air mixedness along with "significant challenge to both operability (combustion efficiency and dynamics) and durability." The latter requires the use of high-temperature ceramic matrix composite and advanced liner cooling. Because of the continuous tradeoffs among low NO_x, cruise combustion efficiency, combustion instability, and autoignition, the TAPS2 technology development effort started with the down-select process from the 13 configurations tested in the single-element flame tube rig.

In order to make TAPS technology suitable for products, there were two additional objectives that were clearly met as shown in Fig. 27. First, make successive technology development for increasingly higher engine pressure ratios without corresponding increase in takeoff NO_x. The data from the CFM-TAPS engine technology demonstrator and full-scale annular rig data for the AST-TAPS clearly demonstrate the viability of the first objective. It should be reminded that all the three applications use approximately 70 % combustion air.

The second objective was contrary to the prior lean dome low-emission technology development experience, in that the number of nozzle tips increased relative to the rich domes in the first generation lean domes. Each nozzle tip of the rich domes was replaced by two tips in DAC's including the CFM56 and GE90 DAC products. The dry low-emission LM6000 product has 75 tips compared to 30 with the water-injected diffusion flame combustor (Fig. 53) which will be discussed later. Other OEMs have also used more nozzle tips than rich-domes in their lean dome technology programs (viz. Bruce et al. 1977, 1978, 1981; Dubiel 1986; Rizk and Mongia 1990; Roberts et al. 1977; Sanborn et al. 1983; Segalman et al. 1993). As expected, the lean dome fuel/air mixing devices are more complex than rich domes, viz. Figs. 5, 6, 25 and 26. Therefore, the former's unit production cost should be higher than the latter until new cheaper manufacturing technology (viz. additive manufacturing technology) matures. Therefore, the second objective of TAPS technology development was to have fewer lean mixers than the swirl cups used in the comparable engines and simultaneously emit 50 % lower LTO NO_x emission. For example, GEnx with 22 tips in higher thrust class engine (viz. 255–345 kN) than its predecessor CF6-80C2 engine (viz. 231–273 kN) with 30 swirl cups. We were glad that this objective was met as shown in Fig. 27. However, for the technologist there is one more point to remember: the CF6-80C2 engine's takeoff $EINO_x$ ranges between 22 and 29 whereas that of GEnx is between 15 and 48. Combustion product person can take pride in claiming victory in regard to LTO

NO_x because GEnx average margin to CAEP8 is approximately 50 % compared to 5 % for the CF6-80C2; but the technologist has to worry about the slope of takeoff NO_x for the GEnx ($EINO_x = 1.08 \times 10^{-5}\ OPR^{3.97}$) compared to the CF6-80C2 given by $EINO_x = 0.112\ OPR^{1.59}$. However, based on published TAPS technology development experience, the first generation CFM-TAPS technology's takeoff NO_x is approximately $2 \times EINO_{xGEnx}$ whereas that of the GE $N + 2$ Tech is given by an unbelievable number $0.2 \times EINO_{xGEnx}$, as shown in Fig. 27 by two dotted lines identified, respectively, as 2 and 0.2.

But what does it mean in regard to future NO_x stringency regulations, especially the long-term? This is discussed in Fig. 28. The aviation engine combustion product community has worked together in proposing and consenting to future ICAO stringy regulations because they have not been used to try to eliminate qualified competitors. Combustion technologists have pursued the long-term goal of 70 % reduction in takeoff NO_x since the early 1970s, Bruce et al. 1977, which along with gains in propulsion system efficiency gave approximately 80 % reduction from the base CAEE regulation at 30 OPR, namely 20 g NO_2/kN.

If we follow the trendline of stringencies set by CAEP4, CAEP6, and CAEP8 with the corresponding regulatory values of 67, 59, and 50.1 g NO_2/kN at 30 OPR along with the corresponding effective dates of December 31 of 2003, 2007, and 2013, we assume that the next five rounds of NO_x stringencies, each 4 year apart would require the LTO NO_x at 30 OPR not to exceed, respectively, 42.6, 36.2, 30.8, 26.2, and 22.2 gm NO_2/kN. With a 30 % margin from the regulatory standard of 2033, the $N + 2$ product goal should be $0.7[-37.763 + 2OPR]$ as shown in Fig. 28 which also summarizes the current status of the best products, namely TALON II, Trent1000, and GEnx in addition to the current regulation (CAEP/8) and GE $N + 2$ technology as reported by Lee et al. (2013). We can safely assume that the next round of TALON-X-based P&W combustion product will be at least as good as

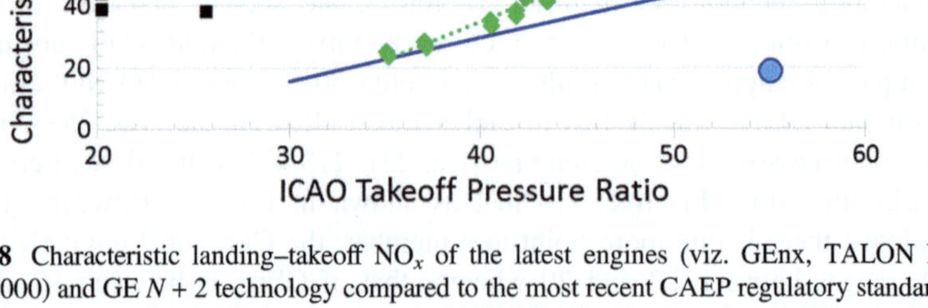

Fig. 28 Characteristic landing–takeoff NO_x of the latest engines (viz. GEnx, TALON II, and Trent1000) and GE $N + 2$ technology compared to the most recent CAEP regulatory standard (viz. CAEP/8) and potential CAEP long-term goal, CAEP LT. The $N + 2$ product goal is set with 30 % margin to CAEP LT specified as $-9.884 + 2\pi$

Talon1000 which has 15 % margin from CAEP/8. In summary, the LTO NO_x of the $N + 2$ product at 30 OPR will be 15.6 g NO_2/kN giving approximately 85 % reduction from the first generation modern aviation engines listed in the ICAO engine emissions database. The low-OPR GEnx meets the $N + 2$ product goals; however at ≥50 OPR, it needs to reduce LTO NO_x by approximately 50 % in order to meet the proposed CAEP/18 goals, namely—37.763 + 2π. However, if the GE $N + 2$ technology can be transitioned successfully into product, then it will have approximately 75 % margin from the future NO_x regulatory standard. In other words, no new NO_x technology innovations are required other than the general challenges with transitioning it into products effectively.

Low-power combustion efficiency, CO, and hydrocarbons were other major concerns of the first generation lean domes which have been successfully resolved in the second generation as summarized in Figs. 29 and 30 in terms of average LTO HC and CO. The data for TALON II and Trent1000 are also shown in order to facilitate a back-to-back comparison between the best rich domes and GEnx. For all practical purposes, GEnx performs as well as the rich domes; both have significant margins from the CAEE baseline standards, namely 19.6 g CH_4/kN and 118 g CO/ kN. The comparison between the smoke regulatory standard in terms of maximum SAE smoke number $\leq Min\left[\frac{83.6}{F^{0.274}}, 50\right]$, is presented in Fig. 31 for both the rich and lean domes; the latter are expected to produce considerably lower values than the former. But that is not the case with the GEnx engine tested in 2009. Since the characteristic smoke numbers of GEnx and the best of the Rolls Royce rich domes including RR Phase 5 and Trent1000 are essentially comparable (Fig. 32), we have plotted the average smoke numbers of Trent1000 and GEnx, respectively, in Figs. 33 and 35 for each of the four parts of the landing–takeoff operation, namely takeoff, climbout, approach, and idle.

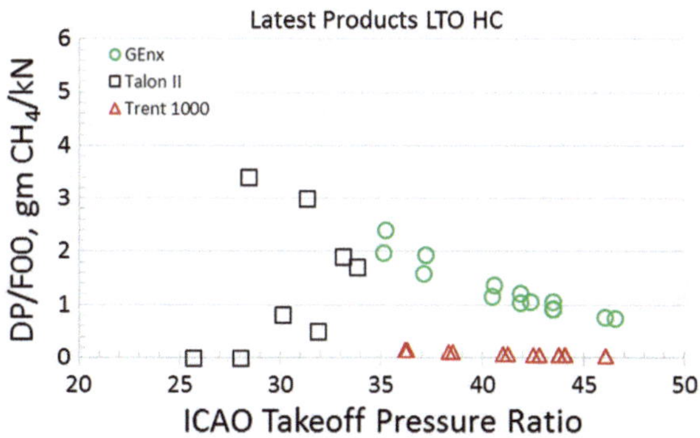

Fig. 29 Landing–takeoff LTO HC of the latest engines (viz. GEnx, TALON II and Trent1000) versus ICAO takeoff pressure ratio. Average landing–takeoff unburned hydrocarbons may be compared to the standard, 19.6 g of CH_4/kN rated takeoff thrust

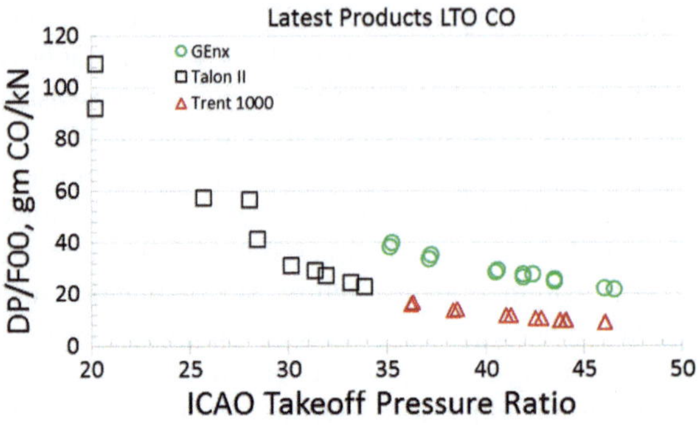

Fig. 30 Landing–takeoff LTO CO of the latest engines (viz. GEnx, TALON II and Trent1000) versus ICAO takeoff pressure ratio. Average landing–takeoff CO may be compared with the standard, 118 g of CO/kN of rated takeoff thrust

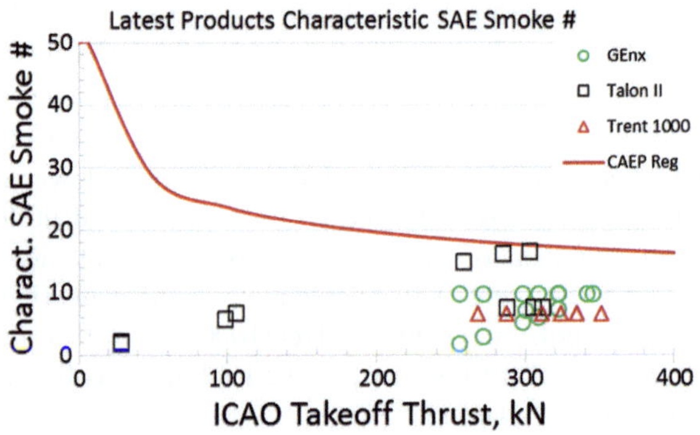

Fig. 31 Characteristic SAE smoke number versus takeoff thrust of the selected engines (viz. GEnx, TALON II, and Trent100) in order to show margin from the ICAO CAEP regulatory standard, namely $\leq \mathrm{Min}\left[\frac{83.6}{F^{0.274}}, 50\right]$

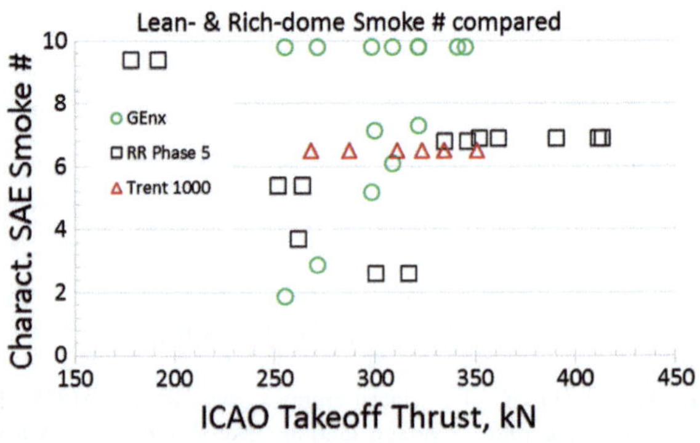

Fig. 32 Characteristic SAE smoke number versus takeoff thrust and back-to-back comparison between lean dome (GEnx tested in 2009) and the several Rolls Royce rich-domes, namely RR Phase 5 and Trent 1000

Fig. 33 Average SAE smoke numbers of the rich-dome Trent1000 engine models for each of the four modes of the landing–takeoff cycle

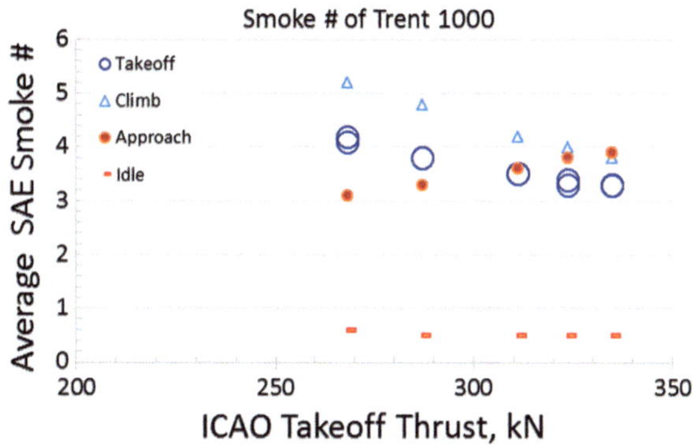

The modern N and $N + 1$ generation Trent1000 rich domes have consistently performed very well in regard to reducing NO_x and simultaneously bringing maximum smoke number in single digit as shown in Fig. 33 indicating very low values at idle, increases by a factor of 5–10 during approach, climb, and takeoff modes of the landing–takeoff cycle. It even retains the classical pattern of having smoke number slightly higher during climb than takeoff; the difference reduces as the engine runs closer to its highest rated level. The data presented for rich domes in Figs. 31, 32, and 33 clearly show that all the three OEMs have the empirical know-how for keeping smoke considerably lower than the first generation aviation engines in which the small engines could have smoke numbers approaching 50. Similarly with more experience on lean domes, it would be equally easy to achieve even lower smoke emission as illustrated in Figs. 34 and 35, respectively, for the first and second generation lean domes. The first generation lean domes include 10 models of the CFM56-5B, 5 models of the CFM-7B, and 5 models of the GE90; whereas the second generation lean domes include 15 and 11 models of the GEnx tested, respectively, in 2009 and 2012. As shown in Fig. 34, the maximum smoke

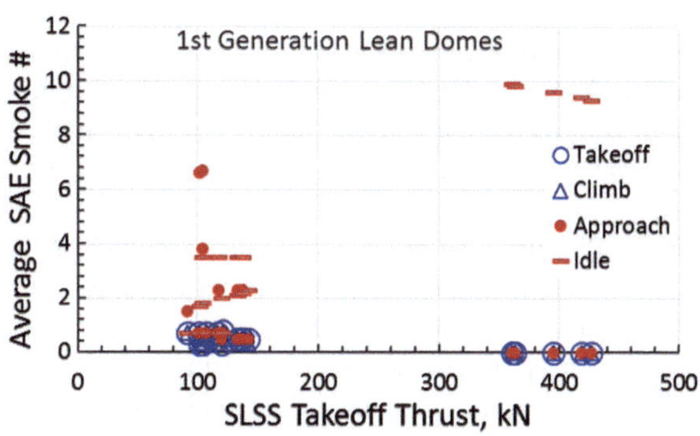

Fig. 34 Average SAE smoke numbers of the first generation lean domes (viz. CFM DAC and GE90 DAC1) for each of the four modes of the landing–takeoff cycle, namely takeoff, climbout, approach, and flight idle

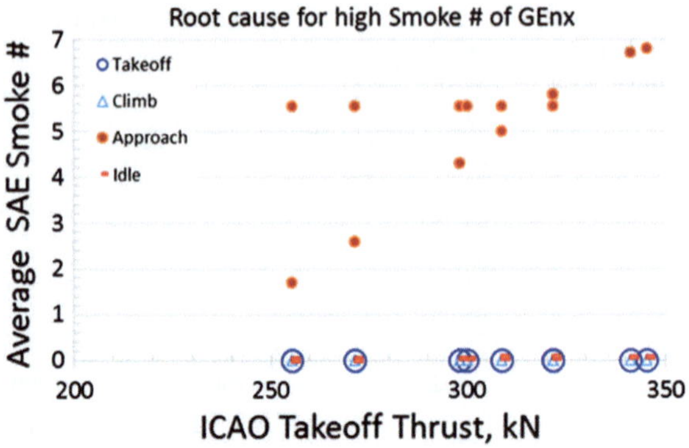

Fig. 35 Average SAE smoke numbers of GEnx engine models for each of the four modes of the landing–takeoff cycle, namely takeoff, climbout, approach, and flight idle. Data are from the engines tested in 2009 and 2012

number (ranging between 9.3 and 9.9) of the GE90 occurs during idle for the DAC I design which one could attribute to the pilot zone which is fueled at the idle operating condition with the generally used hypothesis number one, namely, the primary zone is too rich and the intermediate temperature levels are not high enough to oxidize soot produced in the primary zone. Since the GE90 DAC I fuel staging details are not available publicly, reduction in its smoke emission at approach could be due to hypothesis number two (here the intermediate zone temperature is high enough to oxidize most of the soot formed in the primary zone) or hypothesis number three, namely the primary zone is lean enough because of fuel staging and therefore its soot formation rate is negligible. Both the residence times and the degree of mixedness play a very important role in determining the level of smoke emissions in addition to the fuel properties especially aromatics and attendant fuel hydrogen content as discussed in Sect. 5. These simple hypotheses help technologists to synthesize their rich and lean domes' data and take appropriate steps for combustor modifications to achieve low smoke emission. The author has not run into any protracted development effort due to high smoke since the late 1980s and therefore does not depend on any new breakthrough in fundamentals and/or modeling of soot formation and oxidation. The high smoke data on the first generation CF56 lean domes can similarly be explained by one or more of the hypotheses proposed for explaining the GE90 data.

However, the TAPS fuel staging is somewhat complicated because it could employ pilots with different atomizing mechanisms including pressure atomizers with single and/or dual orifices followed by the spray optimization by the concentric swirlers which may be co- or counter-swirling. It could also use pure or piloted airblast nozzles. It may even have dual fuel injection systems as shown previously in Fig. 26. Because the pilot and main flames are concentric, the pilot fuel fraction does not have to be a simple function of power setting. Nevertheless, the same set of smoke formation and reduction hypotheses discussed in the previous paragraph are applicable except that the intermediate zone here is lying in a domain in between the pilot and main flames; and that the main flame by itself can become the source of higher smoke. However, the TAPS product team has done a good job in modifying the engine tested in 2009 (which had the highest smoke number of 8) to produce only 0.6 smoke number in the engine tested in 2012, as summarized in Fig. 35. In summary, the second generation lean domes produce exhaust smoke number in an order of magnitude lower than the rich domes.

3.2.3 LEAP-X and GE9-X Engines' TAPS II

LEAP-X is expected to use TAPS II combustion system (M'Bengue 2010), with pressure ratios considerably higher than its predecessor CFM56 engine. It may be recalled that the TAPS technology demonstration was done on a CFM demonstrator engine with engine NO_x data presented previously in Fig. 27 which shows that its takeoff was approximately twice of the GEnx. We do not know whether the LEAP-X NO_x will fall on the dotted black or green curves shown in Fig. 27; but the product results for takeoff NO_x and the characteristic LTO NO_x (viz. Fig. 28) will add another significant milestone on the viability of the second generation lean domes especially considering that "more than 7,500 total LEAP engine orders to date" a claim by the CFM press release dated July 17, 2014, http://www.cfmaeroengines.com/press/cfm-signs-21-4-billion-in-new-deals-at-farnborough-air-show/774.

The GE9X a new engine for the next Boeing 777x airplane with approximately 12 % lower fuel burn will use "a 3rd generation TAPS, and advanced fuel nozzles using additive manufacturing" technology according to the GE Aviation press release dated November 19, 2013, http://www.geaviation.com/press/ge90/ge90_20131119a.html. The author looks forward to adding data points of LEAP-X and GE9X to Figs. 27, 28, 29, 30 and 31 to illustrate maturity of the second generation lean domes for medium, large and very large aviation engines for a very large range of operating pressure ratios, namely 25–60.

4 NO_x Entitlement

Figure 11 of Mongia (2011c) gives description and NO_x data on a unique accelerating swirler passage (ASP) module tested behind the NASA iconic lean premixing vaporizing multiventuri injectors (Tacina 1990). The NASA test data was conducted at the High Speed Research (HSR) supersonic cruise point of P_3 = 150 psi (1,034 kPa) and T_3 = 1100°F (866 K) for combustor outlet temperature up to 2,100 K giving $EINO_x \leq 1.0$. Mongia (2011c) assumed that ASP NO_x data can be used as NO_x entitlement for aviation engines. Figure 3 of Mongia (2008) gives $EINO_x$ data taken in a NASA flametube rig at fixed P_3 = 400 psi (2,758 kPa) and T_3 = 1,000 °F (811 K) and FAR = 0.027 of an Advanced Multi-Swirlers (AMS)

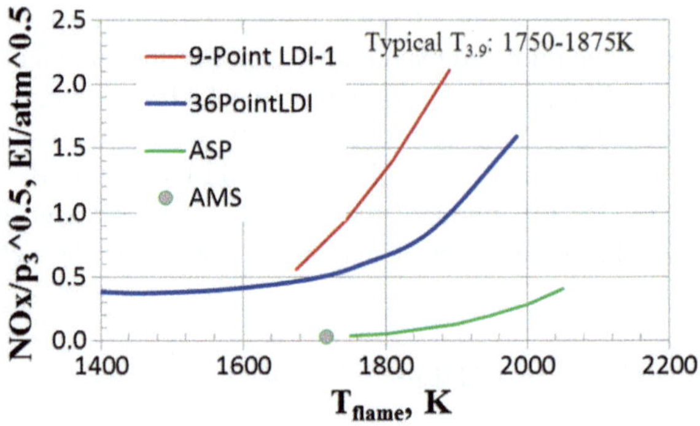

Fig. 36 NO_x entitlement ($EINO_x/P^{0.5}$) represented by an accelerating swirl passage mixer (*ASP*) as a function of adiabatic flame temperature can be compared with an advanced multi-swirlers mixer (*AMS*) and LDI-1 from Mongia (2011e)

mixer. Its minimum $EINO_x$ value along with the data of NASA LDI-1 and ASP were pulled together in Fig. 20 of Mongia (2011e) to propose it as the current status of the promising lean dome technologies. This figure is reproduced here as Fig. 36 in order to facilitate comparison with the most recent products' takeoff NO_x as summarized in Fig. 37. At least two comments are relevant here. First, we have assumed that the pressure exponent of 0.5 is valid for comparing data from different lean dome concepts, in addition to an assertion that it can be represented as a function of average adiabatic flame temperature. Second, the products' takeoff $EINO_x$ data are presented as a function of SLSS (ICAO) takeoff pressure ratio and therefore only indirect comparison can be made between Figs. 36 and 37; we have assumed that the pressure exponent of 0.5 for the products' $EINO_x$ is reasonable.

The P_3 normalized NO_x data of Fig. 37 compared to the earlier figures (viz. Figs. 13 and 27) does not provide any additional insight but a means to make comparison with rig data at different design conditions, namely Fig. 36. As mentioned by the TAPS $N + 2$ technology team in Lee et al. (2013), their selected configuration was not necessarily the lowest NO_x, one driven by other

Fig. 37 Takeoff $EINO_x/PR^{0.5}$ of several rich-domes, GEnx and GE $N + 2$ technology versus takeoff pressure ratio PR

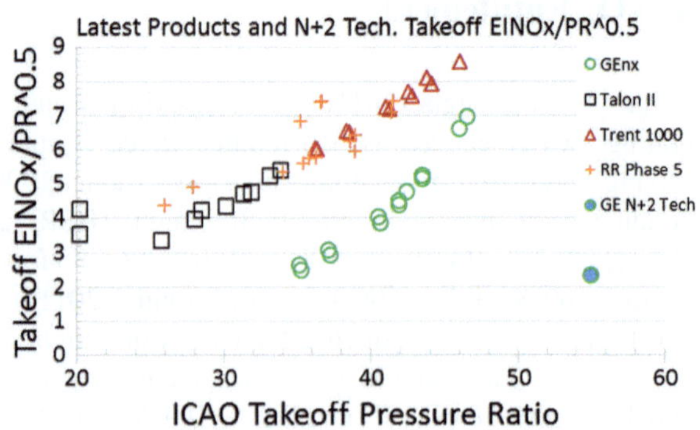

considerations including durability. Clearly, there is a potential for reducing lean dome's NO_x by a factor of four over the GE $N + 2$ technology; but there is no need to push for further reduction based on the discussion presented in the previous section.

5 Emissions with Alternative Fuels

There are several outstanding papers dealing with critical aspects of alternative fuels which have significantly lower aromatic and sulfur contents, including a few mentioned here (viz. Bester et al. 2009; Bulzan et al. 2010; Chisty et al. 2011; Corporan et al. 2004; Carter et al. 2011; Davison et al. 2011; Longwell 1977; Moses et al. 2008; and reference provided therein). Therefore, we have very limited objective here on this topic in that the effect of FT fuel on $EINO_x$ is very small (Fig. 38) but its benefits in regard to particulate emissions are enormous in terms of both the number density and mass emissions; typical results for the former are shown in Figs. 39 and 40 for the first generation of the modern turbofan engine for single aisle engine, namely the CFM56-2C discussed earlier in Figs. 11 and 13. Effect of alternative fuels on the first and second generation LDI as measured by Wey and Bulzan (2013), Dolan et al. (2013) and Tacina et al. (2014) show similar beneficial results.

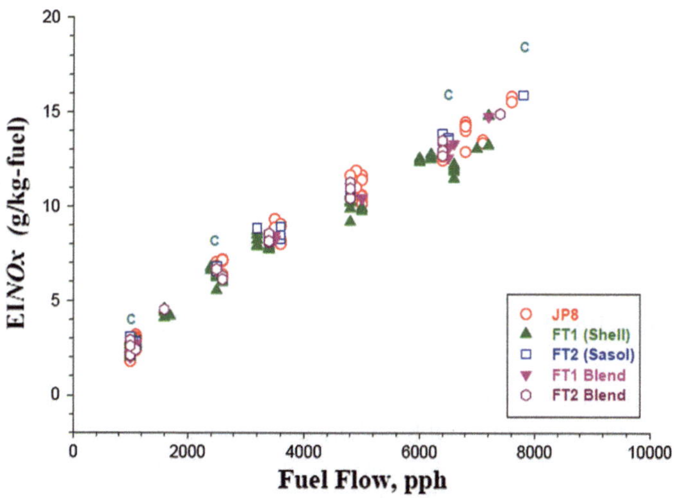

Fig. 38 $EINO_x$ versus fuel flow rate of the CFM56 engine with the base fuel (JP8), two FT fuels and their blends; reproduced from Bulzan et al. (2010)

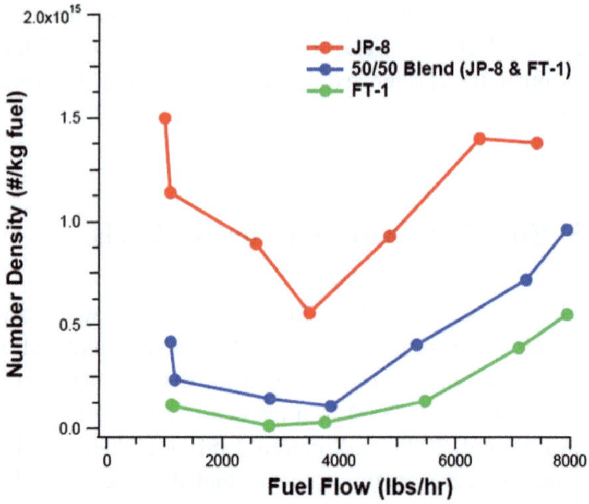

Fig. 39 Particle number density versus fuel flow rate of the CFM56 engine with the base fuel (JP8), FT-1, and its blend; reproduced from Bulzan et al. (2010)

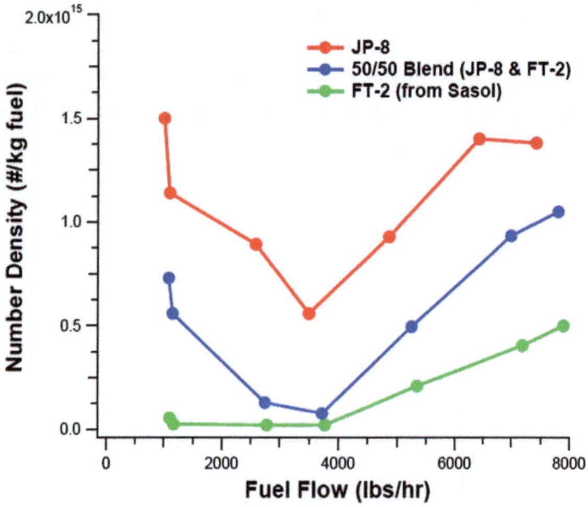

Fig. 40 Particle number density versus fuel flow rate of the CFM56 engine with the base fuel (JP8), FT-2 and its blend; reproduced from Bulzan et al. (2010)

6 Second Alternative to Rich Domes SV-LDI

The first generation lean dome aviation engines have been in revenue service since the middle 1990s for up to 40 OPR with no flameholding, flashback, or autoignition incidence (F^3) reported publicly; but the author calls them as LDI. The same statement is valid for the recently introduced second generation partially premixing lean domes flying in both the Boeing 747 and 787 airplanes. The premix aeroderivative industrial engines with up to a pressure ratio of 30 since the middle 1990s and the most recently introduced 40 pressure engines have not reported these F^3 problems normally associated with lean domes. This does not imply that these problems are not encountered during technology development (viz. Bruce et al. 1977, 1978). Nevertheless, the robust lean domes are F^3 safe; otherwise, they will not be introduced as reliable aviation engines where safety is the primary design criteria without any exceptions. The technologists and combustion product teams have developed strong guidelines on the lean domes for safety; but that does not mean that we should not look for alternatives to current premixing designs in revenue service. Only a "fool" should ignore the fact that increasing takeoff pressure ratio from 45 to 60 (viz. Table 5) may reduce the allowable mixing time by a factor of 3 in addition to propensity to flameholding and flashback. Therefore, this section and the next two sections deal with lean dome concepts considered alternative to TAPS.

6.1 Swirl-Venturi Lean Direct Injection (SV-LDI)

The origin of the modern generation of single venturi lean direct injection concepts (SV-LDI) can be traced to Tacina (1990) who discussed different options and narrowed down to the basic configuration investigated by many including the use of axial swirlers by Tacina et al. (2005a); radial swirlers by Tacina et al. (2005b); and discrete jets by Tacina et al. (2004). It is educational to see the similarity of these three types of swirling flows to the primary swirlers of swirl cups presented earlier in Figs. 3, 4 and 5; but one should not make strong statements about the apparent similarity because the geometrical details of the subsystems including the swirler, venturi, injector, and its location can have big impact on overall emissions and operability. The two most recent extensions to the baseline LDI configurations of Tacina et al. (2004, 2005a), known by LDI-2 are described in Villalva et al. (2013) and Tacina et al. (2014) which we will describe briefly here.

As shown in Fig. 41, typical SV-LDI fuel/air mixer comprises of two elements, viz. a converging/diverging venturi fed by a fuel injector with concentric axial swirler with the vane angle optimized for meeting the requirements.

As shown in Fig. 42, the baseline or LDI-1 has nine identical SV-LDI fuel/air mixers arranged in a 3 × 3 grid for testing in an adiabatic combustor with a square

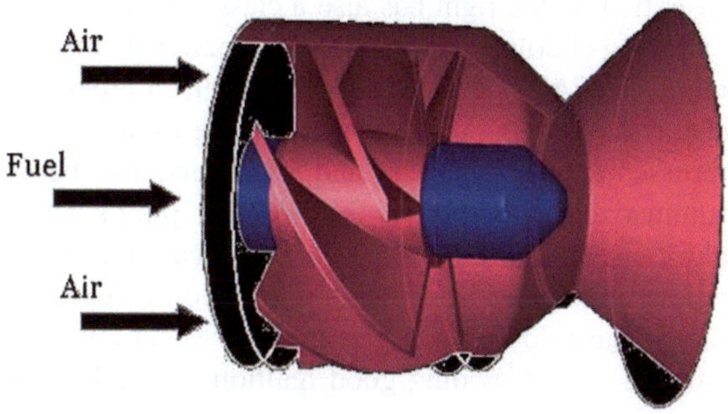

Fig. 41 First generation of swirl venturi lean direct injection mixer (SV-LDI); from Tacina et al. (2005a)

Fig. 42 The baseline 9-point SV-LDI design; from Tacina et al. (2005a)

cross-section of 76.2 mm × 76.2 mm. Tacina et al. (2005a) summarize test results with three different helical vane angles, viz. 45°, 52.5°, and 60° showing advantage of the 45° swirlers for lower NO_x but limited by operability range. The measured effective area of this module was 955 mm^2 giving a combustor area expansion ratio of 6.1, a design parameter of great significance.

The second generation SV-LDI (LDI-2) comprises of a "pilot" surrounded by 12 main/intermediate (M) fuel/air mixers for a combustor of 114.3 mm × 114.3 mm square cross-section. The M mixers were divided into three discrete groups of four identical mixers identified as m_1, m_2, and m_3 (Fig. 43) in two very different arrangements. The configuration on the left (to be called flat dome LDI-2) has classical fuel staging pattern used in diffusion flame annular combustors (viz. Bruce et al. 1977), namely alternately flowing fuel nozzle at idle with attendant reduction in lean blowout fuel/air ratio (LBO FAR), CO, and hydrocarbons (HC). The configuration on the right has also a classical fuel staging pattern called sector burning involving alternate pairs of flowing nozzles which might lead to further reduction in idle LBO FAR, CO, and HC. However, in order to further improve its operability characteristics, the configuration on the right has both the pilot and the m_1 mixers recessed relative to the m_2 and m_3 mixers; and therefore called recessed LDI-2. As summarized in Table 6, the baseline pilot's swirl vane angle of 55° counter clockwise (CCW) with a longer increased exit diameter diverging section and pressure atomizer was designed to give higher low-power stability. However, m_1 mixers have 45^0 vane angles in order to get lower NO_x, but with simplex pressure atomizers for providing good ignition source for both the m_2 and m_3 mixers. The

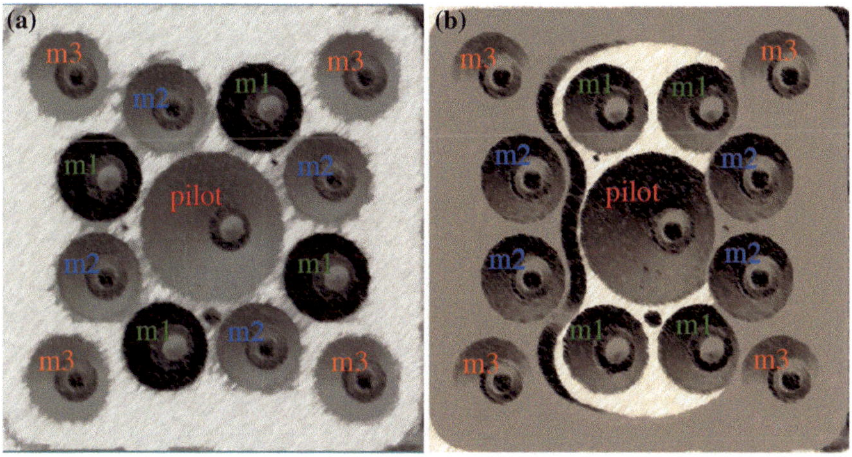

Fig. 43 Second generation SV-LDI for two configurations, left flat dome and right 5-point recess dome; reproduced from Tacina et al. (2014)

same m_1 set of mixers configuration is also used for the recessed LDI-2. In order to offset supposedly higher NO_x emission from the recessed configuration, its pilot injector and swirler arrangement was different from that of the flat dome. It uses Woodward's expertise in pure airblast nozzle with inner swirler air (identified as IAS in Table 6) vanes of 57° clockwise (CW), and outer air swirler (OAS) 57° CCW vanes with the attendant low swirl air atomizing pilot; and the flame front being barely visible. In order to further reduce NO_x of the recessed LDI-2, the eight main mixers (m_2 and m_3) were similarly designed as pure airblast nozzles but with 45° CW/45° CCW arrangement in contrast to the flat dome with 45° CW/45° CW swirlers. In other words, the flat dome uses higher swirl number m_2 and m_3 mixers than the recessed dome, again to offset higher NO_x of the latter. Extensive testing was done in the NASA facility and the first paper on the results are summarized by Tacina et al. (2014), with several more to follow to deal with many aspects of these designs. Effective areas of the flat and recessed modules are 1,338 and 1,326 mm^2, respectively, with the corresponding combustor area expansion ratios of 9.77 and 9.85 compared to 6.1 for the 45° LDI-1.

Tacina et al. (2014) have provided extensive database and correlation effort on the both LDI-2 configurations. Here we share only one aspect of the design, namely back-to-back comparison on $EINO_x/P_3^{0.5}$ versus adiabatic flame temperature T_{flame} between the recessed dome (at T_3 = 1,024 °F, P_3 = 253 psi, their standard deviations σ = 14 °F and 7 psi), and the flat dome at T_3 = 1,014 °F, P_3 = 255 psi, σ = 14 °F, and 7 psi. The results are summarized in Fig. 44; here P_3 is in atmosphere and T_{flame} is in degrees Kelvin. The left part of the figure shows the entire datasets covering T_{flame} range between 1,500 and 2,100 K for the recessed dome in addition to the flat dome for T_{flame} up to 1,900 K; the right part is the zoomed-in data for the both configurations for T_{flame} up to only 1,900 K. Combustion efficiency of all the data presented

Table 6 Second Generation SV-LDI configurations. For each stage, the table gives the type of fuel injector and the air swirler angle(s); reproduced from Tacina et al. (2014)

Configuration	Pilot injector	Pilot swirler	Main 1 Injector	Main 1 swirler	Main 2 injector	Main 2 Swirler	Main 3 Injector	Main 3 Swirler
Flat dome	Simplex	55° C CW	Simplex	45° CCW	Airblast	IAS: 45° CW OAS: 45° CW	Airblast	IAS: 45° CW OAS: 45° CW
5-point recess	Airblast	IAS: 57° CW OAS: 57oCCW	Simplex	45° CW	Airblast	IAS: 45° CW OAS: 45° CCW	Airblast	IAS: 45° CW OAS: 45° CCW

OAS outer air swirler, *IAS* inner air swirler, *CW* clockwise, *CCW* counterclockwise

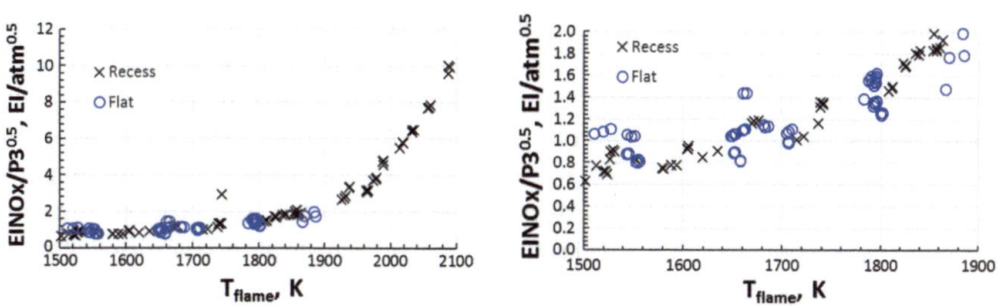

Fig. 44 $EINO_x/P_3^{0.5}$ versus adiabatic flame temperature for the LDI-2 flat and recessed domes; refer to Tacina et al. (2014) for more details

here is more than 99.9 %. Several important observations can be made including: (1) NO_x increases slowly from ~ 1.0 EI/atm$^{0.5}$ at 1,500 K to ~ 2.0 EI/atm$^{0.5}$ at 1–900 K T_{flame}; (2) no noticeable difference between the recessed and domes, except the latter shows slightly bigger data scatter band; (3) rapid increase in NO_x for $T_{flame} > 1{,}900$ K; (4) and finally the simple correlation (viz. NO_x as a function of T_{flame}) needs further enhancements to take into account other secondary effects of the flat dome LDI-2. The development of good and useful engineering correlation is a long process which will be shared for these designs in future publications. The Woodward/NASA LDI-2 team should be very pleased to see that their relevant data (viz. right part of Fig. 44) is pretty competitive with the GE $N + 2$ technology point (Fig. 37) which the author feels is a proud milestone for the TAPS team that has worked hard continuously since 1995, and spent more than $500 million in developing the TAPS technology and GEnx products.

7 Modeling and Correlations

It should be obvious from Figs. 41, 42, and 43, and Table 6 that both LDI-2 configurations were significant departures from the baseline LDI-1 design and that no use was made of the CFD models in making the LDI-2 design selections. The LDI-1 empirical correlation was not applicable either and therefore the selection process was driven by combustion fundamentals tempered by the guidelines provided by the LDI-1 design experience. The author is a strong believer in using CFD in the design process (Mongia and Smith 1978, Mongia et al. 1986, 2004a, b, c, d, e, f; Mongia 1993, 2001a, b, 2008). His teams have spent 10–25 % of their resources in the development and calibration of CFD-based combustion models applicable in the design process since the middle 1970s. Subsequently, many publications have been shared by his former team members contributing significantly toward the usefulness of CFD in the design process. The author has shared in several of these publications the applications and limitations of the models in the design process; the latter is covered by complimenting with simpler models as explained briefly. The author is also a strong believer in the formulation and calibration of semi-empirical correlation followed by their applications (Rizk and Mongia 1990; Mongia 2010a, b). The conceptual design of the combustion concepts and their execution through engineering-based testing process are equally important for meeting the design goals. An optimum combination of the three was coined as empirical/analytical design methodology (viz. Mongia and Smith 1978) which has continuously improved through the years to the best process currently followed by several including Rida et al. (2012), Sen et al. (2012), Sekar et al. (2014), Sripathi et al. (2014).

Our new CFD modeling approach is required for managing combustion dynamics in addition to predicting CO and NO_x. Therefore, we started with the National Combustion Code (NCC) that solves the turbulent two-phase fluid transport equations coupled with compact kinetic scheme including turbulence chemistry interaction. Both Reynolds-averaged Navier–Stokes (RANS) and time-filtered Navier–Stokes (TFNS) simulations are planned to be calibrated with the LDI-1 and LDI-2 data before the models can be applied for screening promising new concepts. The reader may want to refer to the publications (viz. Ajmani et al. 2013a, b, c, 2014; Ajmani and Breisacher 2014a, b) for details; here we will discuss only one figure from Ajmani et al. (2014).

Figure 45 shows the current status in terms of experimental versus predicted NO_x for the flat dome (identified as baseline) and recessed dome, identified as

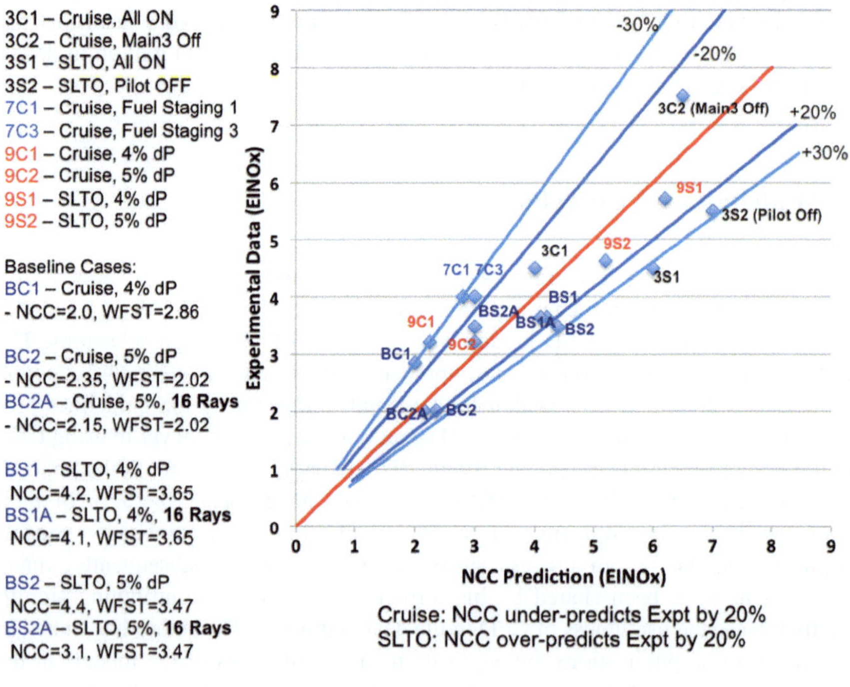

Fig. 45 Summary of $EINO_x$ predictions (NCC) and Experimental Data (WFST) at P_3 = 130 psia and various cycle conditions for four different LDI-2 geometries; from Ajmani et al. (2014)

configurations 3, 7, and 9. Details on the configurations, simulations, and data are given by Ajmani et al. (2014); here we will provide only the summary of the results and future plans.

The RANS simulation of the 13 mixer modules (Fig. 43) by using a tetrahedral mesh with more than 17 million elements to resolve important features of the geometry. A two-equation turbulence model was used, namely cubic k-ε model with variable Cμ along with dynamic wall functions with pressure gradient effects. The governing equations for the liquid phase were solved based on a Lagrangian formulation along with Jet-A fuel properties and turbulence transport effects. Initial spray conditions came from measured data and empirical knowledge with the spray divided into 10 droplet groups, discretized into 32 spatial streams along the 360^0 circumference. A compact 14 species with 18 chemical reaction steps for Jet-A fuel combustion and NO_x predictions was used along with the other transport equations. The results shown in Fig. 45 have provided motivation to develop and calibrate the next generation of computationally affordable kinetics mechanism with ≤20 species to be spearheaded by Prof. Fred Dryer based on his most recent outlook (Dryer 2014).

The data and calculations in Fig. 45 cover a broad range of operating conditions, viz. configuration 3 identified as 3C1, 3C2, 3S1, and 3S2; here C and S are abbreviations, respectively, for the cruise and sea-level takeoff operating conditions; and 1, 2, and 3 as three fueling options as described. Similar short description given as legends in Fig. 45 helps one to appreciate the highlights of configurations 7 (viz. 7C1, 7C2), 9 (viz. 9C1, 9C2, 9S1, 9S2), and several baseline (flat) dome operating conditions listed as BC1, BC2, BC22A, BS1, BS1A, BS2, and BS2A. The range of NO_x data is between 2 and 8EI. Comparison between data and predictions can be broadly be made by noticing that the ideal match is along the red line, followed by two bounding lines of ±20 % accuracy and the outer band of ±30 %. This level of agreement is considered a reasonable start for the RANS simulations because Kumud Ajmani plans to continue with the TFNS calculations with his current kinetic scheme as well as with the improved kinetic scheme to be developed by Prof. Fred Dryer in the coming years.

The author fully appreciates the critical importance of CFD in the design process even when we have several examples of successfully implementing advanced technology concepts through empirical means only, e.g., the CFM-TAPS first engine demonstration, and the two LDI-2 configurations of Fig. 43. He does not need any convincing or discouragement from the summary given in Table 3. However in order to get the acceptance of the turbulent combustion models by the "nonbelievers" (Table 3), models should meet the accuracy metric derived from two directions, namely the measurement repeatability (σ) and semi-empirical correlation σ levels. The average σ values of the 128 samples from the ICAO engine emissions database of the landing–takeoff hydrocarbon, CO, and NO_x emissions (DP/F_{00}) are, respectively, 0.89, 2.61, and 1.33 g of pollutants/kN or 13.4, 5.0 and 2.8 % of the average DP/F_{00}. Therefore, the gas turbine prediction models' long-term σ goal proposed by Mongia has been: 3 % of takeoff $EINO_x$, 7.5 and 15 %, respectively, of idle CO and HC emission indices.

Similarly, the modern semi-empirical simple emission models should provide guidance for the engine operating lines from sea-level to the maximum-cruising altitude; and they should be as reliable as the engine performance models which are used routinely for on-board mission fuel burn calculations. Therefore, CO and NO_x emission characteristics along the sea-level operating lines were presented by Mongia (2008) of the 12 combustors that included rich and lean domes of the N-1 and N-generation engines. Simple engineering correlations for both sea-level and altitude engine operating lines have given promising results worth pursuing for future investigations; details are given in Mongia (2010a, b) with summary figures reproduced as Figs. 46, 47 and 48 for the selected five modern rich domes along their sea-level engine operating lines. These correlations are good for the range of combustor inlet temperatures (T_3) and pressures (P_3); the adiabatic stoichiometric flame temperature (T_{stoich}); HC, CO, and NO_x in EI in terms of maximum,

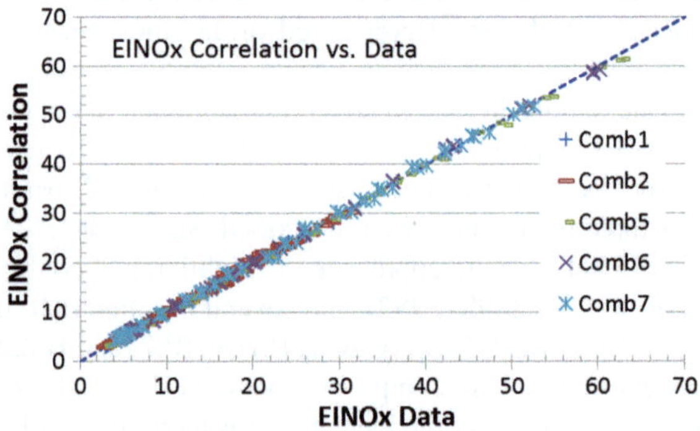

Fig. 46 Comparison between $EINO_x$ correlations versus data from idle to maximum takeoff thrust for the modern rich-domes, namely Comb 1, Comb 2, Comb 5, Comb 6, and Comb 7. The range of the root-mean-square errors is 0.23–0.47 EI and 1.2–3.1 %; *Source* Mongia (2010a, b)

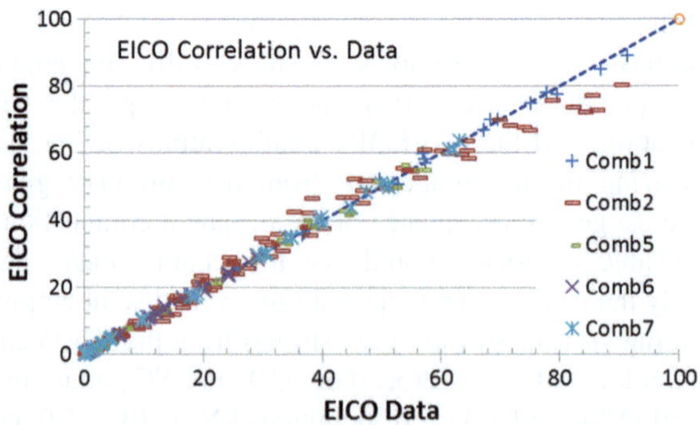

Fig. 47 Comparison between EICO correlations versus data from idle to maximum takeoff thrust for the modern rich-domes, namely Comb 1, Comb 2, Comb 5, Comb 6, and Comb 7. The range of the root-mean-square errors is 0.7–2.5 EI; *Source* Mongia (2010a, b)

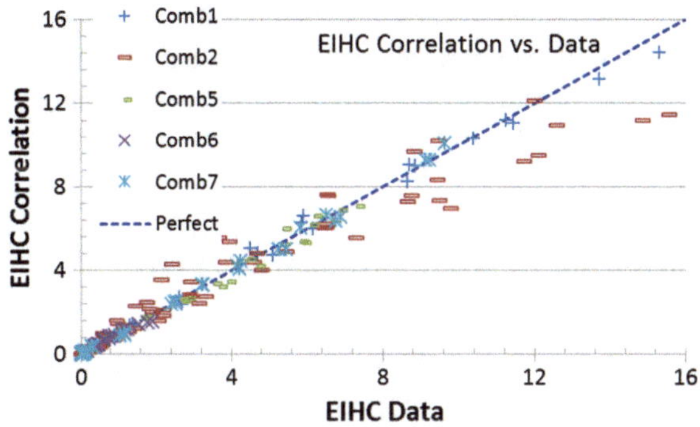

Fig. 48 Comparison between EIHC correlations versus data from idle to maximum takeoff thrust for the modern rich-domes, namely Comb 1, Comb 2, Comb 5, Comb 6, and Comb 7. The range of the root-mean-square errors is 0.07–0.74 EI; *Source* Mongia (2010a, b)

Table 7 Range of P_3, T_3, NO_x, CO, HC and T_{stoich} for applicable correlations of Figs. 46, 47 and 48

	T_3, ^{0}R	P_3, psi	HCEI	COEI	NO_xEl	Tstoich, ^{0}R
Max	1,662	650	43	100	63	4,806
Min	733	36	0	0	3	4,234
Max/Min	2.3	17.9	~430[a]	~1,000[a]	22.3	1.135

[a] Arbitrary, minimum value assumed 0.1 El

minimum, and the ratio max/min in Table 7. In order to avoid diving by the zero values of minimum HC and CO to calculate the max/min ratio, we assumed an arbitrary minimum value of 0.1 EI for HC and CO. It is interesting to note that with the max/min ratios of 2.3 and 17.9 for T_3 and P_3, we get the corresponding ratios of 1.135, 22.3, 430, and 1,000 for T_{stoich}, NO_x, HC, and CO, respectively. This clearly illustrates the complexity of achieving reliable engineering correlations. Nevertheless, it is very interesting and encouraging to see the level of agreement for all the three species (NO_x, CO and HC) in terms of the root-mean-square and maximum errors for EI and %EI as summarized in Table 8. We feel that eventually CFD

Table 8 Root-mean-square (RMSE) and maximum errors for correlating sea-level NO_x, CO, and HC along engine operating lines of the selected five combustors

Combustor		1		2		5		6		7	
		El	%	El	%	El	%	El	%	El	%
No_X	RMSE	0.23	1.21	0.444	3.11	0.34	1.63	0.32	1.47	0.47	1.86
	Max	1.129	3.85	1.404	8.5	1.069	5.5	0.972	3.1	1.726	7.57
CO	RMSE	0.738	3.25	2.474	[a]	0.864	[a]	0.327	[a]	0.488	[a]
	Max	3.063	12.48	13.098	[a]	3.153	[a]	1.4	[a]	1.363	[a]
HC	RMSE	0.188	12.9	0.741	[a]	0.164	[a]	0.068	[a]	0.122	[a]
	Max	0.868	[a]	4.36	[a]	0.621	[a]	0.383	[a]	0.453	[a]

[a] Values too close to zero's making ratios in terms of % not relevant

models will reach the desired accuracy levels. On the downside, the author feels strongly that without CFD support, the technologist and designers' process will be directed toward wrong directions by "experts" so frequently that it reduces down to containing the "brush fire" almost every week; and that is not a pleasure for anyone.

8 Third Alternative to Rich Domes: AGT101

This brief section with four figures (Figs. 49, 50, 51, and 52) is written to share with the reader the feasibility of a low-NO_x combustion concept successfully rig demonstrated for a regenerative automotive gas turbine AGT101 sponsored by a NASA/DOE effort in the early 1980s because two of the three reasons. First, its calculated ignition delay time of 0.19 ms at the most critical operating point (viz. $T_3 = 1{,}328$ K and $P_3 = 201$ kPa) is comparable to 0.27 ms for a 60 OPR hot-day takeoff condition summarized earlier in Table 5. Second, an interesting discussion on the two very different approaches was presented in Mongia (2011b). The author and his counterpart colleague at the Allison Gas Turbines for the AGT100 (Ross et al. 1983) used two very different approaches. Further details on the AGT100 liquid premixed prevaporized combustion system and its CFD analysis are provided by Rizk and Mongia (1990). On the other hand, limited description and data on the AGT101 are provided by Sanborn et al. (1983) with more details and back-to-back comparison for the two approaches given in Mongia (2011b). Third, the AGT101 injector has subsequently been studied by several in the research community (viz. Fureby 2012; Iudiciani et al. 2009; Li and Gutmark 2005, 2006) that should be of interest to combustion technologists.

The AGT101 fuel injector hardware (Fig. 49) and its general layout cross-section, mixer swirler (with high turning radial inflow vanes) followed by a combustor of 4 inch diameter are shown in Fig. 50. Two types of mixer swirlers (viz. radial or combination of smaller radial swirler followed by a dome mounted axial

Fig. 49 A "lean-direct injection" concept based fuel injector from Mongia (2011b); known popularly in the research community as Triple Annular Research Swirlers (*TARS*)

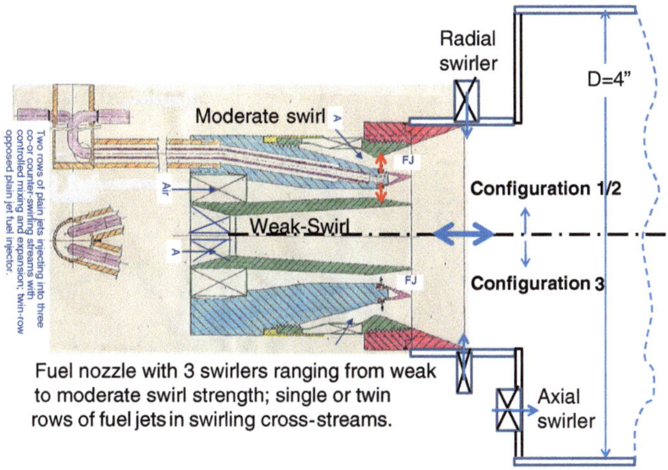

Fig. 50 LDI-based combustor cross-section from Mongia (2011b)

swirlers) were tried; radial inflow swirler closer to exit plane of the premixing tube gave the best results. Typical airflow distribution at the simulated max-max power (T_3 = 1158 K and P_3 = 490 kPa) is given in Fig. 51 along with its CO and NO_x EI characteristics in Fig. 52. Two interesting observations need to be made here. First, its $NO_x/P_3^{0.5}$ value ranges between 0.6 and 1.5 EI/atm$^{0.5}$ which is very competitive with the lowest values discussed earlier concerning Figs. 37 and 44. Second, the mixer survived during extensive rig testing at temperatures exceeding 1328 K with the corresponding ignition delay times less than 0.2 ms even when the mixer may be classified as a premixing and not LDI which we do not agree for the reasons outlined in the following paragraph.

First, we used liquid fuel injection at the throat of two concentric venturis involving both radially inward and outward directions normal to cross swirling flows, JIC. Second, two separate diverging concentric passages followed axially downstream of JIC. This was followed by continuous fuel/air mixing in a single diverging passage until the radial swirler. The swirl strength of the three concentric swirlers was optimized, which starting from center increased from very weak (≤15°), to weak (≤30°), and moderately strong (≤45°) for the outer-mounted injector swirler, as shown in Figs. 49 and 50. The point of disagreement could be

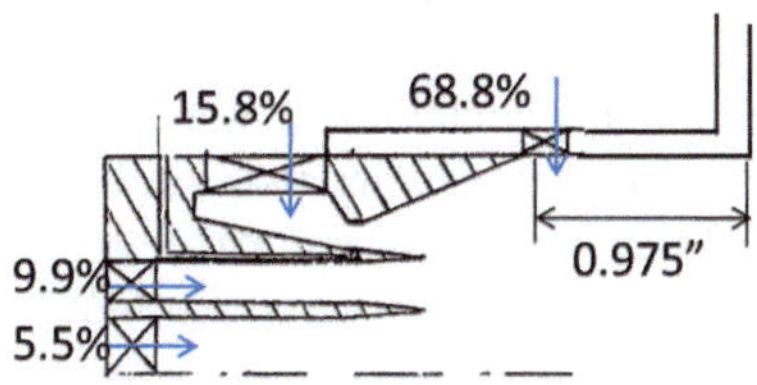

Fig. 51 Air flow distribution of the mixer configuration tested at the simulated max-max operating point; from Mongia (2011b)

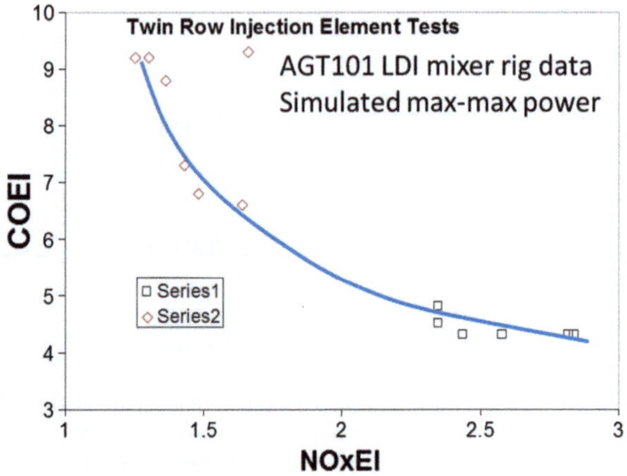

Fig. 52 Measured CO versus NO_x characteristics of the mixer configuration for the airflow distribution shown in Fig. 51 at the simulated max-max operating point; from Mongia (2011b)

the mixing length (viz. 0.975″ in Fig. 51) that prevents it from calling an LDI; but the same argument can be levied against LDI-2 of Fig. 43. We can call it by any name; but it surely gave very low NO_x (by more than an order of magnitude compared to diffusion flame combustors) and survived very high T_3 values.

9 Operability and Dynamics

No article on modern aviation engine combustion products is considered complete without making some comments about operability and combustion dynamics. We have covered operability topic already; and only a few summary statements will be made regarding combustion dynamics in this section. Even though the author finished his graduate studies on unsteady combustion dynamics (Murthy and Mongia 1969; Mongia and Ambs 1974), he was fortunate enough not to run into gas turbine combustion dynamics until 1994; even his first experience on a small industrial engine dry NO_x combustion product before 1994 did not get him to face combustion dynamics. It is good karma because he believes in the statement frequently made by one of his most able co-workers, Tim Held, "No one is expert in dynamics; we all are victims." Several of author's close friends have made seminal contributions; and he is grateful for the several outstanding books on the subject matter (Culick et al.1993; Lieuwen and Yang 2005; Lieuwen 2012), because they have added considerably to his fundamental knowledge base on combustion dynamics. He is happy to see the continuing commitment in the research community including the 2013 Best Paper of the AIAA propellant and Combustion Technical Committee, Caux-Brisebois et al. (2013), and a recent paper by his colleague, Prof. Bill Anderson, Gejji et al. (2014); we will discuss two figures from this paper in this article after summarizing author's perspective on dynamics as a former gas turbine combustion technologist and product leader in the following paragraph that starts by duplicating a statement lifted off from Mongia et al. (2003).

"A fundamental issue in designing gas turbine combustors is the late stage of the development process at which combustion dynamics phenomena become apparent. Many, if not most of the critical performance parameters of a combustor can be determined analytically or through a combination of analysis and component testing. Although some indications of the susceptibility of a combustion system to high levels of dynamics can sometimes be inferred from component tests, the behavior of the full system cannot be predicted on the basis of either analysis or component testing. The severity and character of the problem is not determined until an engine test is conducted, at which point significant changes to component design are very expensive and are likely to have major schedule and development cost impacts. Various acoustic control strategies, both passive and active, are then applied to deal with the unacceptable dynamics levels. However, the application of these strategies is a very empirical process with little assurance of success." Some additional insight is provided in Mongia et al. (2005). Nothing substantial has changed during the last 10 years. The basic methodology first successfully demonstrated during the 1994–1996 time frame for solving dynamics on the industrial engines LM2500 and LM6000 DLE, Fig. 53, as shown schematically in Fig. 54 continues to evolve to meet significantly more demanding requirements of aviation engines combustion products including the CFM DAC and GEnx; and in coming years for the LEAP-X and GE9X. Computationally expensive models have shown some success in identifying potential resonant frequencies that can be used as design guide. Fundamental experiments coupled with computational efforts (Gejji et al. 2014), show enormous complexity of the phenomenon (viz. Figs. 55 and 56), not exactly a morale booster for the technologists and designers. In the meantime, the author along with his colleague Prof. Galen King and three PhD students are working on a research project called proactive state selection in order to bypass the annoying dynamics for the $N + 3$ technology elements (Mongia 2013e).

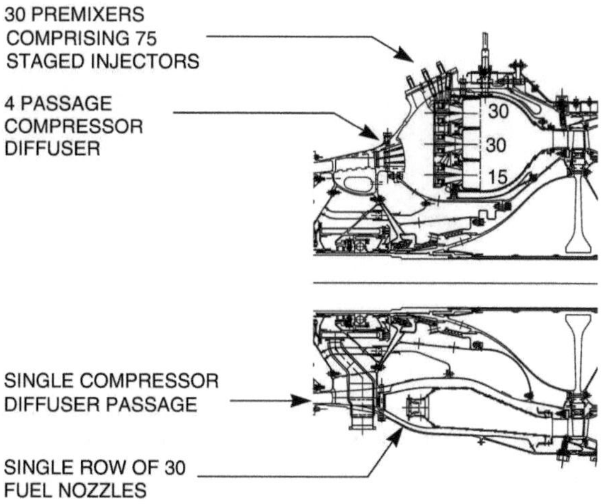

Fig. 53 Schematic comparison between the dry low emissions (*DLE*) (*top*) and rich-dome (*bottom*) aeroderivative industrial engines; adopted from Joshi et al. (1998)

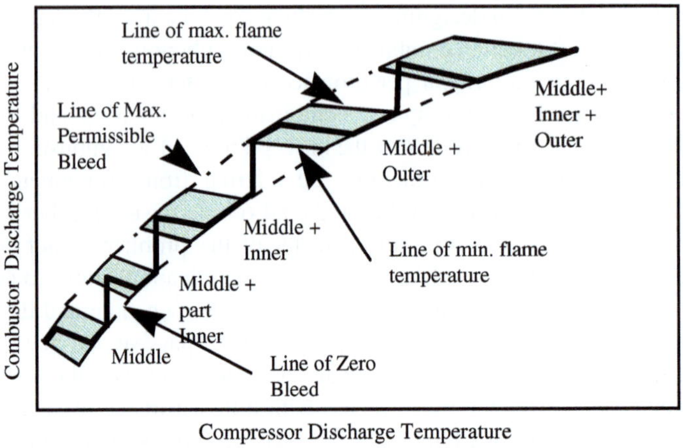

Fig. 54 Staging scheme for the LM2500, LM2500 + and LM6000 combustors and resulting domain of operation; from Joshi et al. (1998)

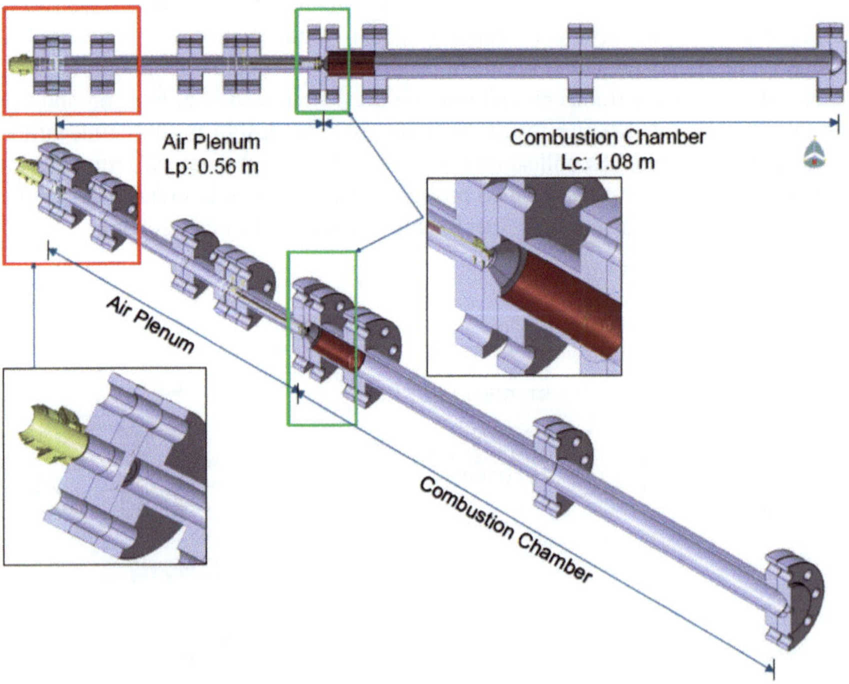

Fig. 55 Schematic of the LDI combustor set-up for conducting controlled dynamic studies conducted by Prof. Bill Anderson's group; reproduced from Gejji et al. (2014)

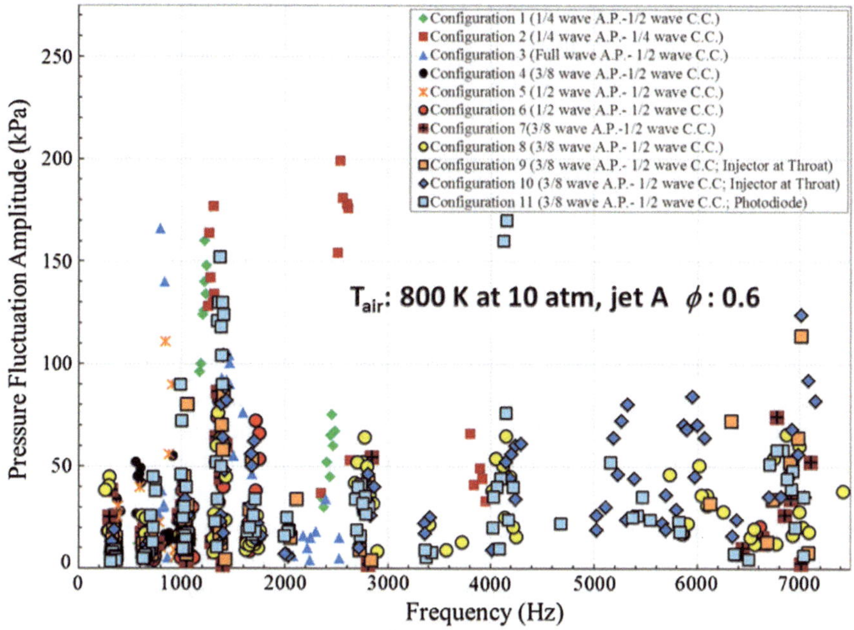

Fig. 56 60^0 single element SV-LDI dynamic investigations in a flametube conducted by Prof. Bill Anderson's group; reproduced from Gejji et al. (2014)

10 Summary

This article gives an overview of the current rich-dome combustion system design, requirements, challenges, three alternatives to rich domes, lean dome NO_x entitlement, alternative fuels, operability, dynamics, and accuracy level expectations from modeling and engineering correlations. Limited by the page count normally associated with any article, the most of the material presented here is from the author's personal experience; and this should not be misinterpreted by the reader for lack of relevance and/or respect for the outstanding quality works being conducted by the numerous combustion research, technology, and product development teams. Several large low-NO_x rich domes' takeoff $EINO_x$ is reproduced well by a simple correlation $EINO_{x\text{RD}-\text{L}} = 0.0303\text{PR}^{1.9722} w/R^2 = 0.9906$ including Talon II and Trent1000. However, the LTO NO_x is correlated well by a similarly good quality curve only for the group of combustors without Trent1000, $\text{DP}/F_{00} = 0.6793\text{PR}^{1.2241}$ compared to lower value expression for Trent1000 alone $\text{DP}/F_{00} = 0.1292\text{PR}^{1.6327}$. Consistent with the NO_x stringency pattern set by CAEP4, CAEP6, and CAEP8 and longtime goal for achieving 85 % reduction in takeoff NO_x at 30 OPR, we propose the long-term LTO regulatory standard of $CAEP/18 = -37.763 + 2\pi$ effective December 31, 2033.

The combustor inlet temperatures for desired overall pressure ratio at sea-level, standard-day, static condition can be estimated by using $T_{3,SSS} = 317.544\ \text{OPR}^{0.2704}$; $320.1955\ \text{OPR}^{0.272}$, respectively, for the $N + 1$ and $N + 2$ generation engines. This along with generally accepted requirements for combustor operability, we have to manage significantly increased range of P_3, T_3, and fuel/air ratio, viz. T_3: 216.67–1,084 K; P_3: 0.33–60 atm; ΔP: ~(0.1–1.2) ΔP_{design}; and FAR: FAR_{min}–FAR_{max}. The numerical values of the dome design pressure drop (ΔP_{design}), and the minimum and maximum fuel air ratios (FAR_{min} and FAR_{max}) depend upon the combustion system design and its potential applications. We will assume typical values of these variables, respectively, 3–5 %, 0.005–008, and 0.025–0.040 for the aviation engines.

The first generation of lean dome products met their original objectives of achieving lower NO_x within the specified design constraints including cooling technology. They were immediately followed by the second generation lean dome products known popularly as TAPS in GEnx in addition to planned LEAP-X and GE9X. The takeoff $EINO_x$ of GEnx is given by $EINO_{xGEnx} = 1.079 \times 10^{-5}\ \text{OPR}^{3.971} w/R^2 = 0.991$. These products will be able to meet the proposed long-term LTO NO_x regulatory standard within the generally accepted design modification and refinement process. TALON-X and recently introduced P&W concept (viz. Lee et al. 2013) will provide credible alternatives to TAPS; an interesting technological competition. However, the second generation lean domes produce an order of magnitude lower exhaust smoke number than the rich domes. For all other design requirements, both the lean and rich domes have comparable characteristics; the dynamics topic will be covered separately. For future technologists, lean dome NO_x entitlement as given in Figs. 36 and 37 are proposed. The effect of FT fuel blends on combustion efficiency, $EINO_x$ is very small; but its benefits in regard to particulate emissions are enormous in terms of both the number density and mass emissions due primarily to significantly lower aromatic and sulfur contents. Future CFD and semi-empirical model should be developed with the proposed long-term accuracy goal expressed in term of the standard deviation σ goal: 3 % of takeoff $EINO_x$, 7.5 and 15 %, respectively, of idle CO and HC emission indices.

Acknowledgments The author would like to express his gratitude for the support and encouragement he has received from his colleagues, friends, customers, bosses, and mentors since 1972.

References

Acosta WA (1987) Liner cooling research at NASA Lewis Research Center. AIAA paper 1987-1828

Ajmani K, Breisacher K (2014) Simulations of NO_x emissions from low emissions discrete jet injector combustor tests. AIAA paper 2014-3524

Ajmani K, Mongia HC, Lee P (2013a) Evaluation of CFD best practices for combustor design: part I—non-reacting flows. AIAA paper 2013-1144

Ajmani K, Mongia HC, Lee P (2013b) CFD best practices to predict NO_x, CO and lean blowout for combustor design. GT2013-9566
Ajmani K, Mongia HC, Lee P (2013c) Evaluation of a time-filtered Navier stokes (TFNS) CFD approach for LDI combustor analysis. AIAA paper 2013-3688
Ajmani K, Mongia H, Lee P (2014) CFD computations of emissions for LDI-2 combustors with simplex and air blast injectors. AIAA paper 2014-3529
Anonymous (1982) Design documentation report, counterflow film-cooled combustor program. Report 21-4007-A, Garrett Turbine Engine Co., Phoenix, AZ, June 1982, (NASA CR-167922)
Arvai ES (2011). http://airinsight.com/2011/11/09/comparing-the-new-technology-narrow-body-engines-gtf-vs-leap-maintenance-costs/#.U8KShPldU00. Accessed on 9 Nov 2011
Bahr DW, Gleason CC (1973) Experimental clean combustor program, phase I—final report. NASA CR-134737, July 1973.
Bahr DW, Gleason CC (1975) Experimental clean combustor program, phase I—final report. NASA CR-134737
Bester N, Yates A (2009) Assessment of the operational performance of fischer–tropsch synthetic–paraffinic kerosene in a T63 gas turbine compared to conventional Jet A-1 fuel. GT2009-60333
Bruce TW, Davis FG, Kuhn TE, Mongia HC (1977) Pollution reduction technology program small jet engines (PRTP) phase I final report. NASA CR-135214, Sept 1977
Bruce TW, Davis FG, Kuhn TE, Mongia HC (1978) PRTP phase II final report. NASA CR-159415, Sept 1978
Bruce TW, Davis FG, Kuhn TE, Mongia HC (1981) PRTP phase III final report. NASA CR-165386, Dec 1981
Bulzan D et al (2010) Gaseous and particulate emissions results of the NASA alternative aviation fuel experiment (AAFEX). GT2010-23524
Burrus DL et al (1984) Energy efficient engine combustion system component technology performance report. NASA CR-168274, July 1984
Carter NA et al (2011) Energy and environmental viability of select alternative jet fuel pathways. AIAA 2011-5968
Chishty WA et al (2011) Emissions assessment of alternative aviation fuel at simulated altitudes. GT2011-45113
Caux-Brisebois V et al (2013) Thermo-acoustic coupling in swirl-stabilized flames with helical vortices. AIAA Paper 2013-3650
Corporan E, DeWitt MJ, Wagner M (2004) Evaluation of soot particulate mitigation additives in a T63 engine. Fuel Process Technol. 85:727–742
Culick F, Heitor MV, Whitelaw JH (eds) (1993) Unsteady Combustion. In: Proceedings of the NATO Advanced Study Institute. Series: NATO science series E, vol 306, Praia da Granja, Portugal, 6–17 Sept 1993
Danis AM, Pritchard BA, Mongia HC (1996) Empirical and semi-empirical correlation of emissions data from modern turbo propulsion gas turbine engines. ASME paper 96-GT-0186
Davison CR, Chisty WA (2011) Altitude performance of a turbojet with alternate fuels. GT2011-45132
Diehl LA, Biaglow JA (1974) Measurement of gaseous emissions from a turbofan engine at simulated altitude conditions. NASA TM X-3046, April 1974
Dolan B et al (2013) Biofuel emissions of a multipoint staged low NO_X combustion system at intermediate pressures. AIAA paper 2013-3651
Dryer FL (2014) Chemical kinetic and combustion characteristics of transportation fuels. In: Plenary paper, combustion (international) symposium
Dubiel DJ (1986) Energy efficient engine combustion system component technology performance program. NASA CR-179533, Sept 1986
Fear JS (1976) The NASA pollution-reduction technology program for small jet aircraft engines—a status report. NASA-TM-X-73419, July 1976
Foust MJ, Thomsen D, Stickles R, Cooper C, Dodds W (2012) Development of the GE aviation low emissions TAPS combustor for next generation aircraft engines. AIAA-2012-0936

Fureby C (2012) A comparative study of flamelet and finite rate chemistry LES for a swirl stabilized flame. J Eng Gas Turbines Power 134: 041503-1–041503-13

Gejji R, Huang C, Yoon C, Anderson W (2014) A parametric study of combustion dynamics in a single-element lean direct injection (LDI) gas turbine combustor. AIAA paper 2014-133

Gleason CC, Bahr DW (1979) Experimental clean combustor program, phase III—final report. NASA CR-135384, June 1979

Gleason CC, Rogers DW, Bahr DW (1975) Experimental clean combustor program, phase II—final report. NASA CR-134971, July 1975

Gupta AK et al (1991) Burner geometry effects on combustion and NO_X emission characteristics using a variable geometry swirl combustor. J Propul 7(4):473–480

Iudiciani P et al (2009) Characterization of a multi-swirler fuel injector using simultaneous laser based planar measurements of reaction zone, flow field, and fuel distribution, 2009. GT2009-60278

Johansen KM, Johnston BH, Mongia HC, Sanborn JW (1977) Combustion process testing for reduced wall temperature gradients. In: Proceedings 1977 DARPA/NAVSEA ceramic gas turbine demonstration engine program review

Joshi ND, Mongia HC, Leonard G, Steggmaier JW, Vickers EC (1998) Dry low emissions combustor development. ASME paper 98-GT-310

Kumar V, Panda P, Mongia H, Naik S (2012a) Innovative approaches for reducing CO_2 emissions of aviation engines part 1: selection of promising approaches. AIAA Paper 2012-3951

Kumar V, Panda P, Mongia H, Naik S (2012b) Innovative approaches for reducing CO_2 emissions of aviation engines part 2: NPSS calibration with existing gas turbine engines. AIAA paper 2012-4228

Kumar V, Panda P, Mongia H, Naik S (2012c) Innovative approaches for reducing CO_2 emissions of aviation engines part 3: advanced Brayton cycle optimization. AIAA paper 2012-4175

Kurzke J (2009) Fundamental differences between conventional and geared turbofans. GT2009-59745

Lee CM, Chang C, Kramer S, Herbon JT (2013) NASA project develops next generation low-emissions combustor technologies. AIAA paper 2013-540

Li G, Gutmark E (2005) Effect of exhaust nozzle geometry on combustor flow field and combustion characteristics. Proc. Comb. Inst. 30:2893–2901

Li G, Gutmark E (2006) Experimental study of boundary conditions effects on non-reacting and reacting flows in a multi-swirl gas turbine combustor. AIAA J 44(3):444–456

Lieuwen T (2012) Unsteady combustor physics. Cambridge Press, Cambridge

Lieuwen TC, Yang V (Eds) (2005) Progress in Astronautics and Aeronautics, vol 210

Longwell JP (1977) Synthetic fuels and combustion. Prog Energy Combust Sci 3:127–138

M'Bengue L (2010) Toward ACARE 2020: innovative engine architectures to achieve the environmental goals? In: 27th international congress of the aeronautical sciences

McKinney RG, Sepulveda D, Sowa W, Cheung AK (2007) The Pratt &Whitney TALON X low emissions combustor: revolutionary results with evolutionary technology. AIAA 2007-386

Mongia HC (1993) Application of CFD in combustor design technology. AGARD CP-536, pp 12-1/12-18

Mongia HC (1997) Recent progress in low-emissions gas turbine combustors. In: 13th international symposium on air breathing engines, ISABE 1997

Mongia HC (2001a) Gas turbine combustor liner wall temperature calculation methodology. AIAA 2001-3267

Mongia HC (2001b) A synopsis of gas turbine combustor design methodology evolution of last 25 years. ISABE-2001-1086

Mongia HC (2003) TAPS—a 4th generation propulsion combustor technology for low emissions. AIAA paper 2003-2657

Mongia HC (2008) Recent progress in comprehensive modeling of gas turbine combustion. AIAA paper 2008-1445

Mongia HC (2010a) On initiating 3rd generation of correlations for gaseous emissions of aero-propulsion engines. AIAA paper 2010-1529

Mongia HC (2010b) Correlations for gaseous emissions of aero-propulsion engines from sea-level to cruise operation. AIAA paper 2010-1530

Mongia HC (2011a) Engineering aspects of complex gas turbine combustion mixers part I: high ΔT. AIAA paper 2011-107

Mongia HC (2011b) Engineering aspects of complex gas turbine combustion mixers part II: high T_3. AIAA paper 2011-106

Mongia HC (2011c) Engineering aspects of complex gas turbine combustion mixers part III: 30 OPR. AIAA paper 2011-5525

Mongia HC (2011d) Engineering aspects of complex gas turbine combustion mixers part IV: swirl cup. AIAA paper 2011-5526

Mongia HC (2011e) Engineering aspects of complex gas turbine combustion mixers part V: 40 OPR. AIAA paper 2011-5527

Mongia HC (2013a) N+ 3 and N+ 4 generation aeropropulsion engine combustors part 1: large engines' emissions. GT2013-94570

Mongia HC (2013b) N+ 3 and N+ 4 generation aeropropulsion engine combustors part 2: medium size rich-dome engines and lean-domes. GT2013-94571

Mongia HC (2013c) N+ 3 and N+ 4 Generation aeropropulsion engine combustors part 3: small engine emissions and axial staging combustion technology. GT2013-94572

Mongia HC (2013d) N+ 3 and N+ 4 generation aeropropulsion engine combustors part 5: NO_x, CO, HC and smoke emissions. AIAA paper 2013-3653

Mongia HC (2013e) N+ 3 and N+ 4 generation aeropropulsion engine combustors part 6: operating conditions, target goals and lifted jets. AIAA paper 2013-3654

Mongia HC, Ambs LL (1974) A model for the combustion and extinction of composite solid propellants during depressurization. Combust Flame 22:59–69

Mongia HC, Smith KF (1978) An empirical/analytical design methodology for gas turbine combustor. AIAA Paper 78-998

Mongia HC, Reynolds RS, Srinivasan R (1986) Multidimensional gas turbine combustion modeling: applications and limitations. AIAA J 24(6):890–904

Mongia HC, Held TJ, Hsiao GC, Pandalai RP (2003) Challenges and progress in controlling dynamics in gas turbine combustors. J Propul Power 19:822–829

Mongia H, Hsiao G, Burrus D, Sreedhar PSVS, Rao A, Naik P (2004a) Combustor diffuser modeling part I: inlet profiles and 2-D calculations. AIAA paper 2004-4168

Mongia H, Hsiao G, Burrus D, Sreedhar PSVS, Rao A (2004b) Combustor diffuser modeling part II: inlet profiles and 3-D calculations. AIAA paper 2004-4169

Mongia H, Hsiao G, Burrus D, Habeeb K, Sreedhar PSVS (2004c) Combustor diffuser modeling part III: validation w/typical separating single passage diffusers combustor diffuser modeling. AIAA paper 2004-4170

Mongia H, Hsiao G, Burrus D, Sreedhar PSVS, Rao A (2004d) Combustor diffuser modeling part IV: effect of cowling geometry, mixer size and nozzle blockage. AIAA paper 2004-4171

Mongia H, Hsiao G, Ravindra MVV, Sreedhar PSVS (2004e) Combustor diffuser modeling part V: validation with a three passage diffuser rig data. AIAA paper 2004-4172

Mongia H, Hsiao G, Mueller M, Dai Z, Ravindra MVV, Sreedhar PSVS, Arvind P (2004f) Combustor diffuser modeling part VI: validation with a four passage diffuser rig data. AIAA paper 2004-4173

Mongia HC, Held TJ, Hsiao GC, Pandalai RP (2005) Incorporation of combustion instability issues into design process: GE aeroderivative and aero engines experience. In: Lieuwen TC, Yang V (eds) Progress in astronautics and aeronautics, vo 210. pp 43–64

Moses CA, Roets PN (2008) Properties, characteristics and combustion performance of SASOL fully synthetic jet fuel. GT2008-50545

Murthy SNB, Mongia HC (1969) Progressive acoustic and weak shock waves in AI, Be and B flames. AIAA paper 1969-540

Nealy DA, Reider SB, Mongia HC (1985) Alternate cooling configuration for gas turbine combustion systems. In: AGARD CP-390, pp 25-1–25-15

Panda P, Kumar V, Mongia H, Naik S (2012a) Innovative approaches for reducing CO_2 emissions of aviation engines part 4: turbine exhaust driven thermal cycle TED-T. AIAA paper 2012-4230

Panda P, Kumar V, Mongia H, Naik S (2012b) Innovative approaches for reducing CO_2 emissions of aviation engines part 5: turbo reaction-controlled internal combustion system TRCS. AIAA paper 2012-3952

Paskin MD, Ross PT, Mongia HC, Acosta WA (1990) Composite matrix cooling scheme for small gas turbine combustors. AIAA paper 1990-2158

Paskin MD, Mongia HC, Acosta WA (1993) An efficient liner cooling scheme for advanced small gas turbine combustors. AIAA paper 1993-1763

Rida S et al (2012) Imprinted effusion modeling and dynamic CD calculation in gas turbine combustors. GT2012-68804

Rizk NK, Mongia HC (1990) Lean low NO_x combustion concept evaluation. In: 23rd international symposium on combustion. The Combustion Institute, pp 1063–1070

Roberts R, Fiorentino A, Greene W (1977) Experimental clean combustor program phase III—final report. NASA CR-135253, Oct 1977

Ross PT, Williams JR, Anderson DA (1983) AIAA J Energy 7:429

Sanborn JW, Mongia HC, Kidwell JR (1983) Design of a low-emission combustor for an automotive gas turbine. AIAA paper 83-0338

Segalman I, McKinney RG, Sturgess GJ, Huang LM (1993) Reduction of NO_x by fuel-staging in gas turbine engines-A commitment to the future. In: AGARD Conference Proceedings

Sekar J et al (2014) Liquid jet in cross flow modeling. GT2014-26124

Sen BA et al (2012) Pratt and Whitney gas turbine combustor design using ANSYS fluent and user defined functions. GT2012-70145

Snecma (2014). http://www.snecma.com/-leap-x,381-.html?lang=en. Accessed 13 July 2014

Sung CJ (2014) Personal communication

Sripathi M et al (2014) Laminar flamelet based NO_x predictions for gas turbine combustors. GT2014-27258

Stearns EM et al (1982) Energy efficient engine core design and performance report. NASA CR-168069, Dec 1982

Tacina RR (1990) Low NO_x potential of gas turbine engines. AIAA paper 90-0550

Tacina R, Mao C-P, Wey C (2004) Experimental investigation of a multiplex fuel injector module with discrete jet swirlers for low emissions combustors. AIAA-2004-0135

Tacina R, Lee P, Wey C (2005a) A lean-direct-injection combustor using a 9 point swirl-venturi fuel injector. ISABE-2005-1106

Tacina R, Wey C, Laing P, Mansour A (2005b) Low-NO_x lean-direct injection, multi-point integrated module combustor concept for advanced aircraft gas turbines. NASA/TMX 2002-211347

Tacina KM, Chang CT, He ZJ, Lee P, Dam B, Mongia H (2014) A second generation swirl-venturi lean direct injection combustion concept. AIAA paper 2014–3434

Therkelsen PL, Littlejohn D, Cheng RK (2012) Parametric study of low-swirl injector geometry on its operability. GT2012-68436

Villalva R et al (2013) Medium pressure emissions of a multi-point low NO_x combustion system. AIAA paper 2013-1044

Wey C, Bulzan D (2013) Effects of bio-derived fuels on emissions and performance using a 9-point lean direct injection low emissions concept. GT2013-95135

Spectroscopic Methods and Visualization Applied to Combustion Diagnoses

Kuniyuki Kitagawa

Abstract Recently, it has well been realized that from the viewpoints of depletion of fossil fuels and global warming effect along with increasing demand for electricity and thermal energy, the energy issue is currently among the most important ones. The key strategies in this regard include the development of renewable energy and increase in energy system efficiency.

Keywords Optical diagnostics · Spectroscopic temperature profile · Soot temperature profile · High-temperature air combustion · Premixed and diffusion flames

1 Introduction

Recently, it has well been realized that from the viewpoints of depletion of fossil fuels and global warming effect along with increasing demand for electricity and thermal energy, the energy issue is currently among the most important ones. The key strategies in this regard include the development of renewable energy and increase in energy system efficiency.

Combustion systems are nowadays one of the key technologies for energy conversion from chemical potentials in fossil fuels and biomass as well to heat and/or electric power. Two important issues for combustion systems to be environmentally benign are how to increase the conversion efficiency and to decrease toxic emission. A key issue to this is to unravel combustion. However, its phenomena are extremely complicated with many chemical reactions as well as mass and heat transfers. Numerical simulations are therefore sometimes inadequate, and experimental results are necessary for their validations.

The key parameters involved in combustion phenomena include temperature, mass (chemical molecules and particles) and heat flows, mass concentrations.

K. Kitagawa (✉)
Nagoya University, Furo-Cho, Chikusa-Ku, Nagoya 464-8603, Japan
e-mail: a41596a@yahoo.co.jp

A.K. Agarwal et al. (eds.), *Novel Combustion Concepts for Sustainable Energy Development*, DOI 10.1007/978-81-322-2211-8_8

In addition, spatial distributions, or profiles of these parameters, and their temporal changes are very important for unraveling combustion phenomena and improving the system efficiency and suppressing toxic emission.

For this purpose, measurement methods are inevitable for monitoring spatial distributions of temperature and intermediate chemical species in flames during combustion.

Spectroscopic methods and 2D visualization are powerful and promising tools to meet the requirement described above. In this chapter, several different spectroscopic methods and visualizations, and their applications to combustion systems will be described. The spectroscopic applications here include spontaneous optical molecular emission spectroscopy or spectrometry (OMES or MES), planar laser-induced fluorescence spectroscopy (PLIF), and laser ionization mass spectrometry (LI-MASS).

1.1 Spontaneous Emission Spectroscopy and Visualization and Application to Ecological Industrial Furnace High-Temperature Air Combustion (HITAC) (Hino et al. 2004; Shimada et al. 2005)

Flames are chemical reaction fields to maintain high temperatures and excite chemical species to higher energy levels. Consequently, the excited species, or molecules, and atoms emit spontaneously radiation when they return to their stable lower or ground states. In other word, flames at high temperatures are regarded as thermal excitation source and have been used for flame photometry for trace analyses of alkali elements, alkaline elements, sulfur, phosphorus, etc. On the other hand, spontaneous emission from intermediate chemical species such as OH, CH, and C_2 gives us useful information for combustion conditions. Gray body emission from soot in combustion flames enables us to measure flame temperatures spectroscopically. Since the measurement systems for spontaneous emission spectrometry, or molecular/atomic emission spectrometry, simply consist of a converging lens and spectrometer, they are suitable for on-site measurements of, e.g., industrial furnaces. When the spectrometer is replaced by a ***spectro-camera*** (Kubota et al. 1998), a monochromator or an optical bandpass filter attached to a CCD camera, spatial resolution of spectral information becomes available. Furthermore, a ***spectro-video camera*** is used instead, and time resolution is given. Thus, we can obtain spectral, spatial, and time resolutions. In this chapter, such a systems were applied to measurements of combustion temperatures and chemical species during combustion in a regenerative industrial furnace used for reheating of steal slabs. The furnace is based on a high-temperature air combustion technology, enabling us to save energy consumption up to 30 %, leading to less CO_2 emission. This new type of combustion had not been unraveled in conjunction with temperature uniformity and low NO_x emission.

1.1.1 Spectroscopic Temperatures

Regarding with heat transfer and toxic emission, temperature is among the most important physical properties for combustion science. Temperature distributions in combustion flames have conventionally been measured by thermocouples placed in different positions. However, several disadvantages are encountered in measurement, such as disturbance of combustion flows, errors due to gray body emission, low spatial resolution, due to the limited number of thermocouples etc. In addition, thermocouples have no capability to discriminate temperatures of combustion and non-combustion fields, e.g., inside flames and surrounding air.

From those viewpoints, spectroscopic measurements have advantages over thermocouple measurements, as well as no limited temperature ranges. However, careful selection of spectral lines and/or bands should be made to obtain accurate results.

In this chapter, spectroscopic measurements of soot temperature and molecular temperature of an intermediate species, C_2, during combustion will be described.

1.1.2 Soot Temperature

The temperature on the surfaces of soot particles is determined by the solid lattice vibration, or graphite lattice typically, and the gray body emission intensity at a wavelength λ is well known to be expressed as a function of soot temperature, T_s by the Planck's law as follows:

$$I_\lambda = (8\pi hc\,\varepsilon(T_s,\lambda)\lambda^{-5})/\{\exp(ch/\lambda kT_s) - 1\} \tag{1}$$

where h is the Planck's constant, c the light velocity, ε the emissivity, and k the Boltzmann's constant. Taking a ratio of the intensities at two different wavelengths, or two colors, we obtain the expression for well-known two-color thermometry.

$$I_{\lambda 1}\big/I_{\lambda 2} = \{\varepsilon(T_s,\lambda_1)/\varepsilon(T_s,\lambda_2)\}(\lambda_2/\lambda_1)^5\{\exp(ch/\lambda_2 KT_s) - 1\}\big/\{\exp(ch/\lambda_1 kT_s) - 1\} \tag{2}$$

The temperature is calculated from this equation.

The issue here is to cancel the emissivity term, $\varepsilon(T,\lambda_1)/\varepsilon(T,\lambda_2)$. If the difference in the wavelength between the two colors, the two emissivity is not identical in magnitude and the ratio is variable dependent on the temperature, leading to the errors in the obtained temperature. If the wavelengths are too close, the measurement precision is low since the exponential terms becomes insensitive to the temperature. Thus, optimized selection should be made, also avoiding the structured spectra from chemical species, such as molecular bands from OH, CH, and C_2 by using narrowband optical filters. We named this method taking these issues into account, as a narrowband two-color method (Kubota et al. 1998). The filter pair should be selected after careful examination of structured spectra from combustion fields. Soot temperature measurement is suitable for radiant flames of high equivalence ratios, and/or high-carbon fuels such as heavy oil.

1.2 C_2 Vibrational Temperature (Kitagawa et al. 2003)

We have developed a method for spectroscopic temperature measurements of gas temperature during hydrocarbon combustion on the basis of two-color principle by using the two-vibrational branches of the Swan band emission from C_2 radical (Hino et al. 2004). The obtainable temperature is vibrational temperature of this species.

The ratio of the intensities integrated over the two-color bands, I_1/I_2, is given by the following expression.

$$I_1/I_2 = \sum_i (A_i/\lambda_i)/\exp(-E_i/kT_{vib}) \Big/ \sum_j (A_j/\lambda_j)/\exp(-E_j/kT_{vib}) \quad (3)$$

Here, A is the Einstein's probability for spontaneous emission of vibrational transition, T_{vib} the vibrational temperature, and the subscripts i and j refer for the individual vibrational spectrum line of the two bands, respectively. This method is suitable for spectroscopic temperature measurements of non- /semi-radiant combustion flames (blue flames or green flames during HITAC) of low equivalence ratios and low carbon fuels such as natural gas.

2 Experimental

Figures 1 and 2 show the schematic diagrams of the regenerative test furnace developed for preheating of steel slabs prior to rolling process and the measurement system for spectroscopic temperature profiles, constructed in front of the furnace, respectively.

The test furnace had been developed as a new type system in an NEDO project and located at a steel making plant site of NKK Co. (currently JEF Co.) and has a pair of burners which is alternatively switched at an interval, e.g., 1 min. The exhausted gas, from the burner B, is fed to the heat accumulator of the partner burner A. The fresh air and part of the exhausted gas are preheated through the accumulator up to 1,200 °C prior to combustion in the furnace. This high-temperature air combustion (HITAC) (Kitagawa et al. 2003) has a great advantage of capability to attain energy saving up to 30 % that is a green technology. The issues taken into account include thermal NO_x emission expected due to the excess enthalpy during combustion. However, the HITAC combustion can be maintain with low oxygen concentration down to 5 %, thus lowering NO_x was expected. Two types, diffusion and premixed combustions, were tested. Spectroscopical profilings of the soot and C_2 temperature and chemical species were made in conjunction with the NO_x emission.

The image of the combustion flame is converged by 1/18 through the lens, divided into two paths, and each beam pass through the two-color (or wavelength) bandpass

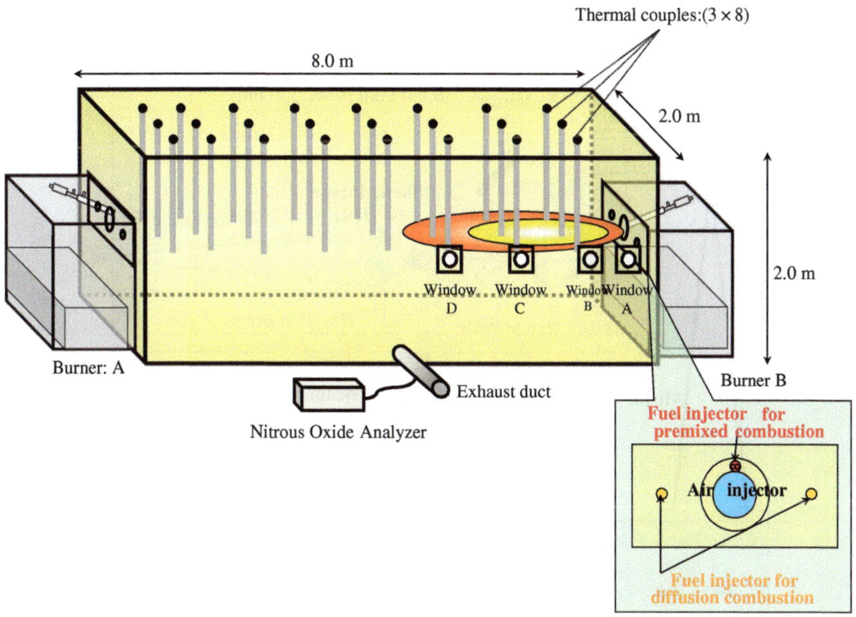

Fig. 1 Schematic diagram of regenerative test furnace for steel slub preheating. Fuel: heavy oils or natural gas

filters, merged again in the view field of the video camera through another half mirror after reflected. Either mirror was slightly tilted to separate the two-color images in the view field. Each pixel data obtained from the video camera was processed to obtain temperature by post-numerical processing based on Eqs. (1) and (2).

3 Result and Discussion

Figures 3 and 4 show the one-shot profiles of soot temperature observed through the window A (upstream) and B (middle stream) during the premixedand diffusion combustions, respectively.

It is clearly demonstrated in the figure that during the premixed combustion the high-temperature zone (shown in white) is observed around the left edge of the window A. This high-temperature zone was found to be very steady, leading to the high level of NO_x emission over 120 ppm. Near the injector, premixed gas undergoes very fast combustion reaction, leading to the high-temperature zone, leading to high NO_x emission.

On the other hand, the high-temperature zone is not encountered in the upstream, during the diffusion combustion. It was found from a series of profile data that the

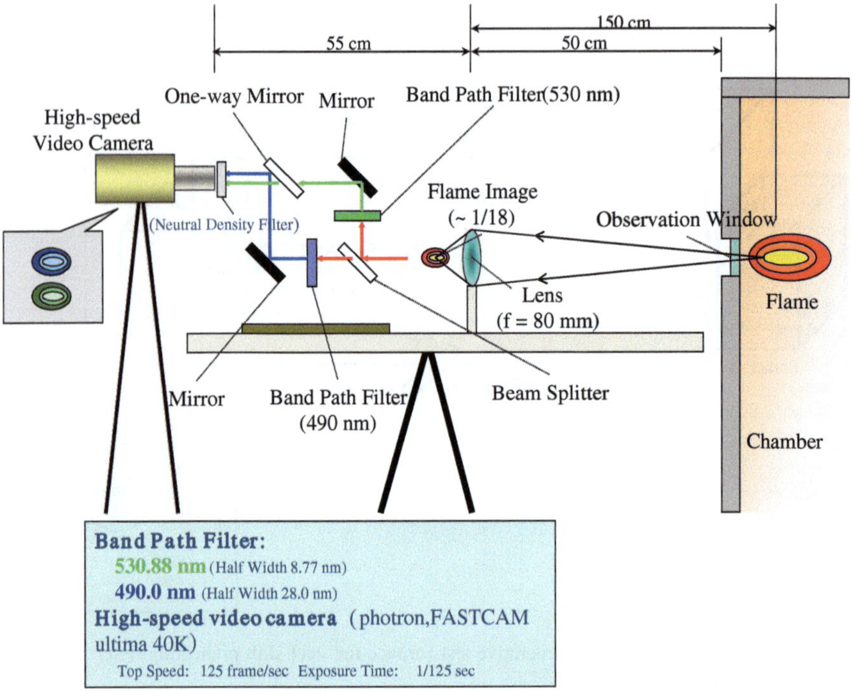

Fig. 2 Schematic diagram of the temperature profile measurement

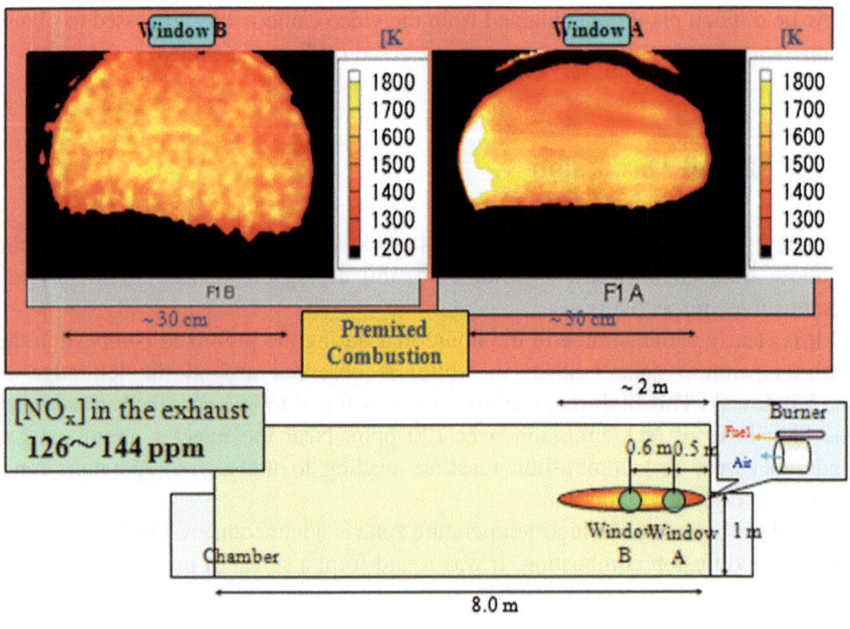

Fig. 3 Soot temperature profiles during premixed HITAC. Fuel: heavy oil

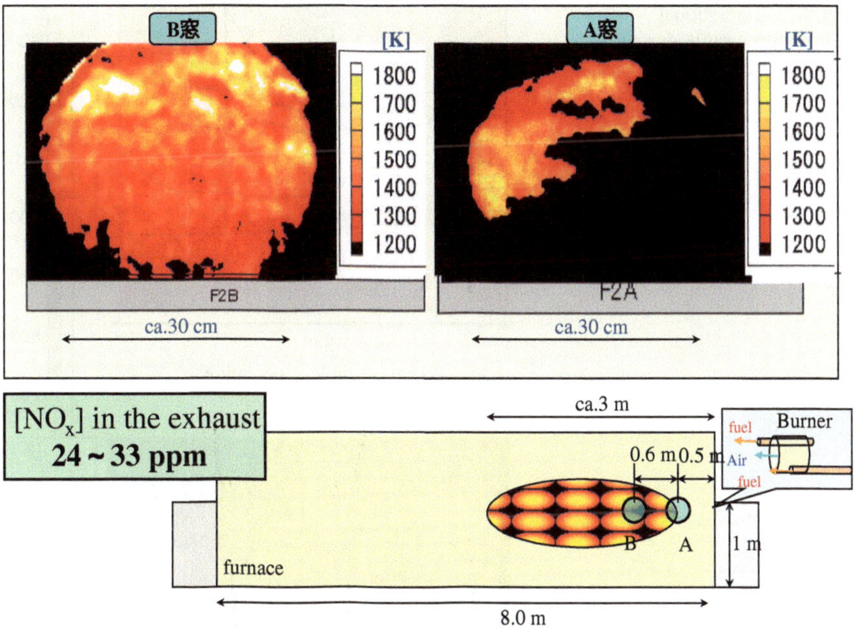

Fig. 4 Soot temperature profiles during diffusion HITAC

high temperature island was not steady but instantaneous leading to the low NO_x level. During the diffusion combustion, the fuel and air are injected from distant nozzles, followed by slow reactions due to a low oxygen concentration along with mixing the gases.

Spectroscopic temperature measurements were made for non- or semi-radiant combustion flames. Figures 5 and 6 show the C_2 vibrational temperature profiles during coal gas premixed and diffusion combustions together with the thermocouple temperature profiles, respectively.

The spatial resolutions of the spectroscopic temperature profiles are much higher than those obtained with the thermocouples. As in case of the heavy oil combustion, highly structured temperature distribution is encountered during the premixed HITAC, high-temperature zone localized around the injector while the temperature uniformity is seen during the diffusion one.

The origin of the difference in the NO_x level between the premixed and diffusion HITACs was also surveyed by directly measuring of spontaneous emission of radiation from thermally excited NO molecules in the HITAC field. Figure 7 shows the spectral intensity profiles of NO spontaneous emission of radiation.

The combustion conditions are the same as in Figs. 5 and 6.

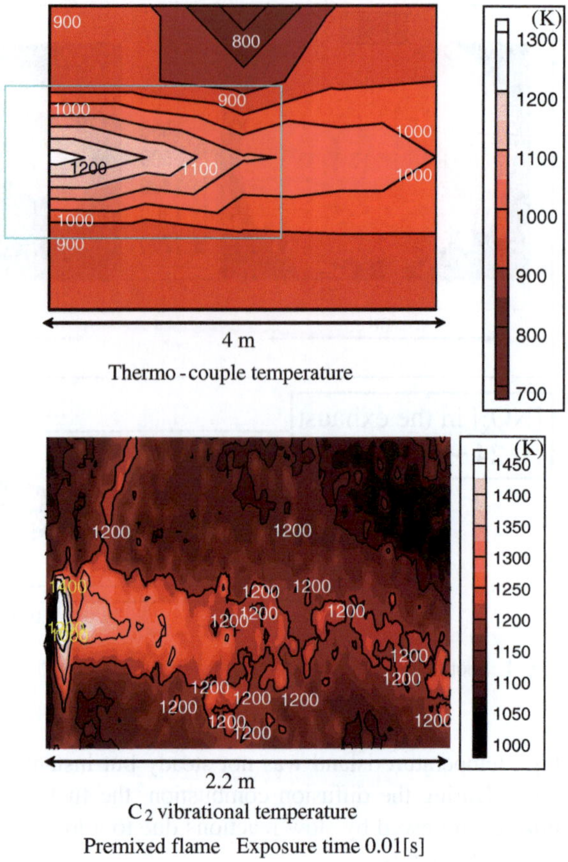

Fig. 5 C_2 vibrational temperature profile during premixed HITAC. Fuel: coal gas

3.1 Applications of Planer Laser-Induced Fluorescence Spectroscopy (PLIF)

Spontaneous emission spectroscopy and visualization is a simple method as described above and suitable for on-site measurements of industrial combustion system. For more scientific research, this has such disadvantages as the measured image is constructed from integrated emission intensity of radiation along the view line and also dependent on the combustion temperature. Laser-induced Fluorescence Spectroscopy (LIF) is a powerful tool to overcome these problems though the systems are more sophisticated and expensive. When Q-branch lines of vibrational spectra are selected for the excitation lines, the temperature dependence is minimized. The excitation laser beam is made in a plane and induces cross-sectional fluorescence in combustion flames. This method is called as planar laser-induced fluorescence (PLIF). In this chapter, two application will be described: 2D monitoring of OH for

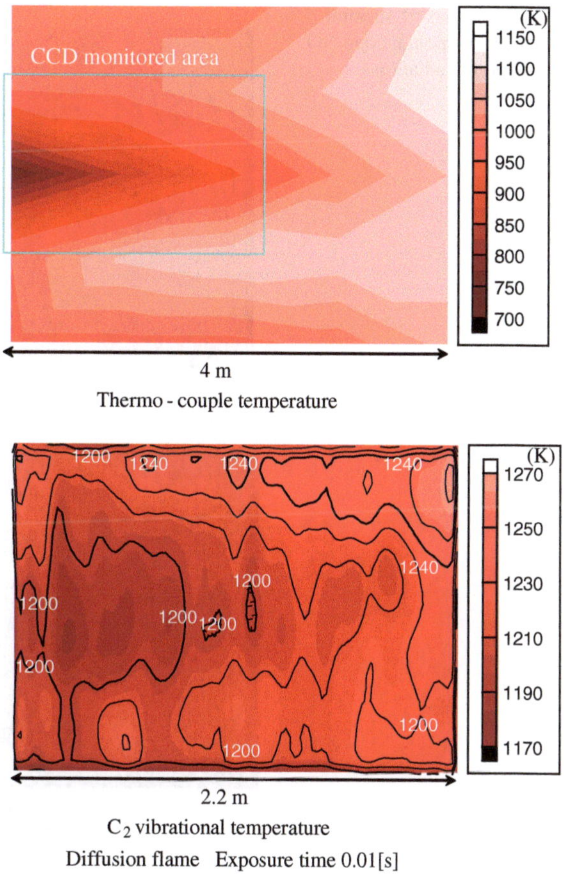

Fig. 6 C_2 vibrational temperature profile during diffusion HITAC. Fuel: coal gas

the evaluation of a steam addition effect during methane–air combustion by utilizing isotope shift and 2D monitoring of redox (reductive-oxidative) environment during acetylene–air combustion.

3.1.1 2D Monitoring of OH for the Evaluation of a Steam Addition Effect During Hydrogen–Air Combustion by Utilizing Isotope Shift PLIF (Kajimoto et al. 2013)

Mixing of fuels and oxidizer is among the key issues to attain high-combustion efficiency for decreasing soot and CO emission. In this research, Mixing conditions was monitored by PLIF measurements of OH and OD. Figure 8 shows the schematic diagram of the PLIF system utilizing isotope effect. The PLIF system is a commercially available conventional apparatus. For monitoring the mixing condition, heavy water (D_2O) steam is introduced in the airflow entering in the outer ring of the coaxial diffusion burner. OD radicals are produced form thermal dissociation

Fig. 7 Profiles of spectral intensity of spontaneous NO emission of radiation

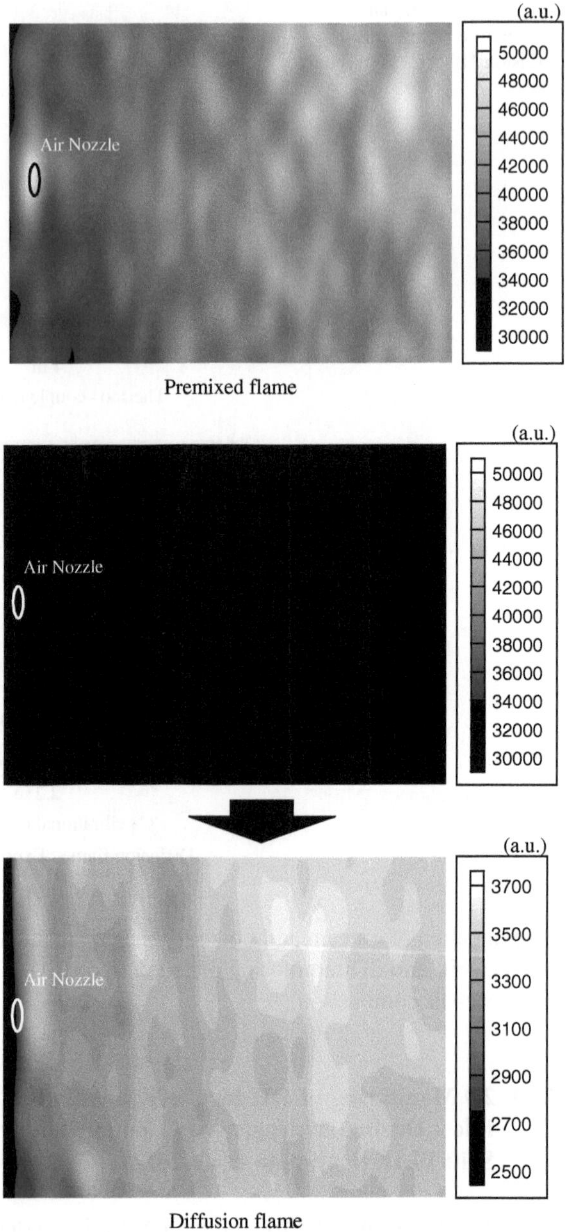

of D_2O. On the other hand, the fuel H_2 flows out of the inner tube, followed by combustion reaction to produce OH. Thus, OH and OD are the indicator of the two flows, and mixing condition is visualized by PLIF monitoring of OH and OD.

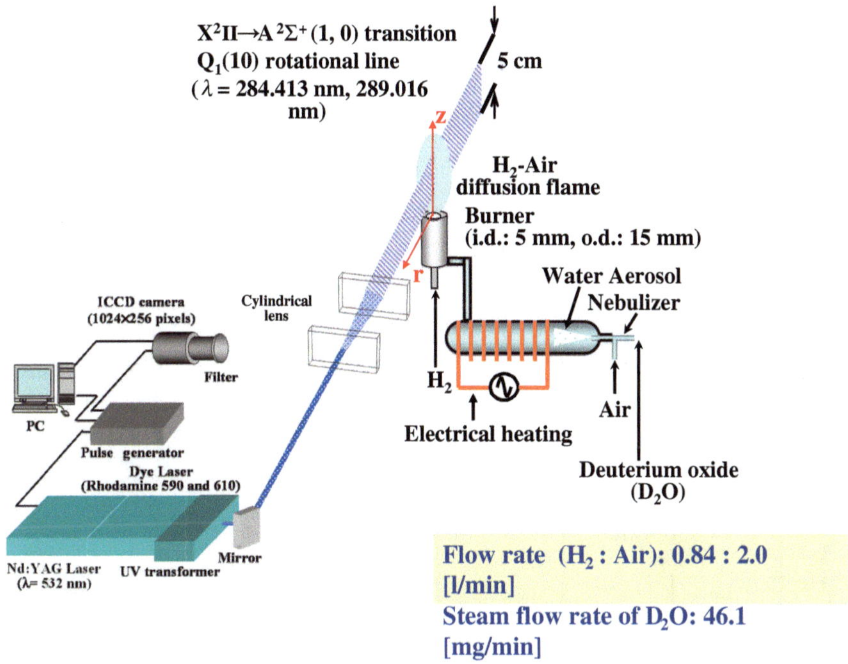

Fig. 8 The schematic diagram of the PLIF system utilizing isotope effect

Figure 9 shows the isotope effect for the rotational lines $Q_1(10)$ of OH and OD caused by the difference in atomic mass between H and D. When the dye laser is tuned in either excitation line, the PLIF detector can selectively measure OH or OD fluorescence followed by reconstruction of either image in the flame.

Figure 10 demonstrates the OH and OD LIF images. The sectional profile along with the combustion axis clearly visualized for OH and OD radicals in the flame. This images are proportional to their concentrations.

Figure 11 shows the flame temperature profiles reconstructed by thermocouple measurements. The PLIF signal is almost independent on this flame temperature.

However, the dissociations of H_2O or D_2O molecules are dependent on the temperature. Therefore, the PLIF images do not directly reflect the mixing conditions. We, therefore, estimated a ratio of D_2O to H_2O concentrations on the basis of the chemical equilibrium equation as a function of the OH and OD concentration (Fig. 10) and the temperature (Fig. 11), followed by its image reconstruction (Fig. 12).

$$\frac{[D_2O]/[H_2O]}{[[D_2O]/[H_2O]]_{max}}\,[-]$$

It is clearly indicated that the ratio is higher in the peripheral zones. This suggests that the mixing is inadequate even in the hydrogen flame where the mutual

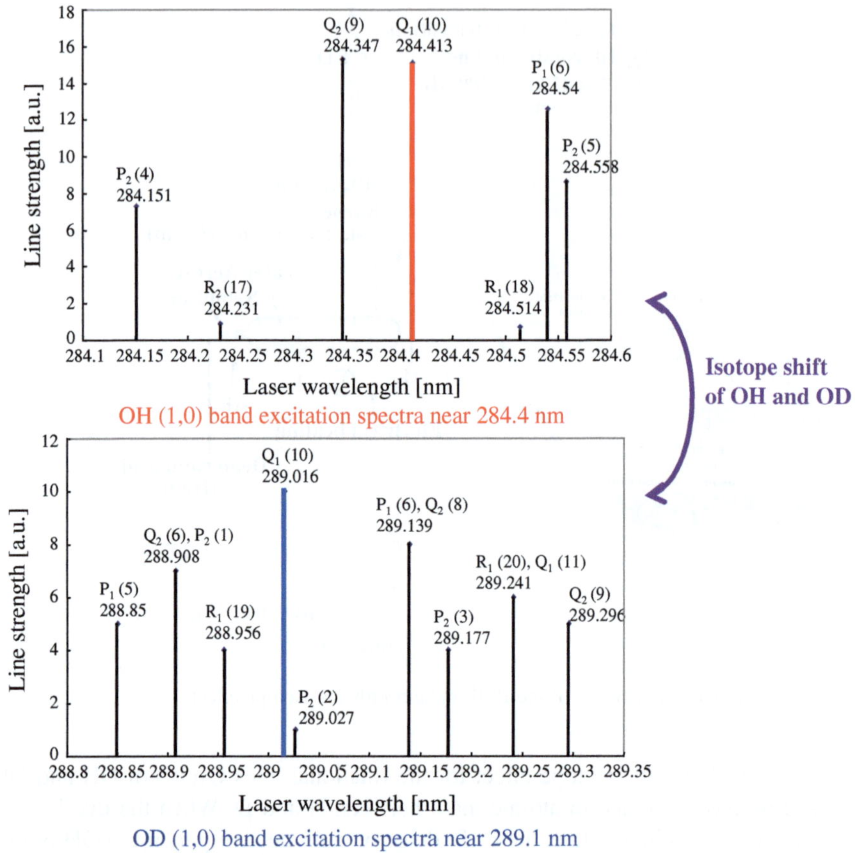

Fig. 9 Isotope shift between OH and OD lines for LIF excitation

molecular diffusion is higher due to the light mass of hydrogen. Swirling is required also for diffusion flames of hydrogen.

3.1.2 Profiling of Redox Environment in Combustion Flame by PLIF with Chemical Seeding (Itoh et al. 2006)

It is an important issue to determine whether the flame has a reducing or oxidizing (***redox***) environment in the material processing (the processing of steel, other metals, and ceramics). The material surface is greatly affected by the combustion atmosphere. Furthermore the atmosphere in flames is not uniform and often contains spatial distributions in practical flames.

To analyze the ***redox*** environment, the flame is estimated by measuring spatial PLIF distributions of Fe and FeO: FeO + Red. ⇔ Fe + O + Ox.

Higher Fe concentration (LIF intensity) reflects "reductive".

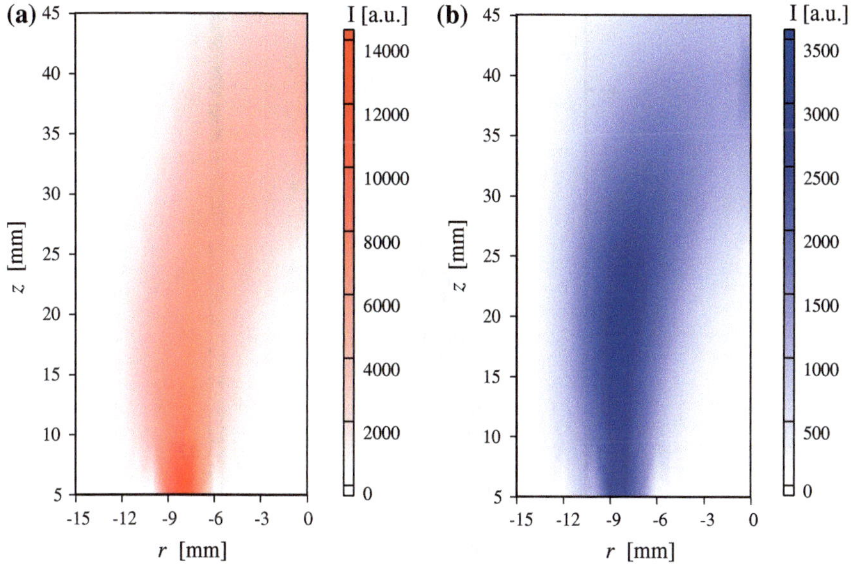

Fig. 10 PLIF images of **a** OH and **b** OD

Fig. 11 Flame temperature profiles reconstructed by thermocouple measurements

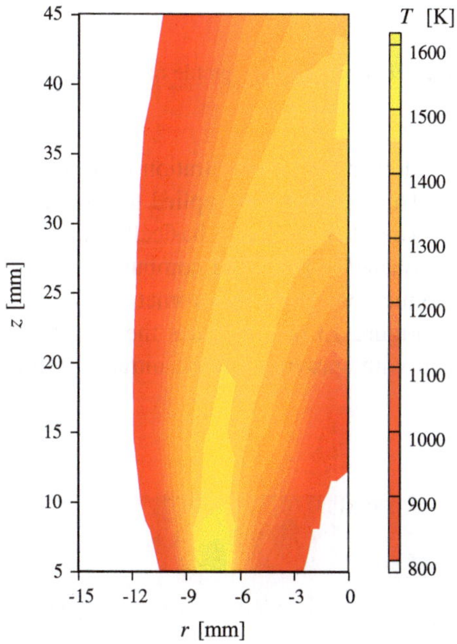

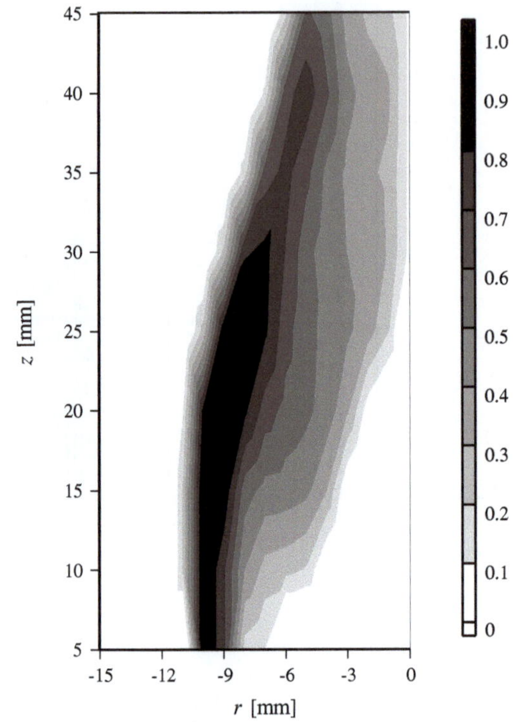

Fig. 12 Dimensionless profile of [D_2O/H_2O] ratio

Higher FeO concentration (LIF intensity) reflects "oxidative".

This method was applied to a premixed acetylene/air combustion flame.

The PLIF system is basically same as that described above. The burner used for acetylene/air premixed combustion is a commercially available slit burner used in atomic absorption spectrometry for trace element analyses.

Figure 13 demonstrates the acetylene/air premixed combustion flames without and with spraying a ferric nitrate solution as a chemical seed.

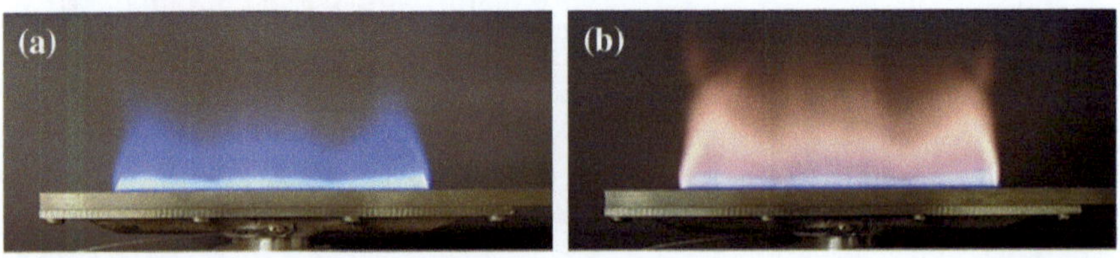

Fig. 13 Photographs of acetylene/air premixed combustion flames. **a** Without chemical seeding of $Fe(NO_3)_3$ and **b** with chemical seeding of $Fe(NO_3)_3$

The blue color of the intact flame (a) is mainly due to molecular emissions of radiation from thermal excited CH and C_2 molecules, while the pink color of the flame with seeding is attributable to those from FeO and Fe [ref].

Figure 14 shows the LIF spectra of Fe atoms and FeO molecules in the flames with seeding.

In this experiment, the rotational temperature of the flames were measured by comparing temperature dependent LIF lines of OH, or two-line method of LIF. Figure 15 shows the rotation temperature profiles under the different conditions of equivalence ratio.

Figure 16 shows the profiles of redox index calculated from the LIF intensity ratio of Fe/FeO.

It is clearly indicated from Fig. 15 that the temperature reaches maximum when the equivalence ratio is unity. On the other hand, Fig. 16 demonstrates that the redox index is highest in the fuel-rich combustion flame, independent of the temperature.

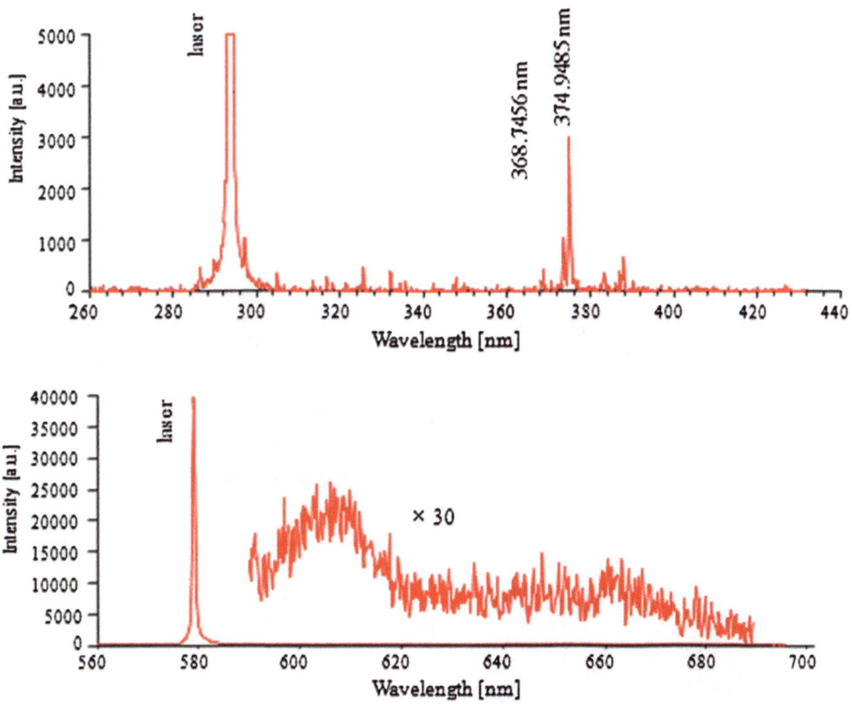

Fig. 14 LIF spectra of Fe atoms (*upper*) and FeO (*lower*) molecules in the acetylene/air premixed combustion flames with chemical seeding

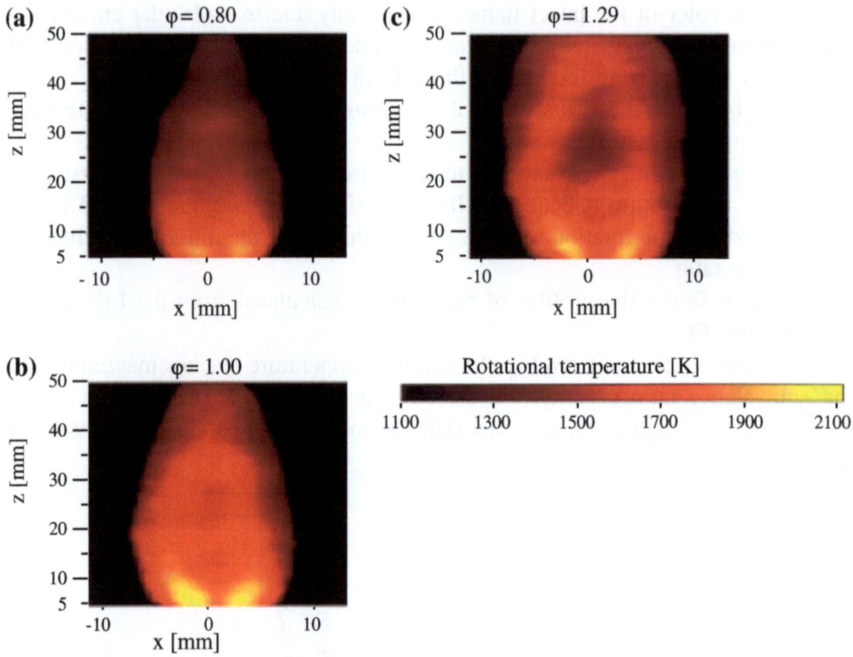

Fig. 15 Profiles of rotation temperature of OH by two-line PLIF

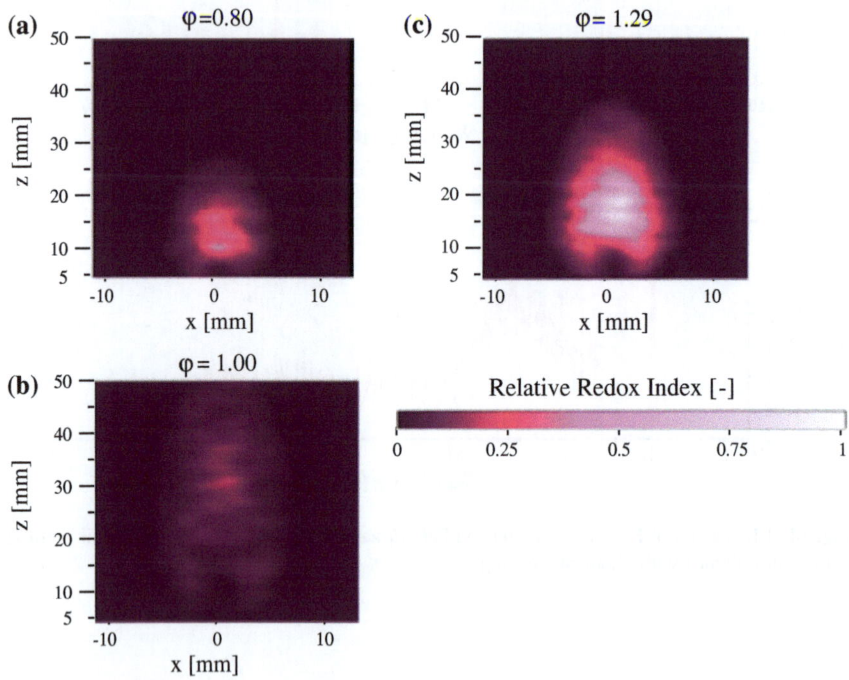

Fig. 16 Profiles of redox index with different equivalence ratio

3.1.3 Laser Ionization Mass Spectrometry (LIMS) and Application to Monitoring Intermediate Chemical Species During Combustion (Nishimura et al. 2010)

Spectroscopic visualization methods describe above are advantageous in that spatial, spectral, and time resolutions are concurrently attainable. However, the detectable chemical species are relatively limited in the number, typically NO, CH, OH, C_2, H_2O etc. Nevertheless, more intermediate chemical species are expected from elementary reactions during the combustion. Mass spectrometry (MS) is known to be a very powerful tool to determine molecular weight with high sensitivity to trace levels. A key issue is how to introduce the combustion flows and efficiently ionize the chemical species to be analyzed. For this purpose, a laser-induced ionization source (LI) with a sampling cone has been developed as a promising technology and LIMS successfully applied to visualization of intermediate chemical species in a methane/air premixed combustion flame. Figure 17 shows the schematic diagram of the experimental apparatus, and Fig. 18 shows the photographs of the combustion flame-MS interface.

The combustion flow is introduced into the laser-induced ionization (LI) chamber through the narrow orifice of the coupling interface. The Nd^+-YAG laser beam of frequency-doubled, tripled, or quadrupled radiation focussed on the Al target tip, consequently very tiny plasma is induced (laser-induced plasma) where intermediate chemical species from the combustion flame are ionized and partially decomposed. The key issue here is how to increase the ionization efficiency and lower the decomposition degree, leading high yield of ionized species to give high sensitivity and intact composition of chemical species in the combustion flame.

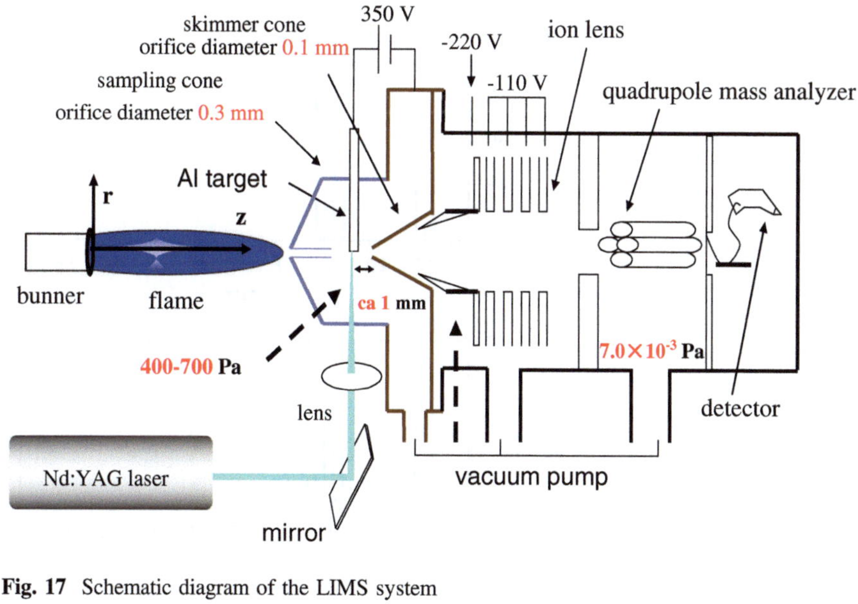

Fig. 17 Schematic diagram of the LIMS system

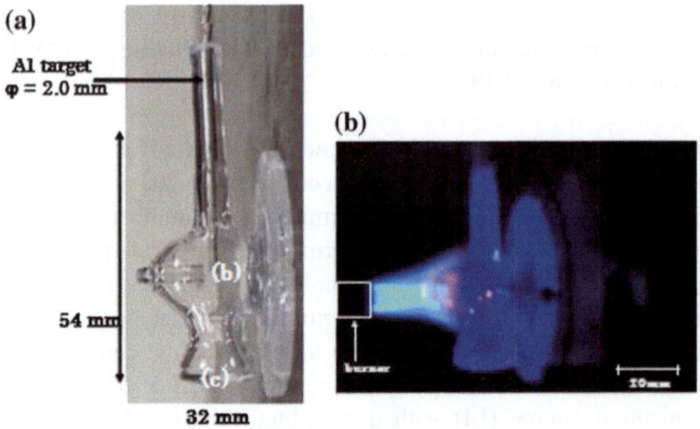

Fig. 18 Coupling cone interface between combustion flame and laser-induced ionization (LI) chamber of MASS. **a** Without flame, **b** with flame

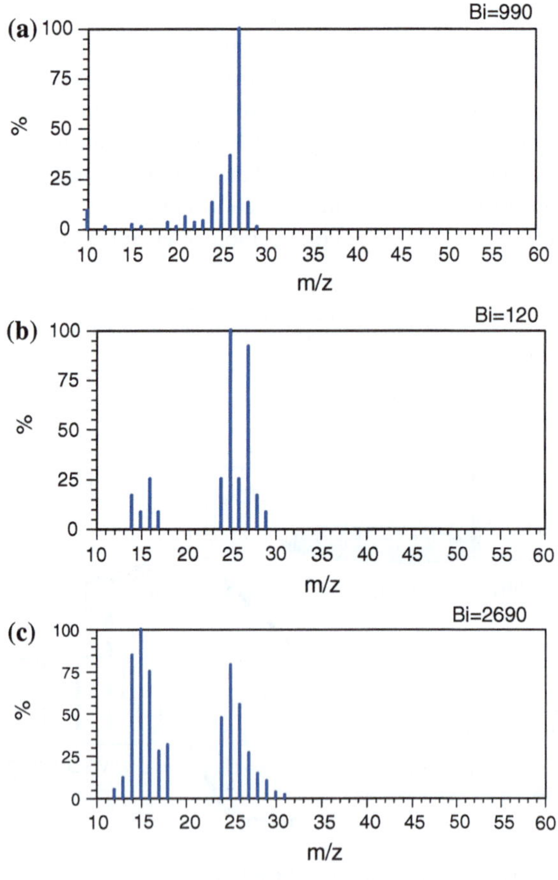

Fig. 19 Dependence of mass spectrum intensity on laser frequency. **a** 532 nm, **b** 355 nm, **c** 266 nm

Table 1 Peak assignment for the MS spectra in Fig. 19

m/Z	Chemical formula
12	C
13	CH
14	CH_2
15	CH_3
16	CH_4
17	OH
18	H_2O
24	C_2
25	C_2H
26	C_2H_2
27	C_2H_3
28	C_2H_4
29	C_2H_5
30	C_2H_6
31	CH_3O

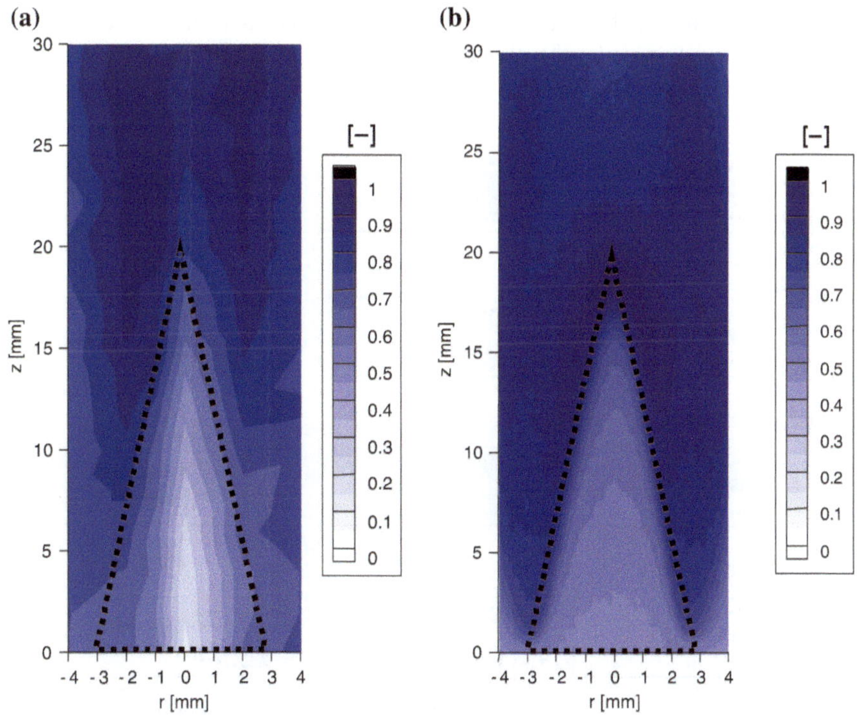

Fig. 20 Comparison of OH profiles obtained by LIMA (**a**) and PLIF (**b**)

Figure 19 shows the dependence of the typical mass spectrum on the laser frequency. It indicates that the wavelength 532 nm is better for low degree of fragmentation though 355 nm gives highest spectrum intensity. The peak assignment for the spectra in Fig. 19 is listed in Table 1. Except for CH, OH, C_2, and H_2O, all other intermediate species is difficult by spectroscopical methods described above. (Our most recent challenges include visualization by infrared spectroscopy to overcome this, which is not described here.) Most of important intermediate species during methane/air combustion are successfully detected. These includes the species dehydrogenated during oxidation of the fuel methane, CH_n, such as CH_3, CH_2, and CH as well as unburnt CH_4, and C_2H_m formed by coupling of C_1H_n. Another interesting species detected LIMS is CH_3O, or methoxy radical which clearly reflects the beginning pathway of fuel oxidation.

The next step is to evaluate the quantitative property of LIMS. Figure 20 shows the comparison of the OH profiles obtained by LIMS mapping and PLIF. These profiles are almost same geometrically. The intensity difference is probably due to the flow disturbance caused by the sampling cone interface (Fig. 18b).

Figure 21 shows the typical profiles of CH_n intermediate species for methane/air premixed combustion flame under the different equivalence ratios. Under the lean fuel condition, all the reductive (able to be oxidized) or combustible species, CH_n, do not exist around the peripheral zones. This is attributable to the very fast combustion reaction under the premixed condition.

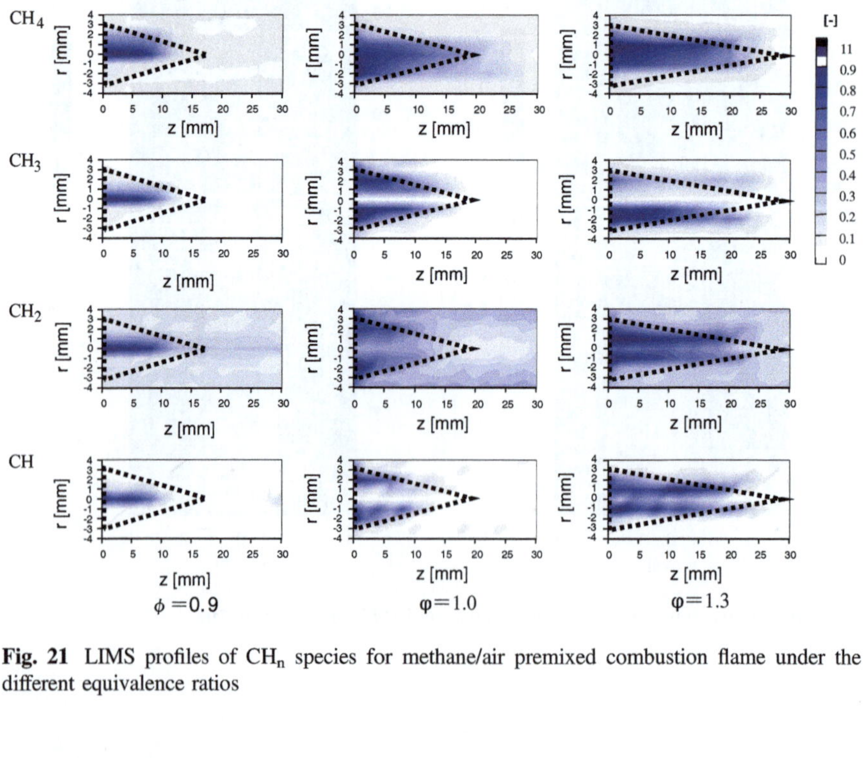

Fig. 21 LIMS profiles of CH_n species for methane/air premixed combustion flame under the different equivalence ratios

On the other hand, along with the increase of the equivalence ratio, their number density is enhanced in the outer zone. The results indicate clearly that the diffusion of ambient air into the flame becomes dominant for the oxidation process during the combustion.

4 Conclusion

In this chapter, the developments and applications are described on the three different kinds of spectroscopy or spectrometry for visualization of intermediate chemical species. Other spectroscopic methods have also been challenged for such visualization, including laser-induced plasma spectroscopy (LIPS) (Itoh et al. 2001) and infrared spectroscopy (Morita et al. 2012). These methods play very important roles in controlling and optimizing combustion to increase the efficiency and suppressing toxic emission, leading to establishing environmentally friendly, or green energy systems.

References

Hino Y, Sugiyama S, Suzukawa Y, Mori I, Konishi N, Ishiguro T, Kiatagwa K, Arai N, Gupta AK (2004) Two-dimensional spectroscopic observation of non-luminous flames in a regenerative furnace using coal gas. J Eng Gas Turbines Pow 126:20–25

Itoh S, Shinoda M, Kitagawa K, Arai N, Lee YI, Zhao D, Yamashita H (2001) Spatially resolved elemental analysis of a hydrogen-air diffusion flame by Laser-induced Plasma Spectroscopy (LIPS). Microchem J 70:143–152

Itoh S, Kitagawa K, Arai N (2006) Profiling of redox atmosphere in flames by chemical seeding/planar laser induced fluorescence. J Eng Gas Turbines Pow 128:765–772

Kajimoto T, Yamada E, Shinoda M, Desmira N, Kitagawa K, Gupta AK (2013) Analysis of flame structure by isotope shift-planar laser-induced fluorescence spectrometry of trace OH and OD radicals. Microchem J 106:334–339

Kitagawa K, Konishi N, Arai N, Gupta AK (2003) Temporally resolved 2-D spectroscopic study on the effect of highly preheated and low oxygen concentration combustion. J Eng Gas Turbines Pow 125(1):326–327

Kubota M, Tsuge S, Kitagawa K, Arai N, Ushigome N, Kato Y (1998) Analysis of degradation processes of carbon/carbon composites in a high temperature chemical flame by a spectrovideo camera. Carbon 36:1783–1790

Morita S, Yatsuhashi A, Kitagawa K (2012) Infra-red imaging of combustion flames. Bunseki Kagaku (Anal Chem) 61:275–279 (Japanese)

Nishimura T, Yamamoto A, Torii T, Matsumoto K, Kitagawa K, Gupta AK (2010) Develop of laser ionization mass spectrometer for detection of unstable species in flames. Microchem J 95:50–56

Shimada T, Akiyama T, Fukushima S, Mitsui K, Jinno M, Kitagawa K, Arai N, Gupta AK (2005) Time-resolved temperature profiling of flames with highly preheated/low oxygen concentration air in an industrial size furnace. J Eng Gas Turbines Pow 127:464–471

Lean Blowout Detection Techniques for Partially Premixed Flames in a Dump Combustor

Swarnendu Sen, Rajendra R. Chaudhari and Achintya Mukhopadhyay

Abstract Lean premixed or partially premixed combustion in gas turbine combustors is one of the popular strategies for achieving very low NO_x levels. However, lean combustion is extremely susceptible to lean blowout (LBO). LBO is a significant safety hazard for aeroengines and necessitates an expensive and time-consuming shutdown and restarting procedure for stationary engines. To avoid blowout, usually current combustors are therefore typically operated with a fuel–air mixture that is significantly richer than the LBO limit (Φ_{LBO}). These techniques found to be good in terms of safety margin but produces higher NO_x due to high temperature of flame. Online LBO prediction techniques are required for real-time control of combustors to maintain a continuous flame with smaller margins from the LBO limit. Experiments carried out on a gas-fueled swirl-stabilized dump combustor at different degrees of premixing and flow rates. LPG and methane are used as fuel for the combustor. Combustion behavior at different degrees of premixing has been studied. Existing LBO detection techniques are evaluated, which necessitates the development of new LBO techniques. Two optical signal-based detection techniques have been discussed here—flame color-based technique and symbolic time series analysis (STSA)-based technique. Both the methods performed well for a wide range of parametric variation.

Keywords Environmental degradation · Lean blowout detection · Liquid petroleum gas · Stoichiometric combustion · Anomaly measure

The monograph is based on a part of the doctoral work of R. R. Chaudhari and published papers by the authors in *International Journal of Spray and Combustion Dynamics*, 5, 49–66, 2013 and *Journal of propulsion and Power*, 29, 950–960, 2013.

S. Sen (✉) · R.R. Chaudhari · A. Mukhopadhyay
Department of Mechanical Engineering, Jadavpur University, Kolkata 700 032, India
e-mail: drssen@mech.jdvu.ac.in; sen.swarnendu@gmail.com

A.K. Agarwal et al. (eds.), *Novel Combustion Concepts for Sustainable Energy Development*, DOI 10.1007/978-81-322-2211-8_9

Nomenclature

$I(x, y)$	Image intensity at pixel (x, y)
$I_r(x, y)$	Red intensity of the image at pixel (x, y)
$I_g(x, y)$	Green intensity of the image at pixel (x, y)
$I_b(x, y)$	Blue intensity of the image at pixel (x, y)
L_f	Premixing length
N	Data length captured in experiment
Q	Volume flow rate
γ	Ratio of integrated red to blue intensities
Φ	Equivalence ratio
Φ_{LBO}	Equivalence ratio at LBO; LBO limit
Σ	Number of symbols

1 Introduction

Flames, in different industrial combustion applications, such as gas turbine, aircraft engine, and burners, are in general partially premixed in nature. These flames are stable and reliable. Major disadvantage of this type of combustor is production of unacceptable level of thermal NO_x. NO_x is found to be a harmful atmospheric pollutant and causes much environmental degradation such as acid rain, ozone layer depletion, and human health (Turns 2000; Correa 1993). During combustion process, atmospheric nitrogen, in high temperature, combines with oxygen to form NO_x. The formation in the combustion process is due to the oxidation of atmospheric nitrogen in the high-temperature regions of the post-flame gases, which is chemically described by thermal Zeldovich mechanism (Zeldovich 1946).

Due to harmful effects of NO_x, stringent environmental pollution norms have been devised, which recently motivated engine manufactures to develop low NO_x combustors that meet various regulatory emission standards (Correa 1993). New concept for combustion technology like lean premixed (LPM) combustion (or lean premixed prevaporized [LPP] combustion when liquid fuels are employed) has been introduced to the gas turbine industry (Lefebvre 1995; Correa 1998). Lean premixed combustion appears to be a most promising technology recently because it offers practical solution to reduce the emissions of nitrogen oxides (NO_x) and its simplicity in implementation to practical combustors. In LPM combustion, the temperature of the flame is significantly reduced due to excess air present in combustion zone which consequently eliminates the production of thermal NO_x (Zeldovich 1946). In addition, the lean operating condition elevates low maintenance requirements because the lower combustor exhaust temperature increases the lifetime of turbine blades and other mechanical components (Lefebvre 1999).

However, the combustors are operated at lower equivalence ratios in the LPM approach. The range of operation of such LPM combustors is much narrower compared to conventional combustors (Glassman 1996). So this operation results in low NO_x production, at the cost of overall stability of the combustion process.

Flame stabilization in lean premixed gas turbine combustor is a challenge due to weak reaction rate and thereby reduced flame speed against the high-speed flow of combustor. The popular stabilization mechanisms used for sustaining the flame inside the combustor are bluff-body flame holders (Cheng and Kovitz 1958; Hertzberg et al. 1991; Chaudhuri and Cetegen 2008), dump plane, swirlers (Ateshkadi et al. 2000; Schefer et al. 2002; Muruganandam et al. 2005), and pilot flame (Jensen and Shipman 1958). If the flame blowout from the combustor due to insufficient stabilization mechanisms, the phenomenon is called lean blowout (LBO). The equivalence ratio (Φ), at which blowout takes place, is called LBO limit (characterized by the limiting equivalence ratio Φ_{LBO}). Further small disturbances in combustor operating conditions such as velocity fluctuations or equivalence ratio perturbations caused due to fuel/air unmixedness (Barnes and Mellor 1998) elevate the risk of blowout. The major risk develops when the combustor faces power setting changes (Shashvat et al. 2005; Ateshkadi et al. 2000; Chao et al. 2000), which causes loss of power.

Blowout is a serious problem in both land-based engines and aeroengines. The blowout in stationary power station gas turbines may cause shutdown of the power station. This leads to an expensive as well as lengthy process of restart, which increases the maintenance cost and reduces productivity. Severe problem occurs for LBO in aircraft engines where combustion is the ultimate source of the vehicle thrust. During rapid deceleration, sudden changes in throttle setting rapidly reduce the fuel flow rate. But the transience of air flow occurs at a slower rate for the rotational inertia of the compressor (Rosfjord and Cohen 1995). This leads to lean fuel–air mixture, and the resulting blowout prevents the engine from recovering from the compressor stall event (DuBell and Cifone 1989). This puts a more severe situation in high-altitude vehicles where the stability limits are narrowed.

To avoid blowout, usually conventional combustors are therefore operated with a wide margin above the LBO limit as the LBO limit is uncertain and does not follow well-defined curve on the operating map of combustor (see Fig. 1). For these reasons, the designers have to design the combustor operability (denoted by solid line) to assure that the worst possible state (state 2) maintains a proper safety margin from LBO limit. The designer should also take care while drawing the operating curve of combustor to include dynamic power setting changes that occur during transit condition of combustor like throttling of turbine for reducing the power output (dotted curve adjacent to solid line) and similarly maintain an optimal safety margin from LBO limit.

These requirements keep all other states (like state 1) in a regime of high equivalence ratio (rich condition) which is not optimal state from emission point of view. The best operating condition in terms of NO_x for state 1 will be the leanest condition possible for that velocity of the combustor (i.e., state 1). But due to design constraint, the combustor operated at higher equivalence ratio which produces more NO_x.

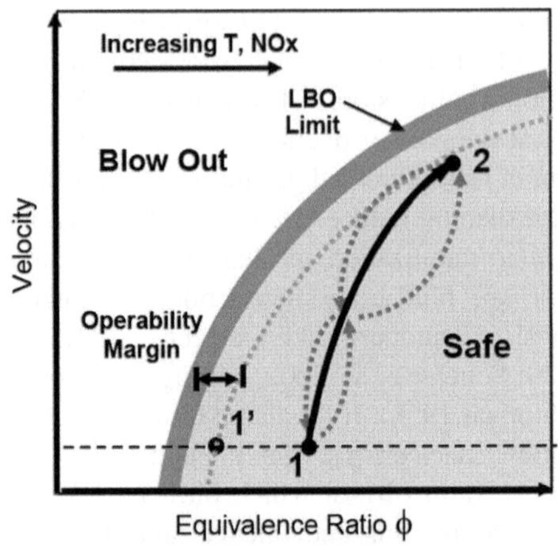

Fig. 1 Detailed schematic of combustor operating regime and transients. The operating regime (*solid line*) is designed such that the worst possible state (*2*) lies with safe margin from *LBO limit*. As a result, the other operating states (such as *1*) become non-optimal state produce more NO_x (Chaudhari 2011)

Thus, presently, the scheme of avoiding LBO by operating the combustor in rich condition regime with a wide margin above the LBO limit tends to increase the NO_x formation from engine. Enhanced performance can therefore be obtained with a reduction in this margin. So, the ability to sense the proximity of combustor to incipient LBO can provide significant payoff in terms of operation economy and reduce the overall emission.

1.1 Flame Dynamics Toward Lean Blowout

Before LBO, the flame shows some large-scale unsteadiness. Many studies characterize this flame behavior toward LBO. Nicholson and Field (1949) reported that the flame undergoes large-scale pulsation as the combustor approaches blowout. They also observed the repeated detachment and reattachment of flame to the flame holder before extinction. Hertzberg et al. (1991) reported similar pulsation of flame. Increased ignition time and local extinction of flame due to high instantaneous strain rate are considered as the major causes of blowout. Chao et al. (2000) observed the jet flame for characterizing the blowout dynamics. They observed the flame blowout transience turning off the acoustic excitation. They postulate that the blowout transient process can be divided into four regions: the pulsating, onset of receding, receding and extinction. They found the flame base pulsate in pulsating region before the onset of liftoff and eventually blowout. Through LIPF and PIV, they found that high strain rate encountered by the flame base which is more than extinction strain rate is the primary cause of blowout process. Hedman et al. (2002), characterized the blowout phenomenon using OH PLIF. They also reported the same pulsation and extinction of flame toward blowout. Norton (2003), in their CFD study, observed oscillations near extinction. De Zilwa et al. (2000) investigated flame dynamics close to blowout in dump combustors with and without low swirl. They reported very low-frequency oscillations as the rich and lean extinction limits were approached.

Muruganandam et al. (2002, 2003, 2005) and Muruganandam and Seitzman (2005), at Georgia Tech., characterize LBO phenomenon in methane-fueled, swirl-stabilized, laboratory-scale gas turbine combustor using optical (OH chemiluminescence) signature of flame. They observed extinction and reignition events near blowout regime, which are called precursor events. They reported that the frequency of those precursor events increases monotonically as the extinction limit is reached. During extinction event if the flame packets are unable to reignite the reactant, the flame finally blows off. They also report that the power in low-frequency fluctuation increases dramatically toward LBO. The authors assert that the reason behind increasing the low-frequency power is the combination of localized extinction timescale (~ 10 ms) and the mean time between such events (~ 1 s). Similarly, the parallel study was carried out by Prakash et al. (2005), Prakash (2007), on the same swirl-stabilized combustor where the focus was on acoustic emission rather than the optical emission. Similar to Muruganandam's observation in optical signature, they also observed dominant low-frequency oscillations in acoustic signal near LBO. They also found that the acoustic signal exhibits the same pattern of localized extinction events of optical signal, which they extracted using a set of wavelets whose shapes are similar to the pattern. Nair et al. (2004) and Nair and Lieuwen (2005) extended the characterization of blowout phenomenon in terms of acoustic emission on different configurations of combustion setup such as natural gas-fueled simple piloted burner, laboratory-scale swirl-stabilized dump combustor, and bluff-body-stabilized combustors, and their findings in terms of dominance of low-frequency spectrum and increased presence of time-localized and intermittent events in acoustic signal match with the earlier observation reported by Prakash.

Recently, Chaudhuri and Cetegen (2009) and Chaudhuri et al. (2010) characterized LBO phenomenon in a bluff-body-stabilized premixed combustor and reported that there is a significant reduction in flame speed toward LBO and flame shape changes from conical to columnar and high local stretch rates that exceed the extinction stretch rates are the main cause for final blowoff.

1.2 Existing Lean Blowout Sensing Approaches

The advantage of lean premixed or partially premixed combustion technique in practical systems and its susceptibility to blowout put a major challenge to the designers to reduce the emission as well as to maintain the power and reliability of engine. The development of strategies for early detection of imminent blowout and adoption of appropriate measure to mitigate it would allow the combustor to run at leaner equivalence ratio without loss of flame. Several studies have been reported in the past where the research community has exploited the quantifiable flame behavior preceding blowout for the development of lean blowout precursors, used for early sensing the proximity of combustor to imminent LBO. Accurate theoretical prediction of LBO is a major challenge due to vast complexity of the subject. Many fundamental mechanisms of flame ignition, stabilization, and propagation are still undiscovered despite the wide research conducted in the past. Phenomenological detection of LBO is found to be more promising technique in the present time. Lefebvre (1999) explored the LBO detection technique based on correlation function which was found to be very informative from design consideration of combustor but may not be suitable in real-time prediction of LBO.

The very first among LBO-predicting strategies relied on monitoring pressure in the combustion chamber and involved the rate of pressure drop and compressor rotating speed (Barnum and Bell 1993; Lucenko et al. 1996) or monitoring the rate of pressure fluctuation (Snyder and Rosfjord 1998). The major limitation of those techniques is speed detection and complexity of algorithm.

A non-intrusive diagnostic system is, however, preferred to avoid the harsh condition inside the combustor, and this has motivated the use of optical (Schefer et al. 2002; Muruganandam et al. 2003) and acoustic (Nair et al. 2004; Prakash et al. 2005; Shashvat et al. 2005) sensor-based techniques for LBO prediction. A comparison between the two in the same experimental setup (Muruganandam et al. 2002; Prakash et al. 2005) portrays optical methods as the superior choice on account of faster response time.

Muruganandam et al. (2005) analyzed OH chemiluminescence to identify LBO precursors in the flame of a premixed, swirl-stabilized, methane-fueled idealized combustor. Thresholds were set as a percentage of the local time-averaged intensity level for the particular flame condition. The crossing of this threshold due to fluctuations was identified as a "precursor event." An increase in the frequency of these events marked proximity to LBO conditions. Moreover, the authors implement this threshold-based sensing technique in active control system for mitigating the LBO. They used redirecting main fuel to the pilot line as control action, which enhances the equivalence ratio and helps to stabilize the flame. In a parallel study on the same swirl-stabilized dump combustor, Shashvat et al. (2005) developed the active control system based on acoustic signal for LBO mitigation, thus allowing for lean operation at lower temperature and lower emission. The authors observe increased acoustic power in low-frequency fluctuation of acoustic signature (<200 Hz) which is the result of localized extinction and reignition events. The authors used filtration of acoustic signal to produce the intermittent events, and the events crossing the set threshold generate the control action which is again in the form of diverting the main fuel to pilot line, which enhances the stabilization of flame in the combustor. Later, Nair and Lieuwen (2005) extended the development of fast diagnostic technique to monitor the proximity of combustor to blowout using acoustic signature from flames on three different combustors with different flame-holding mechanisms such as pilot, swirl, and bluff-body-stabilized flames. They observed similar dominance of low-frequency oscillation and time-localized intermittent events in acoustic data toward lean condition. Based on this observation, spectral, statistical, and thresholding signal processing schemes were developed for detecting blowout precursors with varying levels of time response, sensitivity, and robustness.

Yi and Gutmark (2007) reported LBO prediction based on OH emissions in partially premixed, liquid-fueled, multiswirl-stabilized gas turbine combustor. A statistical analysis of the time series is used to justify two detection indexes, namely the normalized chemiluminescence root mean square and the normalized cumulative duration of LBO precursor events for blowout prediction.

Mukhopadhyay et al. (2008) used moving kurtosis and threshold-based techniques applied on combustor pressure time series for early detection of flame extinction. More recently, Li et al. (2007) reported the use of a novel tunable diode laser temperature sensor to acquire a temperature time series signal from partially premixed, swirl-stabilized dump combustor. An increase in the power of low-frequency temperature fluctuations was identified as a LBO precursor. The authors further implement this LBO precursor as a control variable in active control system and demonstrated prevention of LBO during power reduction and transient fuel variation.

1.3 Objective of the Present Work

Several sensing strategies for predicting the proximity of combustor to LBO have been reported by many researches in the past. In all these studies, the flame dynamics near the blowout limit has been used for LBO prediction at a fixed degree of premixing. However, the degree of premixing influences heat release fluctuations which are the primary cause of dynamic instability and should subsequently affect the LBO limit (Cohen et al. 2001). The optical signal-based techniques reported earlier in literature are, however, reliant upon the fluctuation observed near LBO limit. Williams (1989) reported that these fluctuations are not persistent in diffusion/partially premixed condition compared to observations in premixed flame, which motivates us to check the efficacy of those reported optical and acoustic signal-based technique in the regime of variation of degree of premixing.

The statistical and the spectral approach reported earlier for sensing the proximity of combustor to LBO needs a sample data window (N). A larger sample window increases the accuracy of LBO predictor but decreases the efficiency of sensing technique with respect to time response characteristics. The sensing techniques use the quantitative blowout indicator calculated as an average of the sample window and deviation from mean, which is again computationally inexpensive and time-consuming and may lead to delay in real-time monitoring. Limitation of those approaches motivates us to explore different approaches for sensing the proximity of combustor to LBO, which are computationally inexpensive and economical. The development of the proposed LBO detection techniques is based on the quantifiable flame dynamics observed prior to blowout. So the second objective of this study was to characterize the combustion dynamics near LBO limit, quantify the flame characteristics preceding to blowout in terms of LBO precursor, and further use those precursors for LBO detection.

2 Experimental Details

In the present scheme, a swirl-stabilized dump combustor has been designed and fabricated to investigate the outlined objectives of the present study. The combustor simulates a laboratory-scale model of a generic gas turbine combustor.

All the experiments carried out in the present study are at atmospheric pressure condition, air is supplied at ambient temperature, the fuel primarily used is liquefied petroleum gas (i.e., LPG, 60 % propane, 40 % butane), and some tests carried with fuel methane.

A swirl-stabilized dump combustor was designed as a laboratory-scale model of a generic gas turbine combustor. The assembled view of the combustor is shown in Fig. 2. The combustor consists of a premixing section, the combustor, and an exhaust chamber. All metal parts made of type SS316 stainless steel enable the device to withstand high temperatures. All three sections are aligned coaxially. A swirler is provided at the end of the premixing section to stabilize the reacting mixture. The combustor design with a swirling injection of reactants and a dump plane with rear-facing step produces a swirling flow field with large recirculation zones. The global flow structure for this class of geometry is qualitatively similar, irrespective of combustor dimensions (Williams et al. 2007). The combustor section has an inner diameter of 60 mm and a length of 200 mm. This compares favorably with an actual gas turbine combustor, the "combustor can" of the Rolls Royce Trent 60 DLE (a 58-MW*e* aeroderivative gas turbine), which is approximately 200 mm in diameter and 700 mm long (Tomas 2005).

The premixing section consists of a 375-mm-long SS pipe of outer diameter 43 mm and inner diameter 23 mm. At one end of the pipe, a SS plate of 65 mm diameter and 15 mm thickness has been welded. A 15-mm-diameter internally threaded hole tapped into the plate serves as the air inlet. Five 6-mm-diameter holes have been drilled on the wall of the pipe at intervals of 50 mm. The first hole is at a distance of 20 mm from the upstream end of the premixing tube where plate is welded for air inlet. These holes act as alternative positions for fuel inlet. We name this fuel inlet ports as port 1 to port 5. Variation in the position of the fuel inlet alters the length available for the premixing of fuel and air (characterized by the premixing length L_{fuel}). This allows one to investigate the effects of extent of fuel/air premixing on the combustion characteristics.

The swirler was constructed by placing 15-mm-long and 1-mm-thick blades on a solid rod of diameter 8 mm at an angle of 60° to the axial direction. The blade heights were such that the swirler fits snugly in the 29-mm-diameter slots in the premixing section just prior to dump plane. The swirler is provided to add swirl in

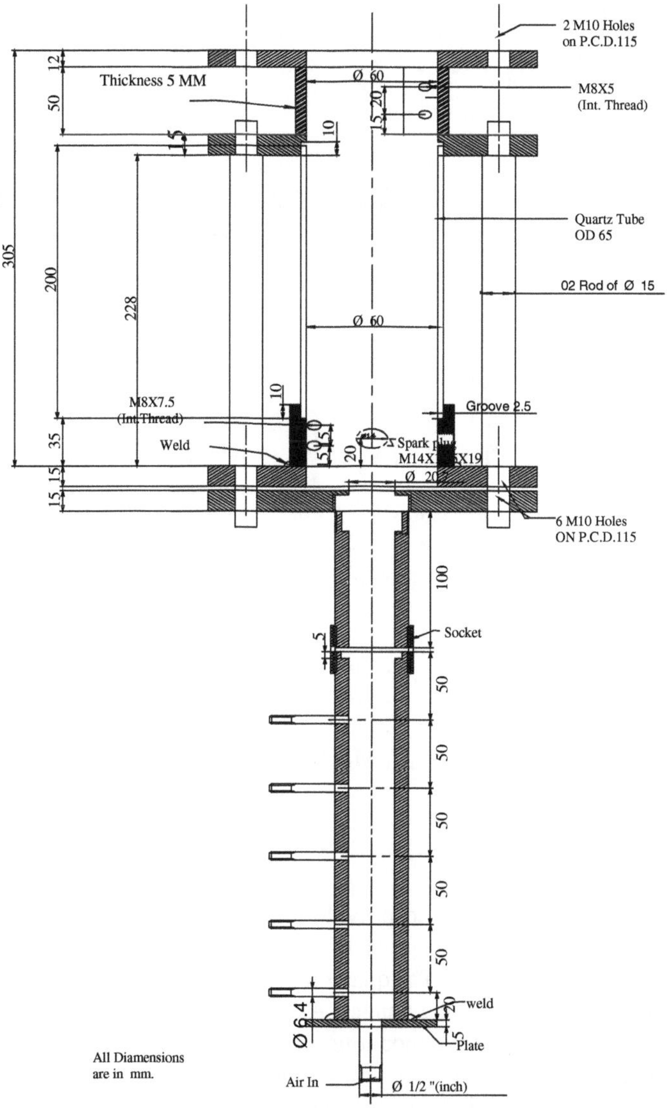

Fig. 2 Assembled view of the laboratory-scale combustor

the flow which increases the residence time as the flow takes helical path instead of straight path in combustor which enhances the combustion efficiency even in short-length combustor. After swirler, the sudden expansion in the flow area (dump plane) slows down the flow velocity and generates recirculation regions which ensure continuous heat–radical feedback to incoming mixture. The combination of swirl–dump combustor is a common method for modern gas turbine engine. The same arrangement is used in the present study in the laboratory-scale model.

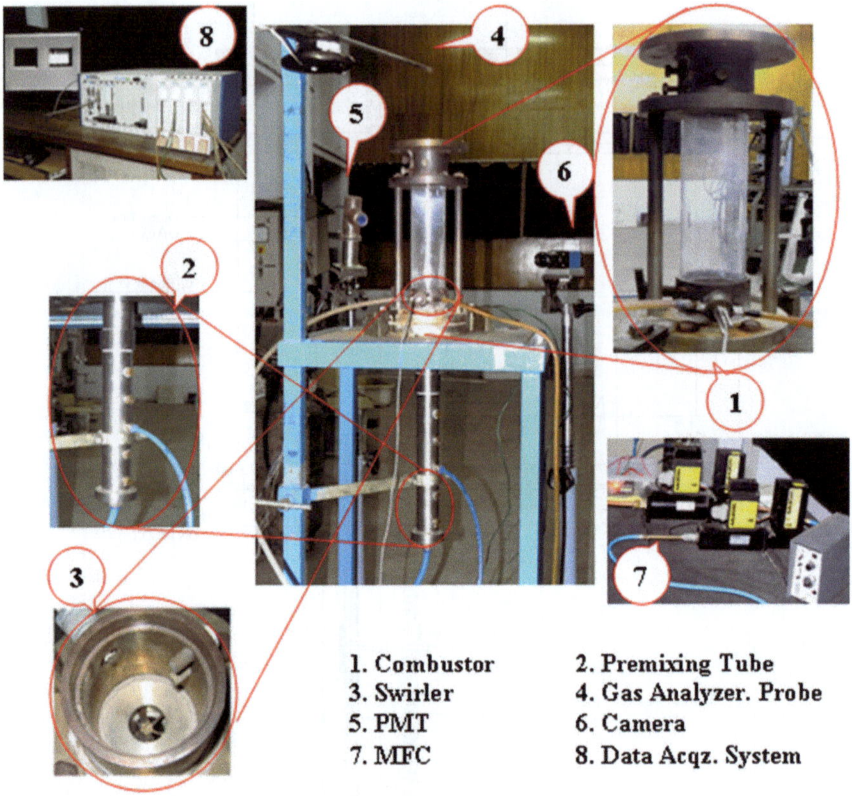

Fig. 3 The combustor setup

The combustor is designed to provide optically accessible swirling flames. It consists of a 200-mm-long quartz tube of inner diameter 60 mm and outer diameter 65 mm to allow optical access. The quartz tube is held between two grooved flanges by two long rods threaded to the flanges. At the base of the combustor and near the top, provisions were made for mounting pressure transducers and spark plug for ignition. The lower flange of the combustor was fastened to the flange of the premixing section. Figure 3 shows the photographs of the experimental rig.

3 Results and Discussion

The experiments were conducted for different air flow rates which give different Reynolds nos. calculated at the exit of swirler section (see Table 1). Five different ports provided on the premixing tube are used for injecting the fuel and thereby vary the premixing length (L_{fuel}) available for fuel/air mixing. The arrangement

Table 1 Experimental details

Mixing length (L_{fuel})		Port number	L_{fuel} (cm)
		1	35
		2	30
		3	25
		4	20
		5	15
Air flow	(low flow rates)	Q_{air} (LPM)	Re
		70	6,645
		80	7,480
		85	7,948
	(high flow rates)	150	13,982
		175	16,312
		200	18,643
Sampling rate from PMT		2 kHz.	
A/D digitization resolution		16 bit, range ±5 V	

allows characterizing the flame dynamics in the regimes of different degrees of premixing. The underline strategy was the higher premixing length (L_{fuel}) that allows greater time for fuel/air mixing and thus improves air–fuel premixing.

The experiment involved first using stoichiometric air–fuel mixture (Φ = 1). Equivalence ratio was then gradually reduced in higher steps (~0.05) initially and then in smaller steps (~0.02) closer to Φ_{LBO}. The reduced equivalence ratio was obtained by reducing the fuel flow rate while keeping the air flow rate constant. This ensured that the incoming velocity is almost constant since the air consists of the bulk of the reacting mixture compared to the low fuel flow rate.

Figures 4 and 5 plot a sequence of images captured at 30 frames per second for a stable equivalence ratio far from LBO, i.e., Φ = 1.0 and Φ = 0.81, and one close to LBO, i.e., Φ = 0.75, respectively. In Φ = 1.0 case, the combustion is seen to be relatively steady and occupies the whole combustor and the shape of the flame is observed to be conical. As the equivalence ratio is reduced, for Φ = 0.81 case, there is a significant change in flame color and flame becomes bluish with reddish tip. Though the flame zone changes from image to image, there is always a well-defined combustion region and the flame looks more stable (Fig. 4b). For conditions close to blowout (Φ_{LBO} = 0.75), the flame shape changes from conical to columnar and the length increases due to reduced reaction rate and burning velocity of flame near LBO (Chaudhari 2011).

Near LBO limit, there are random instances where the flame exhibits oscillation in the combustor and flame lifts off from the dump plane. These oscillations produce near flame loss event in combustor. The combustor initially has a spatially combustion zone. Then, the flame detaches from the center body, showing weak reaction, and moves further downstream from the combustor inlet and stabilizes there. The flame begins to disappear from the field of view, and there is almost complete loss of flame, suggesting extinction event. From downstream, the flame packets are

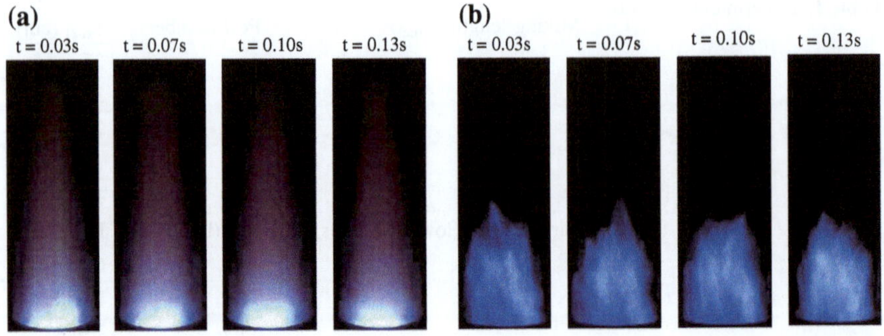

Fig. 4 Still images of the flame at 80 lpm with Δt = 33-ms intervals away from LBO for **a** $\Phi = 1.0$ and **b** $\Phi = 0.81$ $L_{\text{fuel}} = 35$ cm (port 1)

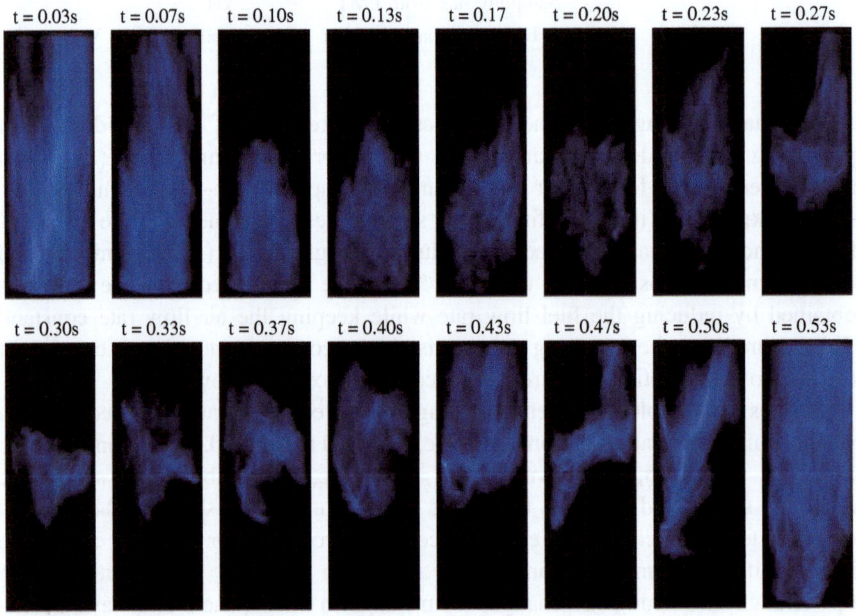

Fig. 5 Still images of precursor event of the flame at 80 lpm with Δt = 33-ms intervals observed near LBO, i.e., $\Phi \approx \Phi_{\text{LBO}} = 0.75$, $L_{\text{fuel}} = 35$ cm (port 1) (Chaudhari et al. 2013)

convected back to inlet, which reignites the unburned fuel which entered the combustor in the previous period. The fuel is rapidly consumed which exhibits intense combustion, and the flame is re-established in the combustor (reignition event). These unique extinction and reignition events span a period of several milliseconds, and they occur randomly in time, prior to LBO. As the combustor approached toward LBO limit, the frequency of these events increases, and thus, the time between two such events decreases closer to LBO. Muruganandam et al. (2005)

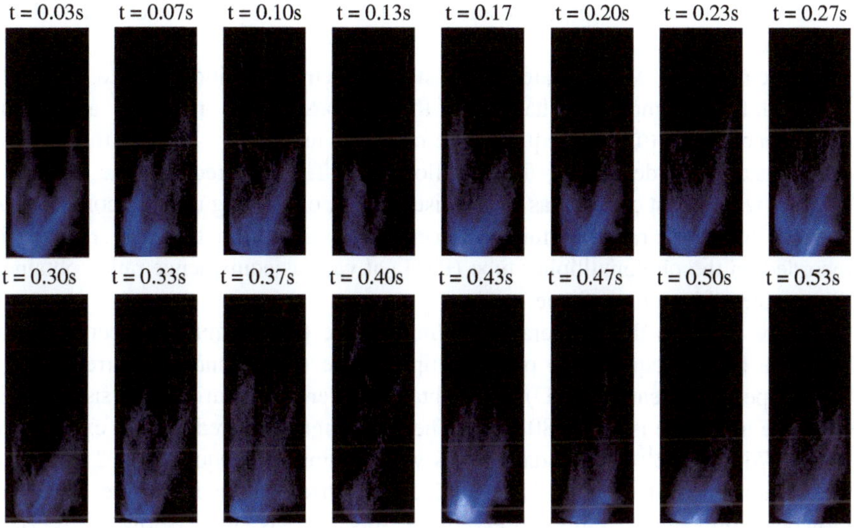

Fig. 6 Still images of the flame at 80 lpm with Δt = 33-ms intervals observed near LBO, i.e., $\Phi/\Phi_{LBO} = 1.0$, $L_{fuel} = 15$ cm (port 5) (Chaudhari et al. 2013)

observed the same extinction and reignition events (precursor to blowout) on the swirl-stabilized dump combustor, which matches with the basic design of the present experimental setup, with lean premixed methane–air flame.

Figure 6 shows the images of the flame at $\Phi/\Phi_{LBO} = 1$ for port 5 case, where the premixing length is minimum ($L_{fuel} = 15$ cm). The low premixing length for fuel–air mixing ensured that the flame obtained in the combustor is with lower degree of fuel–air premixing or mostly partially premixed condition. From Fig. 6, it is revealed that all the time the flame is attached to the dump plane and not shows any liftoff pattern. Though the flame shows some flame loss events where the flame intensity is significantly reduced and later reignition of flame, the flame does not oscillate in the combustor, and the precursor events are not so intense. The flame base is attached to the dump plane all the time, which is in contrast to the observation of port 1 flame characteristics. The flame in low flow rates exhibits more cycles of oscillation in near LBO regime, whereas in higher flow rates, a less number of oscillation cycles were observed.

For sensing techniques, the quantifiable flame characteristics prior to blowout have been explored to obtain LBO metric and use for detecting the LBO. The sensing methodology presented in the current study is based on optical emission, i.e., chemiluminescence and flame color variation analysis from combustor. The present study used chemiluminescence from CH radicals (emitted at 432 nm.) for detecting the lean blowout. The choice of optical sensing allows the optical sensor to be located outside the harsh environment of combustor so that the requirements of non-intrusive sensing can be fulfilled. Further, inherent fast time response characteristic of optical sensing made it the best choice for fast detection of flame instability events.

The combustor was operated from stoichiometric condition up to complete LBO, and the temporal behavior of the combustor was recorded at various equivalence ratios (Φ). The equivalence ratio was lowered by keeping the air flow rate constant and decreasing the fuel flow rate. This ensured that the incoming velocity was almost constant as air consists of bulk of reacting mixture compared to the low fuel flow rate. Chemiluminescence emission from the CH* radicals is recorded on the photomultiplier tube (PMT) at various equivalence ratios obtained till complete blowout of flame occurs.

Figure 7 shows the temporal recordings of the CH chemiluminescence signal from the flame, acquired on photomultiplier tube (PMT) and the corresponding power spectrum density (PSD) plots obtained after fast Fourier analysis. The air flow rate used here is $Q_{\text{air}} = 80$ LPM. The data length captured in each experiment N = 32,768 (i.e., 2^{15} points) data points with a sampling frequency of 2 kHz, and the lowermost port (port 1, $L_{\text{fuel}} = 35$ cm.) is used for injecting the fuel. The blowout equivalence ratio observed in this test is just below 0.75 (i.e., $\Phi_{\text{LBO}} = 0.75$). The data are presented for different equivalence ratio conditions which are normalized by the value at blowout (i.e., Φ/Φ_{LBO}). The normalization facilitates the comparison of same state of combustor with different port conditions, described subsequently below.

It is observed that away from Φ_{LBO} at high equivalence ratios ($\Phi/\Phi_{\text{LBO}} = 1.33$), CH* oscillations are coherent and strong in intensity. The corresponding PSD plot revealed some dominating frequency around 20 Hz with adjacent bunch of frequencies. Apart from this, the spectrum also shows some low-powered high frequencies in a band of 50–60 Hz. From the different CH time series plots, it is observed that as the equivalence ratio is lowered, the mean value of CH chemiluminescence intensity decreases, which is expected as the fuel flow is lowered to obtain lower equivalence ratio which correspondingly gives lower heat release rate. At $\Phi/\Phi_{\text{LBO}} = 1.08$, the strength of oscillation at the dominating combustion frequency $\sim$20 Hz diminishes and some high-powered low frequencies appear in spectrum. As combustor approaches to LBO (i.e., $\Phi/\Phi_{\text{LBO}} = 1.02, 1.01$), the flame becomes weak in intensity with the increase in length due to lower reaction rate and changes shape from conical to columnar and started showing some oscillations in combustor. The corresponding CH intensity also decreases as flame becomes weak and exhibits some relatively high-amplitude bursts where the signal shows high spike in both below and above the mean value which matches with the flame oscillations. The corresponding PSD elucidates that high-powered low dominating frequencies apparently match with flame oscillations. At $\Phi/\Phi_{\text{LBO}} = 1.0$, just before the LBO, the flame lifts off from dump plane and starts oscillating in combustor. The chemiluminescence intensity decreases and shows some relatively high-amplitude bursts in CH time series plot, with the signal going both below and well above the mean value, and occasionally drops to near-zero value. Often, these bursts or events are characterized by an almost complete loss of chemiluminescence signal quickly, followed by intense emission from an imaged region. These bursts

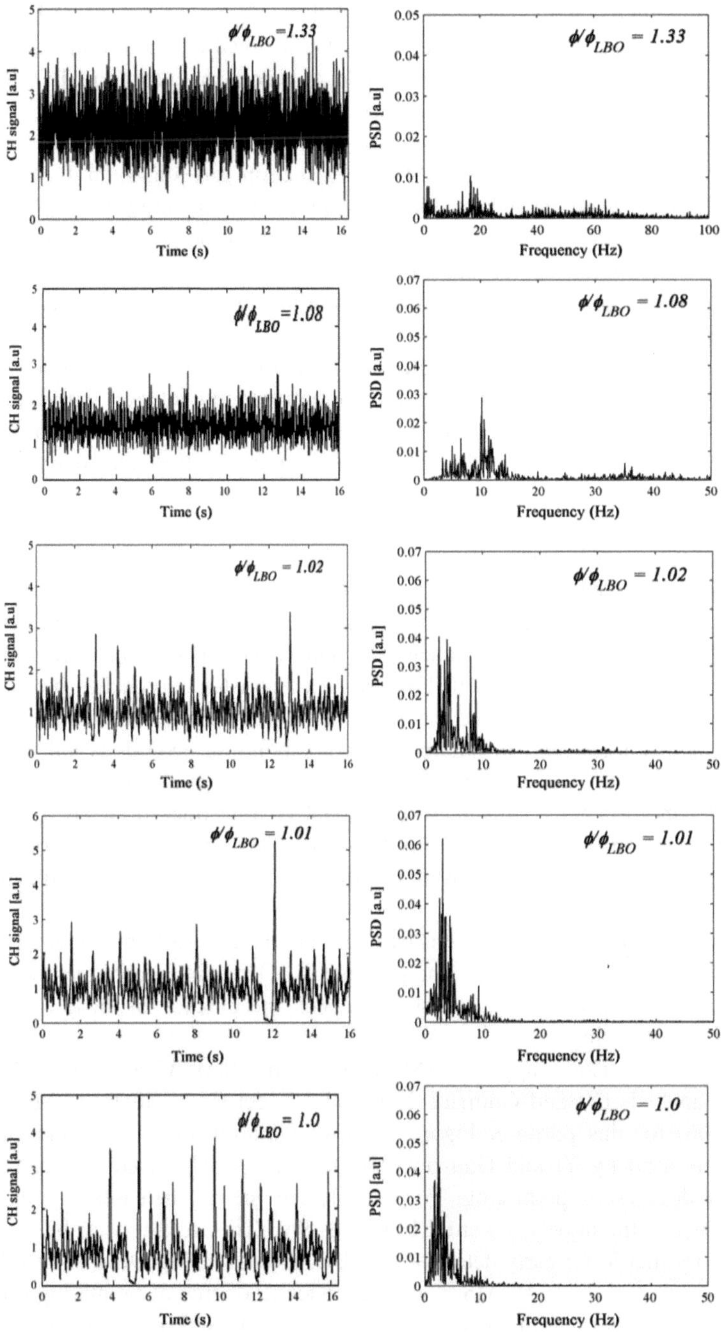

Fig. 7 CH chemiluminescence time series and PSD plots for different equivalence ratios for port 1 (L_{fuel} = 35 cm)

or events coincide with the occurrence of flame loss (extinction) and reignition events. The bursts (unsteady events) arise due to local extinction and strong reignition of flame. These events can be called as precursor events to LBO, and they grow in number near blowout. The corresponding PSD plot reveals some strong dominating high-powered low-frequency events which are likely related to time-scale of occurrence and duration of flame extinction and reignition events.

Similar tests were carried out by changing the fuel-injecting ports. As the fuel entry point is raised, as discussed earlier, the length of the premixing section (L_{fuel}) available for mixing of fuel and air decreases. Thus, the flame obtained using a lower mixing length (i.e., higher port) for fuel injection results in increasingly less premixed flames.

Figure 8 shows the time series data for CH chemiluminescence for port 5 and the corresponding PSD plots for different normalized equivalence ratio conditions. Similar to port 1 and port 3 tests, here also the mean of the chemiluminescence intensity decreases with the decrease in equivalence ratios due to reduced heat release rate.

At higher equivalence ratio condition (i.e., $\Phi/\Phi_{\text{LBO}} = 1.69$), the CH oscillations are strong in intensity and coherent in nature which is confirmed from the corresponding PSD where it exhibits a strong dominating frequency (≈20 Hz). At $\Phi/\Phi_{\text{LBO}} = 1.35$, the 20 Hz dominating combustion frequency diminishes from the spectrum and a few lower frequencies are added. The spectrum shows a bunch of frequencies in a broadband 1–50 Hz. As the combustor approaches blowout condition ($\Phi/\Phi_{\text{LBO}} = 1.06$, 1.03), the CH chemiluminescence time traces are still random and not coherent in nature. The corresponding PSD does not reveal any dominating frequency and exhibits different frequencies in a band of 1–50 Hz. Near blowout condition (i.e., $\Phi/\Phi_{\text{LBO}} = 1.0$), the CH time trace does not exhibit significant burst (unsteady events) from the mean. This corroborates our observation that the flame never lifts off from the dump plane and distinct extinction and reignition events are absent.

From the above observation, it is evident that the near LBO behavior changes as the degree of premixing varies. The previous LBO sensing techniques were tested on premixed flames. So, first we should check the efficacy of those for different degrees of premixing.

We tested statistical approach (Muruganandam 2006; Yi and Gutmark 2007), spectral analysis (Yi and Gutmark 2007), and direct signal analysis (Muruganandam 2006) for this purpose. Figure 9 shows variation of NRMS value as LBO metric (as used by Yi and Gutmark 2007) with normalized equivalence ratio at different degrees of premixing. For port 1, the metric increases rapidly as the equivalence ratio moves toward LBO (Fig. 9a). This feature can be used as a satisfactory metric for early detection of LBO. However, Fig. 9b–e shows that this rapidly rising feature of NRMS is absent for lesser degree of premixing (port 2 to port 5). This shows that the metric is suitable for premixed flames or for the flames with a larger degree of premixing. But, for partially premixed and non-premixed flame, the metric cannot provide satisfactory result. The other methods mentioned

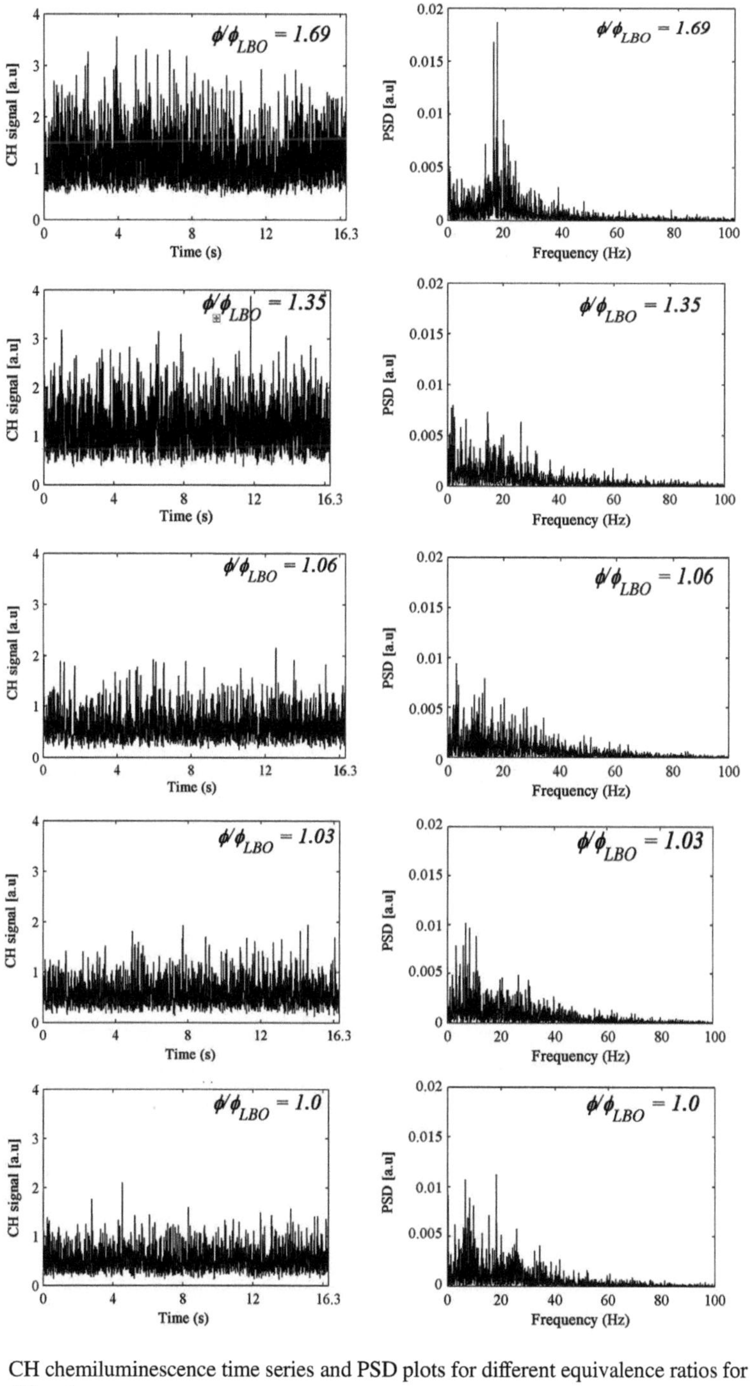

Fig. 8 CH chemiluminescence time series and PSD plots for different equivalence ratios for port 5 (L_{fuel} = 15 cm)

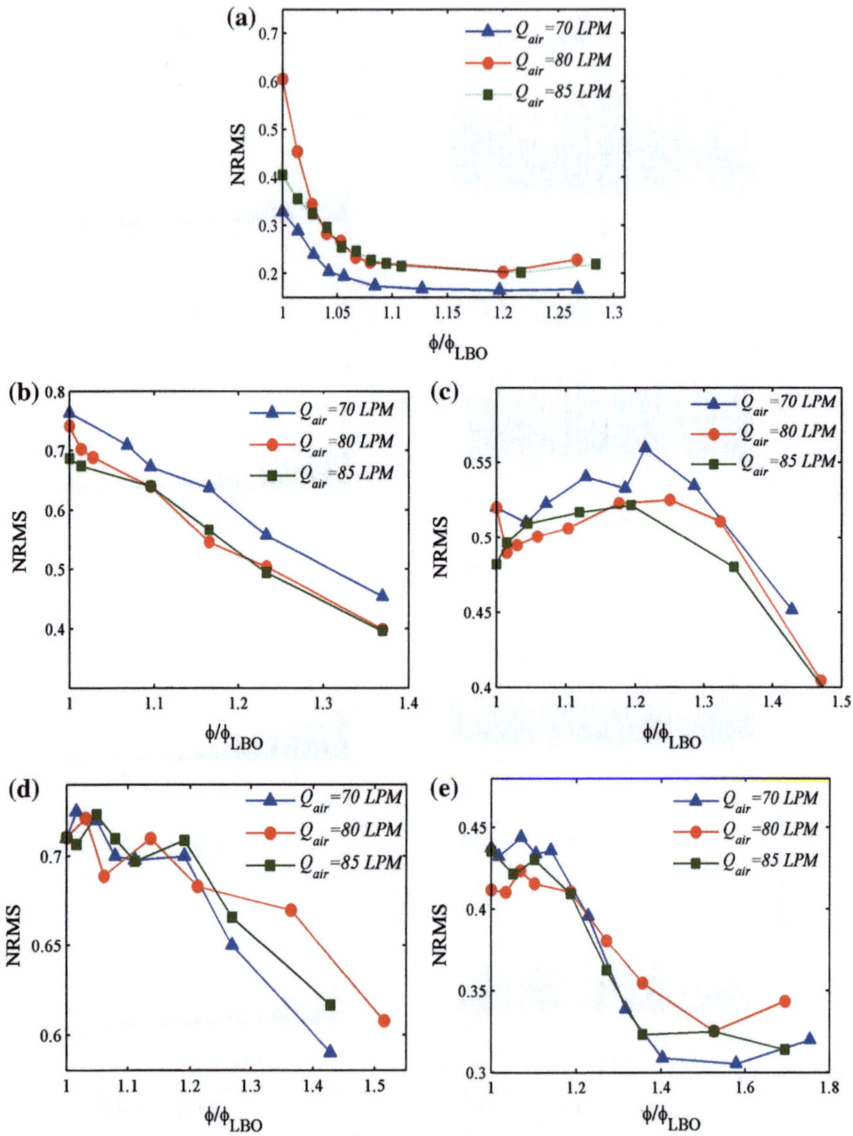

Fig. 9 Variation of NRMS for different flow conditions for **a** port 1, **b** port 2, **c** port 3, **d** port 4, and **e** port 5

have also shown similar trend. For brevity, the results are not included here, and the details may be found at Chaudhari (2011).

3.1 Flame Color: LBO Predictor

A LBO prediction strategy based solely on the flame color variation was proposed by the authors (Chaudhari 2011; Chaudhari et al. 2013). The technique is based on flame images captured by color CCD or CMOS sensor. Data were acquired for three different air flow rates ($Q_{\rm air}$) for fuel supply to each of the five premixing lengths ($L_{\rm fuel}$). At each equivalence ratio (Φ) condition, four photographs of the flame modes were taken by a digital SLR camera. Each color image acquired from the SLR camera was analyzed using MATLAB® to obtain a color metric used for LBO prediction. Each pixel in a color image is mapped as a vector of three intensity components, i.e.,

$$I(x,y) = \begin{Bmatrix} I_{\rm r}(x,y) \\ I_{\rm g}(x,y) \\ I_{\rm b}(x,y) \end{Bmatrix} \tag{1}$$

where $I_{\rm r}$, $I_{\rm g}$, and $I_{\rm b}$ refer to the red, green, and blue components, respectively, of the intensity of the corresponding pixel. This is a result of the presence of only these three colors on the Bayer filter used in conventional digital cameras. Considering that changes in image color range from red to blue in a flame as LBO is approached, which is illustrated in Fig. 10 where $\sum_{x,y} I_{\rm r}(x,y)$ and $\sum_{x,y} I_{\rm b}(x,y)$ represent the mean value of red and blue color intensities determined from our frames captured at each equivalence ratio (Φ) value for $Q_{\rm air}$ = 80 LPM, $L_{\rm fuel}$ = 35 cm. The two quantities, respectively, decrease and increase monotonously with decreasing equivalence ratio (Φ). Based on this color variation, we propose LBO detection metric in the form of

$$\gamma = \sum_{x,y} I_{\rm r}(x,y) / \sum_{x,y} I_{\rm b}(x,y) \tag{2}$$

Determination of γ is computationally inexpensive and can be performed on a real-time basis.

We compared the flame color variation detected by color camera with spectrometer results. Spectrometer captured the electromagnetic emission of flame simultaneously with the color camera at different equivalence ratio conditions. Figure 11 shows the visible electromagnetic spectrum of flame emission for different operating conditions obtained when the combustor operated from stoichiometric condition until blowout takes place. At $\Phi/\Phi_{\rm LBO}$ = 1.49 (represent stoichiometric condition of combustor) the dominating wavelength of flame emission is observed to be around 700 nm which is close to the standard red color

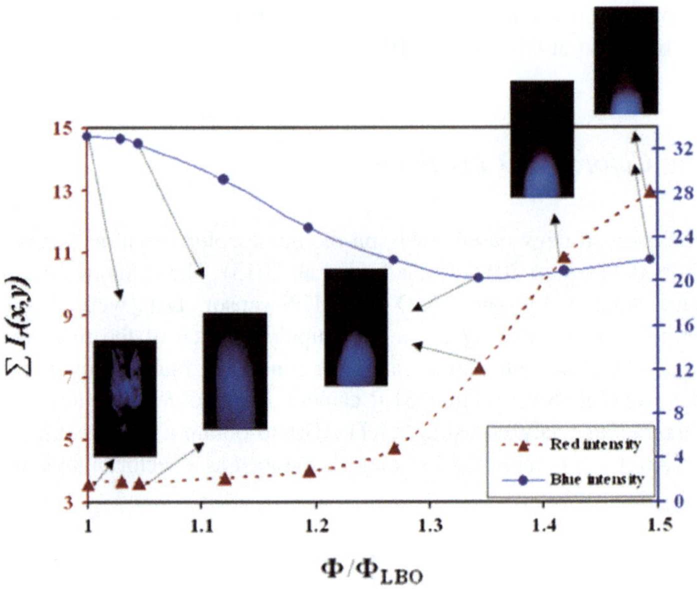

Fig. 10 Variation of red and blue color intensities with flame modes for different equivalence ratios (port 1) (Chaudhari et al. 2013)

emission band in visible spectrum. As the equivalence ratio is further reduced, the dominating 700-nm peak in spectrum diminishes and the spectrum shows peak around 432 nm wavelengths, which is close to blue color emission band wavelength in the visible spectrum. The results of the spectrometer are consistent with the color analysis of flame carried by camera.

We explain this color observation as follows. In the most common hydrocarbon flames, the most important factor determining flame color is oxygen supply and the extent of fuel–oxygen premixing which determines the rate of combustion and thus the temperature and reaction paths, thereby producing different colors. At higher equivalence ratio (Φ), the presence of incandescence of very fine soot particles produced in the flame makes flame reddish. With increasing oxygen supply (lean condition), less soot is produced due to more complete combustion and the reaction creates enough energy to excite and ionize gas molecules in the flame, leading to blue appearance of leaner flame.

Here, we first presents the variation of γ with equivalence ratio for a range of parameters permitted by our setup to illustrate the consistency of the proposed metric stated in Eq. (2) for quantifying LBO proximity.

It is observed that Φ_{LBO} is dependent on L_{fuel}, i.e., the degree of premixing. In order to explore the applicability of our predictor for various degrees of premixing, we conducted different tests for a wide range of air flow rates and L_{fuel} permitted by the setup. The color ratio, i.e., γ, was determined for each decreasing equivalence ratio condition and reported over a normalized equivalence ratio (Φ/Φ_{LBO}) in

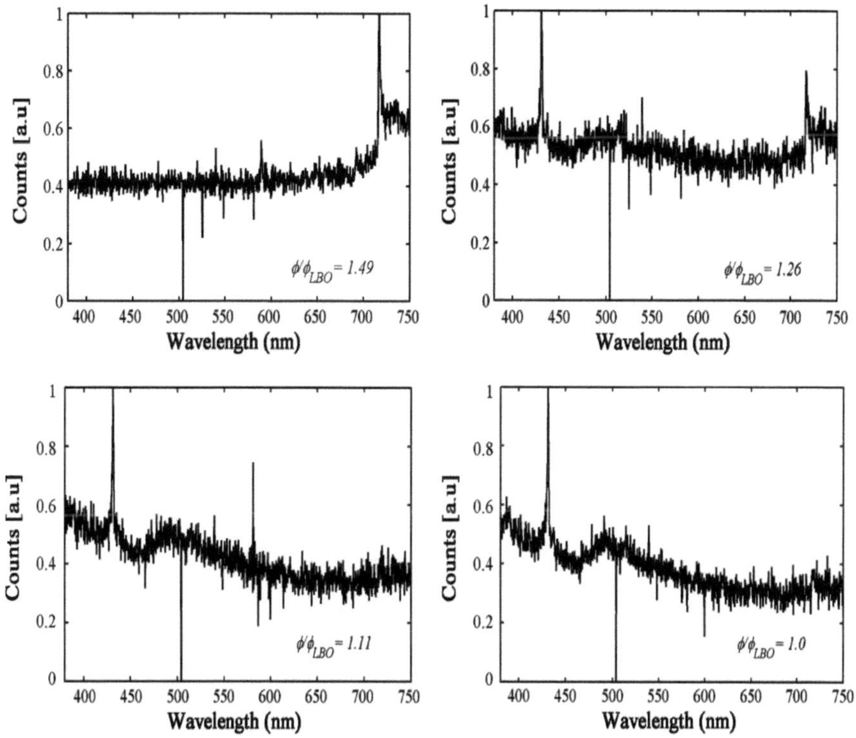

Fig. 11 Spectrometer spectrum for different Φ/Φ_{LBO} clearly shows the variation of colors from their different dominating wavelength peak as Φ approaches to Φ_{LBO} (Q_{air} = 80 LPM, L_{fuel} = 35 cm) (Chaudhari et al. 2013)

Fig. 12. Again, for each selected L_{fuel} and air flow rate (Q_{air}), three different runs have been taken to estimate the percentage error. It is found that error percentage for most of the readings lies within ±10 %.

From Fig. 12, it is clearly observed that irrespective of the L_{fuel} and the air flow condition used, γ decreases monotonously with equivalence ratio, which justifies the choice of γ as LBO detection metric for sensing the proximity of combustor to blowout. As evident from the figure, a set threshold (γ_{thresh}) on the value of γ would be sufficient to detect proximity to blowout for this particular configuration. Setting $\gamma_{thresh} \approx 0.3$ would provide a reasonably safe margin of $\Phi/\Phi_{LBO} \approx 1.2$–1.3 for all the flow rates and premixing lengths considered, with the higher ratio corresponding to the more premixed configuration (L_{fuel} = 35 cm). Since premixed flames are more susceptible to lean blowout, a higher safety margin for this configuration is advantageous. A single threshold value for different configuration and flow condition would simplify the development of controller for preventing blowout. Again, selection of threshold value (γ_{thresh}) is not limited to fixed value and can be varied depend upon the safety margin required. Another advantage of

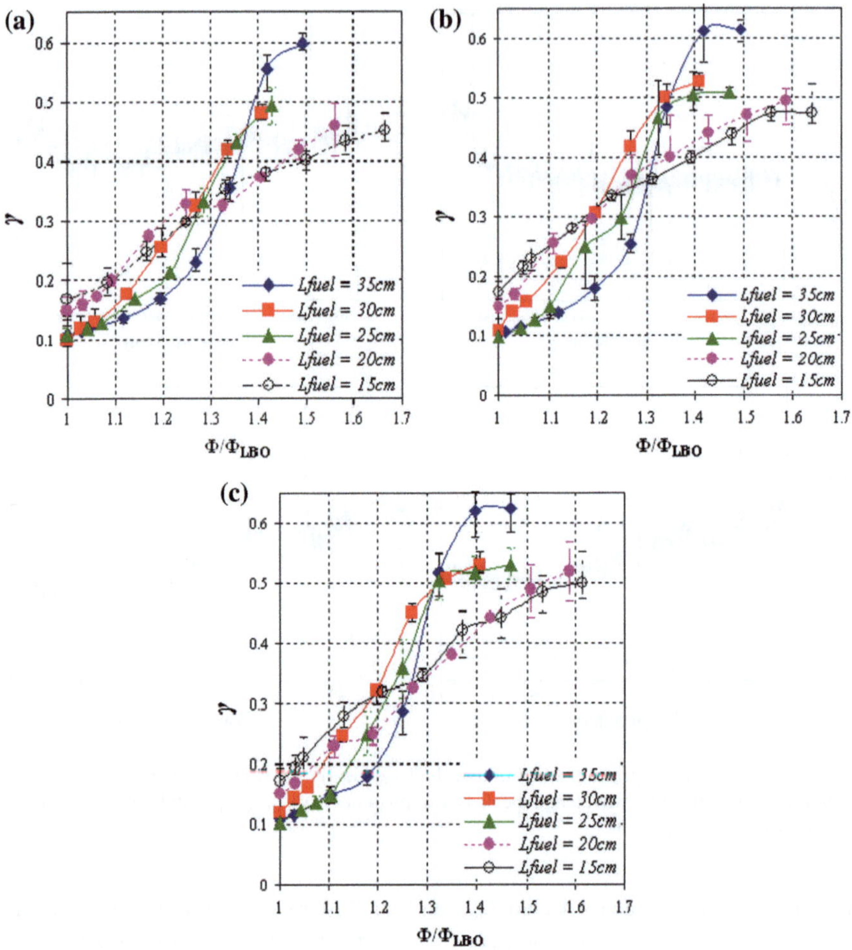

Fig. 12 Variation of γ with Φ/Φ_{LBO} for different air flow rates: **a** Q_{air} = 70 LPM, **b** 80 LPM, and **c** 85 LPM (Chaudhari et al. 2013)

this technique is that the characterization of the flame dynamics itself in terms of this metric does not involve any a priori definition of threshold.

For calculating γ, very few computations are required so the real-time requirement is satisfied. Time can be reduced by processing a single image instead of taking average of four images. This has also been tested and found satisfactory. The LBO warning will be initiated on the set threshold (γ_{thresh}), and the active control system will take appropriate action for enhancing the stabilization of flame in combustor so that the incipient blowout could be avoided and LBO limit can be extended.

From the above results, it is observed that the proposed flame color-based LBO detection technique works consistently over the various degrees of air/fuel

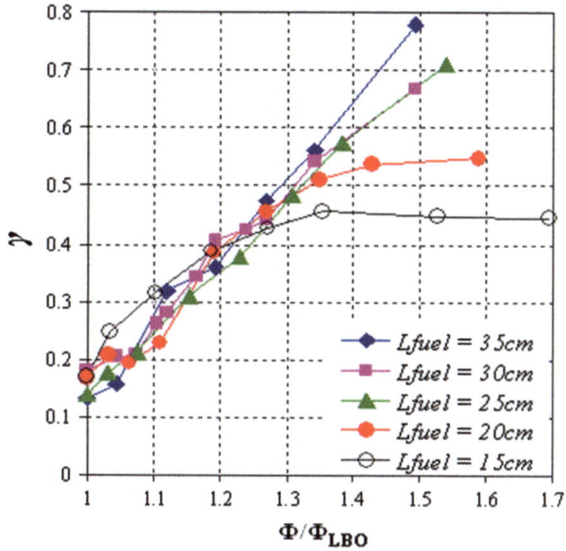

Fig. 13 Variation of γ with Φ/Φ_{LBO} for high air flow rate 175 LPM for different premixing lengths (L_{fuel}). Shown method is capable to detect LBO in high flow rate (Chaudhari et al. 2013)

premixing. The Reynolds numbers for which the technique is tested and discussed above are low compared to the actual flow condition in practical gas turbine system. In order to check the efficacy of the current LBO detection technique in high flow regime, experiments has been carried out where we used Q_{air} = 175 LPM and flow rate is tested for all premixing lengths allowed by the design of premixing tube. Figure 13 plots the variation of γ for equivalence ratio normalized by its value at LBO limit (i.e., Φ/Φ_{LBO}) for all the allowable L_{fuel} and with fixed air flow rate (Q_{air}) 175 LPM.

The trend obtained for γ in high flow rate for different premixing lengths is quite satisfactory and matches with the earlier observed trend of low flow rate. This proves that the flame color-based LBO detecting technique consistently works satisfactorily over a wide range of flow rates.

All the experiments discussed above are carried out with fuel (LPG: 60 % propane, 40 % butane). The proposed method works satisfactorily and can be implemented with active control system for early detection of incipient LBO. However, natural gas is a common fuel in practical gas turbine system. In order to check the efficacy of the proposed LBO predictor, methane is used as fuel. We tested high flow rate 200 LPM on 3 different premixing lengths. Figure 14 shows the variation of γ reported over normalized equivalence ratio (i.e., Φ/Φ_{LBO}). From the plot, it is observed that the γ significantly drops near LBO and matches with the earlier trend observed in LPG fuel. Moreover, the values of the metrics also match closely for the two fuels. This proves that the proposed LBO detection technique can satisfactorily monitor the proximity of LBO in configuration of high flow with methane fuel which is common configuration used in practical gas turbine.

The previously discussed conventional LBO detecting techniques are highly dependent on the extent of fuel/air premixing, and they need a sample length (N) to

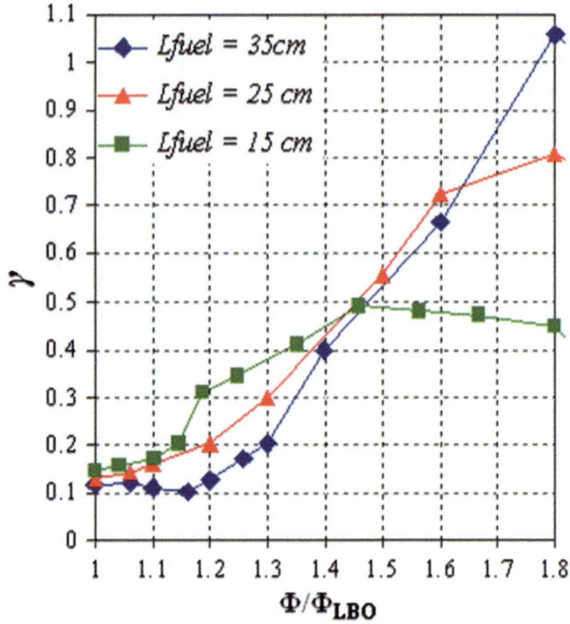

Fig. 14 Variation of γ with Φ/Φ_{LBO} for different L_{fuel}, when methane is used as fuel. Result shows that method is capable to detect LBO (Chaudhari et al. 2013)

produce the robust measure of the proximity to LBO. Consequently, the computation time required for producing the desired LBO metric is higher, which hampers the time response characteristic of sensing technique. This major shortcoming of those discussed methods is considerably reduced in the proposed flame color-based LBO detecting technique as the computation time required for simply reading the images is very low. Apart from this, the proposed method is cost-effective and easily implementable and shows consistency at different degrees of reactant premixing.

3.2 Symbolic Time Series Analysis

In this investigation, the use of STSA of captured heat release fluctuations was explored as a tool for early detection of imminent LBO. This approach has been successfully implemented by Bukkapatnam et al. (1995, 2000), Chin et al. (2005), Gupta et al. (2006, 2007), Chakraborty et al. (2008) for different systems. However, STSA for experimentally obtained time series data in combustion systems has not been previously reported. A major advantage of working with symbols is that the efficiency of numerical computation is significantly enhanced relative to what can be achieved by direct analysis of the original time series data.

In STSA, the range of data is divided into a number of non-overlapping regions. A symbol is assigned to each region. The collection of symbols is referred to as *alphabet*. When data from the time series sequence fall in a particular region, it is coded with the symbol associated with that region. Thus, the time series data are converted to a sequence of symbols, referred to as *symbolic time series*. There are different approaches for partitioning the range of data into regions. The most intuitive approach generates the partitions such that the region rich in data is divided into partitions of smaller size and region with sparse data has partitions of large size. In this work, the entire region is partitioned in such a way that each region contains equal number of data. This maximizes the information entropy associated with the data and is known as *maximum entropy partitioning*. To achieve this, the elements of time series data are sorted in ascending order. Starting from the first data in the sorted list, every consecutive data segment of length $\lfloor N/|\Sigma| \rfloor$ constitutes a distinct element of the partition where N and Σ represent the total number of data and number of symbols used and $\lfloor x \rfloor$ represents the greatest integer less than or equal to x. Once the partitions are generated, the time series data are converted into symbol sequence and the number of data in each segment is calculated. The array containing numbers of data in all the partitions constitutes the *state vector*. For detection of LBO, a condition away from the blowout limit where combustion is stable is identified as the nominal state. The partitions are generated using maximum entropy partitioning of this nominal time series data. In other words, at the nominal state, the number of data in each partition is same. Once the partitions are generated, the same partitions are used for all subsequent operating conditions. As the equivalence ratio is reduced, new time series data are obtained. Each time series data are partitioned using the same partitions as the nominal state, and the number of data in each region is calculated. A new state vector containing the number of data in all the partitions for the new operating condition is computed. Some measure of deviation of the new state vector from the nominal state vector is considered as an anomaly measure. The objective of the present research is to use this anomaly measure for predicting LBO. For this, two methods of generating partitions are used. In one approach, the time series data are directly used. This has been referred to as *simple partitioning* (simple P). In the other approach, complex analytic functions are generated by Hilbert transform of the time series data. Given the Hilbert transform of a real-valued signal $x(t)$, the corresponding complex-valued analytic signal is defined as $A[x](t) = x(t) + i\tilde{x}(t)$ where $\tilde{x}(t)$ is the Hilbert transform of $x(t)$ (Subbu and Ray 2008). The partitioning is generated using both the amplitude and the phase of the complex analytic function. This method of STSA, referred to as analytic signal space partitioning (ASSP), is more effective than simple partitioning and computationally very efficient. The algorithms for simple partitioning and ASSP are given below. Further details could be found in Mukhopadhyay et al. (2013).

Simple Partitioning (Simple P)

Generation of partitions

1. Generate time series data of length N, $x(i)$ for the nominal state
2. Select number of states, N_1, and assign one symbol to each partition (alphabet)

3. Arrange the time series data in ascending order and store the sorted series as $x1(i)$
4. Assign $N_2 = \left\lfloor \frac{N}{N_1} \right\rfloor$ where $\lfloor\rfloor$ denotes nearest lower integer
5. Assign partition(1) = x1(1) and partition $(i + 1) = x1(i * N_2)$ for $1 \leq i \leq (N_1 - 1)$ and partition $(N_1 + 1) = x1(N)$

Construction of state probability vector for nominal state

1. Initialize state probability vector p(i) = 0 for $1 \leq i \leq N_1$
2. for $j = 1 : N$

   for $i = 1 : N_1$

   if $x(j) \geq partition(i)$ & $x(j) < partition(i+1)$,

   $p(i) = p(i) + 1$

   end

   end

   end
3. p0(i)=p(i)/N, for $1 \leq i \leq N_1$

Construction of state probability vector for other states

1. Generate time series data of length N', x(i) for the non-nominal state
2. Initialize state probability vector p(i) = 0 for $1 \leq i \leq N_1$

   for $j = 1 : N'$

   for $i = 1 : N_1$

   if $x(j) \geq partition(i)$ & $x(j) < partition(i+1)$,

   $p(i) = p(i) + 1$

   end

   end

   end
3. p1(i)=p(i)/N, for $1 \leq i \leq N_1$

Use the same partitions for all the states

Determination of Anomaly Measure

1. Compute anomaly measure, $M = \cos^{-1} \frac{\mathbf{p_0} \cdot \mathbf{p_1}}{\|\mathbf{p_0}\| \|\mathbf{p_1}\|}$

ASSP

Generation of partitions

1. Generate time series data of length N, $x(i)$ for the nominal state
2. Generate complex analytic function using Hilbert transform, $x0(i) = Hilbert\,(x(i))$
3. Arrange $abs(\mathbf{x0})$ and $angle(\mathbf{x0})$ in ascending orders and store the sorted series as $x1(i)$ and $x2(i)$
4. Select number of states, N_{11} and N_{12} for $x1$ and $x2$, respectively, and assign one symbol to each partition (alphabet)
5. Assign $N_{21} = \left\lfloor \frac{N}{N_1} \right\rfloor b$ and $N_{22} = \left\lfloor \frac{N}{N_2} \right\rfloor$ where $\lfloor\rfloor$ denotes nearest lower integer
6. Assign partition1(1) = x1(1) and partition1$(i+1) = x1(i * N_2)$ for $1 \le i \le (N_{11} - 1)$ and partition $(N_{11} + 1) = x1(N)$
7. Assign partition2(1) = $x2(1)$ and partition2$(i+1) = x2(i * N_2)$ for $1 \le i \le (N_{12} - 1)$ and partition $(N_{12} + 1) = x2(N)$

Construction of state probability vector for nominal state

1. Initialize state probability vector p(i) = 0 for $1 \le i \le N_{11} * N_{12}$
2. for $j = 1 : N$

    for $i1 = 1 : N_{11} - 1$

      for $i2 = 1 : N_{12} - 1$

        if $\begin{array}{l} x1(j) \ge partition1(i1)\ \&\ x1(j) < partition1(i1+1)\ \& \\ x2(j) \ge partition2(i2)\ \&\ x2(j) < partition2(i2+1)) \end{array}$,

          $p((i1-1)*(N_{12}-1)+i2) = p((i1-1)*(N_{12}-1)+i2)+1$

        end

      end

    end
3. p0(i)=p(i)/N, for $1 \le i \le N_{11} * N_{12}$

Construction of state probability vector for other states

1. Generate time series data of length N', $x(i)$ for the non-nominal state
2. Generate complex analytic function using Hilbert transform, $x0(i) = hilbert\,(x(i))$
3. Arrange $abs(\mathbf{x0})$ and $angle(\mathbf{x0})$ in ascending orders and store the sorted series as x1(i) and x2(i)
4. Initialize state probability vector p(i) = 0 for $1 \le i \le N_{11} * N_{12}$

5. for $j = 1 : N'$

   for $i1 = 1 : N_{11} - 1$

      for $i2 = 1 : N_{12} - 1$

         if $\begin{matrix} x1(j) \geq partition1(i1)\ \&\ x1(j) < partition1(i1+1)\ \& \\ x2(j) \geq partition2(i2)\ \&\ x2(j) < partition2(i2+1)) \end{matrix}$,

         $p((i1-1)*(N_{12}-1)+i2) = p((i1-1)*(N_{12}-1)+i2)+1$

         end

   end

end

6. $\mathrm{p1(i)} = \mathrm{p(i)}/N'$, for $1 \leq i \leq N_{11} * N_{12}$

Use the same partitions for all the states
Determination of Anomaly Measure

1. Compute anomaly measure, $M = \cos^{-1} \frac{\mathbf{p}_0 \cdot \mathbf{p}_1}{\|\mathbf{p}_0\| \|\mathbf{p}_1\|}$

In the present context, the time series of heat release fluctuation of flame (i.e., CH signal) captured on photomultiplier tube (PMT) is used for STSA analysis. The algorithms stated above are applied, and the LBO predictor in the form of anomaly measure has been deduced to monitor the proximity of imminent blowout. The same time traces of CH signal which are captured in different experimental runs on which the preceding optical LBO detecting methods were described are used in the present analysis. We observed that as Φ approaches to Φ_{LBO}, there is a distinct change in the flame dynamics which apparently vary the heat release rate.

Figure 15a shows the variation of anomaly measure as a function of equivalence ratio Φ for port 1 (L_{fuel} = 35 cm), Q_{air} = 85 LPM, and the nominal condition (Φ) taken as 0.82. In this investigation, alphabet size was taken as 8 for simple partitioning, while for ASSP, the amplitude and phase spaces were divided into 8 and 5 regions, respectively. From Fig. 15a, it has been found that the anomaly measure first increases sharply as the equivalence ratio decreases from the nominal value but changes very slowly and acquires a value close to 1 as LBO is approached. Thus, a sudden change in the slope of the anomaly measure, which is observed at a reasonable distance from LBO, may be useful for LBO prediction. On carefully observing the trend of anomaly measure, thresholding of the anomaly measure was found to be very suitable for sensing the proximity of LBO. From Fig. 15a, the threshold of 0.7 can be satisfactorily used, which gives enough margin for sensing the imminent blowout. Once the set threshold value is crossed, the active control system takes appropriate action for enhancing the stabilization of flame in combustor so that the incipient blowout could be avoided and LBO limit can be extended. This can be done either by increasing the pilot fuel or possibly by using small amplitude fuel modulation.

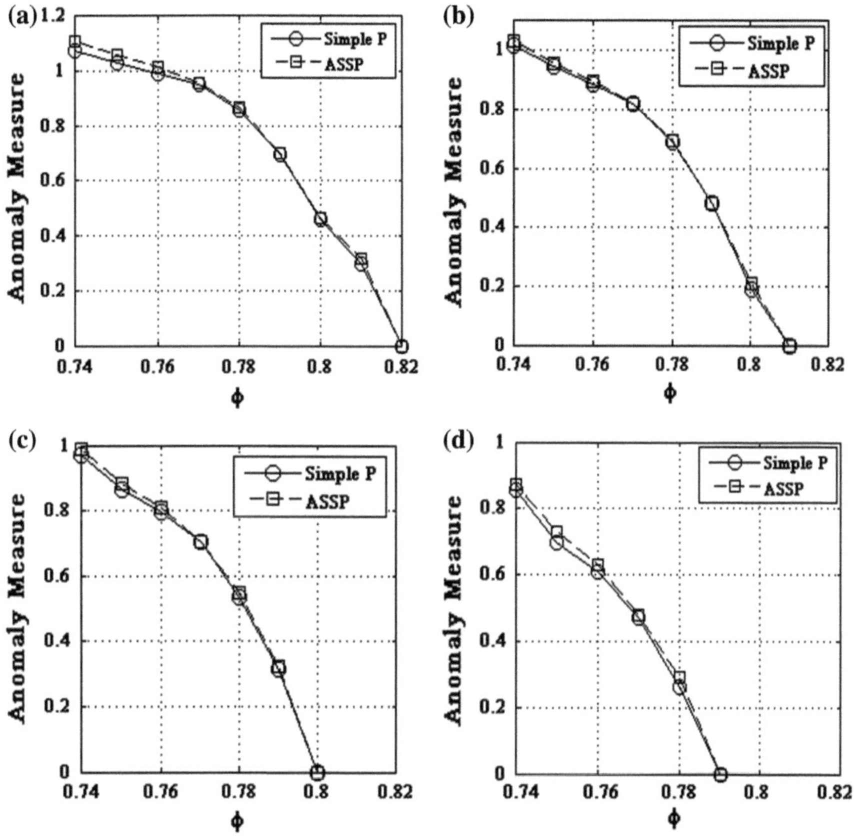

Fig. 15 Variation of anomaly measure with Φ for air flow Q_{air} = 85 LPM, L_{fuel} = 35 cm (port 1) for different nominal conditions taken: **a** Φ = 0.82, **b** Φ = 0.81, **c** Φ = 0.8, and **d** Φ = 0.79

The range of operating condition where the combustor exhibits stable flame is long; it is preferable to select the nominal condition nearer to LBO where flame shows stable operation. We explore the technique by varying the nominal condition with different equivalence ratios, which is shown in Fig. 15a–d, and the results obtained are quite satisfactory and a chosen value of threshold can be used for sensing LBO.

As noted earlier, Φ_{LBO} is observed to be dependent on L_{fuel}, i.e., the degree of premixing. In order to explore the applicability of anomaly measure LBO predictor for various degrees of premixing permitted by our setup (characterized by L_{fuel}), we present the variation of anomaly measure with Φ/Φ_{LBO} calculated with simple partitioning and ASSP for different L_{fuel} with nominal condition close to Φ = 0.80 for different flow conditions shown in Figs. 16 and 17. The implication of this observation is that the anomaly measure calculated with simple partition acquires a value in the range 0.9–1.1 in all the cases irrespective of the premixing length used

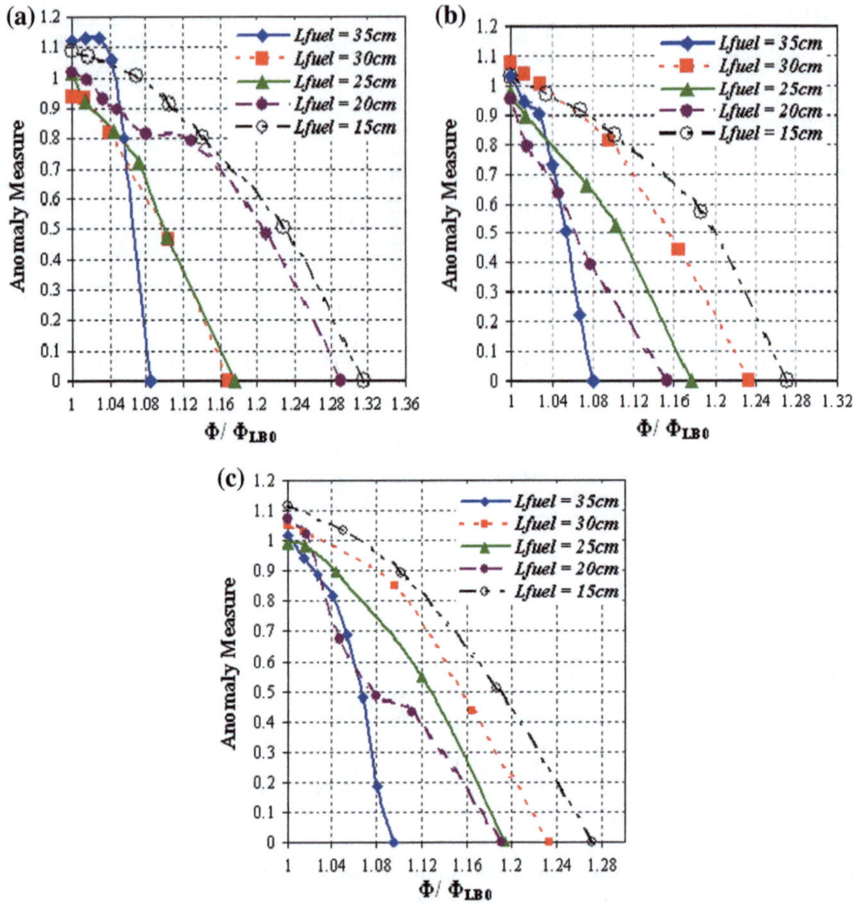

Fig. 16 Variation of anomaly measure calculated by simple portioning with Φ/Φ_{LBO} for different port positions (L_{fuel}) permitted by set up with air flow Q_{air}: **a** 70 LPM, **b** 80 LPM, and **c** 85 LPM

(L_{fuel}); similarly, the anomaly measure calculated with ASSP acquires a value in the range 0.9–1.2 for all experiments which suggest that the set threshold of the anomaly measure can be successfully used for sensing the proximity of LBO. The choice of selecting the anomaly threshold is not fixed and can be varied depending upon the configuration of the combustor and the safety margin required.

Similar to flame color technique, here also we checked the capability of proposed symbolic time series analysis LBO detecting technique at high flow rate which simulates Reynolds number near to practical gas turbine system with three ports correspondingly giving 3 premixing lengths (port 1, port 3, and port 5) for early detection of LBO. The variation of anomaly measure with Φ/Φ_{LBO} calculated from both simple and ASSP partitioning methods, respectively, for different high flow rates has been calculated. The trends of anomaly measure match with the

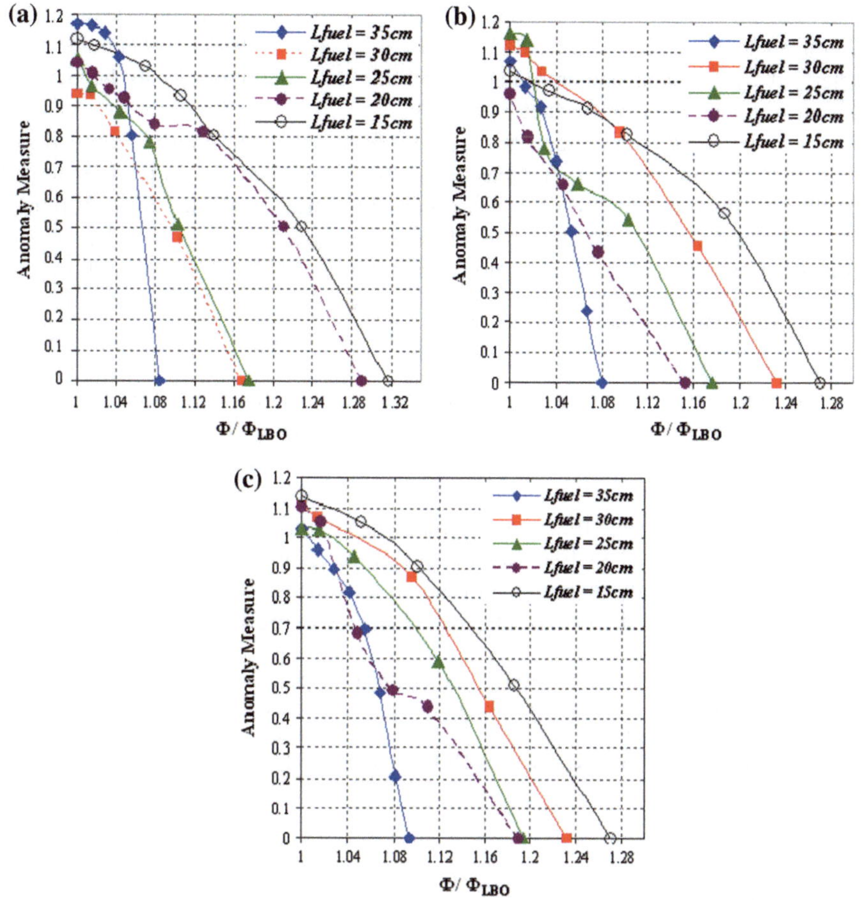

Fig. 17 Variation of anomaly measure calculated by ASSP with Φ/Φ_{LBO} for different port positions (L_{fuel}) permitted by set up with air flow Q_{air}: **a** 70 LPM, **b** 80 LPM, and **c** 85 LPM

earlier trends of lower flow rates plotted in Figs. 16 and 17 (Mukhopadhyay et al. 2013), which proves that the STSA-detecting LBO technique consistently works satisfactorily over a wide range of flow rates.

4 Conclusions

We present an experimental study on a swirl-stabilized, LPG–air, and methane–air-fueled dump plane combustor. Experiments carried out at different flow rates varying from 70 to 200 lpm with varied degree of fuel–air premixing. Equivalence ratios are also varied from stoichiometric to LBO limit. Studies have been carried out to characterize and predict flame behavior near LBO. It is observed that for premixed flames, LBO is preceded by a high level of intermittent behavior characterized by frequent and irregular extinctions and reignitions. This behavior is absent in lower level of air–fuel premixing.

For all levels of premixing, a shift of color from reddish to bluish as equivalence ratio reduces toward LBO. A metric has been developed for early detection of imminent blowout. The ratio of red to blue intensities, integrated over all points, showed a monotonical variation with change in equivalence ratio. This behavior is consistent over a large range of degree of premixing of different fuels and flow rate. The characteristics can be measured using a standard CCD/CMOS sensor and used as a detection metric. The method is cheaper, computationally easier, and easy to implement in real time.

The time series of optical data obtained from PMT has been used to develop another metric for early LBO detection. The time series is first converted to a symbolic time series, by partitioning the data. Two methods of partitioning have been discussed here. As the flame shifts toward LBO, the symbolic time series shows a deviation from the normal operating condition (stoichiometric or near stoichiometric). The deviation can be quantified using a norm called anomaly measure. This anomaly measure can be used as a metric for LBO detection. Both partitioning methods showed satisfactory result regarding LBO detection.

Acknowledgment The authors sincerely acknowledge the help extended by Suvojit Ghosh, Rakesh Prasad Sahu, Tanoy Paul, Souvick Chatterjee, Arkadeep Kumar, and Kamalika chatterjee. The contribution and collaboration from Prof. Asok Ray, Pennsylvania State University, is gratefully acknowledged. We acknowledge DRDO, AICTE, and CSIR for their financial support toward this work.

References

Ateshkadi A, McDonell VG, Samuelsen GS (2000) Lean blowout model for a spray-fired swirl-stabilized combustor. Proc Combust Inst 28:1281–1288

Barnes JC, Mellor AM (1998) Effects of unmixedness in piloted-lean premixed gas turbine combustors. J Propul Power 14:967–973

Barnum BM, Bell RC (1993) Flame failure detection method. US Patent-5235802, filed 17 Aug 1993

Bukkapatnam S, Kumara S, Lakhtakia A (1995) Analysis of sensor signal shows turning on a lathe exhibits low-dimensional chaos. Phys Rev E 52:2375–2387

Bukkapatnam S, Kumara S, Lakhtakia A (2000) Fractal estimation of flank wear in turning. ASME J Dyn Syst Measur Control 122:89–94

Chakraborty S, Gupta S, Ray A, Mukhopadhyay A (2008) Dynamic data-driven detection and estimation of faults in thermal pulse combustors. Proc Inst Mech Eng: J Aerosp Eng 222:1097–1108

Chao YC, Chang YL, Wu CY, Cheng TS (2000) An experimental investigation of the blowout process of a jet flame. Proc Combust Inst 28:335–342

Chaudhari RR (2011) Investigation of thermoacoustic instabilities and lean blowout in a model gas turbine combustor. PhD thesis, Jadavpur University

Chaudhari RR, Sahu RP, Ghosh S, Mukhopadhyay A, Sen S (2013) Flame color as a lean blowout predictor. Int J Spray Combust Dyn 5:49–66
Chaudhuri S, Cetegen BM (2008) Blowoff characteristics of bluff-body stabilized conical premixed flames with upstream spatial mixture gradients and velocity oscillations. Combust Flame 153:616–633
Chaudhuri S, Cetegen BM (2009) Blowoff characteristics of bluff-body stabilized conical premixed flames in a duct with upstream spatial mixture gradients and velocity oscillations. Combust Sci Technol 181:555–569
Chaudhuri S, Kostaka S, Renfro MW, Cetegen BM (2010) Blowoff dynamics of bluff body stabilized turbulent premixed flame. Combust Flame 157:790–802
Cheng SI, Kovitz AA (1958) Theory of flame stabilization by a bluff body. Proc Combust Inst 7:681–691
Chin SC, Ray A, Rajagopalan V (2005) Symbolic time series analysis for anomaly detection: a comparative evaluation. Sig Process 85:1859–1868
Cohen JM, Stufflebeam JH, Proscia W (2001) The effect of fuel/air mixing on actuation authority in an active combustion instability control system. J Eng Gas Turbines Pow Trans ASME 123:537–542
Correa SM (1993) A review of NO_x formation under gas-turbine combustion conditions. Combust Sci Technol 87:329–362
Correa SM (1998) Power generation and aero-propulsion gas turbines: from combustion science to combustion technology. Proc Combust Inst 28:1793–1807
De Zilwa SRN, Uhm JH, Whitelaw JH (2000) Combustion oscillations close to the lean flammability limit. Combust Sci Technol 160:231–258
DuBell TL, Cifone AJ (1989) Combustor influence on fighter engine operability. AGARD meeting on mechanisms of instability in liquid fueled combustors
Glassman I (1996) Combustion, 3rd edn. Academic Press, San Diego
Gupta S, Ray A, Mukhopadhyay A (2006) Anomaly detection in thermal pulse combustors using symbolic time series analysis. Proc Inst Mech Eng: J Syst Control Eng 220(2006):339–351
Gupta S, Ray A, Keller E (2007) Symbolic time series analysis of ultrasonic data for early detection of fatigue damage. Mech Syst Signal Process 21:866–884
Hedman PO, Fletcher TH, Graham SG, Timonthy GW, Flores DV, Haslam JK (2002) Observation of flame behavior in a laboratory—scale premixed natural gas/air gas turbine combustor from PLIF measurement of OH. ASME Turbo Expo, Paper GT-2002-30052, Amsterdam
Hertzberg JR, Shepherd IG, Talbot L (1991) Vortex shedding behind rod stabilized flames. Combust Flame 86:1–11
Jensen WP, Shipman CW (1958) Stabilization of flames in high speed flows by pilot flames. Proc Combust Inst 7:674–680
Lefebvre AH (1995) The role of fuel preparation in low-emission combustion. J Eng Gas Turbines Power 117:617–654
Lefebvre AH (1999) Gas turbine combustion. Taylor and Francis, Philadelphia
Li H, Zhou X, Jeffries JB, Hanson RK (2007) Active control of lean blowout in a swirl-stabilized combustor using a tunable diode laser. Proc Combust Inst 31:3215–3223
Lucenko M, Vanderleest RE, Kenneth J (1996) Method and apparatus for detecting burner blowout. US Patent No. 5581995, filed 10 Dec 1996
Mukhopadhyay A, Datta S, Sanyal D (2008) Effects of tailpipe friction on the nonlinear dynamics of a thermal pulse combustor. J Eng Gas Turbines Pow Trans ASME 130:1–9
Mukhopadhyay A, Chaudhari RR, Paul T, Sen S, Ray A (2013) Lean blowout prediction in gas turbine combustors using symbolic time series analysis. J Propul Power 29:950–960
Muruganandam TM (2006) Sensing and dynamics of lean blowout in a swirl dump combustor. Ph.D. thesis, Georgia Institute of Technology
Muruganandam TM, Seitzman JM (2005) Characterization of extinction events near blowout in swirl-dump combustors. In: 41st AIAA/ASME/SAE/ASEE joint propulsion conference and exhibit. Paper AIAA-2005-4331, Tucson, AZ

Muruganandam TM, Nair S, Neumeier Y, Lieuwen T, Seitzman J (2002) Optical and acoustic sensing of lean blowout precursors. In: 38th AIAA/ASME/SAE/ASEE joint propulsion conference and exhibit, Indianapolis. Paper AIAA-2002-3732, Indiana

Muruganandam TM, Kim B, Olsen R, Patel M, Roming B, Seitzman JM (2003) Chemiluminescence based sensors for turbine engines. In: 39th AIAA/ASME/SAE/ASEE joint propulsion conference. Paper AIAA-2003-4490, Huntsville, AL

Muruganandam TM, Nair S, Scarborough D, Neumeier Y, Jagoda J, Lieuwen T, Seitzman J, Zinn BT (2005) Active control of lean blowout for turbine engine combustors. J Propul Power 21:807–814

Nair S, Lieuwen T (2005) Acoustic detection of blowout in premixed flames. J Propul Power 21:32–39

Nair S, Muruganandan T, Olsen R, Meyers A, Seitzman JM, Zinn BT, Lieuwen T (2004) Lean blowout detection in a single nozzle swirl cup combustor. In: 42nd AIAA aerospace science meeting and exhibit. Paper AIAA 2004-0138, Reno, Nevada

Nicholson HM, Field JP (1949) Some experimental techniques for the investigation of the mechanism of flame stabilization in the wakes of bluff bodies. In: Symposium on combustion and flame, and explosion phenomena, vol 3, pp 44–68

Norton DG (2003) Combustion characteristics and flame stability at the micro scale: a CFD study of premixed methane/air mixtures. Chem Eng Sci 58:4871–4882

Prakash S. (2007) Lean blowout mitigation in swirl stabilized premixed flames. Ph.D. thesis, Georgia Institute of Technology

Prakash S, Nair S, Muruganandan TM, Neumeier Y, Lieuwen T, Seitzman JM, Zinn BT (2005) Acoustic based rapid blowout mitigation in a swirl stabilized combustor. In: Proceedings of ASME Turbo Expo. Paper GT 2005-68589, Reno, Nevada

Rosfjord TJ, Cohen JM (1995) Evaluation of the transient operation of advanced gas turbine combustors. J Propul Power 11:497–504

Schefer RW, Wicksall DW, Agrawal AK (2002) Combustion of hydrogen-enriched methane in a lean premixed swirl-stabilized burner. Proc Combust Inst 29:843–850

Shashvat P, Nair S, Murganandam TM, Neumeier Y, Lieuwen T, Seitzman J, Zinn, BT (2005) Acoustic sensing and mitigation of lean blow-out in premixed flame. In: 43rd aerospace science meeting. Paper AIAA-2005-1420, Reno, Nevada

Snyder TS, Rosfjord TJ (1998) Active gas turbine combustion control to minimize nitrous oxide emissions. U.S. Patent No. 5706643, Filed 13 Jan 1998

Subbu A, Ray A (2008) Space partitioning via Hilbert transform for symbolic time series analysis. Appl Phys Lett 92: 084107-1–084107-3

Turns S (2000) An introduction to combustion. McGraw-Hill, New York

Tomas S (2005) Combustion instability and its passive control. In: Lieuwen TC, Yang V (eds) Combustion instabilities in gas turbine engines: operational experience, fundamental mechanisms, and modeling. Progress in astronautics and aeronautics, vol 210, pp 65–88

Williams FA (1989) Structure of flamlets in turbulent reacting flows and influence of combustion on turbulent fields. In: Borghi R, Murthy SNB (eds) Turbulent reactive flows. Lecture notes in engineering, vol 40. Springer, New York, pp 195–212

Williams TC, Schefer RW, Oefelein JC, Shaddix CR (2007) Idealized gas turbine combustor for performance research and validation of large eddy simulations. Rev Sci Instrum 78:035114-1–035144-9

Yi TX, Gutmark EJ (2007) Real-time prediction of incipient lean blowout in gas turbine combustor. AIAA J 45:1734–1739

Zeldovich J (1946) The oxidation of nitrogen in combustion and explosions. Acta Physicochimica 21:577–628

Fluidized Bed Combustion of Coal, Renewable Fuels, and Waste: Current Status and Developments

Franz Winter

Abstract Coal is a very important fuel because of its high calorific value of up to about 33 MJ/kg and its worldwide distribution (refer also to Lackner et al. 2013). Coal is available on the international market and can be obtained from many mines of the world. However, coal is a fossil fuel, rich in carbon and produces CO_2, a major greenhouse gas. Beside its carbon content, nitrogen and sulfur may be present in the coal as well at significant levels. The nitrogen in the fuel may be converted via complex reaction routes to NO_x which contributes to acid rain formation and shows direct negative effects to human health. The sulfur in the coal is oxidized to SO_2 which subsequently contributes to acid rain formation and direct adverse effects of the human health. Beside high-quality coals which are low in ash, also coals with high amounts of ash and moisture exist. These coals are significantly lower in their energy contents and usually used locally because of their availability and lower costs.

Keywords Fluidized bed technology · Supercritical and ultra-supercritical steam · Biomass · Municipal solid waste · Steam gasification · CO_2-rich flue gas

1 Introduction

Coal is a very important fuel because of its high calorific value of up to about 33 MJ/kg and its worldwide distribution (refer also to Lackner et al. 2013). Coal is available on the international market and can be obtained from many mines of the world. However, coal is a fossil fuel, rich in carbon and produces CO_2, a major greenhouse gas. Beside its carbon content, nitrogen and sulfur may be present in the coal as well at significant levels. The nitrogen in the fuel may be converted via complex reaction routes to NO_x which contributes to acid rain formation and shows

F. Winter (✉)
Institute of Chemical Engineering, Vienna University of Technology, Vienna, Austria
e-mail: franz.winter@tuwien.ac.at

A.K. Agarwal et al. (eds.), *Novel Combustion Concepts for Sustainable Energy Development*, DOI 10.1007/978-81-322-2211-8_10

direct negative effects to human health. The sulfur in the coal is oxidized to SO_2 which subsequently contributes to acid rain formation and direct adverse effects of the human health. Beside high-quality coals which are low in ash, also coals with high amounts of ash and moisture exist. These coals are significantly lower in their energy contents and usually used locally because of their availability and lower costs.

Renewable Fuels like biomass, wood, wood waste, and bark are significantly lower in their calorific values (typically about 15 MJ/kg) compared to coals and are considered CO_2 neutral. They are usually low in ash and sulfur; however, significant nitrogen contents are possible depending on the source of the biofuel. Also high moisture contents are possible depending on the conditions of harvesting and storing. Because of their origin, decomposition and changes in their quality with storage may be of importance.

Waste is a very heterogeneous fuel and may differ in many ways. Some waste which is high in plastics show very high calorific values, for example, up to about 46 MJ/kg if polyethylene or polypropylene is considered. However on the other hand, municipal solid waste (MSW) may be in average as low as about 10 MJ/kg. Also, the calorific value of sewage sludge is low, depending strongly on the water content. Waste may be rich in chlorine forming HCl which is acidic and corrosive and a hazardous pollutant. In addition, heavy metals are usually present in waste in higher quantities and may be found in the fly ash or bottom ash of the combustion system.

2 Fluidized Bed Combustion Technology

The fluidized bed technology is positioned between a fixed bed like a grate furnace where the fuel particles are rather fixed on the grate and an entrained flow where the fuel particles are in a state of pneumatic transport (refer also to Lackner et al. 2010). In fluidized bed technology, the fuel particles are mixed with bed material usually made of ash particles and some additional silica sand particles and they are in a state of fluidization somehow similar to a fluid, similar to a liquid. The combustion air comes from below to fluidize the fuel together with the bed material. Because of the vigorous fluidization, excellent mixing and high heat transfer rates are obtained. This leads to high combustion rates and high efficiencies. The bed material also acts like a heat storage and this characteristic is ideal to handle a wide range of different fuels with different moisture contents and qualities. The combustion temperatures are usually in a lower range about 850 °C, resulting in practically no thermal NO_x formation. Therefore, fluidized bed combustion is usually low in NO_x. The NO_x generated is from the fuel nitrogen as described in the introduction.

If the fuel is rich in sulfur, limestone can be added into the fluidized bed combustor. Then the limestone undergoes calcination and reacts with the SO_2 generated during fuel combustion. This is an in situ desulfurization of the combustion process and decreases SO_2 emissions from the combustion chamber significantly. The in situ SO_2 capture efficiency from the fluidized bed is highest at bed operating temperature of 850 °C.

Potential disadvantages are that a somewhat higher power demand is necessary for the operation of a fluidized bed and that the fuel preparation is very important. There should not be too large and heavy objects in the feeding lines because those are not fluidized in the combustor. Another issue is that low melting mixtures have to be avoided because this may lead to agglomeration and consequently to defluidization of the bed.

Basically there are two different types of fluidized bed combustors (FBC), the stationary or bubbling fluidized bed and the circulating fluidized bed. In a bubbling fluidized bed (BFBC), the fluidized bed stays stationary but fluidized in the combustion chamber, whereas in the case of the circulating fluidized bed combustor (CFBC), the fluidized bed is circulating between the combustion chamber which is in form of a riser then separated usually by a cyclone from the gas and fed back via the return leg into the riser, that is, the fluidized bed is circulating.

Usually in smaller scale, BFBCs are in operation, whereas CFBCs operate up to very large scales. Currently the largest scale in operation of fluidized bed technology is the 600 MWe coal-fired CFBC in Baima, Neijiang City, China.

The CFBC can also handle a somewhat wider range of fuels. In terms of their calorific value, fuels from as low as about 7 MJ/kg up to about 40 MJ/kg can be utilized. In BFBC, typical fuels with calorific values from as low as about 3 MJ/kg up to about 20 MJ/kg are utilized.

3 Current Developments

There are several major developments of fluidized bed technology on-going worldwide.

3.1 One Is the Significant Progress in Scale-up

FBC, that is CFBC, are continuously further developed to increase their capacity (refer to Fig. 1). They operate at high efficiencies and use supercritical steam cycles. In 2014, the largest CFBC worldwide has successfully reached its 600 MWe. It is based on coal and located in Baima, Neijiang City, China (refer to Fig. 2).

In China, there are more than 3,000 units of CFBCs in operation and more than 100 units in the size range of 300 MWe (Yang 2012).

Another significant project of a similar size class is on-going in Korea. In Samcheok, Korea, 4 × 550 MWe CFBCs are in construction (refer to Fig. 3). They are located at the coast location and based on coal with the option of biomass addition. Their full operation is considered in 2016.

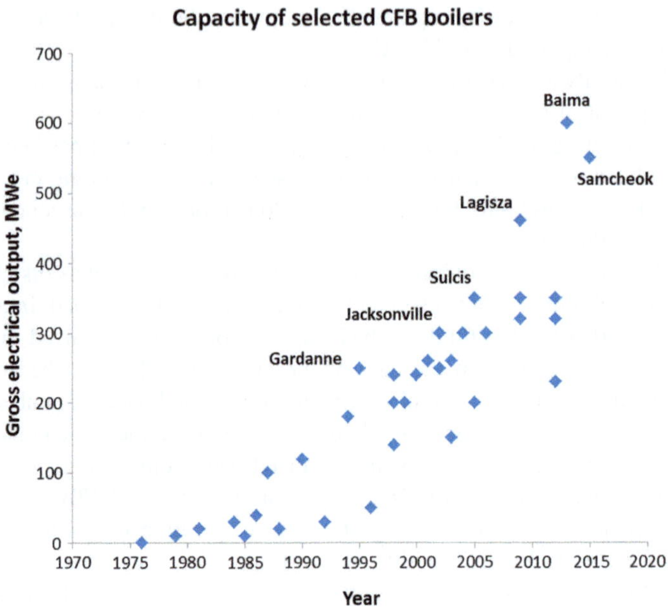

Fig. 1 Continuous development of scale-up of circulating fluidized bed combustors (CFBCs) (IEA-FBC report 2014)

Fig. 2 The 600 MWe CFBC in Baima, Neijiang City, China, with supercritical steam cycle (Yang 2013)

Fig. 3 The Samcheok project, 4 × 550 MWe gross CFBs Korea (KOSPO 2016) with supercritical steam cycle

3.2 *Biomass Addition to Coal Based Large Fluidized Bed Boilers*

There is also one in development that allows some percentage of biomass addition to coal in large scale fluidized bed boilers to gain advantage of using renewable fuels at high boiler efficiencies.

3.3 *Biomass-Based Fluidized Bed Combustors*

Biomass-based FBC are usually in a smaller size range. They typically range from 30 to 130 MWth. BFBCs as well as CFBCs are applied. Usually different biomass fuels are burned at the same time in the same combustor. However coal, sludges or waste is used as well. Such steam generators are used, for example, in the pulp and paper industry. There are on-going projects to increase their availability and to decrease their costs. A special focus is on fuel characterization and to burn the fuel mix in an optimum way.

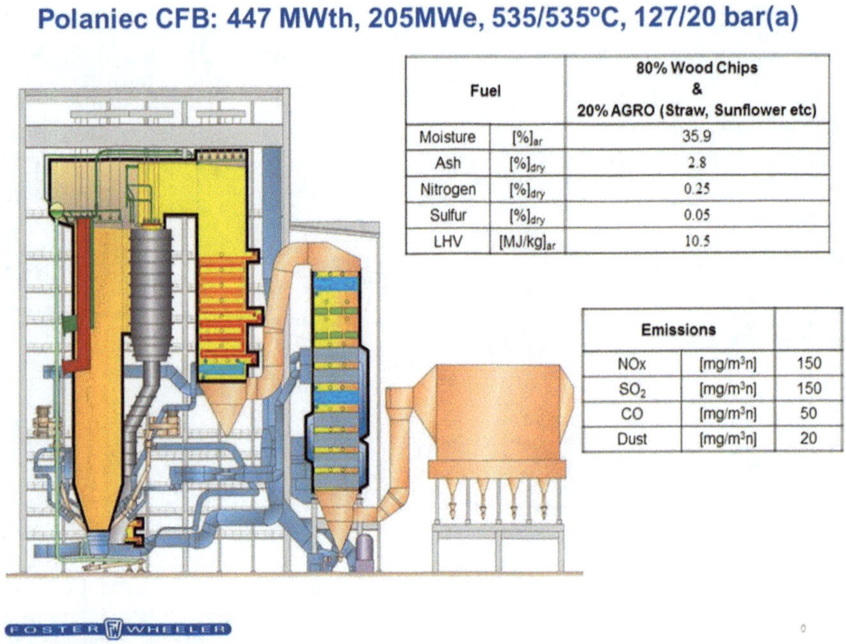

Fuel		80% Wood Chips & 20% AGRO (Straw, Sunflower etc)
Moisture	$[\%]_{ar}$	35.9
Ash	$[\%]_{dry}$	2.8
Nitrogen	$[\%]_{dry}$	0.25
Sulfur	$[\%]_{dry}$	0.05
LHV	$[MJ/kg]_{ar}$	10.5

Emissions		
NOx	$[mg/m^3n]$	150
SO_2	$[mg/m^3n]$	150
CO	$[mg/m^3n]$	50
Dust	$[mg/m^3n]$	20

Fig. 4 World's largest biomass CFB with 447 MWth, 205 MWe in Polaniec, Poland (Foster Wheeler 2014)

On the other hand, there are a few large-scale biomass-based FBC in operation. The world's largest is the Polaniec CFB boiler in Poland with a capacity of 447 MWth (see Fig. 4).

3.4 *Waste-Based Fluidized Bed Combustors*

The fluidized bed technology is very suitable to utilize waste. Most important is that the waste is well prepared to avoid operational problems in the fluidized bed combustor as discussed earlier. A wide range of fuels can be utilized in fluidized beds, for example, assorted MSW (MSW), residue-derived fuels (RDF), and even plastics. Typically the FBCs are in a smaller size range, for example, from 40 to 140 MWth, similar to those FBCs which are based on biomass utilization.

4 Novel Concepts

4.1 Gasification of Biomass—Using Fluidized Bed Technology

Fluidized bed technology is an excellent tool for use in biomass gasification. Because of its very good transportation of the bed material, the fluidized bed can be divided into two zones: A gasification zone where the biomass is gasified and a combustion zone where the residual biomass char is burned and heat is produced from the gasification of the biomass. The hot bed material is then transported into the gasification zone and the two zones are coupled. Steam is used as the fluidizing and gasification agent.

The gas produced is rich in H_2 and CO besides CO_2 and CH_4 and can be used for a range of applications such as Fischer-Tropsch for biodiesel or biofuel production, or synthesis gas which can be used in gas engines for electricity production (see Fig. 5).

4.2 Oxyfuel Combustion—Using Fluidized Bed Technology

In the European Union project Flexi Burn CFB (Tourunen 2013), it was demonstrated that a fluidized bed is able to operate successfully in the oxyfuel combustion mode. In oxyfuel combustion, the fuel is not burned with air but in a gas which is

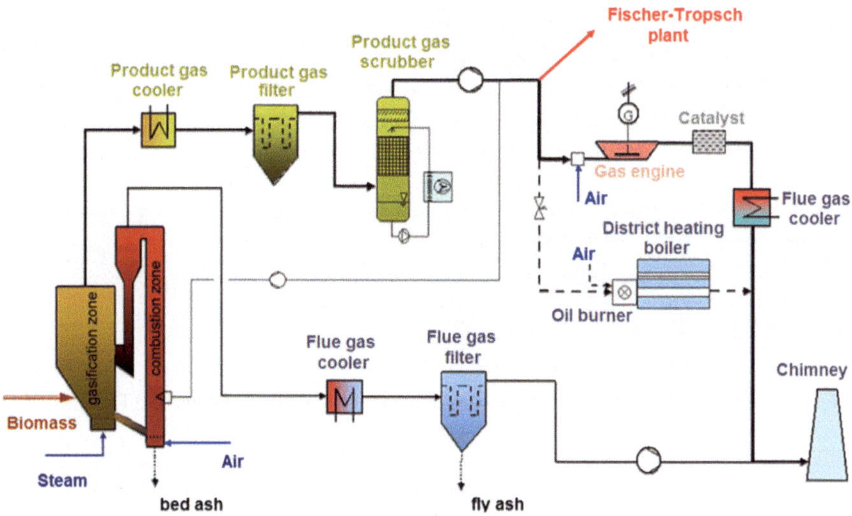

Fig. 5 Concept of the fluidized bed gasifier in Güssing, Austria, using biomass (Hofbauer 2002)

rich of oxygen and CO_2. Because of this technology, the flue gas is almost without nitrogen and the CO_2 is not diluted with nitrogen as it normally combusts with supplied air. These significantly help to remove the CO_2. In Fig. 6, the 30 MWth demonstration CFB unit is shown.

4.3 Chemical-Looping Combustion (CLC)—Using Fluidized Bed Technology

With chemical-looping, it is not necessary to use oxygen. Air is used however not to burn the fuel but to regenerate the metal oxide in the air reactor. The metal oxide is oxidizing the fuel in the fuel reactor. Both reactors are coupled by the mass flow of the metal oxide but the off gases are separated out, see Fig. 7.

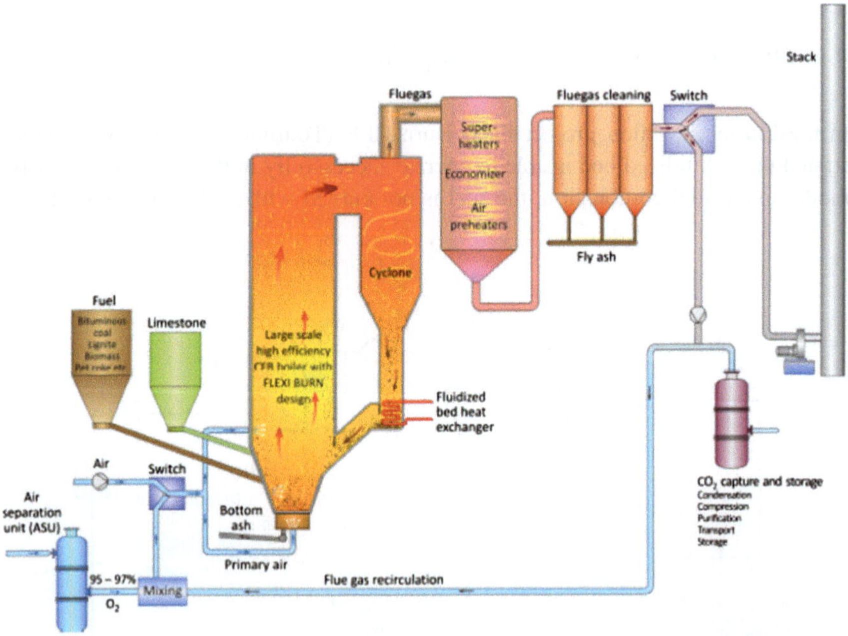

Fig. 6 Scheme of the 30 MWth CFB demonstration unit for oxyfuel combustion (Tourunen 2013)

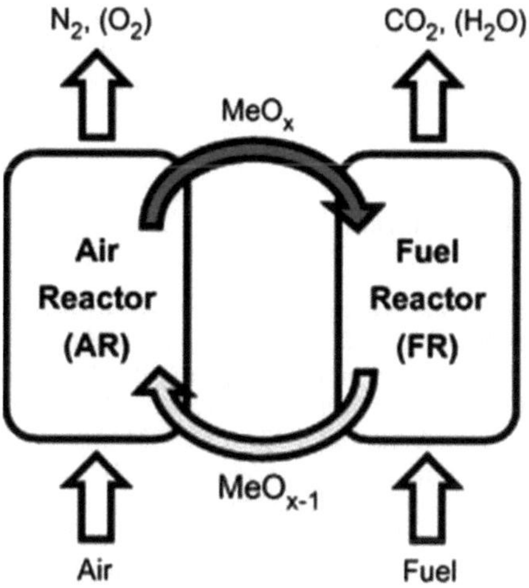

Fig. 7 The concept of chemical-looping combustion (CLC) is based on the application of fluidized bed technology (Penthor et al. 2014)

5 Conclusions

It can be seen that the fluidized bed technology is a very useful technology which has been applied in a wide range of power generation applications. Fuels as different as coals, biomass, sludges, and waste with different qualities can be utilized well in fluidized beds.

Scale-up has been continued and recently successfully demonstrated in the 600 MWe size range. High efficiencies can be achieved using supercritical and ultra-supercritical steam conditions in large-scale CFB boilers. Those units are based on coal.

Biomass is utilized using fluidized bed technology as well. Biomass can be added to large-scale fluidized bed boilers; however, biomass is typically utilized in smaller scale. Those biomass combustors operate usually in a size range from about 30 to about 130 MWth. On the other hand, there are a few large-scale biomass-based FBC in operation. The world's largest at present is the Polaniec CFB boiler in Poland with a capacity of 447 MWth.

The fluidized bed technology is very suitable to utilize wastes, for example, assorted MSW, RDF, and even plastics. Typically, the FBCs are in a smaller size range, which extend from 40 to 140 MWth, similar to those FBCs which are based on biomass utilization.

Novel concepts and successful demonstrations using fluidized bed technology are already on the market or under development, for example, the steam gasification of biomass wherein the syngas produced is rich in H_2 and CO which can be used for synthesis or in gas engines or oxyfuel combustion which produces a CO_2-rich flue gas which significantly helps in CO_2 removal or in chemical-looping combustion (CLC).

Acknowledgements The author thanks the Austrian Ministry for Transport, Innovation and Technology (BMVIT) for the financial support and the International Energy Agency—Fluidized Bed Conversion Group (www.iea-fbc.org) for the fruitful discussions.

References

Foster and Wheeler (2014) Received by personal communication with E. Coda

Hofbauer H (2002) Personal communication

IEA FBC End of Term Report (2014) Unpublished. www.iea-fbc.org

Lackner M, Winter F, Agarwal AK (2010) Handbook of combustion, vol. 4, Solid fuels. Wiley, New York. ISBN 978-3-527-32449-1

Lackner M, Palotas A, Winter F (2013) Combustion: from basics to application. Wiley, New York. ISBN 978-3-527-33376-9

Penthor S, Mayer K, Kern S, Kitzler H, Wöss D, Pröll T, Hofbauer H (2014) Chemical-looping combustion of raw syngas from biomass steam gasification—coupled operation of two dual fluidized bed pilot plants. Fuel 127:178–185

Tourunen A (2013) Development of high efficiency CFB technology to provide flexible air/oxy operation for power plant with CCS, poster, Flexi Burn CFB—EU FP7. Siehe unter http://www.vtt.fi/sites/flexiburncfb/index.htm

Yang H et al (2012) The influence of the new emission standard on the CFB boiler in China and the solutions, vol 64. In: IEA FBC Meeting, 2012, Neapel, Italien. Siehe unter www.iea-fbc.org

Yang H et al (2013) Design and operation of 600 MW CFB boiler in China, 67. In: IEA FBC meeting, 2013. Daejeon, Korea. Siehe unter www.iea-fbc.org

Using Petroleum and Biomass-Derived Fuels in Duel-fuel Diesel Engines

Suresh K. Aggarwal and Xiao Fu

Abstract There is worldwide interest in developing renewable energy sources in a sustainable manner. Syngas and biogas offer significant potential in this context, as these fuels offer great flexibility with regard to their production and utilization. This chapter provides an overview of research dealing with the combustion and emission characteristics of these fuels, both as stand-alone fuels or by blending with petroleum fuels. Conversion methods for producing these fuels from different biomass sources are also briefly reviewed. While the syngas composition can vary widely, it generally has lower heating value, lower density, higher mass diffusivity, higher flame speeds, and wider flammability limits compared to hydrocarbon fuels. Moreover, its combustion leads to almost zero soot emission, although NO_x emission may be a concern depending upon its composition and operating temperatures. Similarly, biogas has lower heating value compared to hydrocarbon fuels, and its ignition and combustion characteristics can vary noticeably depending upon its composition. While there have been few studies focusing directly on biogas combustion, there is extensive literature on methane combustion, including ignition, extinction, flammability limits, flame speeds, cellular instabilities, and emissions. Fundamental combustion aspects requiring further research include cellular instabilities, flame stabilization and blowout behavior, turbulent flames, and emission characteristics. Such efforts would lead to the development of optimized systems for producing these fuels and provide guidelines for optimizing their composition for a given set of operating conditions. The use of syngas and biogas in dual-fuel diesel engines has been a subject of numerous experimental and computational studies. A general observation is that the engine performance and emission characteristics are significantly modified by the presence of gaseous fuel. While the heat release in a diesel engine generally occurs through a hybrid combustion mode,

S.K. Aggarwal (✉)
Flow and Combustion Laboratory, Department of Mechanical Engineering,
University of Illinois at Chicago, Chicago, IL 60607, USA
e-mail: ska@uic.edu

X. Fu
Flow and Combustion Simulation Laboratory, Department of Mechanical
and Industrial Engineering, University of Illinois at Chicago, Chicago, IL 60607, USA

A.K. Agarwal et al. (eds.), *Novel Combustion Concepts for Sustainable Energy Development*, DOI 10.1007/978-81-322-2211-8_11

involving rich premixed combustion and diffusion combustion, that in a dual-fuel engine also involves a lean combustion mode with a propagating flame. The dual-fuel operation at high-load conditions can provide significant reduction in soot and CO soot emissions, while maintaining the engine efficiency, provided the injection characteristics including the amount of pilot fuel can be optimized. However, the NO_x emission may increase, requiring a suitable strategy for lowering the temperatures. The dual-fuel strategy may be less effective at low load, resulting in lower thermal efficiency and higher UHC and CO emissions. Future work should be directed at optimizing the various parameters, such as injection timing, amount of pilot fuel injected, EGR, and multiple injections.

Keywords Syngas · Biogas · Combustion and emissions · Dual-fuel engines

1 Introduction

There is a worldwide interest in developing renewable energy sources in a sustainable manner. This is motivated by concerns about depleting fossil energy sources, climate change due to greenhouse gas emissions, and growing energy needs due to emerging economies and population growth. The climate change concerns are illustrated in Fig. 1, which provides an historic perspective on the concentrations of three most significant greenhouse gases and their contributions to the increase in radiative forcing (W/m^2). As indicated, the CO_2 concentration in atmosphere has increased by 31 %, accounting for 1.46 W/m^2 change in radiative forcing since 1800. Clearly, a sustainable and carbon-neutral energy future will require a significant broadening of our energy portfolio and reducing reliance on non-renewable sources. The non-food and regional fuel sources, especially biomass, can potentially play a major role in this regard. Biomass represents one of the primary energy resources in the world after coal and oil, particularly in developing countries (Hall et al. 1991). It refers to a broad variety of feedstock ranging from agricultural waste, such as straw, bagasse, rice husks, olive pits, and nuts, to energy crops, such as, *miscanthus* and *sorghum* (Werther et al. 2000). It also includes algae, forestry waste, such as wood chips, bark, and thinnings, and other solid wastes including sewage sludge, as well as municipal solid waste. The use of biomass would not only reduce our dependence on fossil energy sources, but also provide energy in a sustainable and carbon neutral manner.

While a variety of fuels can be produced from biomass, syngas and biogas seem to offer the most potential with respect to flexibility in their production and utilization. Syngas can be produced using a variety of feedstock and conversion processes, particularly gasification, while biogas is generally produced by anaerobic digestion of biodegradable materials in an oxygen-free environment (Wikipedia 2014). Syngas mainly consists of H_2 and CO with varying amounts of CH_4 and CO_2, with the main constituents of biogas include CH_4, CO_2, and N_2 along with small traces of H_2O, O_2, H_2, and hydrogen sulfide. Both of these fuels represent a

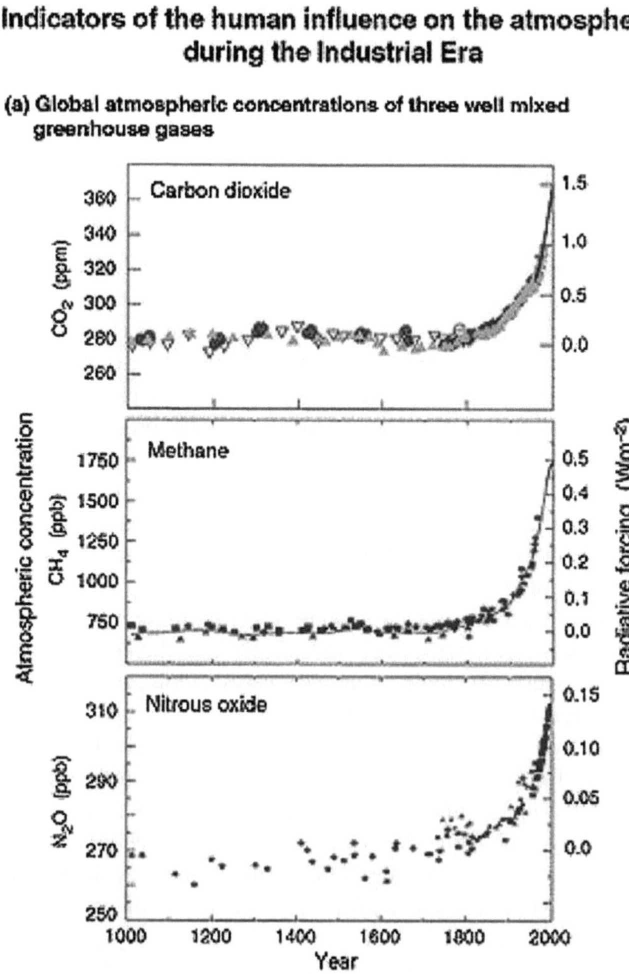

Fig. 1 Global atmospheric concentrations of three most important greenhouse gases

cleaner and renewable energy source and offer great flexibility in their utilization for transportation and power generation. This chapter provides an overview of research dealing with the combustion and emissions of these fuels, both as stand-alone fuel or by blending with petroleum fuels. Clearly, there is a vast body of literature on the fundamental and applied research concerning the production and utilization of these fuels. However, the scope of this chapter is limited to fundamental work concerning their ignition behavior, laminar flame burning these fuels, and the use of these fuels in dual-fuel diesel engines.

The chapter is organized in the following manner. The next section provides a brief summary of processes for converting biomass to more useful fuels, followed by an overview of fundamental studies on the combustion and emission characteristics of syngas and biogas fuels. Studies dealing with the use of syngas and biogas in dual-fueled diesel engines are discussed in Sect. 4, and a summary is provided in the last section.

2 Conversion Methods

Biomass can be converted to more valuable energy forms through a number of processes including mechanical or physical, biological, and thermochemical. A review of various conversion processes and major products (fuels) produced from ligocellulogic biomass is provided by Gill et al. (2011). An example of mechanical process is the extraction of oil from the seeds of biomass crops, such as oilseed rape and groundnuts. This oil can be used directly for the production of energy or processed using esterification to produce biodiesel. Examples of biological methods include the production of biogas from anaerobic digestion or the breakdown of biodegradable material, such as municipal waste and plant material, through microorganisms (Lin and Tanaka 2006), and the fermentation of sugar to produce ethanol and other biofuels.

Thermochemical methods, which include direct biomass combustion, pyrolysis and gasification, have been extensively investigated for the conversion of biomass to a variety of products, including thermal energy, fuels, and chemicals. Direct biomass combustion has traditionally been used to supply heat and power in the process industry. Systems utilizing direct combustion of agricultural waste include kilns and boilers for generating steam used for various industrial applications including electricity production. As discussed by Werther et al. (2000) in their review paper, the typical sequence of events through which a lump of solid fuel undergoes during combustion includes heating up, drying, devolatilization, ignition and combustion of volatiles, and finally the combustion of char. In general, the route involving direct combustion of biomass for electricity generation has low overall efficiency and emits significant pollutants (Caputo et al. 2005). For the details of these processes, and the discussion of many operational and environmental challenges, the reader is referred to their review paper.

Pyrolysis refers to the thermal decomposition of biomass in the absence of oxygen. Depending on the process variables, such as the reactor temperature and residence time, it yields various amounts of gaseous, liquid, and solid products of varying compositions. Conventional pyrolysis, which has been utilized for thousands of years, involves lower temperatures and longer residence times with the principal product being the solid char. In contrast, fast pyrolysis involves high temperatures ($\approx$500 °C) and short residence times ($\approx$2 s), with the main product being a dark brown liquid or bio-oil along with other gaseous, liquid, and solid products, including char. This process is much more commonly used at present compared to conventional pyrolysis. The bio-oil consists of a complex mixture of oxygenated hydrocarbons with varying but appreciable amount of water from both the original moisture and reaction product. Proximate analysis of the bio-oil gives a chemical formula of $CH_{1.9}O_{0.7}$. The typical heating value of bio-oil is about 17 MJ/kg, which is about 40–45 % of that of hydrocarbon fuels. The physical properties of bio-oils are discussed by Czernik and Bridgwater (2004), while Bridgwater (2012) and Mohan et al. (2006) provide reviews on fast pyrolysis and the properties of bio-oils generated from this process. Bio-oil can be utilized in several different ways to produce energy, fuels, and chemicals. It has been used directly as fuels in stationary applications, especially for electricity generation. A more sustainable and value-added approach is based on a bio-refinery concept, similar to a petroleum refinery, for producing energy, conventional fuels, such as syngas and Fischer-Tropsch (FT) fuels, and chemicals. This concept is particularly attractive for biomass because of its chemical heterogeneity and regional variability.

Biomass gasification involves pyrolysis and partial oxidation in a well-controlled oxidizing environment. It is deemed as the most promising technology for producing renewable and carbon-free energy, as it provides great flexibility with regard to feedstock and the fuels produced. In recent years, numerous studies have reported on the different types of reactors used in gasification, and the various gaseous and liquid fuels produced. Wang et al. (2008b) and Gill et al. (2011) provide reviews of these studies. In general, the gasification process converts low-value biomass to a gaseous mixture containing syngas and varying amounts of CH_4 and CO_2. It can also produce hydrocarbons by using lower operating temperatures. The syngas composition, or the relative amounts of CO and H_2 in the syngas, can be varied by using air and steam as the gasification agent (Rapagna et al. 2000). In addition, CO_2 can be used in the presence of a catalyst, such as Ni/Al, to increase the H_2 and CO content (Ollero et al. 2003).

Syngas offers significant flexibility with regard to its utilization. Syngas from coal, known as town gas, was extensively used for lighting and heating during the nineteenth and early part of twentieth centuries. In recent years, there has been renewed interest in the utilization of coal-based syngas for stationary power generation through integrated gasification combined cycle (IGCC) facility (Rodrigues et al. 2003). In addition, it can be used to provide H_2 (Watanabe et al. 2002) or synthesized to produce chemicals and liquid fuels, such as F-T fuels (Tijmensen et al. 2002). The use of syngas in fuel cells, such as solid oxide fuel cells, through the reforming of hydrocarbons and other routes is also being explored (Kee et al. 2005, 2008). Figure 4 in Gill et al. (2011) summarizes the various routes for the utilization of syngas, including the production of F-T and other transportation fuels. Gill et al. (2011) provide an overview of technologies, including Biomass-to-Liquid (BTL) and Coal-to-Liquid (CTL) Gas-to-Liquid (GTL) processes, for producing various fuels through gasification and F-T processes. It should be noted, however, that some of these routes may be more energy intensive, and their cost effectiveness and environmental benefits need to be examined, as it may be more economical to use syngas directly as fuel or for electricity generation.

Biogas or landfill gas (LFG) is typically produced from anaerobic decomposition of organic matter (Gunaseelan 1997). It can also be produced through pyrolysis and gasification processes. Primary sources include biomass, green waste, plant material, manure, sewage, municipal waste, and energy crops. While its composition can vary significantly depending on the source and production process, the main constituents include CH_4 (50–75 % by volume), CO_2 (25–40 %), N_2 (0–10 %), and small traces of H_2O, O_2, H_2, and hydrogen sulfide, as well as some contaminants such as volatile organic compounds, and halogenated hydrocarbons. Some representative biogas mixtures based on the two common sources, namely agricultural waste and household waste, are provided by Quesito et al. (2013). Like natural gas and syngas, biogas can be used as a transportation fuel in IC engines, or for power generation in gas turbines and boilers. It can also be used as compressed natural gas, or reformed to produce syngas, or in solid oxide fuel cells to generate electricity.

3 Syngas and Biogas Combustion and Emission Characteristics

3.1 Syngas Combustion and Emission Characteristics

Considerable work has been reported on syngas combustion and emission characteristics (Lieuwen et al. 2009; Cheng 2009; Aggarwal 2013). Fundamental studies have focused on various aspects, including the development of thermo-transport and kinetic models, and examining the ignition and combustion characteristics in both laboratory flames and practical devices. A major challenge identified in these studies is due to substantial variation in its composition and heating value. This requires that the fundamental properties, such as adiabatic flame temperature, laminar burning velocity, flammability limits, flame stability, extinction, and blowout, need to be determined for a wide range of syngas composition. This also presents challenges while designing syngas combustors, requiring optimization for locally available fuels. Table 1 lists the heating values and adiabatic flame temperatures (T_{ad}) of various syngas-air mixtures at $\phi = 1.0$. As indicated, T_{ad} is nearly independent of the syngas composition. However, diluents, such as CO_2, H_2O, and N_2, can be used to modify its value. Syngas typically has lower heating values compared to biogas and other hydrocarbon fuels. For example, the higher heating value for syngas with 50 %CO and 50 %H_2 by volume is 18,943 kJ/kg or 284,139 kJ/kmol, while the corresponding values for methane (representative of biogas) are 55,500 kJ/kg and 888,000 kJ/kmol, respectively.

Detailed mechanisms for syngas oxidation have been developed by Davis et al. (2005), Li et al. (2007), Wang et al. (2007), and Kéromnès et al. (2013). Since the oxidation chemistries of H_2 and CO are fundamental to those of hydrocarbon fuels, the mechanisms developed for latter fuels, such as the GRI-3.0 (Smith et al.) and San Diego (2002) mechanisms, have also been used for syngas combustion. Validations for the mechanisms have been provided using ignition delay and laminar flame speed data. Ignition delays have been measured using a variety of devices, including shock tube (Petersen et al. 2007), rapid compression machine (RCM) (Walton et al. 2007), and constant volume (or constant pressure) reactor, while

Table 1 Heating values and adiabatic flame temperatures of syngas mixtures

H_2 mole fraction	CO mole fraction	Mol. weight (kg/kmol)	Heating value kJ/kg	Heating value kJ/kmol	Adiabatic flame temp (ϕ = 1.0) (K)
0	1	28.0	10,101	282,814	2,394
0.2	0.8	22.8	12,428	283,365	2,382
0.4	0.6	17.6	16,130	283,892	2,378
0.5	0.5	15.0	18,943	284,139	2,377
0.6	0.4	12.4	22,933	284,368	2,378
0.8	0.2	7.2	39,539	284,684	2,382
1	0	2.0	141,794	283,588	2,387

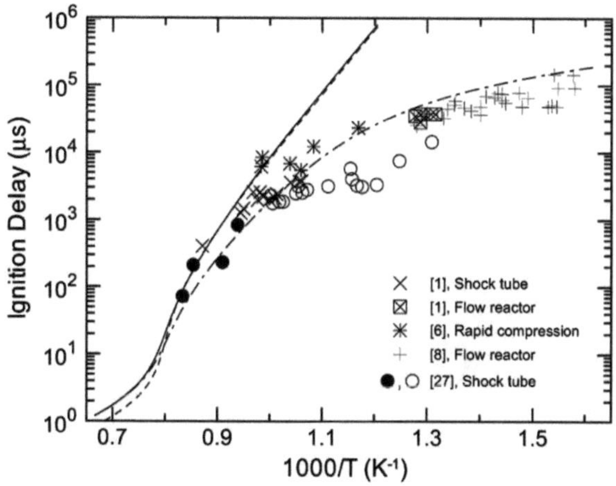

Fig. 2 Ignition delays for various syngas mixtures under different pressure and temperature conditions. *Filled* and *open circles* correspond to strong and weak ignition events, respectively. All experimental data have been normalized to 20 atm assuming $p - 1$ proportionality. *Lines* correspond to ignition delay predictions using the Li et al. mechanism at 20 atm; the *solid line* corresponds to the syngas mixture used in shock tube experiments

computations have often employed a homogeneous reactor configuration (Aggarwal et al. 2011). Figure 2 from Dryer and Chaos (2008) presents some measured and predicted ignition delay data for different syngas mixtures. Additional data can be found in Kéromnès et al. (2013) and other references cited above.

Laminar flame speed represents a fundamental property of a fuel-air mixture. It is of critical importance with regard to burning rate, flame stabilization, flashback, and blowout in practical systems. Laminar burning velocities for H_2–CO mixtures have been measured using different systems, including flat-flame burner (Yan et al. 2001), Bunsen burner (Natarajan et al. 2007, 2009), and expanding spherical flames (McLean et al. 1994; Prathap et al. 2008; Kishore et al. 2011). Figure 3 presents measured and predicted laminar flame speeds versus equivalence ratio and pressure,

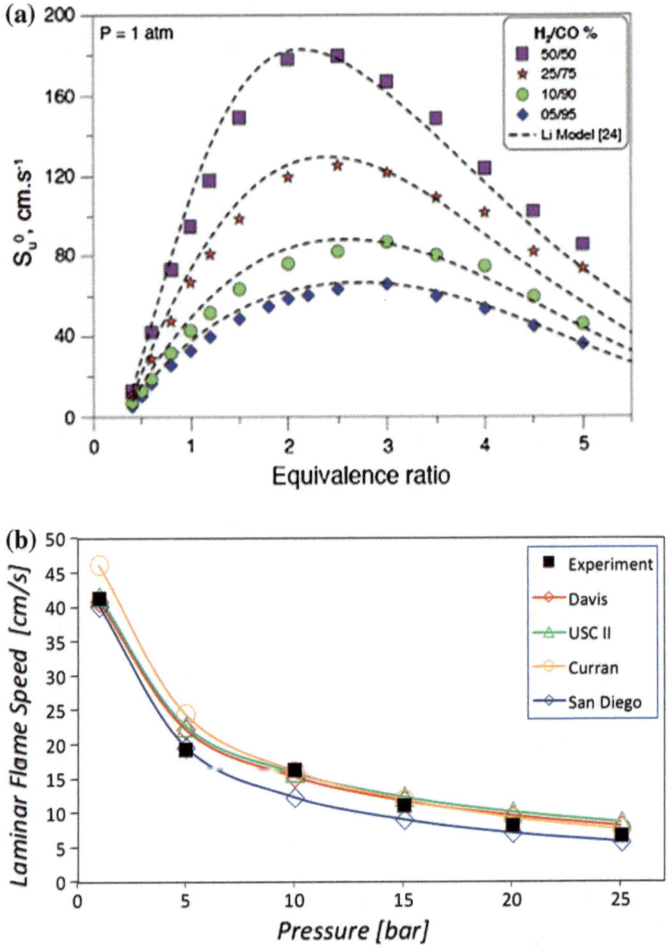

Fig. 3 **a** Measured and predicted laminar flame speeds for various H_2/CO spherically expanding premixed flames. **b** Measured and predicted laminar flame speeds versus pressure at $\phi = 2.5$ for argon-diluted syngas mixture with CO and H_2 mol fractions of 0.265 and 0.029, respectively

respectively. Figure 3a from Bouvet et al. (2011) shows the effect of syngas composition on the burning velocity. As expected, with the increase in H_2 fraction, the flame speed increases, flammability limits become wider, and the peak in flame speed occurs progressively at higher equivalence ratios, ϕ. These effects can be attributed to the higher diffusivity and reactivity of H_2. Figure 3b from Quattrocchi (2014) presents the effect of pressure on the burning velocity. Results are shown for argon-diluted syngas mixture with CO and H_2 mol fractions of 0.265 and 0.029, respectively, at $\phi = 2.5$. Measurements are from Burke et al. (2010), and predictions are performed with the CHEMKIN software using the San Diego (2002), Davis (Davis et al. 2005), USCII (Wang et al. 2007), and Curran (Kéromnès et al. 2013) mechanisms. McLean et al. (1994) and Vagelopoulos and Egolfopoulos (1998) reported premixed flame speeds at pressures from atmospheric to a few atmospheres. Kéromnès et al. (2013) reported a more extensive study on high-pressure premixed flames. Additional data on syngas flame speeds can be found in the references cited above. In addition to unstretched flame speed, the flame response to stretch and the resulting cellular instabilities are important considerations. These phenomena are of fundamental relevance for flame extinction, turbulent flame propagation, flame stabilization, blowout, and transition to detonation. The flame response to stretch is characterized in terms of the Markstein length or non-dimensional parameter, Markstein number. Spherically expanding flames have been commonly used to determine this parameter, and details can be found in Aggarwal (2013) and Kishore et al. (2011).

Since syngas generally contains other species in addition to CO_2 and H_2O, it is important to examine the effects of various diluents on syngas combustion and emissions. Moreover, dilution is often used to lower the flame temperature and thereby limit NO_x emissions. Several researchers have examined the effects of diluents on laminar flame speed, stability, and emissions (McLean et al. 1994; Natarajan et al. 2009; Kishore et al. 2011; Burke et al. 2007; Burbano et al. 2011; Das et al. 2011). It is important to note that only NO_x emission is relevant in syngas flames, while both soot and NO_x emissions are important for hydrocarbon flames. NO_x in hydrocarbon flames can be formed due to four mechanisms, namely thermal (Zeldovich), prompt (Fenimore), N_2O, and NNH mechanisms (Briones et al. 2007; Guo and Smallwood 2007; Fu et al. 2012). Thermal NO involves reactions: $O + N_2 \Rightarrow N + NO$, and $N + O_2 + NO + O$, and $N + OH \Rightarrow NO + H$, while prompt NO formation is initiated through the reaction $CH + N_2 \Rightarrow$ NCN (or HCN) + H (or N). Thus, the prompt mechanism is absent in syngas flames, since it is linked to hydrocarbon combustion chemistry, which produces CH radicals. The prompt NO, however, may be important for syngas mixtures containing CH_4. The N_2O-intermediate mechanism involves $N_2 + O + M \Rightarrow N_2O + M$ as the initiating reaction and is important for lean mixtures and high pressures. Finally, the NO formation through NNH route involves reactions: $N_2 + H \Rightarrow NNH$ and $NNH + O \Rightarrow NO + NH$ (Bozzeli and Deam 1995). Ding et al. (2011) investigated NO_x formation in lean premixed syngas counterflow flames and observed that the NO was formed predominantly through the NNH and N_2O intermediate routes. The contribution of thermal NO was small due to the low-flame temperatures. In addition, increasing the CO fraction in syngas was found to increase the amount of NO formed.

Combustion in many practical devices involves non-premixed (diffusion) and partially premixed flames (PPFs) (Giles et al. 2006; Som et al. 2008). While numerous studies exist of such flames with hydrocarbon fuels, relatively few investigations have focused on syngas fuel. Studies on non-premixed syngas flames have been reported by Giles et al. (2006), Hui et al. (2007), and Park et al. (2004), while those on PPFs have been reported by Som et al. (2008) and Ouimette and Seers (2009). Giles et al. (2006) also examined the effects of diluents (N_2, H_2O, and CO_2) on NO formation in non-premixed flames. Som et al. (2008) examined the effects of strain rate, equivalence ratio, and syngas composition on the detailed

Fig. 4 Images of syngas (50 %H_2/50 %CO)-air partially premixed flames established at $\phi = 6$ (**a**) and $\phi = 16$ (**b**) in a counterflow burner. Strain rate is 35 s^{-1}

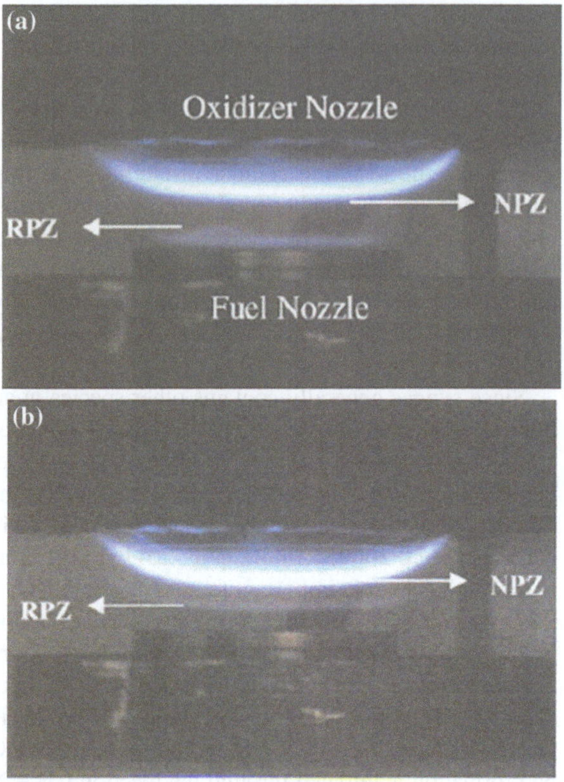

structure and extinction characteristics of PPFs in an opposed-jet burner. Figure 4 from this study depicts images of two such flames established at $\phi = 6$ and 16. For $\phi = 6$, which is just above the rich flammability limit of 50 %H_2/50 %CO syngas-air mixture, the flame exhibits a typical double flame structure with a weak rich premixed reaction zone (RPZ) established close to the fuel nozzle and a non-premixed reaction zone (NPZ) on the oxidizer side near the stagnation plane. As ϕ is increased, the RPZ moves away from the fuel nozzle, and RPZ and NPZ are much closer to each other. Som et al. also examined the effect of syngas composition and partial premixing on NO emissions.

Ouimette and Seers (2009) reported an experimental investigation on syngas partially premixed jet flames. The effects of ϕ, CO_2 dilution, and H_2/CO ratio on the flame structure and NO_x were reported. Figure 5 from this study presents flame images at different ϕ. As expected, the flame length is strongly influenced by the level of partial premixing. As ϕ decreases from the non-premixed to premixed regime, the flame length decreases monotonically. This has important consequences for the emissions of NO_x, greenhouse gases, and other pollutants, since the flame length directly influences the reacting volume and residence time. In addition, images at 2.0 and 1.6 indicate the existence of two reaction zones, with the NPZ enveloping the RZP. Regarding the NO_x formation, results indicated that $EINO_x$ first increases as ϕ

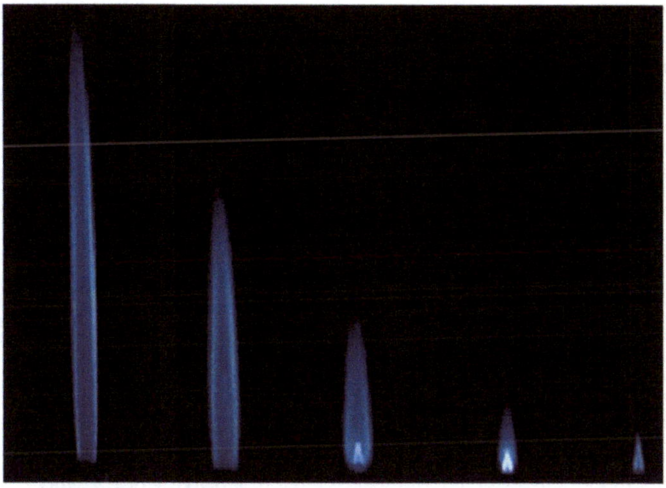

Fig. 5 Images of laminar partially premixed 45 %H_2/35 %CO/20 %CO_2-air flames at different levels of partial premixing. Reynolds number is 1,400

is increased from 1.0 to 1.6, then remains nearly constant for $1.6 < \phi < 3.85$, and subsequently decreases slowly as ϕ is increased to the diffusion limit ($\phi \Rightarrow \infty$).

Studies on turbulent syngas flames have focused on the determination of turbulent flame speeds (S_T); (Daniele et al. 2011; Venkateswaran et al. 2011; Monteiro et al. 2011). Venkateswaran et al. (2011) reported measurements of global turbulent flame speeds using a Bunsen burner and examined the effects of ϕ, syngas composition, mean flow velocity, and turbulence intensity. Consistent with other studies, the flame speed was found to exhibit sensitivity to fuel composition over a wide range of turbulence intensity, increasing significantly with the increase in H_2 content. The data were further analyzed to develop flame speed correlations, indicating the effects of thermo-diffusive instabilities through negative Markstein lengths.

3.2 Biogas Combustion and Emission Characteristics

There have been few studies focusing directly on biogas combustion. However, there is a large body of literature on methane combustion, including ignition, extinction, flammability limits, flame speeds, cellular instabilities, and emissions. Consequently, detailed thermo-transport and kinetic models have been developed to simulate and analyze methane flames in a variety of configurations. Considerable research has also been reported on the extinction and blowout of methane-air flames using various diluents, such as CO_2, N_2, H_2O, and chemical agents (Linteris et al. 2008; Aggarwal 2009). Most of these studies and the associated models can be readily used for analyzing the combustion and emission characteristics of biogas, whose main

Fig. 6 Computed adiabatic flame temperatures of methane-air and two biogas-air mixtures. Pressure = 1 atm and initial temperature = 500 K

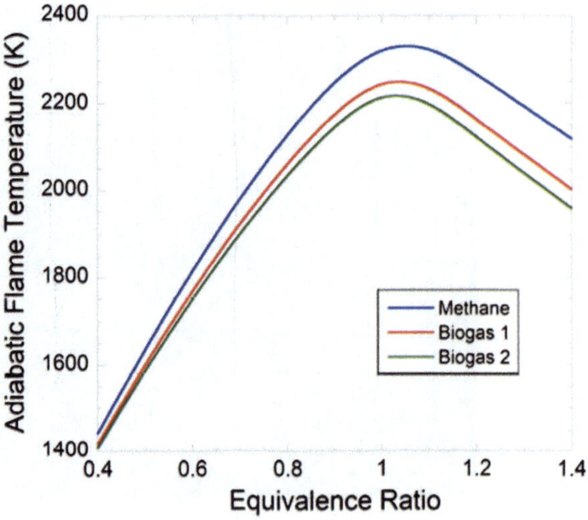

constituents are CH_4 and CO_2 with small traces of H_2O and N_2. This section provides a brief overview of the fundamental combustion and emission properties of biogas.

Biogas has lower energy content compared to natural gas. For example, the volumetric heating values of natural gas (94 %CH_4) and biogas (60 %CH_4/40 % CO_2) are 38.6 and 25 MJ/m^3, respectively. This has consequences for using biogas in natural gas-fired combustion devices, since lower heating value implies higher feeding rates and lower flame temperatures. Figure 6 compares the computed adiabatic flame temperatures for methane-air and two representative biogas-air mixtures. The compositions of these two mixtures, which represent the common feedstock, are 68 %CH_4/26 %CO_2/5 %H_2O/1 %N_2 and 60 %CH_4/33 %CO_2/6 % H_2O/1 %N_2, respectively. As indicated, the biogas flame temperature is about 100–200 K lower than that of methane, depending upon the feedstock biogas. Lower temperatures imply lower flame speeds and thermal NO for biogas flames compared to those for methane flames. The comparison of laminar burning speeds for freely propagating methane and biogas flames is shown in Fig. 7, which shows the flame speed as a function of equivalence ratio and pressure. The flames were computed using the CHEMKIN software and the GRI-3.0 kinetic mechanism. As expected, flame speeds are lower for biogas-air mixtures compared to those for methane-air mixtures. The effect of pressure on flame speed is qualitatively similar for all the three cases shown, with the flame speed first decreasing sharply and then increasing relatively slowly with increase in pressure.

Since biogas is potentially a cleaner and sustainable alternative to natural gas, it is relevant to analyze the emission characteristics of methane and biogas flames in different combustion regimes. Figure 8 from Quesito et al. (2013) compares the computed NO, NO_2, and C_2H_2 profiles in methane and biogas-air PPFs. These flames were simulated in a counterflow geometry at fuel stream equivalence ratio

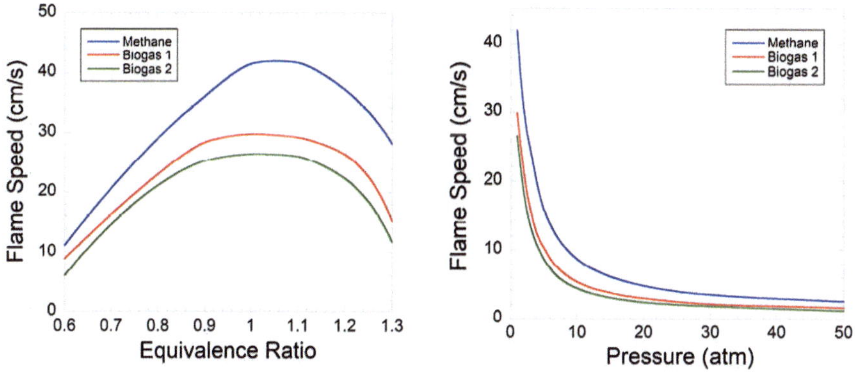

Fig. 7 Computed laminar flame speeds of methane-air and two biogas-air mixtures. Flame speed is plotted versus ϕ (*left*) and pressure (*right*)

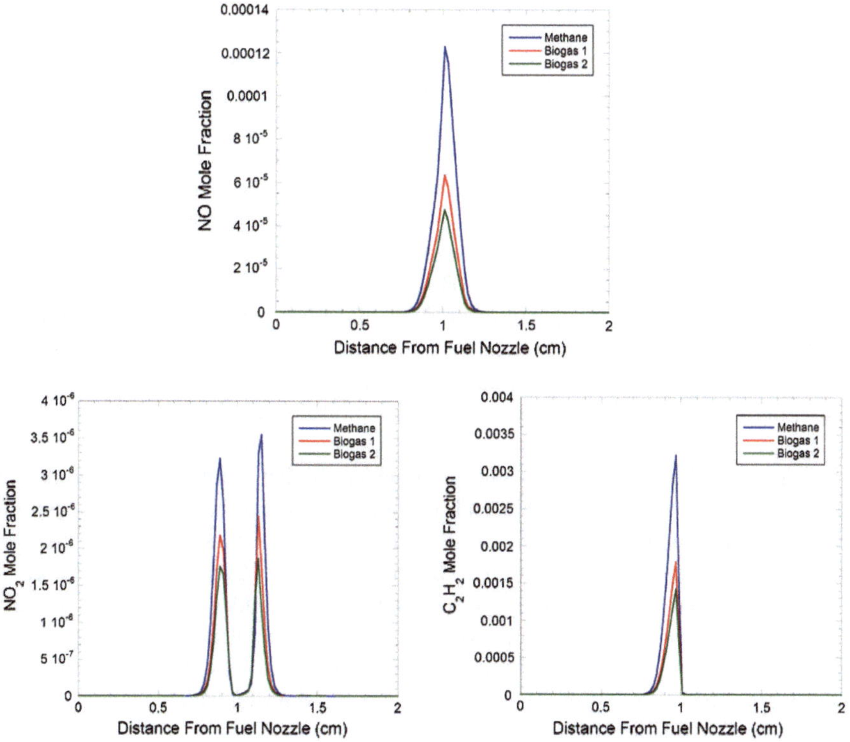

Fig. 8 NO, NO_2, and C_2H_2 mole fraction profiles in counterflow partially premixed flames burning methane and two biogases

$\phi = 3$, pressure = 1 atm, and strain rate = 200 s^{-1}, using the Chemkin software and GRI-3.0 mechanism. For all three cases, the flames exhibited a double flame structure with a RPZ located on the fuel side (at about 0.9 cm from the fuel nozzle) and a NPZ on the oxidizer side near the stagnation plane (at about 1.05 cm from the fuel nozzle). These two reaction zones are clearly seen in the NO_2 profiles. The NO and NO_2 profiles indicate significantly lower level of NO_x formation in biogas flames compared to that in methane flames. This may be attributed to the less thermal NO and prompt NO formed, indicated by lower temperatures and C_2H_2 peaks, in biogas flames compared to those in methane flames. The lower C_2H_2 peaks also imply lower prompt NO and soot formation in biogas flames, although soot emission is not a major concern in natural gas and biogas flames.

3.3 Syngas and Biogas Combustion in Practical Devices

There have been numerous studies in recent years dealing with various aspects of syngas-fueled combustion systems. Many of these studies are discussed in a special issue of Combustion Science and Technology (2008). There is significant interest in using syngas in gas turbine engines (Luessen 1997; Lieuwen et al. 2008; Dam et al. 2011), especially in the context of an IGCC facility, for efficient, low-emission power generation, and for carbon capture and storage. Extensive literature exists dealing with the natural gas (NG)-fired combustors, which is beyond the scope of this review. Some research in this area has also focused on using syngas in natural gas-fired combustors (Colantoni et al. 2010). The viability of using syngas and biogas in spark ignition (SI) and compression ignition (CI) engines has been a subject of numerous investigations (Boehman and Le Corre 2008; Sahoo et al. 2012; Bika et al. 2011; Korakianitis et al. 2011). With regard to CI engines, much of the research has focused on dual-fuel operation, which is discussed in the next section. The use of syngas in SI engines offers many advantages, such as better anti-knocking properties and operation with leaner mixtures. Improved knock resistance is due to the presence of CO and enables operation at higher compression ratio, leading to higher thermal efficiency. However, higher burning rate due to the presence of H_2 can lead to higher end gas temperature and increased propensity to knocking. The presence of H_2 can also increase NO_x emissions, which may be controlled by using leaner mixtures (Boehman and Le Corre 2008). Bika et al. (2011) examined such issues by performing single-cylinder experiments for different syngas compositions, compression ratios, and equivalence ratios. For a given ϕ and spark timing, the knock-limited compression ratio was observed to increase with increasing CO fraction. The burn duration and ignition lag also increased with increasing CO fraction.

Compared to CI engines, the NG-fueled SI engines for automobiles have been used with reasonable success, especially in South America. For instance, there are nearly 3.4 million natural gas vehicles (NGVs) in operation in Brazil and Argentina. A number of studies have examined the performance and emission characteristics of biogas- and NG-fueled SI engines (Korakianitis et al. 2011; González et al. 2013). The advantages of NG or biogas are due to their higher octane number, lower cost, and reduced GHG and particulate emissions. Thus compared with conventional gasoline engines, SI engines using natural gas can run at higher compression ratios, thus producing higher thermal efficiencies and lower CO_2 emissions. However, NO_x, CO, and UHC emissions may be higher, and such engines produce less power than gasoline-fueled engines due to a combination of factors, including lower volumetric efficiency due to natural-gas injection in the intake manifold, lower stoichiometric fuel/air ratio of natural gas compared to gasoline, and the lower equivalence ratio at which these engines may be run in order to reduce NO_x emissions. In addition, high-pressure storage tanks are required to replace the liquid-fuel tank of a conventional vehicle, and fuel injectors or fuel induction systems need to be modified due to higher flow rates required to overcome the low density of natural gas. Despite these modifications, NGVs cannot operate over the same distances as conventional vehicles due to the much lower energy content of NG per unit volume.

In summary, more experimental and computational studies are needed to optimize the engine performance and emissions for various operating conditions and syngas and biogas composition. Research should also focus on examining the use of syngas and biogas in new engine designs, such as Homogeneous Charge Compression Ignition (HCCI) and low temperature combustion.

4 Use of Syngas and Biogas in Dual-Fuel Diesel Engines

Numerous studies have examined the viability of a dual-fuel or blended-fuel strategy using a variety of liquid and gaseous fuels for transportation and power generation. Such blends include bio/petroleum (Hansen et al. 2005), H_2-gasoline (Changwei and Shuofeng 2009), H_2-NG (Ma et al. 2007; Wang et al. 2008a; Das et al. 2000; Kahraman et al. 2009; Dimopoulos et al. 2008; Morrone and Andrea 2009), H_2-CH_4 (Bauer and Forest 2001a, b), H_2-diesel (Shirk et al. 2008), and syngas-diesel (Boehman and Le Corre 2008; Sahoo et al. 2012). The use of NG or biogas in dual-fuel mode has also been investigated using spark SI (Cho and He 2007; Chandra et al. 2011), CI (Duc and Wattanavichien 2007; Ryu 2013), and HCCI (Naber et al. 1994) engines. In diesel engines, a gaseous fuel-air mixture is introduced through the intake valve, while the diesel fuel is injected and compression ignited. Research has focused on the effect of a gaseous fuel on the performance and emission characteristics of gasoline or diesel engines. The effects of various parameters, such as fuel composition, blending ratio, injection timing, EGR, and gaseous fuel-air ratio, have been characterized at different loads. While a number of aspects have been examined (Liu et al. 2003, 2013; Shah et al. 2011; Sahoo et al. 2009), a common observation is that a dual-fuel strategy with NG-diesel at high loads leads to lower PM, CO_2, and volatile organic compounds (VOC) emissions, but somewhat higher CO and UHC emissions, and that the engine power may be reduced due to lower volumetric energy content of NG, although similar thermal efficiency can be maintained. The benefits of dual-fuel operation become less clear at low loads, and issues concerning reduced engine power and higher UHC and CO emissions need to be addressed by optimizing other parameters, i.e., injection timing, EGR, amount of liquid (pilot) fuel injected, and gaseous fuel equivalence ratio (Chatlatanagulchai et al. 2010a, b; Tomita et al. 2009). Engine experiments using syngas-diesel combination (Boehman and Le Corre 2008; Sahoo et al. 2012) indicated that the engine performance and emissions are strongly influenced by the syngas composition, depending upon the load and other conditions. In general, increasing H_2 fraction in syngas was found to improve engine performance, reduce CO and hydrocarbon emissions, but increase NO_x emissions.

The literature review indicates extensive experimental research dealing with dual-fuel (NG-diesel and syngas-diesel) CI engines, but relatively few computational studies on this topic (Singh et al. 2006; Shah et al. 2011). In particular, many details of the two-stage ignition process, different combustion modes, and emission characteristics in dual-fuel engines remain unexplored. Motivated by this consideration, we performed a numerical investigation on the effect of gaseous fuels on the ignition, combustion, and emissions in a dual-fuel diesel engine. The following sections provide a brief description of the computational model, and some representative results focusing on the two-stage ignition behavior, various combustion models including lean premixed burning and extinction, and emissions.

4.1 Computational Model

Simulations were performed using a 3D CFD software CONVERGE (Senecal et al. 2003), which incorporates state-of-the-art models for spray injection, breakup and atomization, turbulence, droplet collision and coalescence, spray-wall interaction, and vaporization. Details of these models are provided elsewhere (Senecal et al. 2007). The mathematical model is based on Eulerian-Lagrangian formulation for the two-phase turbulent reacting flow. The gas-phase flow field is described using the Favre-averaged Navier-Stokes equations along with the RNG k-ε turbulence model. The spray is represented by a stochastic system of a discrete number of parcels, which are tracked computationally using a Lagrangian scheme. The two phases are coupled through the mass, momentum, and energy exchange terms, which are present in both the liquid- and gas-phase equations. The injection process is simulated using a blob injection model, which injects liquid droplet parcels with a diameter equal to an effective nozzle diameter. The subsequent breakup process is simulated by using models based on the Kelvin-Helmholtz (KH) and Rayleigh-Taylor (RT) instabilities (Patterson and Reitz 1998). The droplet evaporation model is based on the Nusselt number and Sherwood number correlations of Chiang et al. (1992). The CFD solver uses an innovative modified cut-cell Cartesian method along with an adaptive mesh resolution technique for grid generation (Richards et al. 2013).

Computations consider a 1.9L 4-cylinder GM light-duty diesel engine, which has been extensively used for experimental studies at Argonne National Laboratory (Ciatti and Subramanian 2011). The engine has a 7-hole common-rail injector in

Table 2 Engine parameters and other properties for the reference case

Engine speed, rpm	1,500
Engine bore diameter, mm	82
Engine stroke length, mm	90.4
Squish height, mm	0.6
Connecting rod length	145.4
IVC (intake valve closing), CAD	−132 ATDC
Gas temperature at IVC, K	350
Gas pressure at IVC, bar	1.23
EVO (exhaust valve opening), CAD	116 ATDC
SOI (start of injection), CAD	−8 ATDC
Fuel injection temperature, K	298
Nozzle diameter, mm	0.141
Injection duration, CAD	8.5
Fuel intake temperature, K	298
Cylinder head temperature, K	500
Cylinder wall temperature, K	450
Piston wall temperature, K	550

each cylinder. Simulations were performed for a 1/7 (51.43°) sector of the cylinder. Table 2 provides the engine parameters and other properties for the reference case. N-heptane is used as the surrogate for diesel fuel, while methane is used to represent biogas. For dual-fuel cases, methane or syngas premixed with air is introduced through the intake manifold. Simulations are started at the intake valve closing. The fuel oxidation chemistry is modeled using the Chalmers mechanism (Golovichev n. d.; Tao et al. 2007), which contains 42 species and 168 reactions. The software also contains models for soot and NO_x formation, which are described by Senecal et al. (2007) and Patterson and Reitz (1998).

While the reaction mechanism has been extensively validated in previous studies, additional validation is provided in Fig. 9, which presents the predicted and measured ignition delays for n-heptane/air, methane/air, and syngas/air mixtures at engine relevant conditions and $\phi = 1$. The constant pressure homogeneous reactor simulations were performed using the CHEMKIN software along with the Chalmers and CRECK (Ranzi 2012) mechanisms. The CRECK mechanism is a comprehensive mechanism with 466 species 14,631 reactions and has previously been validated against various targets, including ignition delays, flame speeds, and species measurements. In the figure, the predicted ignition delays are compared against the measurements for n-heptane/air (Gauthier et al. 2004), methane/air (Huang et al. 2006), and syngas/air mixtures (Mittal et al. 2006). Overall, there is good agreement between the measurements and predictions of the two mechanisms. The Chalmers mechanism somewhat overpredicts ignition delays for n-heptane and underpredicts for methane at high temperatures. For syngas, both the mechanisms significantly underpredict the ignition delay at temperatures below 1,000 K. These differences have been attributed to mixture non-homogeneities present in

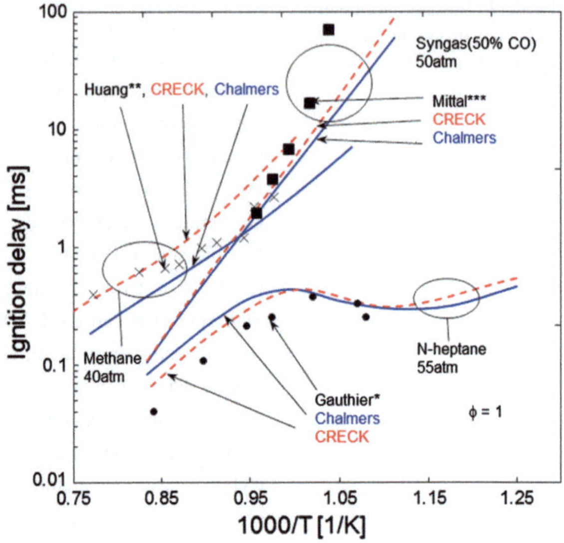

Fig. 9 Predicted and measured ignition delays for n-heptane/air, syngas/air, and methane-air mixtures at high pressures and $\phi = 1$

experiments, but were not duplicated in simulations (Dryer and Chaos 2008; Jain et al. 2013). It is also noteworthy that the Chambers mechanism is able to reproduce the experimentally observed two-stage ignition process and negative temperature coefficient (NTC) region.

4.2 Ignition Characteristics in Dual-Fuel Diesel Engines

In a dual-fuel engine, the ignition is initiated by the pilot diesel. Since ignition represents a critical process with regard to engine performance and emissions, it is important to examine the ignition of pilot fuel in the presence of a combustible gaseous fuel-air mixture. In particular, the ignition chemistry of large hydrocarbons, such as n-heptane, is known to be characterized by the NTC (negative temperature coefficient) and two-stage ignition processes (Curran et al. 1998). Previous studies have not analyzed the effect of gaseous fuels on these processes. Therefore, simulations were performed to examine these aspects by considering homogeneous gaseous mixtures in a constant volume reactor and two-phase mixtures in a diesel engine. Figure 10 depicts the first and second stage ignition processes in diesel engine in terms of the mass profiles of OH and QOOH with respect to crank angle for the single- and dual-fuel cases, as listed in Table 3. As indicated in the table, the start of injection (SOI) is −8° after top dead center (ATDC), and injection duration is 8.5 crank angle degree (CAD).

For n-heptane, the two-stage ignition process is well depicted in the figure, with the first-stage ignition occurring at −4.8° ATDC and the second-stage at −2.4° ATDC. The first-stage ignition is defined by the value of integrated heat release

Fig. 10 Integrated mass profiles of OH and $C_7H_{14}OOH$ (QOOH) for single-fuel (n-heptane), 97 % methane, and 98.5 % syngas dual-fuel cases in diesel engine. The *green rectangles* indicate the first- and second-stage ignitions

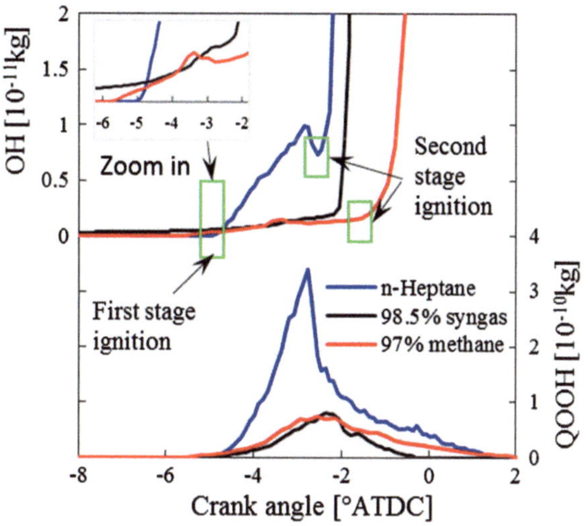

Table 3 Input parameters, and computed performance and emission parameters for single- and dual-fueled engine simulations

	Low load			High load	
Mole fraction of methane	0 %	90 %	0 %	97 %	0 %
Mole fraction of syngas (50 % CO, 50 % H_2)	0 %	0 %	95 %	0 %	98.5 %
Mole fraction of n-heptane	100 %	10 %	5 %	3 %	1.5 %
Total chemical energy per cylinder, J	390			1,014	
Mass of n-heptane per cylinder, mg	8.8	3.3	3.3	3.3	3.3
Mass of methane per cylinder, mg	0	4.8	0	17.3	0
Mass of syngas per cylinder, mg	0	0	13.8	0	50
Mass of air per cylinder, mg	546	542	532	515	456
ϕ based on gaseous fuel	0	0.16	0.17	0.58	0.60
Total volume per cylinder, L	0.448				
Work output per cylinder, J	133	87	95	351	228
Indicated mean effective pressure (*IMEP*), bar	2.78	1.82	1.98	7.36	4.77
CA50 (°ATDC)	1.3	3.6	3.2	5.4	-0.4
Indicated power, kW	1.66	1.09	1.18	4.39	2.85
Engine efficiency (%)	34	22	24	35	22
Specific soot emission (mg/kWh)	15.3	2.2	1.8	1.8	0.6
Specific NO_x emission (g/kWh)	0.51	0.52	0.66	0.89	3.58
Specific UHC emission (g/kWh)	0.21	9.46	0.12	1.81	0.00
Specific CO emission (g/kWh)	0.30	1.76	19.88	0.14	0.00

The emission values are at exhaust valve open (EVO) crank angle

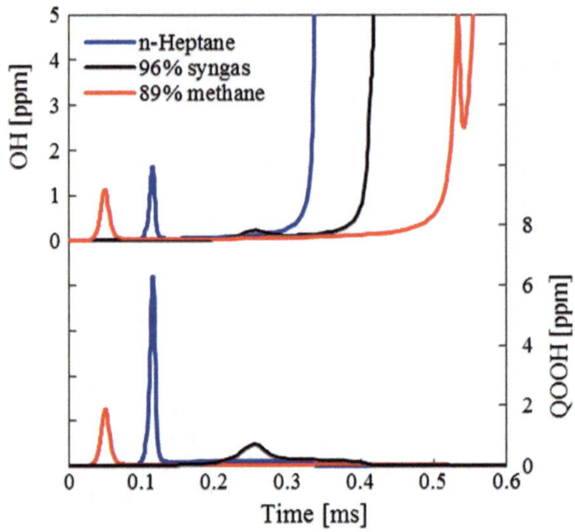

Fig. 11 Temporal profiles of OH, $C_7H_{14}OOH$ (QOOH), and HO_2 mol fraction for single-fuel (n-heptane), 89 % methane and 96 % syngas dual-fuel cases in homogeneous constant volume reactor (CHEMKIN). Simulations for the dual-fuel cases are carried out using the same gaseous equivalence ratios as those for engine simulations discussed in the context of Fig. 10. The n-heptane mole fraction in the initial mixture is specified using a global $\phi = 1$

(IHR) exceeding 0.01 J, while the inflection point in the heat release rate (HRR) profile (not shown) defines the second-stage ignition. These ignition delay values are consistent with those determined from the QOOH and OH profiles shown in Fig. 10, where the first rise in either OH or QOOH indicates the first-stage ignition and the second (sharp) rise (after the first peak) in OH indicates the second stage. The use of QOOH and OH profiles to identify the first and second stage ignition was further confirmed by performing CHEMKIN simulations in homogeneous constant volume reactor. As indicated by the species temporal profiles for n-heptane in Fig. 11, the first sharp rise or peaks in QOOH and OH profiles marks the first stage of ignition, while the second sharp rise in OH marks the second stage ignition. The first and second ignition delays were also confirmed using the 3D contour plots of QOOH and OH mass fractions (not shown). For the dual-fuel case with methane, the first- and second-stage ignition crank angles from Fig. 10 are −5.2° ATDC and −1.2° ATDC. Note, however, that the first ignition is not clearly distinguishable for the engine case (Fig. 10), but can be seen for the homogeneous mixture (Fig. 11). In contrast, the first-stage ignition for the syngas dual-fuel case is not clearly visible for either engine or homogeneous mixture simulations. Another important observation from Figs. 10 and 11 is that for the homogeneous mixture, the presence of gaseous fuel (methane or syngas) increases the ignition delay for the dual-fuel cases compared to single fuel. However, the ignition delay is not affected noticeably by the presence of gaseous fuel in engines, and this can be attributed to the spatially and temporally evolving species and temperature fields in engines sprays.

4.3 Combustion and Emission Characteristics of Dual-Fuel Diesel Engines

Simulations were performed for one single-fuel case (100 % n-heptane), and four dual-fuel cases, two with methane (90, and 97 % by volume), and the other two with 95 and 98.5 % syngas (50 %CO-50 %H_2). As indicated in Table 3, cases with 100 % n-heptane, 90 % methane, and 95 % syngas correspond to low-load condition for which the total (fuel) energy into the cylinder is 390 J. The other two cases with 97 % methane and 98.5 % syngas correspond to high-load condition with total energy into the cylinder being 1,014 J. Also, for the two low-load cases, the gaseous equivalence ratios are $\phi_g = 0.16$ and 0.17, while for the two high-load cases, $\phi_g = 0.58$ and 0.60.

Figure 12 depicts the combustion characteristics for the 0, 90, and 97 % methane cases in terms of the temporal profiles of pressure, HRR, n-heptane vapor mass, and methane mass. For the first two cases corresponding to low-load condition, the amount of energy input is kept fixed, whereas for the 90, and 97 % methane cases, the amount of n-heptane injected is kept fixed. Thus, with the increase in methane mass, the indicated mean effective pressure (IMEP) increases from 1.88 to 7.36 bar, as indicated in Table 3. The increased load for the 97 % methane can also be seen from the pressure and HRR plots in Fig. 12. In addition, the HRR plots indicate that the ignition delay is increased due to the addition of methane, as discussed earlier.

The combustion process in a diesel engine is generally characterized by a hybrid combustion mode involving rich premixed combustion and diffusion combustion (Dec 1997; Som and Aggarwal 2010). In contrast, depending upon operating conditions, the heat release in a dual-fuel engine may also involve a lean combustion mode with a propagating flame (Boehman and Le Corre 2008). The HRR and methane mass profiles in Fig. 12 seem to support this hypothesis, especially for the 97 % methane case. However, the methane mass profile for the 90 % case seems to indicate the extinction of the lean premixed flame due to ultra lean conditions at low load. Various combustion modes and the flame propagation/extinction behavior were analyzed by examining the evolution of some key species during combustion. Figure 13 presents the CO, CO_2, and HO_2 mass fraction contours at different crank angles for the single-fuel case. These plots are cut through at the peak ϕ value location. Black lines in each figure indicate ϕ contours (0.3–3). The corresponding plots for the two duel-fuel cases are shown in Figs. 14 and 15. These plots can be used to identify the different combustion zones. For instance, the high CO region locates the rich premixed zone with ϕ between 2.0 and 3.0, while the high CO_2 region corresponds to diffusion combustion zone with ϕ near 1. This is consistent with previous studies (Azzoni et al. 1999; Aggarwal 2009) dealing with PPFs, indicating that the most of CO is produced in the rich premixed zone, while the most of CO_2 is produced in the diffusion combustion zone, where the temperature is highest. Thus, the rich premixed zone is characterized by n-heptane decomposition and partial oxidation, producing CO, H_2, and intermediate hydrocarbon species. These species are then consumed through oxidation reactions in the diffusion

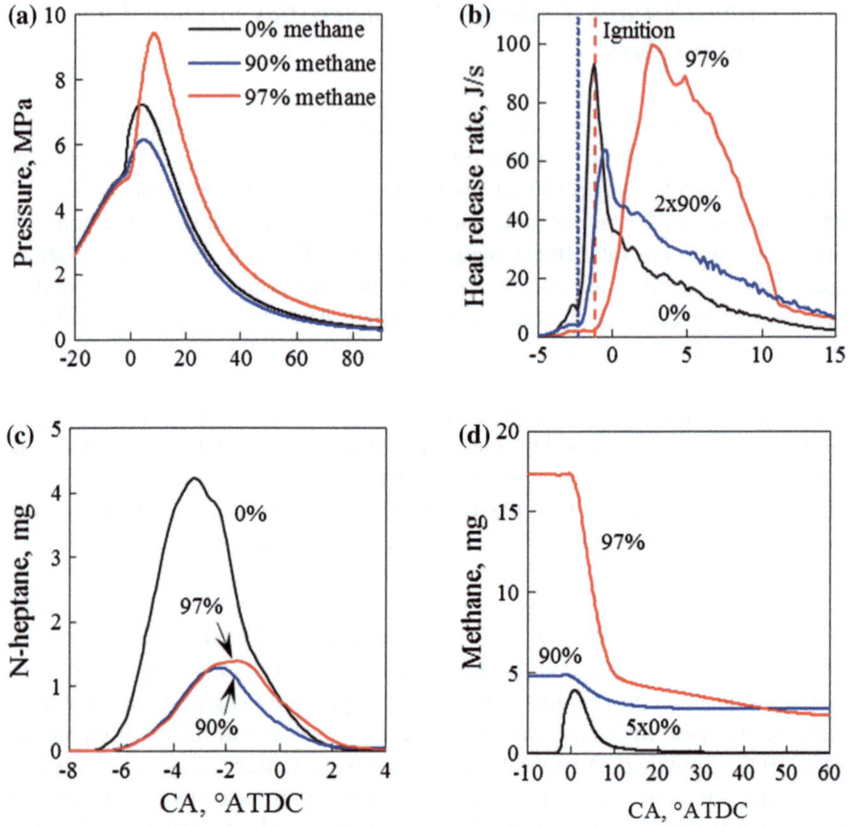

Fig. 12 Pressure, heat release rate, n-heptane vapor, and methane mass profiles with respect to crank angle for the 0 % (100 % n-heptane), 90 and 97 % methane cases. *Vertical lines* in **b** indicate crank angles at ignition

combustion zones. In a similar way, HO_2 plots can be used to examine the lean combustion zone. Thus, for the two dual-fuel cases, HO_2 contours at 0 and 4° ATDC (cf. Figs. 14 and 15) indicate the presence of a lean combustion zone, characterized by a propagating flame. However, the corresponding contours for the single-fuel case do not show this lean combustion zone (cf. Fig. 13).

The three combustion zones for the dual-fuel combustion can be qualitatively identified by deconstructing the HRR diagram based on the temporal mass profiles of some relevant species. Figure 16 presents the mass profiles of several such species for the 97 % methane case. As indicated, the rich premixed combustion mostly occurs between −5° and 3° ATDC and is characterized by the consumption of n-heptane and the production of C_2H_2, CO, and other intermediate species, such as H_2. As indicated in Fig. 16a, C_2H_2 and CO species attain their peak values during this period. Based on this observation, a HRR profile corresponding to rich premixed combustion can be constructed and is shown in Fig. 16b. Similarly, from the

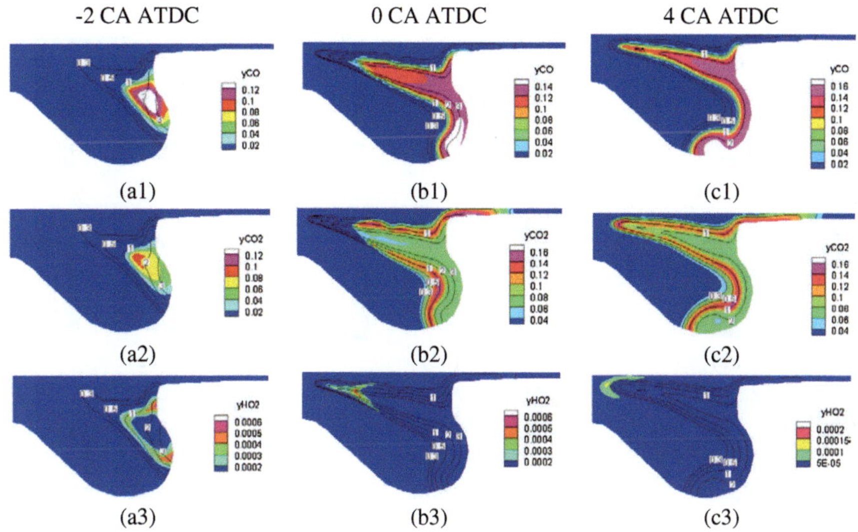

Fig. 13 CO (*a1–c1*), CO_2 (*a2–c2*), and HO_2 (*a3–c3*) mass fraction contours for the n-heptane case at −2, 0, and 40 ATDC. Black lines in each figure indicate ϕ contours (ϕ = 0.3–3)

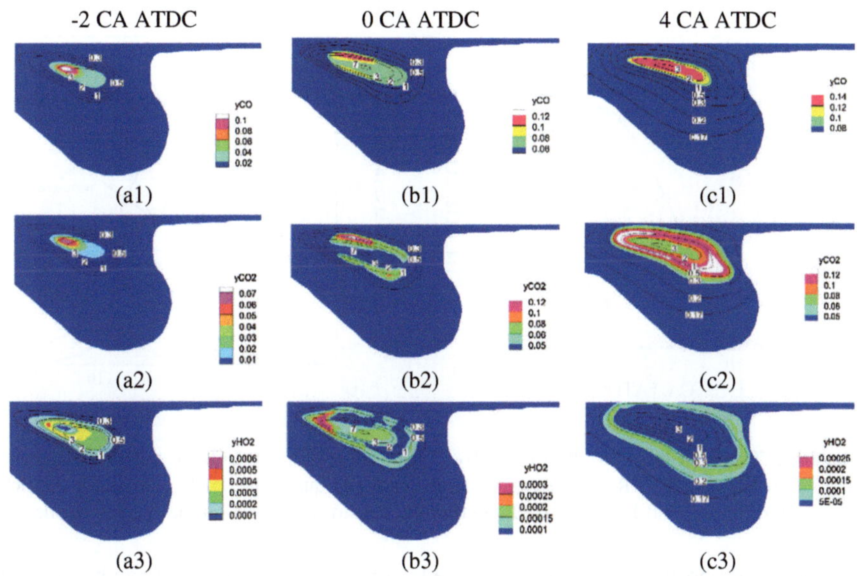

Fig. 14 CO (*a1–c1*), CO_2 (*a2–c2*), and HO_2 (*a3–c3*) mass fraction contours for the 90 % methane dual-fuel case at −2, 0, and 40 ATDC. Black lines in each figure indicate ϕ contours (ϕ = 0.3–3)

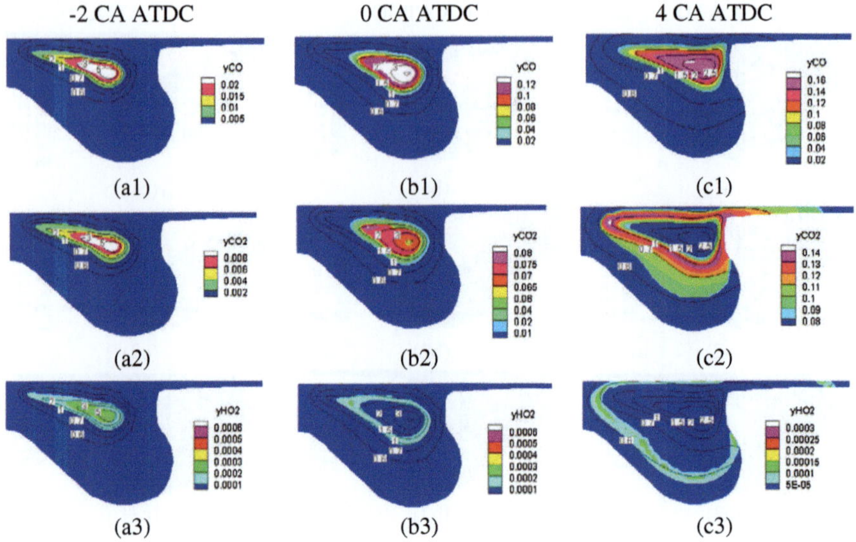

Fig. 15 CO (*a1–c1*), CO_2 (*a2–c2*), and HO_2 (*a3–c3*) mass fraction contours for the 97 % methane dual-fuel case at −2, 0, and 40 ATDC. *Black lines* in each figure indicate ϕ contours (ϕ = 0.3–3)

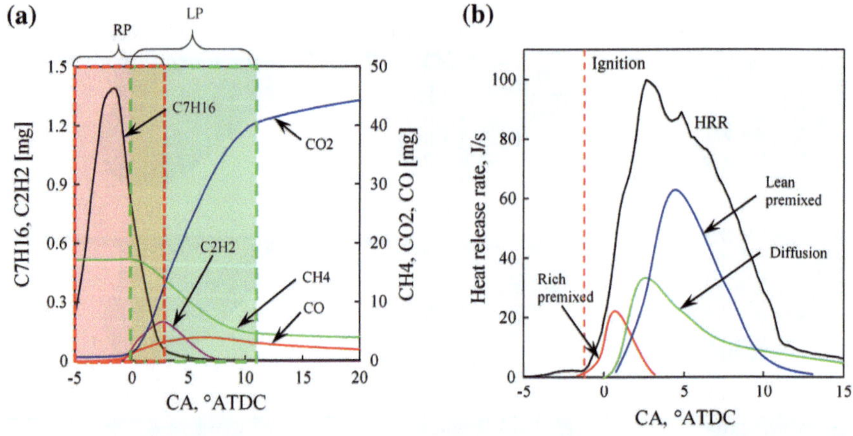

Fig. 16 **a** Temporal mass profiles of n-C_7H_{16}, C_2H_2, CH_4, CO, and CO_2 for the 97 % methane case and **b** Computed HRR for the engine combustion and qualitative HRR profiles for the rich premixed, diffusion, and lean premixed combustion, constructed based on the species profiles

methane mass profile, the duration for the lean premixed combustion can be defined to be between 0° and 10° ATDC. Methane is almost completely consumed during this period, producing a significant amount of CO_2. Accordingly, a HRR profile for lean combustion can be constructed and is also shown in Fig. 16b. Finally, C_2H_2, CO, and other species (such as H_2), which are produced during rich premixed combustion, are consumed in the diffusion combustion zones. Consequently, the

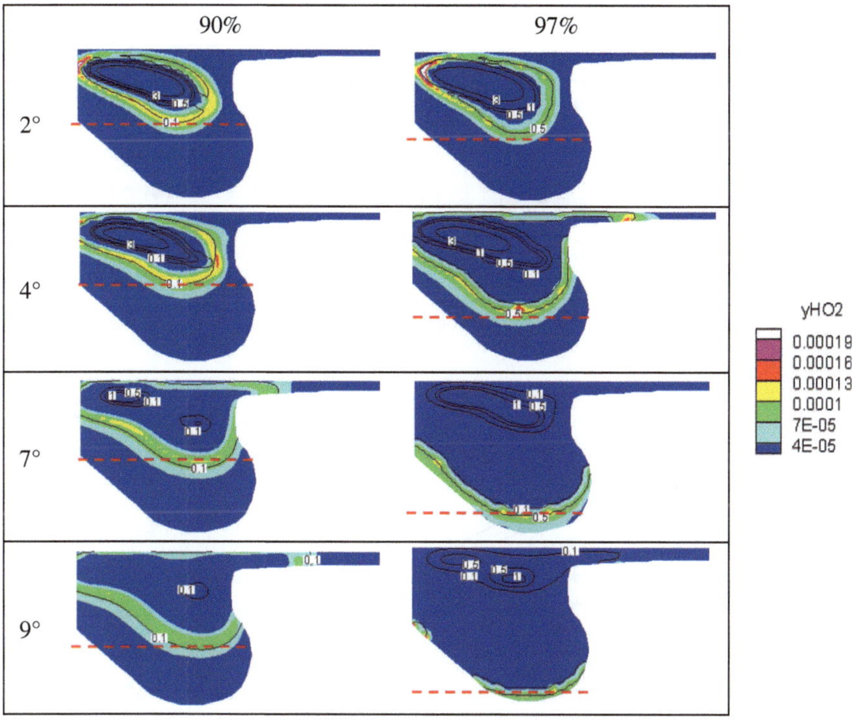

Fig. 17 HO_2 mass fraction contours at different crank angles for the 90 and 97 % methane cases. *Black lines* indicate ϕ contours. *Red line* in these figures shows the flame propagation

HRR profile corresponding to diffusion combustion can be constructed based on this observation and the overall HRR profile. This is also shown in Fig. 16b. The diffusion combustion is estimated to occur between 0° and 60° ATDC, during which time CO_2 (cf. Fig. 16a) and H_2O attain their peak values.

As stated earlier, the methane profiles (Fig. 12) indicate the extinction of lean combustion zone for the 90 % methane case. This is further confirmed by analyzing the HO_2 contours presented in Fig. 17 for the two dual-fuel cases. For both cases, the contours at 2° ATDC indicate the presence of a lean propagating flame in the piston bowl. For the 90 % methane case, however, the HO_2 contours at 7° and 9° ATDC indicate that the flame is extinguished due to ultra lean conditions, with $\phi \approx 0.05$–0.1 ahead of the flame, and due to the fact that pressure and temperature decrease during the expansion stroke. In contrast, the contours for the 97 % case indicate a lean propagating flame with $\phi \approx 0.5$. Consequently, most of methane gets consumed through lean premixed combustion for this case, while a significant amount of methane remains unreacted for the 90 % case. This is further confirmed by the methane profiles discussed earlier in the context of Fig. 12. Note that a small amount of methane remains unreacted for the 97 % case, but it is mostly confined to the crevice and boundary layer regions. These regions also contribute to UHC and CO emissions.

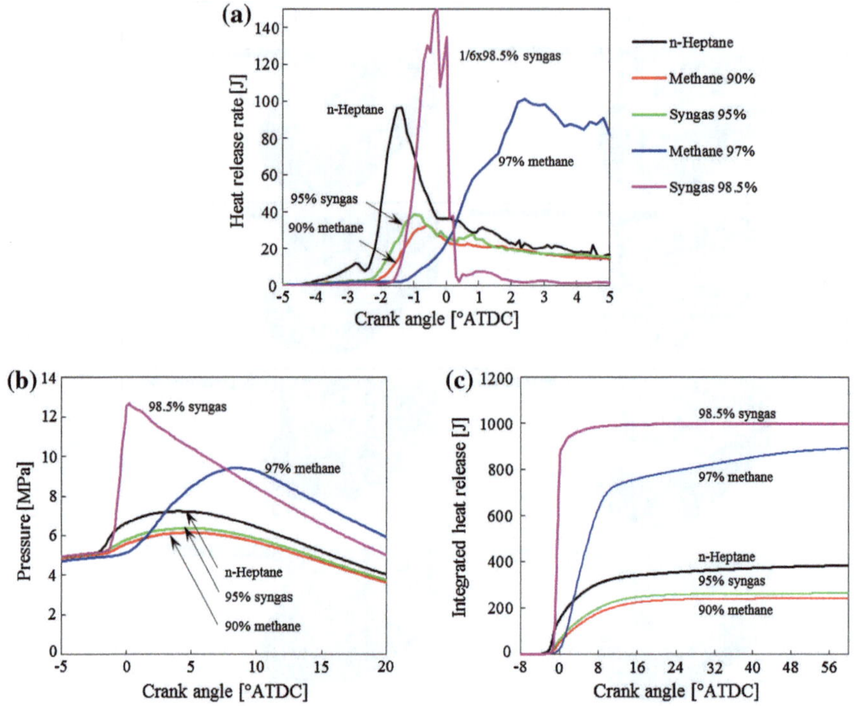

Fig. 18 Heat release rate (**a**), pressure (**b**), and IHR (**c**) profiles with respect to crank angle for the single fuel (n-heptane), and 90 and 97 % methane dual-fuel, and 95 and 98.5 % syngas dual-fuel cases

Results for the syngas dual-fuel cases are shown in Fig. 18, which plots the HRR, pressure, and IHR profiles for all the five cases. The HRR and pressure profiles for the low-load cases (90 % methane and 95 % syngas) indicate more rapid combustion and higher heat release for 95 % syngas compared to that for 90 % methane. Consequently, as indicated in Table 3, IMEP and engine efficiency are higher for 95 % syngas compared to those for 90 % methane. Differences between the syngas and methane become significantly more pronounced at high-load condition (97 % methane and 98.5 % syngas). The HRR and pressure profiles for syngas indicate nearly spontaneous combustion with a very sharp rise in heat release and pressure but a gradual combustion process for methane. The total heat release is also noticeably higher for syngas compared to that for methane. These differences are primarily due to the much faster flame propagation in syngas-air mixture compared to that in methane-air mixture, even though the equivalence ratios for the two mixtures are nearly the same (Table 3).

The flame propagation behavior is illustrated in Fig. 19, which presents the HO_2 contours for the 95 and 98.5 % syngas cases. Similar to the methane case, the contour plots are cut through the peak ϕ value location. As expected, the plots

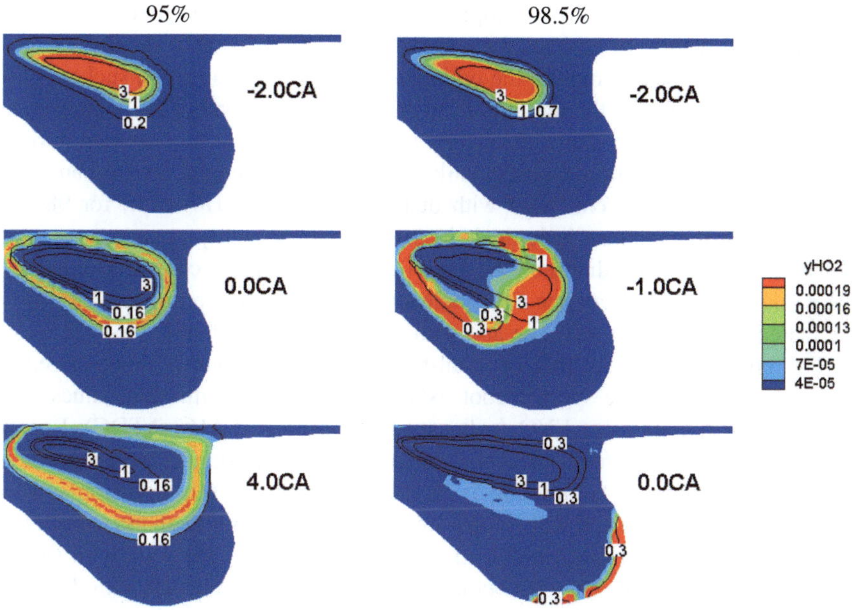

Fig. 19 HO_2 mass fraction contours at different crank angles for the 95 and 98.5 % syngas cases. Black lines indicate ϕ contour ($\phi = 0.16$–3)

indicate significantly faster flame propagation for the 98.5 % syngas case compared to the 95 % syngas, since the gaseous equivalence ratios for the two cases are 0.7 and 0.17 (Table 3). In fact for the latter case, the flame gets extinguished at about 4° ATDC due to ultra lean conditions, and due to decreasing temperature and pressure during the expansion stroke. Furthermore, the comparison of HO_2 contours for the 97 % methane (Fig. 17) and 98.5 % syngas cases indicates significantly faster flame propagation in syngas-air mixture than that in methane-air mixture. However, as indicated in Table 3, the thermal efficiency is noticeably lower for 98.5 % syngas compared to that for 97 % methane. This is further discussed in the next section.

In summary, for certain operating conditions, the combustion process in syngas-based dual-fuel engine may become nearly spontaneous, which can possibly lead to knocking.

4.4 Dual-fuel Engine Performance and Emissions

Two of the performance parameters, i.e., IMEP and engine efficiency computed from pressure profiles (p-v diagram), are listed in Table 3 for all the five cases. At low load, the IMEP and engine efficiency are lower for the dual-fuel (90 % methane and 95 % syngas) cases compared to the single-fuel case, although they are slightly higher for syngas compared to those for methane. The loss of performance at low load can be attributed to the incomplete combustion of gaseous fuel as discussed earlier. Thus, the use of gaseous fuels at low loads may adversely affect the engine performance and would require the optimization of other parameters, such as EGR and injection timing for the liquid fuel. In contrast, at high load (97 % methane and 98.5 % syngas), both methane and syngas are almost completely consumed through lean premixed combustion. Consequently for methane, the engine performance can be maintained or even enhanced with dual-fuel operation. However, for 98.5 % syngas, once the mixture is ignited, it is followed by nearly spontaneous combustion, with the result that the CA50 is −0.4° ATDC for this case. Consequently, the work output, IMEP, and engine efficiency are significantly reduced with syngas compared to those with methane and with single fuel (n-heptane).

The effect of gaseous fuel on dual-fuel engine emissions is summarized in Table 3, which lists the specific soot, NO_x, UHC, and CO emission values (mg/kwh) for the five cases at EVO (exhaust valve opening at 116° ATDC). Results indicate that with dual-fuel operation, the soot emission is significantly reduced at all loads, while the NO_x emission is increased. Moreover, the reduction in soot becomes more pronounced with syngas compared to that with methane at both low and high loads. In contrast, there is greater increase in NO_x emission with syngas than that with methane. The UHC emission is essentially eliminated with dual-fuel operation using syngas. However, the UHC and CO emission can significantly increase with methane at low load due to its incomplete combustion.

5 Summary and Conclusions

There is worldwide interest in developing renewable energy sources in a sustainable manner. Syngas and biogas offer significant potential in this context. These fuels can be produced from a variety of feedstock using different conversion methods and offer great flexibility with regard to their utilization in transportation and power generation systems. This chapter provides an overview of research dealing with the combustion and emissions of these fuels, both as stand-alone fuels or by blending with petroleum fuels. Conversion methods for producing these fuels from different biomass sources are also briefly reviewed.

There are notable differences between the combustion characteristics of these two fuels and those of hydrocarbon fuels. While the syngas composition can vary widely, it generally has lower heating value, lower density, higher mass diffusivity, higher flame speeds, and wider flammability limits. Its combustion leads to almost zero soot emission, although NO_x emission may be a concern depending upon its composition and operating temperatures. Similarly, biogas has lower heating value compared to hydrocarbon fuels, and its ignition and combustion characteristics can vary noticeably depending upon its composition. While there have been few studies focusing directly on biogas combustion, there is extensive literature on methane combustion, including ignition, extinction, flammability limits, flame speeds, cellular instabilities, and emissions. Results from many of these studies can be readily used for analyzing the combustion and emission characteristics of biogas. For both syngas and biogas, a major challenge to combustion system designers is due to the wide variation in their composition. This also provides opportunities for fundamental and applied research on both the production and utilization aspects of such fuels. Fundamental combustion aspects requiring further research include cellular instabilities, flame stabilization and blowout behavior, turbulent flames, and emission characteristics. Such efforts would lead to the development of optimized systems for producing these fuels and provide guidelines for optimizing their composition for a given set of operating conditions.

The use of syngas and biogas in dual-fuel diesel engines has been a subject of numerous experimental and computational studies. In a dual-fuel operation, the gaseous fuel-air mixture is introduced through the intake valve, while the diesel fuel is injected and compression ignited. A general observation from these studies is that the engine performance and emission characteristics are significantly modified by the presence of gaseous fuel. For instance, the two-stage ignition process associated with the diesel fuel can be noticeably influenced depending upon the engine load and the relative amounts of gaseous and liquid fuels. The combustion characteristics are also significantly altered due to the presence of gaseous fuel. While the heat release in a diesel engine generally occurs through a hybrid combustion mode, involving rich premixed combustion and diffusion combustion, that in a dual-fuel engine also involves a lean combustion mode with a propagating flame. The dual-fuel operation at high-load conditions can provide significant reduction in soot and CO soot emissions, while maintaining the engine efficiency, provided the injection characteristics including the amount of pilot fuel can be optimized. However, the NO_x emission may increase, requiring a suitable strategy for lowering the temperatures. The dual-fuel strategy may be less effective at low load, resulting in lower thermal efficiency and higher UHC and CO emissions. Future work should be directed at optimizing the various parameters, such as injection timing, amount of pilot fuel injected, EGR, and multiple injections.

Acknowledgments Prof. Aggarwal's research over the years has been funded by several federal agencies, including NSF, NASA, AFOSR, EPA, Wright-Patterson Air Force Base, and ANL, as well as industry. His work concerning engine combustion and biomass-derived fuels has been supported by ANL. Many results in this monograph were computed by graduate students, Mr. Francesco Quesito and Mr. Xiao Fu.

References

Aggarwal SK (2009) Extinction of laminar partially premixed flames. Prog Energy Combust Sci 35:528–570

Aggarwal SK (2013) Simulations of combustion and emissions characteristics of biomass-derived fuels. In: Dahlquist E (ed) Technologies for converting biomass to useful energy: combustion, gasification, pyrolysis, torrefaction and fermentation. CRC Press, Boca Raton

Aggarwal SK, Awomolo O, Akber K (2011) Ignition characteristics of heptane-hydrogen and heptane-methane fuel blends at elevated pressures. Int J Hydrogen Energy 36:15392–15402

Azzoni R, Ratti S, Aggarwal SK, Puri IK (1999) The structure of triple flames stabilized on a slot burner. Combust Flame 119:23–40

Bauer CG, Forest TW (2001a) Effect of hydrogen addition on performance of methane-fueled vehicles. Part II: driving cycle simulation. Int J Hydrogen Energy 26:71–90

Bauer CG, Forest TW (2001b) Effect of hydrogen addition on the performance of methane-fueled vehicles. Part I: effect on SI engine performance. Int J Hydrogen Energy 26:55–70

Bika AS, Franklin L, Kittelson DB (2011) Engine knock and combustion characteristics of a spark ignition engine operating with varying hydrogen and carbon monoxide proportions. Int J Hydrogen Energy 36:5143–5152

Boehman A, Le Corre O (2008) Combustion of syngas in internal combustion engines. Combust Sci Technol 180:1193–1206

Bouvet N, Chauveau C, Gokalp I, Halter F (2011) Experimental studies of the fundamental flame speeds of syngas (H2/CO)/air mixtures. Proc Combust Inst 33:913–920

Bozzeli JW, Deam AM (1995) O + NNH: a possible new route for NO_x formation in flames. Int J Chem Kinet 27:1097–1109

Bridgwater AV (2012) Review of fast pyrolysis of biomass and product upgrading. Biomass Bioenergy 38:68–94

Briones AM, Som S, Aggarwal SK (2007) The effect of multi-stage combustion on NO_x emissions in methane-air flames. Combust Flame 149:448–462

Burbano HJ, Pareja J, Amell AA (2011) Laminar burning velocities and flame stability analysis of H_2/CO/air mixtures with dilution of N_2 and CO_2. Int J Hydrogen Energy 36:3232–3242

Burke M, Chaos M, Dryer F, Ju Y (2010) Negative pressure dependence of mass burning rates of H_2/CO/O_2/diluent flames at low flame temperatures. Combust Flame 157:618–631

Burke MP, Qin X, Ju Y (2007) Measurements of hydrogen syngas flame speeds at elevated pressures. In: 5th US combustion meeting

Caputo AC, Palumbo M, Pelagagge PM, Scacchia F (2005) Economics of biomass energy utilization in combustion and gasification plants: effects of logistic variables. Biomass Bioenergy 28:35–51

Chandra R, Vijay VK, Subbarao PMV, Khura TK (2011) Performance evaluation of a constant speed IC engine on CNG, methane enriched biogas and biogas. Appl Energy 88:3969–3977

Changwei J, Shuofeng W (2009) Effect of hydrogen addition on the idle performance of a spark ignited gasoline engine at stoichiometric condition. Int J Hydrogen Energy 34:3546–3553

Chatlatanagulchai W, Rhienprayoon S, Yaovaja K, Wannatong K (2010) Air/fuel ratio control in diesel-dual-fuel engine by varying throttle, EGR valve, and total fuel. SAE paper 2010-01-2200

Chatlatanagulchai W, Yaovaja K, Rhienprayoon S, Wannatong K (2010) Air–fuel ratio regulation with optimum throttle opening in diesel-dual-fuel engine. SAE paper 2010-01-1574

Cheng RK (2009) Synthesis gas combustion—fundamentals and applications. CRC Press, Boca Raton, pp 129–168

Chiang CH, Raju MS, Sirignano WA (1992) Numerical analysis of a convecting, vaporizing fuel droplet with variable properties. Int J Heat Mass Transf 35:1307

Cho HM, He BQ (2007) Spark ignition natural gas engines—a review. Energy Convers Manag 48:608–618

Ciatti S, Subramanian SN (2011) An experimental investigation of low-octane gasoline in diesel engines. J Eng Gas Turbines Power 133:092802-1–092802-11

Colantoni S, Gatta SD, Prosperis RD, Russo A, Fantozzi F, Desideri U (2010) Gas turbines fired with biomass pyrolysis syngas: analysis of the overheating of hot gas path components. J Eng Gas Turbines Power 132:061401–061408

(2008) Combust Sci Technol 180(6), May 2008

Curran HJ, Pitz WJ, Westbrook CK, Callahan CV, Dryer FL (1998) Oxidation of automotive primary reference fuels at elevated pressures. Proc Combust Inst 27:379–387

Czernik S, Bridgwater AV (2004) Overview of applications of biomass fast pyrolysis oil. Energy Fuels 18:590–598

Dam B, Love N, Choudhuri A (2011) Flashback propensity of syngas fuels. Fuel 90(2):618–625

Daniele SS, Jansohn P, Mantzaras J, Boulouchos K (2011) Turbulent flame speed for syngas at gas turbine relevant conditions. Proc Combust Inst 33:2937–2944

Das AK, Kumar K, Sung CJ (2011) Laminar flame speeds of moist syngas mixtures. Combust Flame 158:345–353

Das LM, Gulati R, Gupta PK (2000) A comparative evaluation of the performance characteristics of a spark ignition engine using hydrogen and compressed natural gas as alternative fuels. Int J Hydrogen Energy 25:783–793

Davis SG, Joshi AV, Wang H, Egolfopoulos F (2005) An optimized kinetic model of H_2/CO combustion. Proc Combust Inst 30:283–1292. http://ignis.usc.edu/Mechanisms/H2-CO/H2-CO.html

Dec JE (1997) A conceptual model of DI diesel combustion based on laser-sheet imaging. SAE paper 970873

Dimopoulos P, Bach C, Soltic P, Boulouchos K (2008) Hydrogen–natural gas blends fuelling passenger car engines: combustion, emissions and well-to-wheels assessment. Int J Hydrogen Energy 33:7224–7236

Ding N, Arora R, Norconk M, Lee SY (2011) Numerical investigation of diluent influence on flame extinction limits and emission characteristic of lean-premixed H_2–CO (syngas) flames. Int J Hydrogen Energy 36:3222–3231

Dryer FL, Chaos M (2008) Ignition of syngas/air and hydrogen/air mixtures at low temperatures and high pressures: experimental data interpretation and kinetic modeling implications. Combust Flame 152:293–299

Duc PM, Wattanavichien K (2007) Study on biogas premixed charge diesel dual fuelled engine. Energy Convers Manag 48:2286–2308

Fu X, Garner S, Aggarwal SK, Brezinsky K (2012) Numerical study of NO_x emissions from n-Heptane and 1-Heptene counterflow flames. Energy Fuel 26:879–888

Gauthier BM, Davidson DF, Hanson RK (2004) Shock tube determination of ignition delay times in full-blend and surrogate fuel mixtures. Combust Flame 139:300–311

Giles DE, Som S, Aggarwal SK (2006) NO_x emission characteristics of counterflow syngas diffusion flames with airstream dilution. Fuel 85:1729–1742

Gill SS, Tsolakis A, Dearn KD, Rodríguez-Fernández J (2011) Combustion characteristics and emissions of Fischer Tropsch diesel fuels in IC engines. Prog Energy Combust Sci 37:503–523

Golovichev V (n.d.) Mechanisms (combustion chemistry). (Online) Available at http://www.tfd.chalmers.se/~valeri/MECH.html. Accessed 2013

González FOC, Mahkamov K, Lora EES, Andrade RV, Jaen RL (2013) Prediction by mathematical modeling of the behavior of an internal combustion engine to be fed with gas from biomass, in comparison to the same engine fueled with gasoline or methane. Renew Energy 60:427–432

Gunaseelan VN (1997) Anaerobic digestion of biomass for methane production: a review. Biomass Bioenergy 13:83–114

Guo H, Smallwood GJ (2007) A numerical study on the effect of CO addition on extinction limits and NO_x formation in lean counterflow CH_4/air premixed flames. Combust Theory Model 11:741–753

Hall D, Rosillo-Calle F, Woods J (1991) Biomass, its importance in balancing CO_2 budgets. In: Grassi G, Collina A, Zibetta H (eds) Biomass for energy, industry and environment. In: 6th E. C. conference, Elsevier Science, London, pp 89–96

Hansen AC, Zhang Q, Lyne PWL (2005) Ethanol–diesel fuel blends—a review. Bioresour Technol 96:277–285

Huang J, Bushe WK, Hill PG, Munshi SR (2006) Shock initiated ignition in homogeneous methane–hydrogen–air mixtures at high pressure. Int J Chem Kinet 38:221–233

Hui X, Zhang, Z, Mu K, Wang Y, Xiao Y (2007) Effect of fuel dilution on the structure and pollutant emission of syngas diffusion flames. In: Proceedings of GT2007 ASME turbo expo: power for land, sea and air

Jain S, Li D, Aggarwal SK (2013) Effect of hydrogen and syngas addition on the ignition of iso-octane/air mixtures. Int J Hydrogen Energy 38:4163–4173

Kahraman N, Ceper B, Akansu SO, Aydin K (2009) Investigation of combustion characteristics and emissions in a spark-ignition engine fuelled with natural gas–hydrogen blends. Int J Hydrogen Energy 34(2):1026–1034

Kee RJ, Zhu H, Goodwin DG (2005) Solid-oxide fuel cells with hydrocarbon fuels. Proc Combust Inst 30:2379–2404

Kee RJ, Zhu H, Sukeshini AM, Jackson GS (2008) Solid oxide fuel cells: operating principles, current challenges, and the role of syngas. Combust Sci Technol 180:1207–1244

Kéromnès A, Metcalfe WK, Heufer KA, Donohoe N, Das AK, Sung CJ, Herzler J, Naumann C, Griebel P, Mathieu O, Krejci MC, Petersen EL, Pitz WJ, Curran HJ (2013) An experimental and detailed chemical kinetic modeling study of hydrogen and syngas mixture oxidation at elevated pressures. Combust Flame 160:995–1011

Kishore RV, Ravi MR, Ray A (2011) Adiabatic burning velocity and cellular flame characteristics of H2–CO–CO2–air mixtures. Combust Flame 158:2149–2164

Korakianitis T, Namasivayam AM, Crookes RJ (2011) Natural-gas fueled spark-ignition (SI) and compression-ignition (CI) engine performance and emissions. Prog Energy Combust Sci 37:89–112

Li J, Zhao Z, Kazakov A, Chaos M, Dryer FL, Scire JJ (2007) A comprehensive kinetic mechanism for CO, CH_2O, and CH_3OH combustion. Int J Chem Kinet 39:109–136

Lieuwen T, Mcdonell V, Santavicca D, Sattelmayer T (2008) Burner development and operability issues associated with steady flowing syngas fired combustors. Combust Sci Technol 180:1169–1192

Lieuwen TC, Yang V, Yetter RA (eds) (2009) Synthesis Gas combustion: fundamentals and applications. Taylor and Francis, London, 193–222

Lin Y, Tanaka S (2006) Ethanol fermentation from biomass resources: current state and prospects. Appl Microbiol Biotechnol 69:627–642

Linteris GT, Rumminger MD, Babushok VR (2008) Catalytic inhibition of laminar flames by transition metal compounds. Prog Energy Combust Sci 34:288–329

Liu J, Yang F, Wang H, Ouyang M, Hao S (2013) Effects of pilot fuel quantity on the emissions characteristics of a CNG/diesel dual fuel engine with optimized pilot injection timing. Appl Energy 110:201–206

Liu S, Zhou L, Wang Z, Ren J (2003) Combustion characteristics of compressed natural gas/diesel dual-fuel turbocharged compressed ignition engine. In: Proc Inst Mech Eng Part D: J Automobile Eng:217–833

Luessen HP (1997) Gas turbine technology for steel mill gas and syngas applications. ASME paper 97-GT-221

Ma F, Wang Y, Liu H, Li Y, Wang J, Zhao S (2007) Experimental study on thermal efficiency and emission characteristics of a lean burn hydrogen enriched natural gas engine. Int J Hydrogen Energy 32:5067–5075

McLean IC, Smith DB, Taylor SC (1994) The use of carbon monoxide/hydrogen burning velocities to examine the rate of the CO + OH reaction. Proc Combust Inst 25:749–757

Mittal G, Sung CJ, Yetter RA (2006) Autoignition of H_2/CO at elevated pressures in a rapid compression machine. Int J Chem Kinet 38:516–529

Mohan D, Pittman CU, Steele PH (2006) Pyrolysis of wood/biomass for bio-oil: a critical review. Energy Fuels 20:848–889

Monteiro E, Sotton J, Bellenoue M, Afonso Moreira N, Malheiro S (2011) Experimental study of syngas combustion at engine-like conditions in a rapid compression machine. Exp Thermal Fluid Sci 35:1473–1479

Morrone B, Andrea U (2009) Numerical investigation on the effects of natural gas and hydrogen blends on engine combustion. Int J Hydrogen Energy 34:4626–4634

Naber JD, Siebers DL, Di Julio SS, Westbrook CK (1994) Effects of natural gas composition on ignition delay under diesel conditions. Combust Flame 99:192–200

Natarajan J, Lieuwen T, Seitzman J (2007) Laminar flame speeds of H_2/CO mixtures: Effect of CO_2 dilution, preheat temperature, and pressure. Combust Flame 151:104–119

Natarajan J, Kochar Y, Lieuwen T, Seitzman J (2009) Pressure and preheat dependence of laminar flame speeds of $H_2/CO/CO_2/O_2/He$ mixtures. Proc Combust Inst 32:1261–1268

Ollero P, Serrera A, Arjona R, Alcantarilla S (2003) The CO_2 gasification kinetics of olive residue. Biomass Bioenergy 24:151–161

Ouimette P, Seers P (2009) NO_x emission characteristics of partially premixed laminar flames of $H_2/CO/CO_2$ mixtures. Int J Hydrogen Energy 34:4626–4634

Park J, Hwang DJ, Kim KT, Lee SB, Kee SI (2004) Evaluation of chemical effects of added CO_2 according to flame location. Int J Energy Res 28:551–565

Patterson MA, Reitz RD (1998) Modeling the effects of fuel spray characteristics on diesel engine combustion and emission. SAE paper 1998-98-0131

Petersen EL, Kalitan DM, Barrett AB, Reehal SC, Mertens JD, Beerer DJ, Hack RL, McDonell VG (2007) New syngas/air ignition data at lower temperature and elevated pressure and comparison to current kinetics models. Combust Flame 149:244–247

Prathap C, Ray A, Ravi MR (2008) Investigation of nitrogen dilution effects on the laminar burning velocity and flame stability of syngas fuel at atmospheric condition. Combust Flame 155:145–160

Quattrocchi S (2014) Liftoff and blowout characteristics and structure analysis of syngas diffusion flames. MS thesis, Politecnico Di Torino

Quesito F, Santarelli M, Leone P, Aggarwal SK (2013) Biogas combustion in premixed flames or electrochemical oxidation in SOFC: exergy and emission comparison. J Energy Res Technol 135:021202-1–021202021202021202 11

Ranzi E (2012) CRECK modeling—C1-C16 low and high temperature. (Online) Available at http://creckmodeling.chem.polimi.it/index.php/current-version-december-2012/c1c16-low-and-high-temperature. Accessed 2013

Rapagna S, Jand N, Kiennemann A, Foscolo PU (2000) Steam gasification of biomass in a fluidised-bed of olivine particles. Biomass Bioenergy 19:187–197

Richards KJ, Senecal PK, Pomraning E (2013) Converge (version 2.1.0). Convergent Science, Inc., Middleton

Rodrigues M, Walter A, Faaij A (2003) Co-firing of natural gas and biomass gas in biomass integrated gasification/combined cycle systems. Energy 28:1115–1131

Ryu K (2013) Effects of pilot injection timing on the combustion and emissions characteristics in a diesel engine using biodiesel–CNG dual fuel. Appl Energy 111:721–730

Sahoo BB, Sahoo N, Saha UK (2009) Effect of engine parameters and type of gaseous fuel on the performance of dual-fuel gas diesel engines—a critical review. Renew Sustain Energy Rev 13:1151–1184

Sahoo BB, Sahoo N, Saha UK (2012) Effect of H_2:CO ratio in syngas on the performance of a dual fuel diesel engine operation. Appl Therm Eng 49:139–146

San Diego (2002) The San Diego mechanism. (Online) Available at http://web.eng.ucsd.edu/mae/groups/combustion/mechanism.html

Senecal PK, Pomraning E, Richards KJ (2003) Multi-dimensional modeling of direct injection diesel spray liquid length and flame lift-off length using CFD and parallel detailed chemistry. SAE paper 2003-01-0243

Senecal PK, Richards KJ, Pomraning E, Yang T, Dai MZ, McDavid RM, Patterson MA, Hou S, Sethaji T (2007) A new parallel cut-cell Cartesian CFD code for rapid grid generation applied to in-cylinder diesel engine simulations. SAE paper 2007-01-0159

Shah A, Thipse S, Tyagi A, Rairikar S, Kavthekar K, Marathe N, Mandloi P (2011) Literature Review and Simulation of Dual Fuel Diesel-CNG Engines. SAE Paper 2011-26-0001

Shirk MG, McGuire TP, Neal GL, Haworth DC (2008) Investigation of a hydrogen-assisted combustion system for a light-duty diesel vehicle. Int J Hydrogen Energy 33:7237–7244

Singh S, Liang L, Kong SC, Reitz RD (2006) Development of a flame propagation model for dual-fuel partially premixed compression ignition engines. Int J Engine Res 7:65–75

Smith GP, Golden DM, Frenklach M, Moriarty NW, Eiteneer B, Goldenberg M, Bowman CT, Hanson RK, Song S, Gardiner WC, Lissianski VV, Qin Z, GRI Mech-3.0: http://www.me.berkeley.edu/gri_mech/

Som S, Aggarwal SK (2010) Effects of primary breakup modeling on spray and combustion characteristics of compression ignition engines. Combust Flame 157:1179–1193

Som S, Ramirez AI, Hagerdorn J, Saveliev A, Aggarwal SK (2008) A numerical and experimental study of counterflow syngas flames at different pressures. Fuel 87:319–334

Tao F, Reitz RD, Foster DE (2007) Revisit of diesel reference fuel (n-Heptane) mechanism applied to multidimensional diesel ignition and combustion simulations. In: Seventeenth international multidimensional engine modeling user's group meeting at the SAE congress, 15 Apr 2007, Detroit, Michigan

Tijmensen MJA, Faaij APC, Hamelinck CN, van Hardeveld MRM (2002) Exploration of the possibilities for production of Fischer Tropsch liquids and power via biomass gasification. Biomass Bioenergy 23:129–152

Tomita E, Harada Y, Kawahara N, Sakane A (2009) Effect of EGR on combustion and exhaust emissions in supercharged dual-fuel natural gas engine ignited with diesel fuel. SAE paper 2009-01-1832

Vagelopoulos CM, Egolfopoulos FN (1998) Direct experimental determination of laminar flame speeds. Proc Combust Inst 27:513–519

Venkateswaran P, Marshall A, Shin DH, Noble D, Seitzman J, Lieuwen T (2011) Measurements and analysis of turbulent consumption speeds of H_2/CO mixtures. Combust Flame 158:1602–1614

Walton SM, He X, Zigler BT, Wooldridge MS (2007) An experimental investigation of the ignition properties of hydrogen and carbon monoxide mixtures for syngas turbine applications. Proc Combust Inst 31:3147–3154

Wang H, You X, Joshi AV, Davis SG, Laskin A, Egolfopoulos F, Law CK (2007) USC Mech version II. High-temperature combustion reaction model of H_2/CO/C1-C4 compounds. http://ignis.usc.edu/USC_Mech_II.htm, May 2007

Wang J, Chen H, Liu B, Huang Z (2008a) Study of cycle-by-cycle variations of a spark ignition engine fueled with natural gas–hydrogen blends. Int J Hydrogen Energy 33:4876–4883

Wang L, Weller CL, Jones DD, Hanna MA (2008b) Contemporary issues in thermal gasification of biomass and its application to electricity and fuel production. Biomass Bioenergy 32:573–581

Watanabe M, Inomata H, Arai K (2002) Catalytic hydrogen generation from biomass (glucose and cellulose) with ZrO_2 in supercritical water. Biomass Bioenergy 22:405–410

Werther J, Saenger M, Hartge EU, Ogada T, Siagi Z (2000) Combustion of agricultural residues. Prog Energy Combust Sci 26:1–27

Wikipedia (2014) Biogas. (Online) Available at http://en.wikipedia.org/wiki/Biogas. Accessed 2014

Yan B, Wu Y, Liu C, Yu JF, Li B, Li ZS (2001) Experimental and modeling study of laminar burning velocity of biomass derived gases/air mixtures. Int J Hydrogen Energy 36:3677–3769

Mixture Preparation Effects on Distributed Combustion for Gas Turbine Application

Ahmed E.E. Khalil and Ashwani K. Gupta

Abstract Distributed Combustion is now known to provide significant improvements to the performance of gas turbine combustors. Key features of distributed combustion include uniform thermal field in the entire combustion chamber for significantly improved pattern factor and avoidance of hot-spot regions that promote thermal NO_x emissions, negligible emissions of hydrocarbons and soot, low noise, and reduced air cooling requirements for turbine blades. Distributed combustion necessitates controlled mixing between the injected air, fuel, and hot reactive gases from within the combustor prior to mixture ignition. The mixing process impacts spontaneous ignition of the mixture to result in improved distributed combustion reactions. Distributed combustion can be achieved in premixed, partially premixed, or non-premixed modes of combustor operation with sufficient entrainment of hot and active species present in the combustion zone and their rapid turbulent mixing with the reactants. Distributed combustion with swirl is investigated here to further explore the beneficial aspects of such combustion under relevant gas turbine combustion conditions. The near-term goal is to develop a high-intensity combustor with ultra-low emissions of NO_x and CO and a much improved pattern factor and eventual goal of near-zero emission combustor. Different fuel injection scenarios are examined with focus on mixing to achieve distributed reaction conditions and ultra-low emissions. In all the cases, air was injected tangentially to impart swirl to the flow inside the combustor. Ultra-low NO_x emissions were found for both the premixed and non-premixed combustion modes for the geometries investigated here. Results showed very low levels of NO ($\sim$10 PPM) and CO ($\sim$21 PPM) emissions under non-premixed mode of combustion with air preheats at an equivalence ratio of 0.6 and a moderate heat release intensity of 27 MW/m^3-atm. Further enhancement of the mixing process using dilution reduced NO emission to 4.6 PPM which is nearly equivalent to emissions under premixed combustion mode with reduced CO emissions compared to non-premixed combustion mode. Results are also reported on lean stability limits and OH* chemiluminescence under different fuel injection scenarios

A.E.E. Khalil · A.K. Gupta (✉)
Department of Mechanical Engineering, University of Maryland, College Park, MD 20742, USA
e-mail: akgupta@umd.edu

A.K. Agarwal et al. (eds.), *Novel Combustion Concepts for Sustainable Energy Development*, DOI 10.1007/978-81-322-2211-8_12

for determining the extent of distribution combustion conditions. Numerical simulations have also been performed to help develop an understanding of the mixing process for better understanding of ignition and combustion.

Keywords Distributed combustion · Ultra-low emission · High-intensity gas turbine combustion · Colorless combustion · Green combustion turbine · Uniform thermal field

1 Introduction

Increased concerns about pollutants emission and increased stringent regulations concerning emissions from all kinds of propulsion and power systems have motivated combustion engineers to develop novel combustion techniques for achieving ultra-low levels of pollutants emission (such as NO_x, CO, unburned hydrocarbons, and soot) from gas turbine combustors and improve the pattern factor for increased efficiency and longer life of the turbine blades. The thermal field uniformity in contemporary gas turbine combustors is far from adequate as the combustors do suffer from local burnout, thus causing extensive system downtime as well as increased emission of NO_x, CO, unburned hydrocarbons, and soot. Colorless distributed combustion (CDC), which shares some similar principles utilized in high-temperature air combustion (HiTAC) (Tsuji et al. 2003), has shown huge reduction in emissions of NO_x, CO, unburned hydrocarbons, and soot, along with improved thermal field uniformity in the entire combustor (improved pattern factor). Stable combustion, alleviation of combustion instability, and low noise emission were also demonstrated under CDC conditions for gas turbine combustion applications. The flames in distributed combustion do not show any visible flame signatures so that the flame so formed is termed "colorless" due to negligible visible emissions from the flames as compared to conventional flames.

Mixture preparation in gas turbine combustors can be by different means, and the flame is stabilized using swirlers to radially stretch the flow mixture. Swirl provides hot gas recirculation zone for better mixing of the reactants and increased residence of the reactants in the toroidal recirculation zone of the combustor with the subsequent result of compact flame and enhanced flame stability limits over a wide range of operating conditions (Gupta et al. 1984). A recirculation zone is created downstream of the swirler, wherein the flow reversal results in mixing of hot product gases with the incoming fresh stream of reactants. This provides high and uniform temperatures in the combustion zone in the form of a well-stirred reactor. This hot mixture provides the ignition energy for the fuel to ignite and stabilize the flame. In premixed condition, the reaction zone is stabilized in the region of interaction between the product gases and the fresh stream and is often a very thin reaction zone of the order of a millimeter. However, in non-premixed condition, fuel is injected in the shear region formed near the zero streamline boundary and the recirculation region which provides a low-velocity region for flame stabilization with the evolution of high temperatures from the flame. For flames operating in diffusion mode, the reaction zone is stabilized, resulting in large temperature gradients and hot-spot regions in the entire combustion chamber that cause high NO_x levels from the combustion of fuels (Lefebvre 1999).

In previous studies on HiTAC flames for furnace applications (that are characteristic of low combustion intensities of 0.1–1 MW/m^3-atm), avoidance of thin concentrated reaction front in the flame has been achieved by the controlled recirculation and mixing of large amounts of hot and active combustion gases with the fuel and air streams prior to ignition of the mixture to provide distributed mixture reaction zone (Tsuji et al. 2003). Preheating of the air stream has been employed to provide spontaneous ignition of the fuel with volume distributed combustion and stable flame without any flame-holding device. The pressure drop is also very low compared to other flame stabilization devices. HiTAC has been shown to provide significant fuel energy savings in furnaces. In the HiTAC technique, one may envision that high air preheats are required; however, preheating of the combustion air or the fuel is neither necessary nor required to achieve distributed combustion reactions. Furthermore, no catalyst is used in this distributed combustion mode. HiTAC has been successfully demonstrated and is now widely used to achieve low NO_x and CO emissions, stable combustion, and low noise and simultaneously achieve significant energy savings using a range of gas, liquid, and solid fuels for furnace applications (Tsuji et al. 2003; Gupta 2004; Wunning and Wunning 1997; Katsuki and Hasegawa 1999). This work has been known as HiTAC (Tsuji et al. 2003; Gupta 2004), excess air enthalpy combustion, and flameless oxidation (FLOX) (Wunning and Wunning 1997).

In HiTAC technology, high temperature of the air is obtained by preheating with the hot gases from a furnace or reactor. The hot gases can be harvested from within the combustion zone or external to the combustion zone, using, for example, exit gases. The peak temperature in the flame zone is much reduced with the use of diluted low oxygen concentration combustion air even though the air is preheated to high temperatures. This low oxygen concentration or diluted air is obtained from the exhaust gases by recirculating part of the combustion products into the incoming hot combustion air. The effect of air preheat temperature and oxygen concentration on HiTAC flames have been studied, where it was found that pollutants emission, including CO_2 and NO_x, was much lower with highly preheated combustion air at low O_2 concentration than with normal air (Gupta et al. 1999). The HiTAC technology has been also extended to other applications such as thermal destruction of wastes (Gupta et al. 1996) and combustion in microscale combustion devices wherein the combustor dimensions are smaller than quenching distances (Vijayan and Gupta 2010; Shirsat and Gupta 2011).

CDC investigated here is focused on high combustion intensity for stationary gas turbine combustion application, although other applications are also possible. Therefore, the results presented here are also relevant for other power and propulsion applications. Previous investigations of CDC suggest significant improvement in pattern factor, low sound emission levels, and ultra-low emissions of NO_x

and CO using a variety of gaseous and liquid fuels (including biofuels) (Arghode and Gupta 2010a, b, 2011; Khalil and Gupta 2011a, b, 2013, 2014a, b; Arghode et al. 2012). To achieve reactions closer to the distributed regime and avoid the presence of thin reaction zone and hot-spot zones in the flames, controlled mixing between the combustion air and product gases is necessary so as to form a hot and diluted oxidant with rapid mixing with the fuel. High recirculation of hot recirculated combustion gases and its fast mixing with the fuel leads to spontaneous ignition of the fuel with distributed reaction conditions. The recirculated hot products not only play a role in increasing the oxidant temperature while decreasing oxygen concentration, but also the recirculated active radicals within the hot products play an important role in the reaction kinetics. This results in avoidance of thin reaction zone and hot-spot regions in the flame, which helps to minimize or mitigate NO_x emissions (thermal NO_x) produced from the Zeldovich thermal mechanism (Tsuji et al. 2003; Correa 1992).

In CDC, the reaction occurs in a distributed regime due to the volume distributed nature of the mixture of combustion gases, fuel, and oxidizer in the combustion chamber. Depending on the ignition delay time and mixing timescales, the reaction zone is a distributed regime as compared to thin reaction flame front in conventional flames. Such volume distributed combustion can be achieved by air injection at high velocities to avoid stabilization with large thermal gradients in the flame. This may be accomplished by appropriate separation of air and fuel jets and internal recirculation of a large amount of product gases to aid the spontaneous ignition of the mixture with the evolution of distributed reaction zone. The concept of separate injection of fuel and air at a high velocity with desirable and controlled amounts of gas recirculation and mixing between the product gases and fresh reactants can be applied to combustors operating at higher heat release intensities [5–50 MW/m^3-atm (Vincent 1950)] that are commensurable with the gas turbine application. These requirements can be met with different configurations of fuel and air injection into the combustor while using a carefully tailored flow field in the combustor.

The importance of recirculation zone generation and good preparation of the air–fuel mixture for ignition cannot be overstated. One common practice used to create recirculation and stabilize combustion is to utilize swirl flow that entrains and recirculates a portion of the hot combustion products back to the root of the flame. For such combustors, swirl characteristics play a major role in mixing and combustion (Gupta et al. 1984; Archer and Gupta 2004; Leuckel and Fricker 1976). Swirl flows have been widely investigated for several decades because of their extensive use in all kinds of practical combustion systems, including gas turbine combustion. Numerous experiments in swirl flows have been carried out extending from very fundamental isothermal flows and reacting flows to those formed in complex swirl combustor geometries (Gupta et al. 1984). Experimental results have established the general characteristics of swirl flows that reveal the important effects of swirl on promoting flame stability, increasing combustion efficiency, and controlling emission of pollutants from combustion (Gupta et al. 1984). Leuckel and Fricker (1976) conducted a variety of measurements using a non-premixed single swirl burner consisting of an annular swirling air jet and a centrally located non-swirling fuel jet.

Chen and Driscoll (1990) have examined the physical processes that occur within the non-premixed flames by exploring the enhanced mixing characteristics in swirl flows that emanate from the formation of a central toroidal recirculation zone. Gupta et al. (1977) examined multi-annular swirl burner for improved mixing and combustion with the injection of fuel in the shear layer of the surrounding swirling flow to improve radial mixing and achieve compact combustion.

In a previous investigation, the role of swirling air injection into the combustion chamber for distributed combustion reactions was explored (Khalil and Gupta 2011a, b, 2014a). Air was injected tangentially into the combustion chamber at a high air velocity to form swirling motion. This air jet entrains large amounts of product gases forming a recirculation zone. The amount of this recirculation is controlled so that it increases the temperature of the mixture containing air, product gases, and fuel to a temperature that is above the auto-ignition temperature of the fuel. In the non-premixed condition, the fuel is injected at some distance downstream to provide sufficient mixing (desirable mixing time should be less than the ignition delay time). The uniformly mixed fuel/air/product gas will then spontaneously ignite to result in a distributed reaction regime, instead of a thin concentrated reaction flame front formed in conventional flames. Hence, it may be noted that the CDC cases discussed here differ from conventional gas turbine flames in that they do not require a flow reversal or low-velocity region for flame stabilization. This helps to reduce pressure drop across the combustor. The product gases mix with the fresh mixture to increase the temperature of the mixture high enough to cause spontaneous ignition in the entire zone as compared to only a small region of the fresh mixture for flame stabilization as exhibited in conventional flames. Swirl combustors with tangential air entry have been shown to exhibit high swirl intensity, which helps to reduce NO_x emission and enhance flame stability (Yetter et al. 2000). Also, ultra-low NO_x emission along with low CO emission has been demonstrated for swirling CDC combustor where emissions below 3 PPM NO have been demonstrated for a high heat release intensity of 36 MW/m^3-atm at a rather high equivalence ratio of 0.7 (Khalil and Gupta 2011a). Swirling CDC has also been demonstrated using a recirculation ratio (recirculated mass flow rate to inlet mass flow rate) that is three times higher than that for non-swirling configuration (Khalil and Gupta 2014a).

Lean premixed combustion has been inherently used in gas turbine combustors to lower flame temperatures and to avoid stoichiometric non-uniformities that arise due to incomplete mixing of fuel and air. In agreement with this, experimental investigations performed on the test combustor show that premixed combustion demonstrates lower pollutants emission as compared to non-premixed combustion mode over the operational range (Arghode and Gupta 2010a, b, 2011; Khalil and Gupta 2011a, b, 2013, 2014a, b; Arghode et al. 2012). However, operation of these premixed systems can lead to undesired phenomenon of flame flashback. High temperatures in a premixer caused by flashback can result in damage to or shutdown of an engine. Other phenomena relating to premixed combustion have been studied in detail. The initial un-mixedness in a fuel–air premixer has been shown to make the overall lean mixtures to auto-ignite sooner than might be expected based on the overall fuel–air ratio. Also, carbon deposition arising from the flow of liquid fuel over metal surfaces, for example, as found in fuel injectors and swirl cups, has been described as a function of temperature and surface composition (Correa et al. 1996). Hence, it is advantageous for fuel and air to be injected into the combustor separately and then allowed to mix rapidly before they burn. The main means of achieving this is through careful design of the aerodynamics of burners and enclosures, which allows control over the mixing mechanisms and residence time. Under CDC conditions, it is believed that optimum fuel injection and mixing should result in pollutants emission that are lower than premixed case if adequate control on the mixture preparation can be made to achieve the distributed thermal field.

High-velocity coaxial air in jet diffusion flames has been successfully demonstrated to reduce NO_x emission levels in hydrocarbon flames (Kim et al. 2000). Enhanced shear produced by high-velocity coaxial air results in enhanced mixing between fuel and air before combustion and also lowers the residence times resulting in lower emission levels. Mixing in coaxial jets has been studied in details, where the main characteristics of these jets were described (Villermaux and Rehab 2000). Coaxial jet has been also used in gas turbine model combustors demonstrating low pollutants emission over their range of operation (Bobba et al. 2008). However, it is anticipated that mixing between air, fuel, and hot product gases to achieve CDC conditions does not depend solely on the air and fuel injection method, but rather on the interaction between fresh streams, introduction method combined with flow field developed inside the combustor.

In this paper, the effect of different fuel introduction scenarios on the combustor flow field and on pollutants emission is investigated to evaluate the key features associated with each fuel injection scenario on the combustor performance under various operational conditions. A combustor comprising swirling air injection and a helical path for product gases inside the combustor was adopted. A comparison between using premixed air/fuel jet, coaxial air/fuel jets, and separate air and fuel jets was established to identify the most suitable configuration for the previously developed combustor (Khalil and Gupta 2011a, b, 2014a). The effects of burner exit Reynolds number on the relative effects of burner geometry (circular and elliptic with an aspect ratio 2:4) have been investigated (Kamal and Gollahalli 2001). Results show that an increase in Re decreases the benefits of increased air entrainments into the flame from the elliptic burner geometry. Similarly, the effects of changes in NO and CO emission tend to level off at higher burner Re. Consequently, circular inlet jets associated with high inlet velocities have been used in this study

As the impact of fuel injection location is established, fuel dilution effects with different amounts of air (and also air dilution effects with different amounts of fuel) are investigated for enhanced mixture preparation with emphasis on achieving ultra-low emissions comparable to or less than those from novel premixed combustion in distributed combustion without encountering any flashback or combustion instabilities associated with premixed combustion. Previously, different dilution scenarios were examined where the NO emissions were shown to be as low as premixed combustion case (Khalil et al. 2013). It is worth noting that in these dilution cases, the mixture formed (fuel added to air or air added to fuel) had an equivalence ratio that was well outside the fuel flammability limits. Consequently, no flashback was possible. However, the study was limited to a single equivalence ratio (0.8) on an ultra-high-intensity combustor for aviation application (Khalil et al. 2013). One of the air dilution cases (case 2) is further investigated here in detail to evaluate its performance under different equivalence ratios.

2 Experimental Investigations

Air/fuel mixture preparation and mixing with hot recirculated reactive gases are investigated with swirling air flow created through tangential air injection. A nominal combustion intensity of 36 MW/m^3-atm at a constant heat load of 6.25 kW was used as an experimental condition to simulate stationary gas turbine combustion conditions. For fresh streams (air and fuel), injection scenarios are reported here to show the direct role of mixture preparation in achieving distributed combustion conditions and ultra-low pollutants emission. Figure 1 shows a schematic diagram of the arrangements on the fuel injection scenarios presented here with their relative dimensions in terms of D, where D is the air injector diameter. In the first configuration, named "PR," fuel is injected in the air jet upstream of the combustor, and the jet entering the combustor can be considered as perfectly mixed air/fuel jet (see Fig. 1a). The second configuration involves a coaxial nozzle, where fuel is introduced as a center jet, while air is introduced through the annulus, named "CA" (see Fig. 1b). The third and fourth configuration had separate fuel jet introduced in a "cross-flow" behavior. Two different separation distances between the air and fuel jets were investigated, namely "NP0" and "NP1" (see Fig. 1c, d). Changes in the fuel injection method are expected to have a significant effect on mixture preparation between air, fuel, and product gas. Adequate mixture preparation is critical to achieve distributed reactions as discussed earlier.

The configuration NP1 is further investigated here using air dilution to increase the fuel jet momentum for better mixing based on fuel jet interaction with the core flow. Air dilution also offers mixing enhancement as portion of the air is already mixed with fuel, which results in a less complex mixing issues as compared to non-premixed combustion. In the dilution case, air and fuel flow rates are split into two streams. The main "top stream" injects most of the air flow, while the other "cross stream" injects most of the fuel. Fuel is added to the "top stream," and air is added to the "cross stream" such that the equivalence ratio for each stream is well outside of the flammability limits for methane–air combustion (Turns 2006). The cross-flow equivalence ratio was kept at 5. The top flow equivalence ratio varied from 0.2 to 0.12. Both air and fuel flow rates were varied to furnish the combustor with the appropriate air and fuel flows for operation in the desired range of equivalence ratios. Adding air to the fuel stream will increase the jet momentum of the cross-flow. Increased jet momentum will enhance the penetration of the jet in cross-flow to result in better mixing process. In non-premixed case, the jet momentum ratio (top to cross) was 20.4. In air dilution cases discussed here, this ratio varied

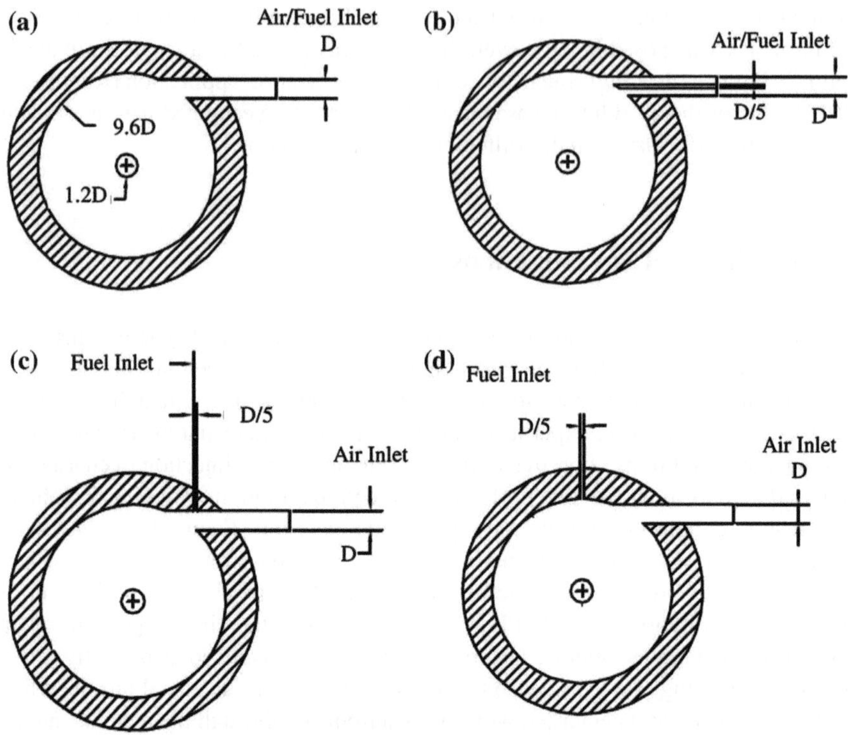

Fig. 1 A schematic diagram of different fuel introduction scenarios, end cross-sectional view at middle location of the combustor **a** configuration "PR" **b** configuration "CA" **c** configuration "NPO" **d** configuration "NPI"

between 2 and 3.7 depending on the equivalence ratio. The higher cross stream jet momentum (or the lower the jet momentum ratio) is expected to significantly enhance the mixing process and also affect the turbulence in this region.

The inlet air temperature was preheated, using an electric air heater, to 600 K to simulate the average air temperature at the gas turbine compressor outlet. The preheating of the inlet air has been shown to affect the combustion kinetics and pollutants emission of this combustor (Khalil and Gupta 2011b). In general, the higher the air inlet temperature, the higher the flame temperature will be. Such high flame temperatures will affect the formation of pollutants, especially NO_x and CO.

2.1 The Test Facility

The combustion chamber under investigation is a cylindrical chamber where air is injected tangentially at half the height of the chamber for all the cases reported here. Fuel is also injected at the same plane as that of air injection. The product gas exit

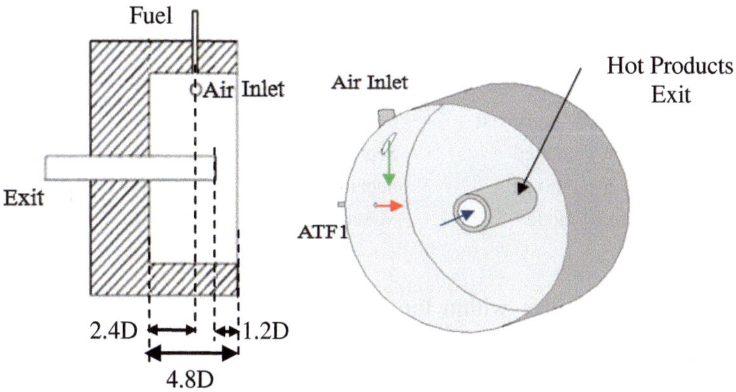

Fig. 2 A schematic diagram for the combustor, longitudinal view along the cylinder axis and a 3D model

Fig. 3 High-intensity CDC cylindrical combustor test rig with optical access (*left*) and product gas exit (*right*)

was extended inside the combustor as shown in Fig. 2. A full description of the experimental facility can be found in previous publications (Khalil and Gupta 2011a, b, 2014a). Figure 3 shows the experimental facility under operation.

Detailed investigations on exhaust emissions and visible emissions were performed for all the experimental investigations reported here. The concentration of NO was measured using a NO–NO_x chemiluminescent gas analyzer; CO concentration was measured using the non-dispersive infrared method, and O_2 concentration (used to correct the NO and CO emissions at standard 15 % O_2 concentration) was measured using the galvanic cell method. The emission readings

Table 1 Experimental parameters investigated

Case no.	Mixing	Fuel introduction	Air temperature (K)
PR	Premixed	With air	600
CA	Partially premixed	Center of air jet	600
NP0	Non-premixed	Separate	600
NP1	Non-premixed	Separate	600
Dil	Mixed mode	Mixed mode	600

were observed to stabilize within three minutes for any change in experimental condition (such as change in equivalence ratio and air preheat temperature for the same configuration). During a single experiment, measurements were repeated three times for each configuration and the uncertainty was estimated to be about ±0.5 PPM for NO and ±10 % for CO emissions. The experiments were repeated at least three times to ensure the consistency and repeatability of each experiment.

For imaging the OH* chemiluminescent intensity distribution, an ICCD camera coupled with a narrow band filter centered at 307 nm wavelength for OH* was used. The gain for all the images obtained was set to 70 with one accumulation per image and f-stop setting of 4.5. The exposure time was set to 50 ms. The resulting OH* intensity distributions were then normalized to have the same scale.

2.2 *Test Conditions*

Detailed investigations were performed on the experimental combustor. The investigations compared the performance of the combustor under different fuel injection scenarios. The facility allowed variation of the injection velocity and nozzle diameter of the fuel and air jets as well as the gas exit diameter. Table 1 summarizes the investigations performed along with the variables manipulated for each investigation. The case name represents the location of fuel injection. The dilution case was studied on NP1 configuration.

3 Numerical Simulations

Numerical simulations were performed to help develop an understanding of the flow field and the mixing process inside the combustor for different air and fuel stream conditions entering the combustor. For these simulations, commercial software FLUENT® code was used. The flow was modeled using finite volume method. For solving turbulence, standard k-ε model with standard wall function was used. For reacting cases, a two-step methane combustion model was adopted to determine mixing between air, fuel, and recirculated gases from within the combustor prior to combustion. The operating pressure was one atmosphere. For any of the numerical models, the volume consisted of more than one million tetrahedral mesh cells.

The model boundary conditions are described in this section. For air inlet, "mass flow inlet" boundary condition was used with inlet turbulence intensity of 10 %, mass fraction of O_2 of 0.233, and inlet temperature of 600 K. For fuel inlet, "mass flow inlet" boundary condition was used with inlet turbulence intensity of 10 %, the mass fraction of CH_4 was set to 1, and the inlet temperature was set to 300 K. For hot product gases to exit, "pressure outlet" boundary condition was used, with zero-gauge back pressure and 10 % turbulence intensity for backflow. For all the numerical simulations, the resultant equivalence ratio was 0.8 with methane as the fuel. For any numerical simulations, the conversion criterion was that the largest residual was less than 10e−4. It is understood that the simulated flow field and mixing behavior will deviate from the actual case; however, the trends are expected to remain consistent as shown through velocity measurements compared to numerical simulations in a previous study (Khalil and Gupta 2014a), thus providing an indication for the causes of pollutants emission behavior for the various cases investigated here.

4 Results and Discussion

4.1 *Injection Location*

The experimental investigation was focused on examining different methods of fuel introduction into the combustor to help develop an understanding of the mixing process so as to enhance and accelerate the mixing and achieve adequate mixing prior to ignition. The distributed combustion process requires that air and/or fuel be mixed with the hot and chemically active gases from the combustion zone prior to the mixture ignition for seeking distributed reactions. In this study, air preheats were employed to simulate gas turbine combustion conditions with elevated inlet combustion air temperatures. The experimental results are reported at equivalence ratios of 0.8, 0.7, 0.6, and 0.5. These upper ends of equivalence ratios are higher than that used in current engines and are used due to higher efficiency that can be achieved at higher equivalence ratios. Emissions of NO and CO were recorded for each arrangement at different equivalence ratios. Figures 4 and 5 show the emission levels at different equivalence ratios with each arrangement.

The results obtained on NO emission indicate that coaxial injection provides highest NO emission levels as compared to other fuel introduction methods. Separate air and fuel introduction exhibited lower emission levels. The premixed combustion mode "PR" revealed the lowest emissions. For instance, at an equivalence ratio of 0.6, coaxial injection provided emission levels of 21 PPM NO and 62 PPM CO. Separate injection mode "NP0" yielded emission levels of 14 PPM NO and 26 PPM CO. Case "NP1" yielded lower emissions at 10 PPM NO and 21 PPM CO.

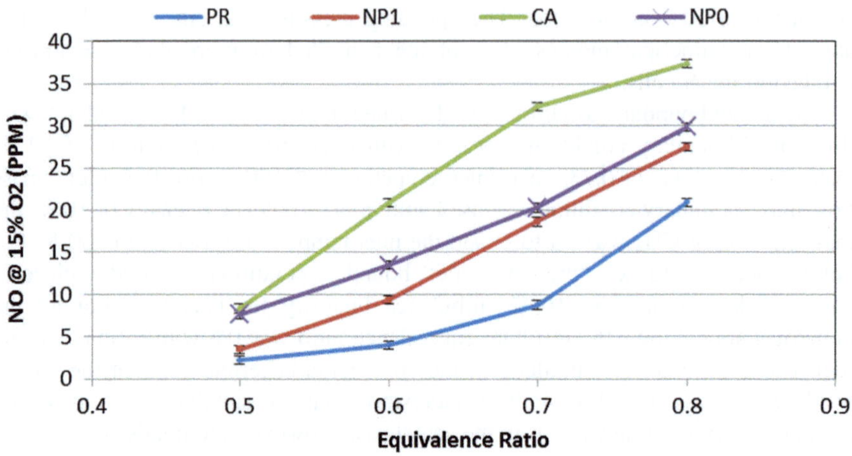

Fig. 4 NO emission for different injection arrangements

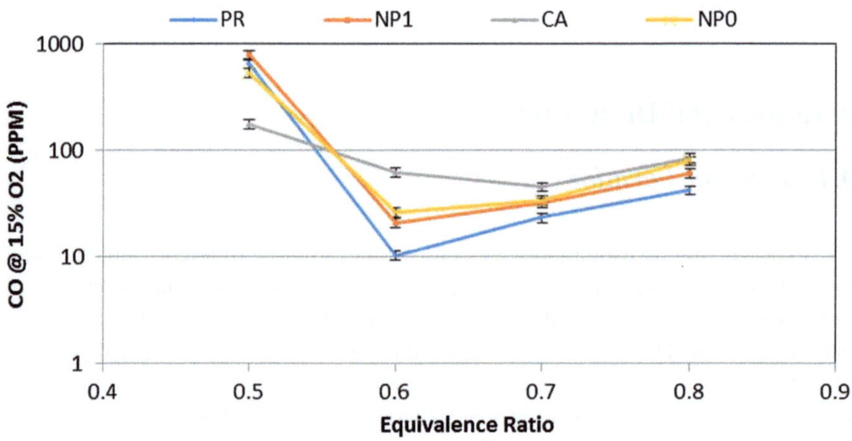

Fig. 5 CO emission for different injection arrangements

However, premixing air and fuel prior to their introduction to the combustion chamber demonstrated ultra-low emission levels of 4 PPM NO and 11 PPM CO. The lower emissions demonstrated from premixed combustion agree with the trend obtained previously for various combustor configurations (Arghode and Gupta 2010a, b, 2011; Khalil and Gupta 2011a, b, 2013, 2014a, b; Arghode et al. 2012). This can be attributed to the fact that the mixing problem in this mode is less complicated. To achieve distributed combustion mode, air and hot product gases mix together to form hot diluted oxidizer. Then, the fuel is mixed with that hot diluted mixture to achieve uniform mixture and spontaneous ignition. However, in premixed mode, air and fuel are well mixed upstream, which helps in mixture preparation for distribution condition.

The NO emissions decreased with the decrease of equivalence ratio as a direct result of lower combustion temperature for all the cases presented here. Note that the trends for CO agree favorably well with the previously reported results (Lefebvre 1995). Emission of CO diminishes with increase in equivalence ratio, reaching a minimum value. Beyond this point, any further increase in the equivalence ratio causes CO emissions to increase. The observed high levels of CO at low equivalence ratios are due to the slow rates of oxidation associated with low combustion temperatures. However, for configuration CA, the observed CO emission at equivalence ratio of 0.5 is lower than the observed emissions for the other configurations examined. Such behavior indicates that the reaction rate is higher at higher temperatures. This suggests that the combustion occurs in a less diluted oxidizer (higher oxygen concentration) so that colorless distributed conditions are not satisfied. NO emission for this configuration, compared to other configurations, was found to be higher to reveal the effect of higher reaction rate and temperature encountered in this configuration.

An increase in the equivalence ratio raised the flame temperature, which accelerated the rate of oxidation to subsequently decrease the CO emissions. However, at temperatures higher than around 1,800 K, the production of CO by chemical dissociation of CO_2 starts to become significant. Thus, at high equivalence ratios, CO levels are high because they attain equilibrium values.

The radical intensity distribution of OH* chemiluminescence showed that the reaction zone is in the shape of a crescent. It can be seen that as the fuel introduction method is changed, the reaction zone moves gradually downstream. Premixed combustion case "PR" resulted in the earliest reaction zone, whereas separate air and fuel introduction as in case "NP1" resulted in a longer ignition delay. OH* zone movement, which can be seen as a flame marker, outlines the difference between mixing and ignition delay in each case discussed here. Figure 6 shows the OH* chemiluminescence intensity distribution at an equivalence ratio of 0.8 for the different cases presented here. A similar trend was also observed at other equivalence ratios.

Numerical simulations were carried out to help develop an understanding of the difference in behavior between each of the fuel injection scenarios discussed herein. The calculated results showed that, for each fuel injection scenario, the ignition of

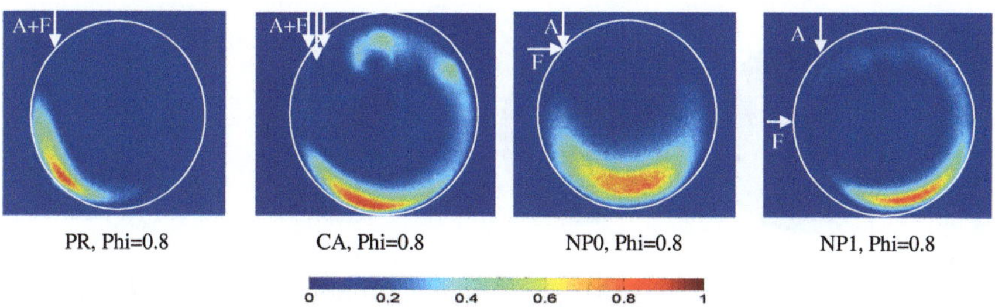

Fig. 6 OH* chemiluminescence intensity distribution for different injection arrangements

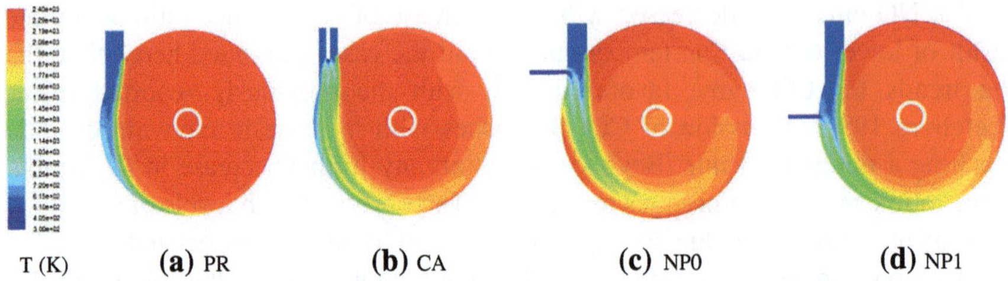

Fig. 7 Temperature distribution at the air/fuel injection plane for the different cases

the mixture is essentially different. Temperature distribution for the air/fuel injection plane is shown in Fig. 7. It can be seen that in case CA (see Fig. 7b), there is a rapid temperature increase at the center of the air/fuel jet. The temperature at the center of the jet approaches 1,700 K which is relatively high as compared to the surrounding jet flow with a temperature of about 1,300 K. This rapid temperature increase indicates that ignition occurs at this point, where the igniting mixture consists of air and fuel with minimal amounts of hot and reactive gases. One should note that, according to temperature plots, recirculated gases having a temperature of about 2,000 K are separated from the ignition zone with cooler air jet (of about 1,300 K). Such oxygen-rich combustion, which fosters thermal NO production, indicates lack of adequate mixing under this scenario. Consequently distributed reaction requirements are not satisfied leading to a "diffusion flame" behavior that should be avoided in combustors for near-zero emissions. A similar behavior can be seen in case "NP0" but to a smaller extent (see Fig. 7c), where combustion occurs only between air and fuel without recirculated hot products. On the other hand, temperature distribution for cases PR and NP1 (see Fig. 7a, d) indicates that such early ignition of only air and fuel does not exist.

Figure 8 shows the fuel jet decay for each of the cases discussed here. By comparing methane decay plots with temperature plots, one can interpret the mixing and ignition process. For case CA, the area at the jet center with high temperature rise as shown in Fig. 7b corresponds with methane depletion shown in Fig. 8b. It is to be noted that for premixed combustion, the plotted results are for fuel mixture fraction relative to the inlet fuel mixture fraction.

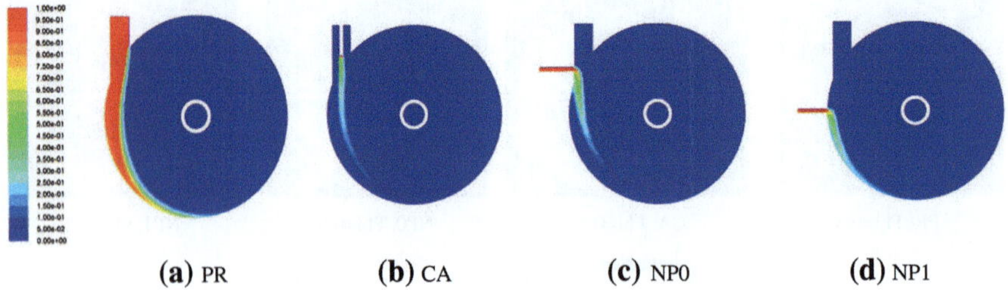

Fig. 8 Methane mass fraction distribution at the air/fuel injection plane for the different cases

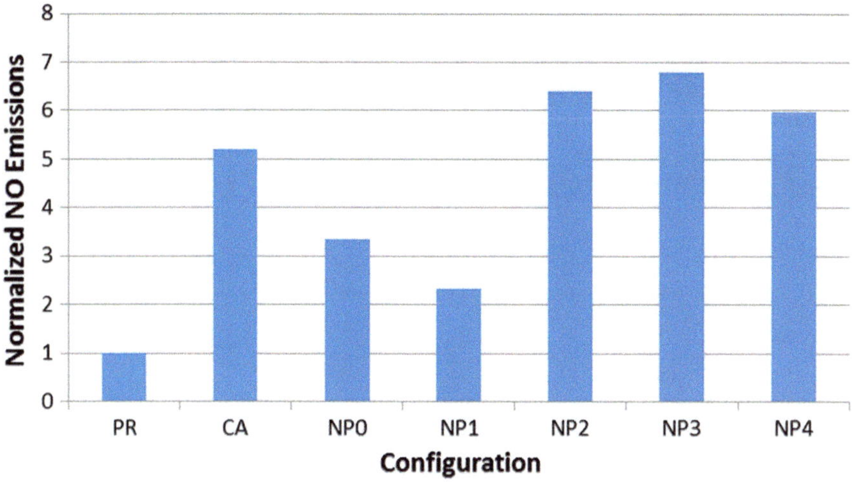

Fig. 9 Normalized NO emissions for different configuration at an equivalence ratio of 0.6

The data obtained here can be combined with precious data for non-premixed combustion to reveal the impact of injection location. NO emissions under different injection scenarios are normalized with respect to NO emission from premixed combustion mode and the results are shown in Fig. 9. These reported values are at an equivalence ratio of 0.6. Early injection as in configurations CA and NP0 resulted in early ignition without proper mixing with hot recirculated reactive gases. On the other hand, late injection cases (NP2, NP3, and NP4) resulted in high emissions. This is attributed to the diffusion flame behavior for these cases were combustion occurs without the fuel properly mixing with air, leading to a locally high equivalence ratio and high emissions regardless of the overall equivalence ratio (Khalil and Gupta 2011a). Configuration NP1 offers a balance of both the issues. Injection of fuel is delayed enough to allow for proper entrainment of recirculated reactive gases and their subsequent mixing with air, yet it is not delayed such that the fuel ignites in a diffusion-like flame without mixing with the hot and diluted oxidizer (i.e., the fresh air and the recirculated reactive species). Hence, configuration NP1 is chosen for further mixing enhancement to reduce the emissions to premixed combustion levels that can bring premixed combustion instability issues.

4.2 Air Dilution

As the emission behavior of the different scenarios explained with NP1 configuration resulted in lower emissions as compared to NP0 and CA, it was sought to further decrease emissions to premixed emission levels through air dilution. Table 2 gives a comparisons between air and fuel distribution for premixed, non-premixed, and dilution cases.

Table 2 Experimental parameters for air dilution

Case	Air (top) stream			Fuel (cross) stream			Overall equivalence ratio
	Air flow (SLPM)	CH_4 flow (SLPM)	Φ	Air flow (SLPM)	CH_4 flow (SLPM)	Φ	
Premixed	137	11.5–7.2	0.8–0.5	–	–	–	0.8–0.5
Non-premixed	137	–	–	–	11.5–7.2	–	0.8–0.5
Dilution	119.9–126.3	2.5–1.6	0.2–0.12	17.1–10.7	9–5.6	5	0.8–0.6

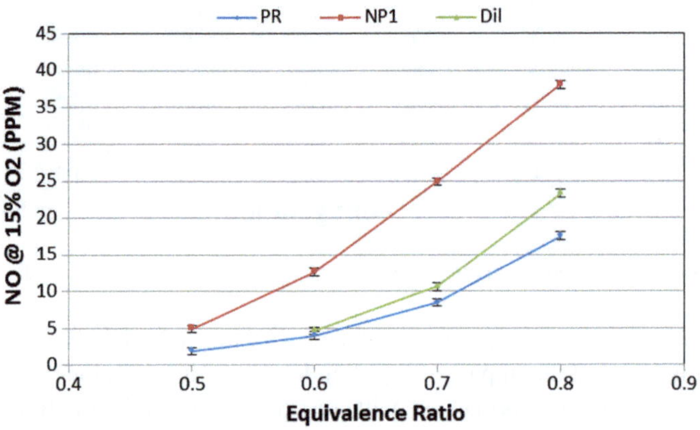

Fig. 10 NO emission for premixed, non-premixed, and air dilution case

Figure 10 shows the measured NO emissions for the premixed, non-premixed, and dilution cases. Air dilution demonstrated favorable performance as NO emissions were substantially lower than that from non-premixed combustion. For instance, at an equivalence ratio of 0.6, NO emission for the dilution case was about 37 % that of the non-premixed case. This percentage changed between 33 and 60 % depending on the equivalence ratio. Comparing these emissions to that of the premixed case, NO emission was almost as low as the premixed case especially at lower equivalence ratio, where NO emission was only 0.5 PPM higher than that of the premixed case (at equivalence ratio 0.6). Previously, this air dilution case had demonstrated emissions equal to those of premixed case (Khalil et al. 2013). However, due to the difference in the flow field, the behavior is different. This difference in the flow field arises from the center exit tube which prevents the fuel (cross jet) from further penetration into the combustor. CO emissions were also reduced under the dilution scenario where the emissions were about 40 % of those under non-premixed conditions. However, these emissions were slightly higher than those of premixed combustion mode (~20 % higher)

To further examine the effect of air dilution, OH* chemiluminescence was used to identify the reaction zone and determine any differences between traditional non-premixed combustion and air dilution cases. The OH* intensity distribution is

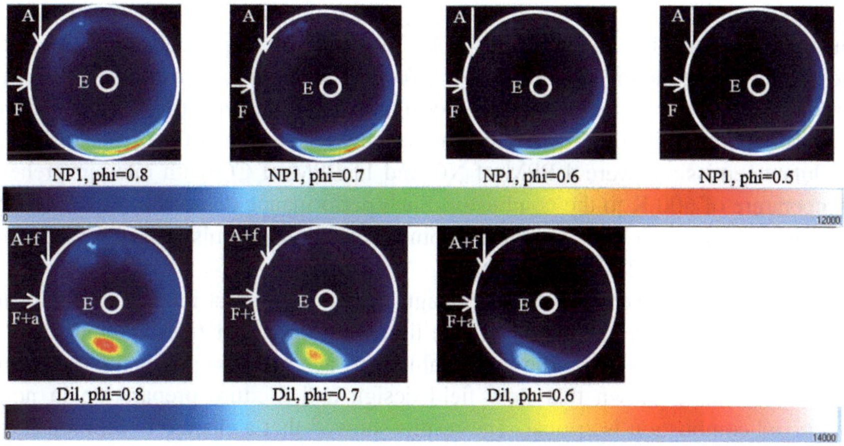

Fig. 11 OH chemiluminescence for non-premixed (*upper*) and air dilution (*lower*) combustion cases

shown in Fig. 11 for both cases, where it can be seen that the air dilution changed the reaction zone location into a larger area compared to a crescent-like zone near the wall of the combustor. This is conjectured to be the direct result of higher jet momentum associated with the diluted fuel jet with air compared to the non-premixed case. For non-premixed case, the top injection momentum was about 0.246 kg m/s, while the cross injection had a momentum of 0.012 kg m/s to result in a jet momentum ratio of 20.4 (top to cross). In the air dilution case, the top injection momentum is about 0.2 kg m/s, while the cross injection momentum is significantly increased to 0.06–0.1 (depending on the equivalence ratio). This change in jet momentums resulted in a much lower jet momentum ratio (2–3.7 depending on the equivalence ratio) which significantly impacts the jet in cross stream penetration and mixing process. The increased momentum of the fuel jet leads to a better mixing process and bends the main air jet to change the reaction zone. Such better mixing process and enlargement in the reaction zone location led to the demonstrated decrease in NO emissions. Also, increased cross jet momentum affects the turbulence level to enhance mixing. Previous investigations showed that any increase in turbulence level enhances mixing (Smith and Mungal 1998) and reduced emissions (Khalil and Gupta 2014c).

5 Conclusions

The distributed combustion with swirl has been shown to provide low emissions of NO and CO under favorable geometrical conditions. The results obtained for the different cases presented here demonstrate the ability of achieving very low pollutants emission with a combustor design incorporating swirl flow in the combustor. Experimental investigations have outlined the importance of flow field design and the relationship between the fuel jet location and resulting pollutants emission. Novel premixed combustion design along with swirling flow has yielded low pollutants emission. At a rather high equivalence ratio of 0.6, the resulting pollutants emissions were 4 PPM of NO and 11 PPM of CO with inlet air preheat temperature of 600 K to the combustor. Such performance demonstrates the ability of this combustor design to produce minimal pollutants emission at a moderate thermal intensity of 27 MW/m^3-atm.

Novel non-premixed combustion design, which is of great interest to eliminate premixed combustion instabilities, flame flashback, and other risks associated with premixed combustion have shown the ability to achieve low pollutants emission. The results have shown that flow field design and mixture preparation in non-premixed combustion is critical. The results showed that a change in fuel injection location greatly affects pollutants emission. For instance, NO emission decreased dramatically from 21 PPM (for coaxial air/fuel injection, case "CA") to 10 PPM (for non-premixed, case "NP1"). CO emission variation was minimal between the different cases due to the fact that inlet air preheats dramatically decreased CO emission due to faster reaction kinetics and complete conversion to CO_2.

The experimental data on emissions and numerical predications revealed that high NO emissions recorded for cases "CA" and "NP0" can be attributed to poor mixing of air, fuel, and hot recirculated reactive gases. Numerical simulations showed that high temperature rise (indicating ignition) was found in areas where only air and fuel existed with minimal amounts of recirculated gases. Such ignition occurring in oxygen-rich mixture fosters thermal NO formation which should be avoided. In order to achieve distributed reaction the air, fuel and hot recirculated reactive gases should be thoroughly mixed prior to ignition and that mixing time to be less than ignition delay. Such good mixing can be found in premixed case "PR" and non-premixed case "NP1."

The interaction between jet mixing ability and the combustor flow field is also of importance. Coaxial injection has been reported to yield favorable mixing between the center jet and surrounding annulus jet. However, in the present case where mixing is required between air, fuel, and recirculated gases, coaxial injection yielded high pollutants emission due to improper mixing between the air/fuel jet and the recirculated flow, leading to the ignition of air/fuel mixture only.

Novel air and fuel mixing, where fuel is diluted with air while portion of the fuel is introduced in the main air jet, demonstrated NO emission comparable to those encountered in premixed combustion mode. These emissions were about 40 % of the NO emissions under non-premixed combustion mode. Fuel dilution, characterized by high cross jet momentum, not only demonstrated NO emission reduction compared to non-premixed mode with comparable emissions to the of premixed mode combustion, but also lower CO emissions compared to non-premixed combustion. Such novel mixing technique eliminates the possibility of flame flashback as the equivalence ratio in the jets is outside the methane air flammability limit. OH* intensity distribution showed that the combustion zone is affected by the cross jet momentum and equivalence ratio. The ability of this mixing technique to demonstrate near-zero emission shows the importance of mixture preparation for achieving colorless distributed combustion for ultra-low NO emissions, without any drawbacks of premixed combustion, such as flame flashback and combustion instability.

Preheated air enhanced the combustion stability. The combustor performance is expected to only improve under more realistic gas turbine conditions of high air preheat temperature and elevated pressures. Global imaging showed no observable unsteadiness in the flame with each fuel injection arrangements.

References

Archer S, Gupta AK (2004) Effect of swirl on flow dynamics in unconfined and confined gaseous fuel flames. In: 42nd AIAA aerospace sciences meeting and exhibit, 5–8 Jan 2004, Reno, Nevada. Paper No. AIAA-2004-0813

Arghode VK, Gupta AK (2010a) Effect of flow field for colorless distributed combustion (CDC) for gas turbine combustion. Appl Energy 87(5):1631–1640

Arghode VK, Gupta AK (2010b) Investigation of forward flow distributed combustion for gas turbine application. Appl Energy 88:29–49

Arghode VK, Gupta AK (2011) Hydrogen addition effects on methane-air colorless distributed combustion flames. Int J Hydrogen Energy 36:6292–6302

Arghode VK, Gupta AK, Bryden M (2012) High intensity colorless distributed combustion for ultra low emissions and enhanced performance. Appl Energy 92:822–830

Bobba MK, Gopalakrishnan P, Periagaram K, Seitzman JM (2008) Flame structure and stabilization mechanisms in a stagnation-point reverse-flow combustor. J Eng Gas Turbine Pow 130(3):031505-1–031505-8

Chen R, Driscoll F (1990) Nitric oxide levels of jet diffusion flames: effect of coaxial air and other mixing parameters. Proc Combust Inst 23:281–288

Correa SM (1992) A review of NO_x formation under gas turbine combustion conditions. Combust Sci Technol 87:329–362

Correa SM, Dean AJ, Hu IZ (1996) Combustion technology for low-emissions gas-turbines: selected phenomena beyond NO_x. J Energy Res Technol 118(3):193–200

Gupta AK (2004) Thermal characteristics of gaseous fuel flames using high temperature air. J Eng Gas Turbines Pow 126(1):9–19

Gupta AK, Beer JM, Swithenbank J (1977) Concentric multi-annular swirl burner: stability limits and emission characteristics. In: Proceedings of 16th symposium (international) on combustion, pp 79–91

Gupta AK, Bolz S, Hasegawa T (1999) Effect of air preheat and oxygen concentration on flame structure and emission. J Energy Res Technol 121:209–216

Gupta AK, Ilanchezhian E, Keating EL (1996) Thermal destruction behavior of plastic and nonplastic wastes in a laboratory-scale facility. J Energy Res Technol 118(4):269–276

Gupta AK, Lilley DG, Syred N (1984) Swirl flows. Abacus Press, Tunbridge Wells

Kamal A, Gollahalli SR (2001) Effects of jet reynolds number on the performance of axisymmetric and nonaxisymmetric gas burner flames. J Energy Res Technol 123(2):167–172

Katsuki M, Hasegawa T (1999) The science and technology of combustion in highly preheated air. In: Proceedings of the twenty seventh symposium (Intl) on combustion, vol 27, pp 3135–4146

Khalil AEE, Arghode VK, Gupta AK (2013) Novel mixing for ultra-high thermal intensity distributed combustor. Appl Energy 105:327–334

Khalil AEE, Gupta AK (2011a) Swirling distributed combustion for clean energy conversion in gas turbine applications. Appl Energy 88:3685–3693
Khalil AEE, Gupta AK (2011b) Distributed swirl combustion for gas turbine application. Appl Energy 88:4898–4907
Khalil AEE, Gupta AK (2013) Fuel flexible distributed combustion with swirl for gas turbine applications. Appl Energy 109:267–274
Khalil AEE, Gupta AK (2014a) Swirling flowfield for colorless distributed combustion. Appl Energy 113:208–218
Khalil AEE, Gupta AK (2014b) Butyl nonanoate as a future biofuel for clean gas turbine engines. Fuel 116:522–528
Khalil AEE, Gupta AK (2014c) Velocity and turbulence effect on distributed combustion for clean gas turbine engines. Appl Energy 125:1–9
Kim S, Yoon Y, Jeung I (2000) Nitrogen oxides emissions in turbulent hydrogen jet nonpremixed flames: effect of coaxial air and flame radiation. Proc Combus Inst 28:463–471
Lefebvre AH (1995) The role of fuel preparation in low-emission. J Eng Gas Turbines Pow 117 (4):617–654
Lefebvre AH (1999) Gas turbine combustion. Taylor & Francis, Philadelphia
Leuckel IW, Fricker N (1976) The characteristics of swirl stabilized natural gas flames. J Inst Fuel 49:103–112
Shirsat V, Gupta AK (2011) Performance characteristics of methanol and kerosene fuelled meso-scale heat recirculating combustors. Appl Energy 88:5069–5082
Smith SH, Mungal MG (1998) Mixing structure and scaling of the jet in crossflow. J Fluid Mech 357:83–122
Tsuji H, Gupta AK, Hasegawa T, Katsuki M, Kishimoto K, Morita M (2003) High temperature air combustion: from energy conservation to pollution reduction. CRC Press, Boca Raton
Turns SR (2006) An introduction to combustion: concepts and application. McGraw-Hill, Boston
Vijayan V, Gupta AK (2010) Combustion and heat transfer at meso-scale with thermal recuperation. Appl Energy 87:2628–2639
Villermaux E, Rehab DH (2000) Mixing in coaxial jets. J Fluid Mech 425:161–185
Vincent ET (1950) The theory and design of gas turbines and jet engines. McGraw-Hill, New York
Wunning JA, Wunning JG (1997) Flameless oxidation to reduce thermal NO formation. Prog Energy Combust Sci 23:81–94
Yetter RA, Glassman I, Gabler HC (2000) Asymmetric whirl combustion: a new low NO_x approach. Proc Combust Inst 28:1265–1272

Flame Characteristics of Vaporized Renewable Fuels and Their Blends with Petroleum Fuels

Subramanyam Gollahalli, Ramkumar Parthasarathy and Arun Balakrishnan

Abstract In order to obtain a fundamental understanding of the thermo-chemical processes and pollutant formation during the combustion of biofuels and their blends, laminar partially premixed flames of palm methyl ester (PME) were studied and compared with those of Jet-A. The fuels were prevaporized and mixed with heated air and injected out of a stainless steel circular tube with a beveled rim. The fuel flow rate was maintained constant and the airflow rates were varied to obtain injector-exit equivalence ratios of 2, 3, and 7. In addition to global CO and NO emission indices and flame radiation, temperature, gas concentration, and soot volume fraction profiles were measured at different flame heights. The measurements highlighted the effects of fuel-bound oxygen in PME. The PME flames had less soot content than the corresponding Jet-A flames; the temperature measurements implied differences in the soot particle size and morphology formed during the combustion of biofuels.

Keywords Global and in-flame combustion characteristics · Partially premixed flames · Radiative fraction of heat release · Equivalence ratio · NO_x formation · Soot formation

1 Introduction

Renewable fuels such as biodiesels and bioalcohols have emerged as attractive alternate energy sources in recent years. These fuels can be produced from renewable feedstocks; they are carbon-neutral and are low in sulfur content. Moreover, they can be produced from locally grown biomass and thus promote energy security. Also, these biofuels can be directly used in existing combustors with minimal modifications. The key to the successful use and consumer acceptance

S. Gollahalli (✉) · R. Parthasarathy · A. Balakrishnan
School of Aerospace and Mechanical Engineering, University of Oklahoma,
212 Felgar Hall, 865 Asp Avenue, Norman, OK 73019, USA
e-mail: gollahal@ou.edu

A.K. Agarwal et al. (eds.), *Novel Combustion Concepts for Sustainable Energy Development*, DOI 10.1007/978-81-322-2211-8_13

of these biofuels is to ensure that their environmental impact is minimal and that their performance in practical combustion devices is positive.

Although the performance of practical devices such as internal combustion engines and gas turbines running on these fuels has been extensively investigated in the past by several researchers, a fundamental understanding of the science of thermochemical processes and pollutant formation during the combustion of biofuels and their blends, particularly the coupling effects of fuel chemistry and combustion conditions, is lacking. In many situations, these fuels burn in partially premixed flame configurations.

This article begins with an overview of the research program on combustion of renewable fuels at the Combustion and Flame Dynamics Laboratory of the University of Oklahoma, USA, and is followed by the specific details of the investigations on vaporized biodiesel fuel combustion. Results include thermochemical characteristics, such as temperature and concentration fields, radiation emission and extinction phenomena, and formation of environmental pollutants such as NO and CO.

2 Background and Motivation

With a motivation to understand the basic thermochemical processes, combustion characteristics, and environmental impact of increasing biofuel use in recent years, a comprehensive research program was initiated in the Combustion and Flame Dynamics Laboratory of the University of Oklahoma, USA. This program was designed with a systematic approach consisting of several phases. The initial phase consisted of simple laminar flames of vapor-phase fuels (biofuels as well as petroleum fuels and their blends) whose characteristics depended mostly on the thermochemical properties of reactants (Love et al. 2009a; Singh et al. 2013). Since air was used as the oxidizer in all the studies, the flame characteristics depended on the fuel properties. The second phase consisted of turbulent flames of the same vapor-phase fuels in which the effects of fluid dynamics were introduced (Dhamale et al. 2011). The third phase consisted of liquid spray flames in which the additional effects of atomization and phase change were added (Erazo et al. 2010). The fourth phase included testing these fuels in intermittent ignition and combustion devices such as internal combustion engines and continuous combustion devices such as gas turbine engines and furnaces (Habib et al. 2010; Sequera et al. 2011; Morton et al. 2013). The fifth phase examined the application in novel combustion burners such as porous media-supported flames (Barajas et al. 2012). The sixth phase consisted of flame extinction characteristics in counterflow burner configurations (Grisanti et al. 2013). The seventh phase dealt with biofuel pool fires, with a focus on fire safety and handling characteristics of biofuels (Tran et al. 2014). Both performance characteristics (in-flame temperature, in-flame concentration, radiation, and exhaust emissions of pollutant species such as NO_x, CO, and soot in-flame configurations), output power/thrust, specific fuel consumption, and emissions (in the studies with practical devices) were recorded. By comparing and contrasting results of different phases, the delineation of the effects of different variables was possible.

The studies of the first phase were highly significant since they dealt with the effects of chemical molecular structure and properties of fuels. A comparison of the flame characteristics of biofuels and their counterparts of petroleum fuels yielded first-order projections of the effects on performance and emissions of combustion devices when the latter were replaced by the drop-in biofuels, of course if the other parameters such as device design and operational characteristics were kept constant. This phase was also designed to facilitate the requirement of small quantities (of the order of 150 ml) of fuels produced in the research stages of fuel production and to rapidly yield results on a comparative basis (Love et al. 2009b). The data could be provided as feedback to the fuel production researchers to help them modify the chemical structure and fuel production process algorithms.

The fuels that have been tested so far in this program include canola methyl ester, soy methyl ester, palm methyl ester, their counterparts of petroleum fuels, Jet A fuel and ASTM No. 2 diesel fuel, and their blends with different concentration of biofuel. The recent work on palm methyl ester (PME) in the first phase of the research program is presented in this paper which consists of the literature pertinent to the combustion of PME, materials and methods, results and discussion, summary and conclusion comparing the flames of Jet A fuel and PME. The fatty acid composition of PME was 1.9 % meristic acid, 44.2 % palmitic acid, 4.5 % stearic acid, 39.6 % oleic acid, and 9.8 % linoleic acid by weight.

3 Flame Studies of Vaporized Biofuels (Jet A and Palm Methyl Ester)

3.1 Relevant Literature

The study by Masjuki et al. (1995) focused on the use of palm oil for its lubricant properties. Masjuki and co-workers targeted Malaysian crude palm oil which when successfully converted to methyl esters, also known as palm oil diesel (POD), was readily combustible in diesel engines. The addition of POD to commercial lubricating oils was found to enhance the performance of such oils. Results obtained from this study showed that the power output and brake specific fuel consumption of the engine lubricated with commercial SAE 40 oil blended with POD were comparable to those obtained with 100 % SAE 40 oil. Wear debris analysis showed that blends of POD and SAE 40 commercial lubricating oil increased the anti-wear characteristics of the engine when compared to those with 100 % SAE 40 lubricating oil.

Karavalakis et al. (2009) studied regulated and unregulated exhaust emissions and fuel consumption with diesel fuel and palm-based biofuel blends at proportions of 5, 20, and 40 % in a Euro 3 compliant light duty vehicle and documented the influence of the driving cycle on NO_x and CO emissions; it was found that the emissions were reduced with the use of biofuels over certain parts of the cycle. Hashimoto et al. (2008) studied the spray combustion of PME and diesel fuel at atmospheric pressure. Both emission indices, NO_x per unit mass of fuel (g-NO_x/kg-fuel) and NO_x emission per unit heating value (g-NO_x/MJ fuel) for PME, were lower than those for diesel fuel. It was also found that the NO_x emission level was reduced with decreasing fuel kinematic viscosity for both fuels and was attributed to the decrease in Sauter mean diameter (SMD) of the spray.

Studies of pollutant emissions from engines fueled with other biofuels and biofuel blends have been carried out by several researchers (McCormick et al. 2001; Canakci and van Gerpen 2003; Tsai et al. 2010). These engine studies do not shed light on the effects of chemical structure of the fuel on the combustion due to a number of complexities, such as atomization, vaporization, and turbulent mixing that occur in an engine. Hence, for this study, we adopted the technique developed in the first phase of our program mentioned above.

4 Methods and Materials

4.1 Experiment Setup

A schematic diagram of the setup is presented in Fig. 1. The experiments were conducted in a large steel combustion chamber (76 cm by 76 cm and 150 cm in height). The burner used for the experiments was housed within the chamber at its bottom center. The walls of the chamber contained windows provided with removable slotted metal sheet covers measuring 96 cm × 25 cm to allow optical access. The top of the combustion chamber was open to atmosphere through an exhaust duct. The ambient pressure of the laboratory was maintained at slightly above the atmospheric pressure ($\sim$20 Pa) to provide a positive draft inside the test chamber to prevent leakage of the combustion products into the laboratory. A stainless steel circular tube (ID of 9.5 mm and OD of 12.7 mm, Fig. 2) with a beveled rim served as the burner. This burner provided a stable, laminar, and repeatable flame and has been used in previous studies (Love et al. 2009a; Singh et al. 2013).

In order to vaporize the fuel completely without liquid-phase pyrolysis that could lead to coking of the fuel, the liquid was injected into a high-temperature airflow. The airflow was provided in a 12.7 mm (OD) steel feed line tubing with heating tape wrapped around it. The heating tape was connected to a proportional temperature controller which was continuously monitored; the air temperature upstream and downstream of the fuel injection location was measured with K-type thermocouples embedded in the feed line and was also monitored. The airflow temperature at the fuel injection location was maintained at 390 °C, which was sufficiently high above the final boiling point of the fuels so as to completely

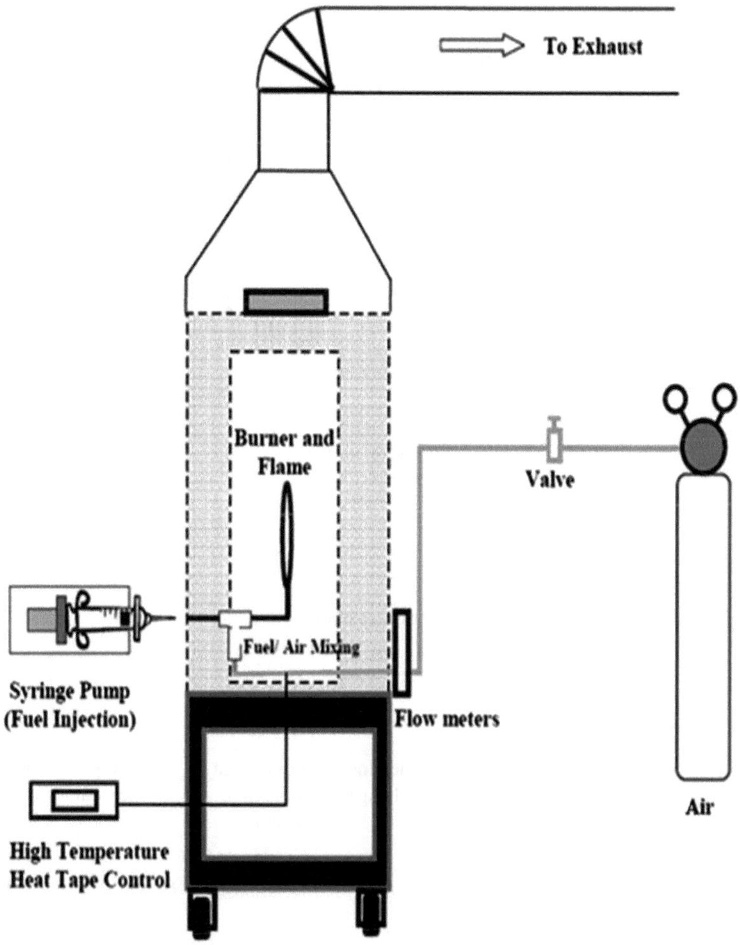

Fig. 1 Diagram of experimental setup

vaporize the injected fuel and low enough to prevent coking in the feed lines. The heated line was long enough (230 cm) to ascertain that the liquid fuel was completely vaporized in the air stream before exiting the burner. The liquid fuel was delivered to the heated air (supplied from a compressed air tank) through a high-temperature silica-based septum with a 50-cm^3-capacity syringe attached to a syringe pump. A periodic examination of the tube walls indicated the absence of any coking. Also, experiments with an air/fuel ratio analyzer indicated that the entire mass flow of liquid fuel injected into the heated air stream exited the burner in vapor state (based on the carbon balance). The volume flow rate of air was monitored using a calibrated rotameter. The fuel–air mixture was ignited at the exit of the burner with an external propane pilot flame which was removed after ignition.

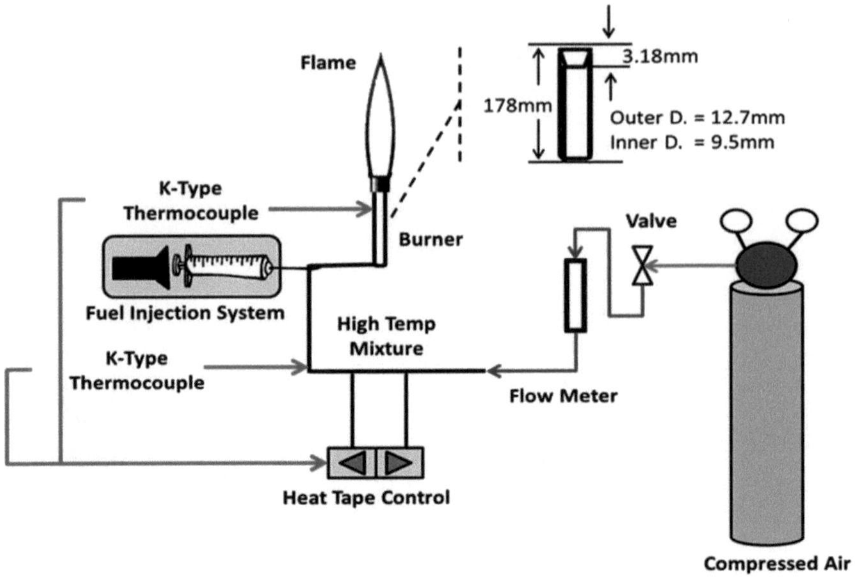

Fig. 2 Schematic diagram of fuel/air injection system

Table 1 Properties of Jet A and PME

Fuel	Density (kg/m^3)	Kinematic viscosity (at 27 °C) (cSt)	Lower heating value (MJ/kg)	Equivalent molecular formula	Oxygen content (%wt.)	Boiling point (°C)
Jet A	802	1.79	42.8	$C_{13} H_{23}$	0.00	145–300
PME	868.5	4.71	36.77	$C_{17.05} H_{32.09} O_2$	11.9	350–354

The properties of the tested fuels are presented in Table 1. Neat PME, designated as P100, has significant oxygen content compared to Jet A. The lower heating value of PME is about 10 % less than that of Jet A. The airflow rate was held constant, and the fuel flow rate was altered to result in burner exit equivalence ratios (Φ) of 2, 3, and 7. The test conditions are presented in Table 2. The Reynolds number (based on the bulk velocity of fuel–air mixture, mixture density, viscosity, and inner diameter of the injector) was in the range of 200–900, indicating laminar flow which was confirmed by the visual appearance of the flame.

Table 2 Test conditions

Fuel type	Equivalence ratio (Φ)	$(A/F)_{\text{stoic by mass}}$	Fuel flow rate (m^3/s)	Air flow rate (m^3/s)
Jet A	2	14.38	3.67×10^{-8}	17.4×10^{-5}
	3			11.6×10^{-5}
	7			4.98×10^{-5}
P100	2	12.37		16.4×10^{-5}
	3			10.9×10^{-5}
	7			4.67×10^{-5}

4.2 Instrumentation

4.2.1 Flame Visualization

The images were obtained at similar lighting and exposure conditions with a dark background at different shutter speeds such as 1/50 s, 1/25 s, and 1 s for all the conditions tested. Images captured at one-second exposure were used for flame length measurement. The flame length was calculated by measuring the number of pixels between the burnet exit and the farthest visible point of the luminous flame, and the number of pixels was converted into equivalent length scale using a known reference. Three images per condition were captured and the resultant flame lengths were averaged.

4.2.2 Global Emission Indices

A Pyrex funnel with a height of 27 cm, bottom diameter of 16 cm, and top diameter of 4 cm was mounted above the flame, where the flue gases were collected and guided to an uncooled quartz probe with a 1-mm-inner-diameter orifice that rapidly expanded to 4 mm (inner diameter). These gas samples were passed through a water condenser immersed in an ice bath in order to report all the emission results on a dry basis and to remove any moisture and subsequently were directed through a fiber filter to trap particulate matter. Measurements of the volumetric concentration of CO, CO_2, and NO_x in the exhaust were carried out using a portable gas analyzer. The analyzer consisted of a built-in infrared detector for CO_2 concentration measurement and electrochemical sensors for the measurement of O_2, CO, and NO_x concentrations.

The measurements were converted into emission indices on a mass basis (g of species/kg of fuel). The emission index for a species is the ratio of the mass of species to the mass of fuel burned by the combustion process. The emission index is particularly useful because it unambiguously expresses the amount of pollutant formed per unit mass of fuel, independent of any dilution of the product stream (Turns 2011). The emission index is expressed as

$$\mathrm{EI}_i = \left(\frac{\chi_i}{\chi_{\mathrm{co}} + \chi_{\mathrm{co}_2}}\right)\left(\frac{N * \mathrm{MW}_i}{\mathrm{MW}_{\mathrm{fuel}}}\right) * 1{,}000 \quad (1)$$

where χ_i, χ_{co} and χ_{co_2} are the mole fraction of the species, CO and CO_2, respectively, N is the number of atoms of carbon in a mole of fuel, and MW_i and MW_{fuel} are the molecular weights of the species, i and fuel, respectively. It is assumed that all the carbon in the fuel is converted into CO and CO_2 with negligible amounts of soot. This assumption was found to be valid since the flames tested were not smoking enough to produce significant amount of solid carbon in the exhaust.

4.2.3 Radiative Heat Fraction

A wide view angle (150°) high-sensitivity pyrheliometer was used to measure the total radiation from the flame. The pyrheliometer had a linear output with a responsivity of 44.56 mV per kW/m^2 and was located far enough (50 cm) from the burner so that its view angle covered the entire flame length and the flame could be assumed as a point source. The measured radiative heat flux was sampled at 1 Hz for time duration of 3 min using LabView software. The background radiation was subtracted from the total radiation to obtain the flame radiation and was expressed as the radiative fraction of heat release, F:

$$F = \frac{4\pi L^2 Q}{\dot{m}\mathrm{LHV}} \quad (2)$$

Here, L is the distance from the flame centerline to the pyrheliometer, Q is the corrected radiative heat flux measured, $\dot{m}$ is the mass flow rate of the liquid fuel, and LHV is the lower heating value of the liquid fuel tested. The radiative fraction of heat release is the fraction of the heat content of the fuel that is lost as radiation from the flame and is an indirect measure of the soot content in the flames.

4.2.4 In-flame Species Concentration

The in-flame gas concentration measurements were performed using a stainless steel sampling probe (1.75 mm ID and 3.2 mm OD); the diameter of the sampling probe was large enough to prevent clogging due to soot accumulation. The gas samples were treated to remove the moisture and particulates before sending them into a portable flue gas analyzer that was used for the global emission measurement. The sampling probe was mounted on a two-dimensional linear traverse mechanism which facilitated the axial and radial movement of the probe across the flame field. Measurements were taken at 2 mm radial distance intervals at three different heights corresponding to 25, 50, and 75 % of the visible flame length.

4.2.5 In-Flame Temperature Measurement

The in-flame temperature profiles were measured using an R-type (Pt–Pt/13 % Rh) thermocouple with a bead diameter of 0.4 mm. Catalytic action was reduced by coating the tip of the thermocouple with a fine layer of silica. The thermocouple was positioned along the length of the flame using a manually guided traverse mechanism. Data acquisition was accomplished using LabView software. The temperature readings were averaged over a period of 30 s with 1 Hz of sampling rate and corrected for radiation, conduction, and convection losses (Jha et al. 2008).

4.2.6 Soot Volume Fraction

The path-integrated soot volume fraction was measured using laser attenuation along with the application of Beer's law and Mie scattering theory, as presented by Yagi and Iino (1962). The soot volume fraction was computed using Eq. (3):

$$F_v = -\frac{\ln\left(\frac{I_0}{I_s}\right)\lambda}{k_\lambda \delta} \tag{3}$$

where I_s and I_o are the incident and attenuated laser intensities, respectively; k_λ is the spectral extinction coefficient based on the refractive indices of the soot (Bryce et al. 2000); λ is the employed laser wavelength; and δ is flame thickness (beam path) obtained from photographs. The spectral extinction coefficient (k_λ) was assumed to be constant, corresponding to that of diesel soot. Previous measurements have indicated that the refractive index of diesel soot was not significantly different from that of the soot formed in soy methyl ester flames (Choi 2009). A 5-mW helium–neon laser (λ = 632.8 nm) was used as a light source with a power detector. The beam attenuation due to the presence of soot was obtained by measuring the intensity of light with and without flame field. The burner remained stationary, with the laser and power detector aligned on a traversing mechanism to obtain radial and axial profiles. The voltage readings from the power detector were digitally sampled at the rate of 10 Hz for 30 s using LabView software.

5 Results and Discussion

5.1 Flame Visualization

The flame photographs (exposure time of 1/25 s) are presented in Fig. 3. The Jet A flames were longer than the corresponding P100 flames at all equivalence ratios. The highest flame length was recorded for both Jet A and P100 flames at $\Phi = 3$, corresponding to the maximum sooting height. In the near-burner region, a blue region corresponding to the homogenous gas-phase reaction zone was observed. The size of this region decreased as the equivalence ratio was increased due to the reduction in the amount of air injected. Also, the size of the blue region was longer for the P100 flames at all equivalence ratios due to the increased oxygen content in the fuel (Table 1).

5.2 Global Emission Indices

The global CO and NO emission indices of the Jet A and P100 flames are presented in Figs. 4 and 5, respectively. The CO emission index increased with equivalence ratio (a dramatic increase was observed at the equivalence ratio of 7) due to the reduction in the amount of injected air. The CO emission index of the P100 flames was smaller than that of the Jet A flames due to the presence of oxygen in the fuel. The NO emission index decreased with equivalence ratio due to a reduction in the flame temperature (as will be seen subsequently). The NO emission index of the P100 flames was lower than that of the Jet A flames, with the difference becoming smaller at the equivalence ratio of 7. The in-flame temperature and gas concentration profiles will help understand the reasons for this behavior.

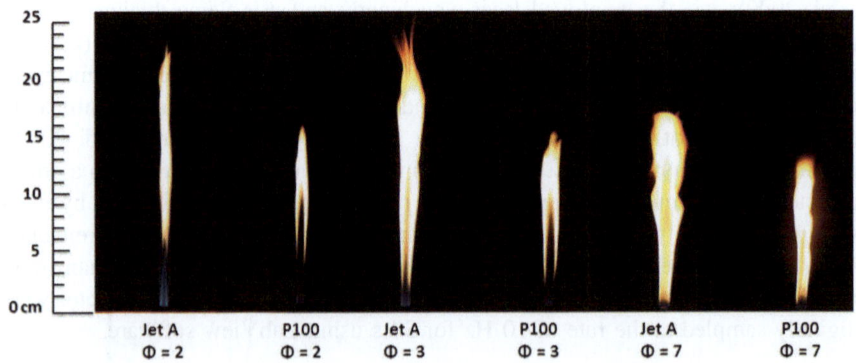

Fig. 3 Flame images at equivalence ratios of 2, 3, and 7 (exposure time of 1/25 s)

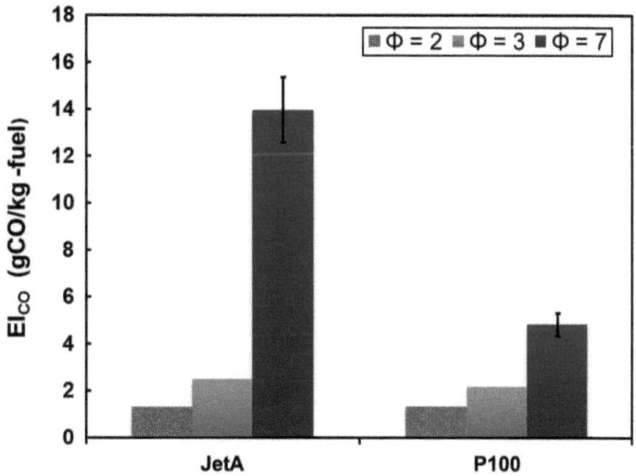

Fig. 4 Global CO emission indices of Jet A and P100 flames

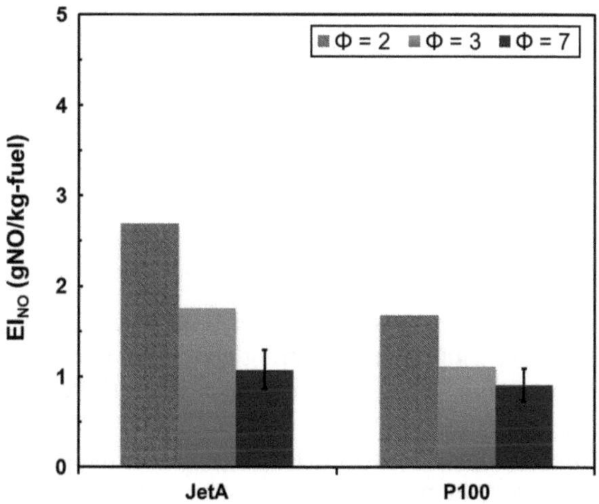

Fig. 5 Global NO emission indices of Jet A and P100 flames

5.3 Radiative Heat Fraction

The radiative heat fraction values for the Jet A and P100 flames are displayed in Fig. 6. The flame radiation depends on the gas and soot concentration, temperature, and gas and soot emissivity. At the equivalence ratios of 2 and 3, the radiative fraction of heat release of the Jet A and PME flames was comparable. At these conditions, though the soot volume concentration in the PME flames was smaller

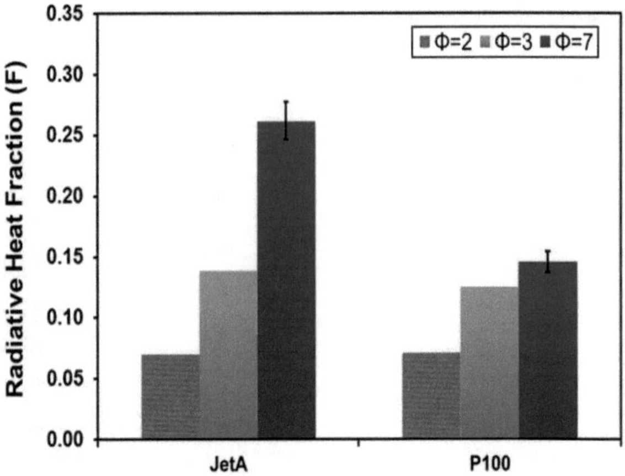

Fig. 6 Radiative fraction of heat release of Jet A and P100 flames

than that in the Jet A flames, the CO concentration was higher at 25 and 50 % flame heights for the PME flames, resulting in more gas radiation. The radiative fraction of heat release in the PME flame was lower (almost one-half) than that of the Jet A flames at the equivalence ratio of 7, because of lower soot concentration and comparable gas concentration.

5.4 In-Flame Temperature Profiles

The in-flame temperature profiles (Fig. 7a–c) displayed a double-hump profile, with the peak temperature occurring about 6–8 mm away from the centerline at 25 % flame height (FL). The maximum value shifted to the centerline by mid-flame height due to entrainment and mixing. The peak temperatures at 25 % flame height were comparable in the Jet A and P100 flames. As the equivalence ratio was increased, the peak temperature was reduced due to mixing with the entrained cool room air. The temperature at 75 % flame height in the PME flames was lower than that of the Jet A flames at equivalence ratios of 2 and 3; the values were comparable at the equivalence ratio of 7. The differences in the diameter of soot particles formed in the PME and Jet A flames and the opposing effects of soot burning and radiative loss from the soot particles could be a reason for this behavior. Previous studies (Merchan-Merchan et al. 2012) have shown that the soot particles formed in canola methyl ester flames were one-half the size of diesel soot and those formed in

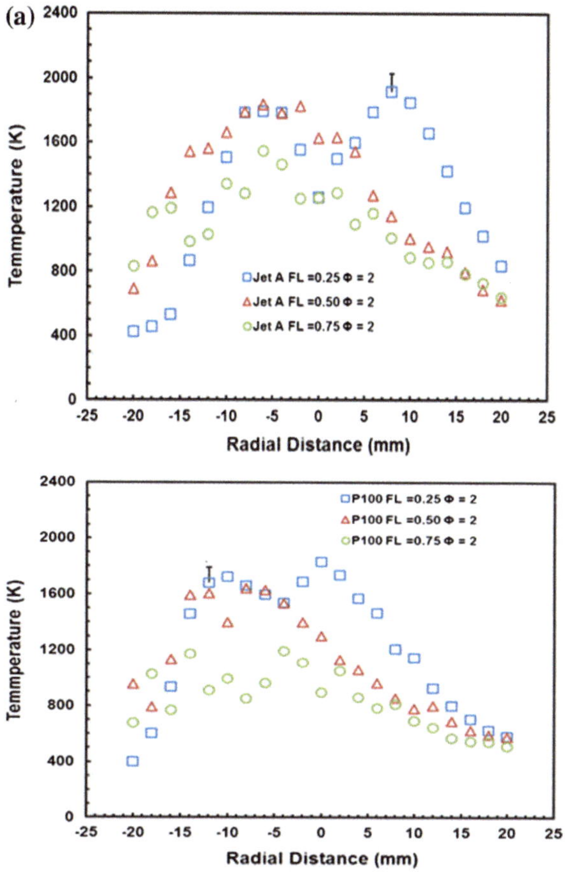

Fig. 7 **a** Radial temperature profiles in Jet A and P100 flames, Φ = 2. **b** Radial temperature profiles in Jet A and P100 flames, Φ = 3. **c** Radial temperature profiles in Jet A and P100 flames, Φ = 7

soy methyl ester flames were about 70 % the size of diesel soot. More knowledge of the soot particle properties formed during PME combustion is needed to better understand the properties of PME flames.

5.5 In-Flame Species Concentration

The oxygen concentration in the flames at 25, 50, and 75 % flame height (FL) is presented in Fig. 8a–c. The oxygen concentration was low near the centerline and increased to the ambient value toward the edge. In general, the oxygen concentration near the centerline increased with downstream distance due to entrainment

◀Fig. 7 (continued)

(b)

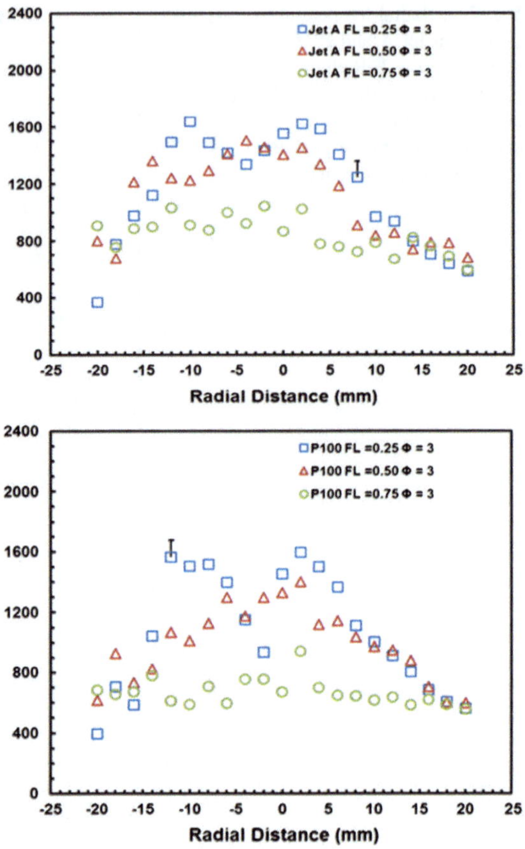

of surrounding air. The oxygen concentration near the centerline for the PME flames was higher than that of the corresponding Jet A flames due to the increased oxygen content of the fuel.

◀**Fig. 7** (continued)

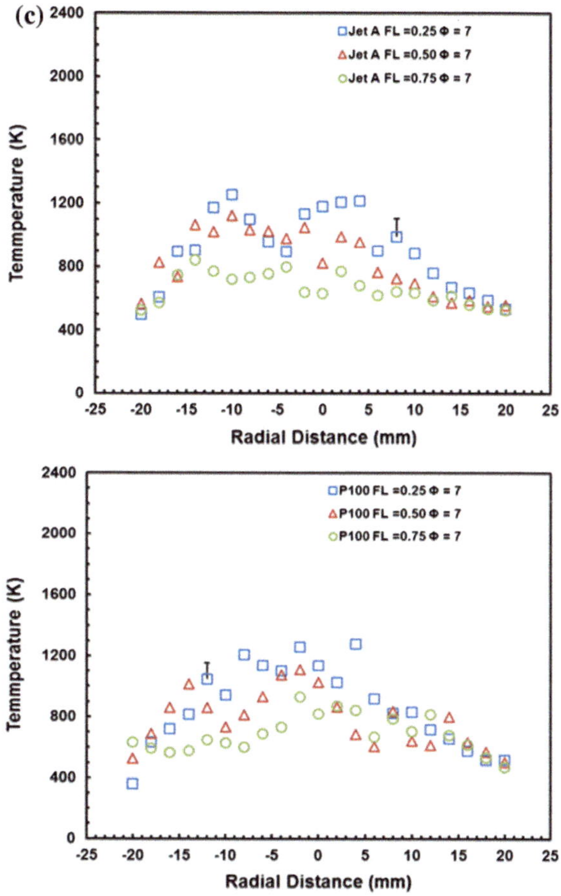

The CO_2 concentrations (Fig. 9a–c) were comparable in the Jet A and PME flames at all equivalence ratios. The CO concentrations at 25, 50, and 75 % flame height (FL) are presented in Fig. 10a–c. Carbon monoxide is produced from the partial oxidation of carbon-containing compounds. The peak CO concentration occurred at 25 % flame height and decreased with flame height. The CO concentration was higher and wider in the PME flames than in the corresponding Jet A

Fig. 8 **a** O_2 concentration profiles in Jet A and P100 flames, $\Phi = 2$. **b** O_2 concentration profiles in Jet A and P100 flames, $\Phi = 3$. **c** O_2 concentration profiles in Jet A and P100 flames, $\Phi = 7$

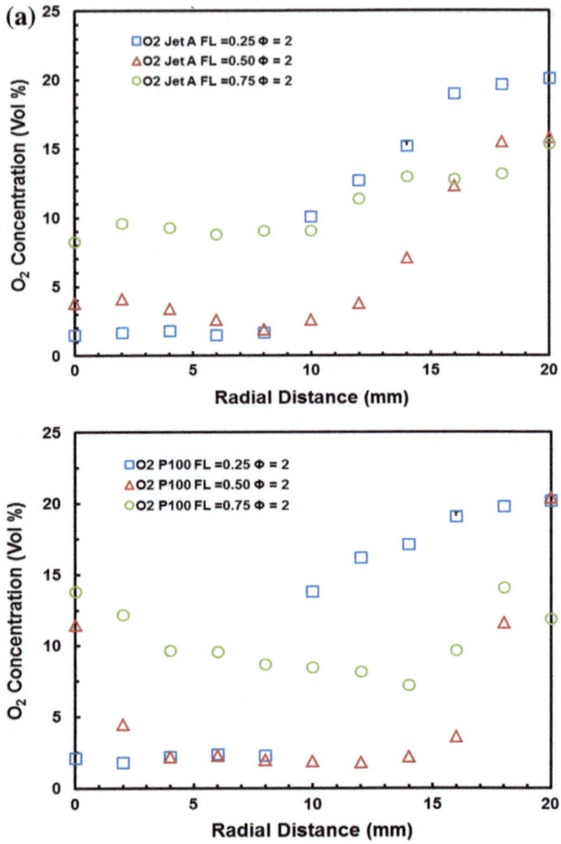

flames due to the significant oxygen content in the fuel molecule. At 75 % flame height, the CO concentration was comparable in both flames and the total CO remaining at the end of the PME flames was smaller than that in the Jet A flames (as seen in the global CO emission index).

The NO concentration profiles, displayed in Fig. 11a–c, indicate that the peak NO concentration occurred at 25 % flame height, where the peak flame temperature was reached (Fig. 6a–c). The NO concentration was higher, and the radial profile was wider in the PME flames than in the Jet A flames due to the increased

◄**Fig. 8** (continued)

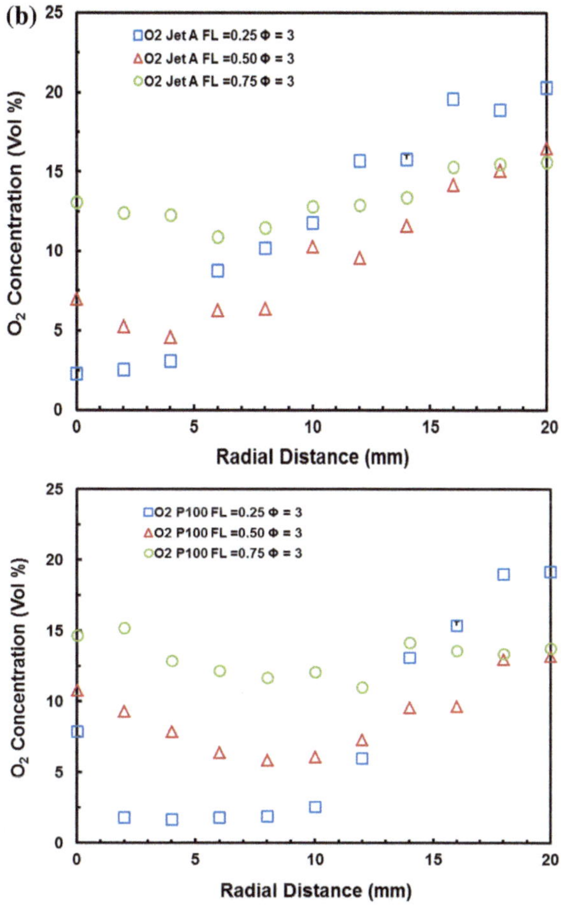

availability of oxygen. At the equivalence ratios of 3 and 7, the NO concentration in the PME flames at 75 % flame height was comparable (slightly higher) than that of the corresponding Jet A flames even though the temperatures were lower. Detailed CH and OH radical measurements are needed to further understand the formation of NO in these flames.

◂**Fig. 8** (continued)

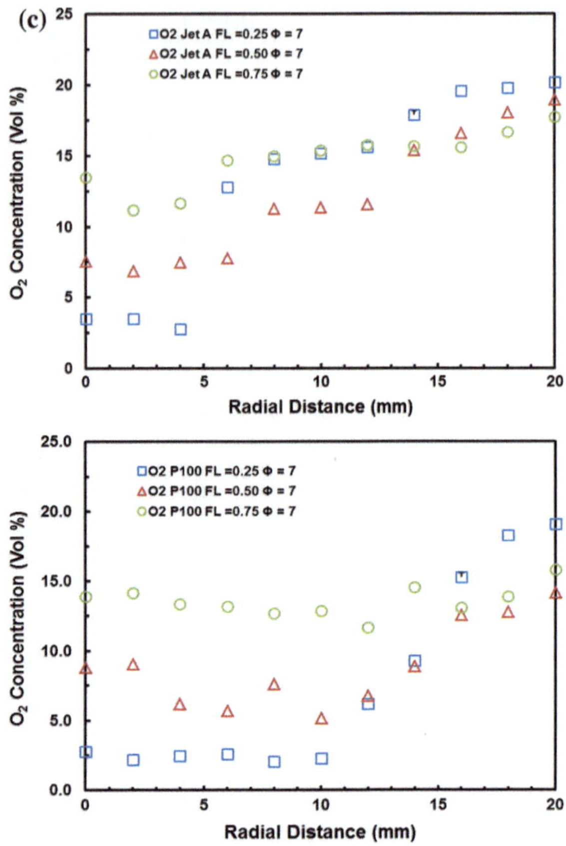

5.6 Soot Volume Concentration

The path-integrated soot volume concentration distribution at 25, 50, and 75 % flame height is presented in Fig. 12a–c. At $\Phi = 2$, the soot volume concentration levels observed at 75 % of flame height were significantly higher than that at 25 and 50 % flame height due to significant particle agglomeration and growth. The peak soot volume concentration in the Jet A flame was significantly higher than that in

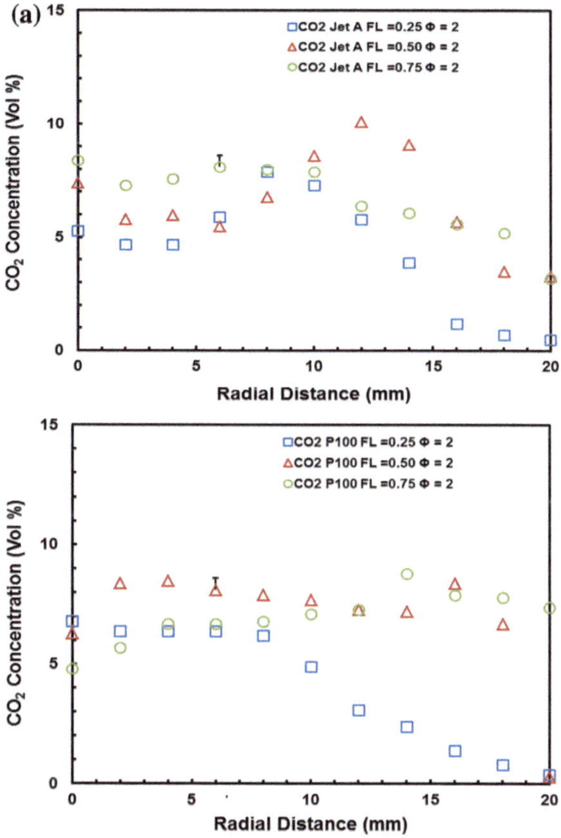

Fig. 9 **a** CO_2 concentration profiles in Jet A and P100 flames, $\Phi = 2$. **b** CO_2 concentration profiles in Jet A and P100 flames, $\Phi = 3$. **c** CO_2 concentration profiles in Jet A and P100 flames, $\Phi = 7$

the P100 flame. At $\Phi = 3$, the soot volume fraction increased toward the edge. The soot volume concentration in the P100 flame was higher than that of the Jet A flame at 25 % flame height; however, the soot volume concentration of the Jet A flame was higher than that of the P100 flame at 75 % flame height. At $\Phi = 7$, the soot volume concentration in the P100 flame was lower than that of the Jet A flame at all heights. Overall, the soot content in the PME flames was smaller than that in the Jet A flames.

◀**Fig. 9** (continued)

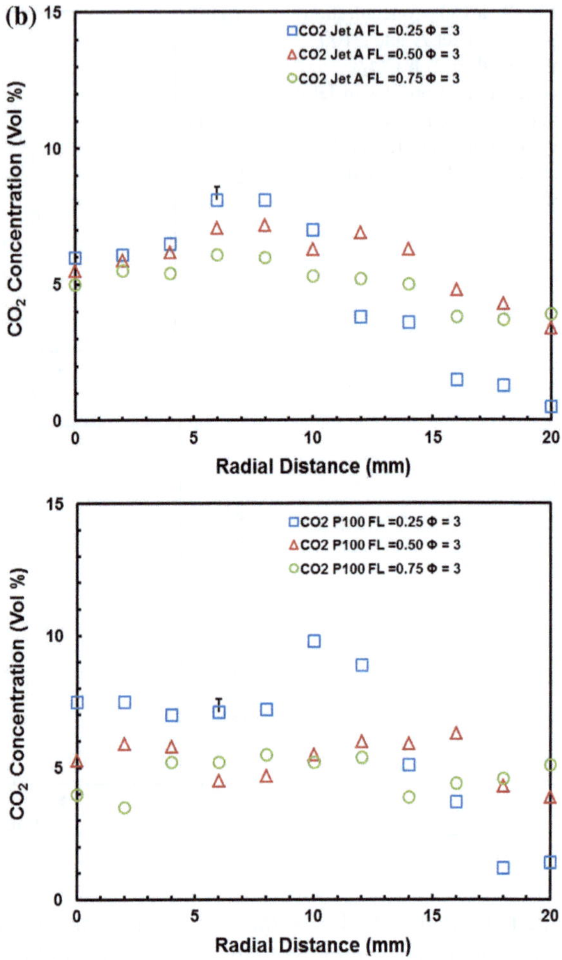

◀**Fig. 9** (continued)

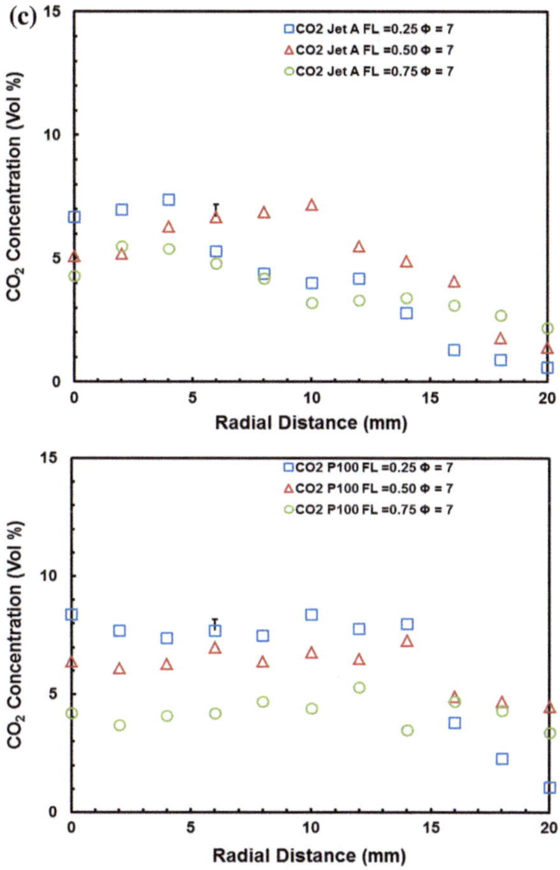

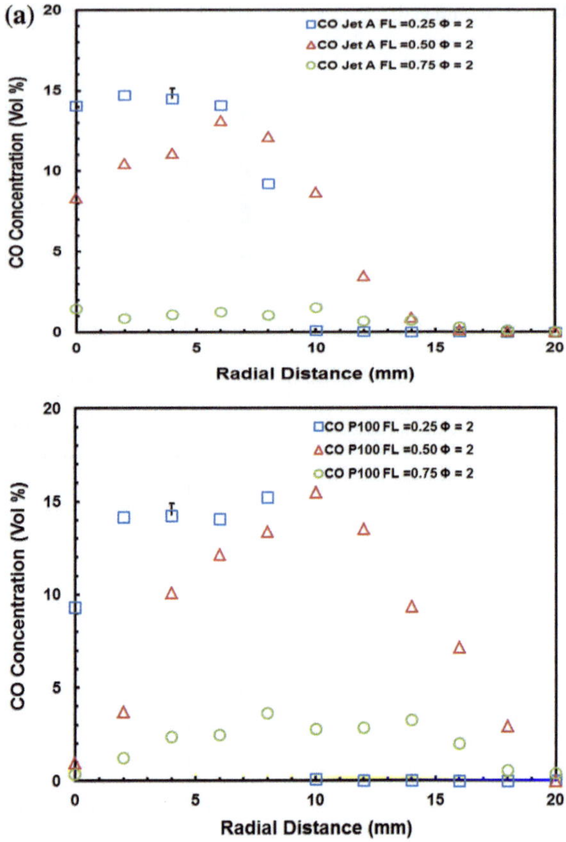

Fig. 10 **a** CO concentration profiles in Jet A and P100 flames, $\Phi = 2$. **b** CO concentration profiles in Jet A and P100 flames, $\Phi = 3$. **c** CO concentration profiles in Jet A and P100 flames, $\Phi = 7$

◄Fig. 10 (continued)

(b)

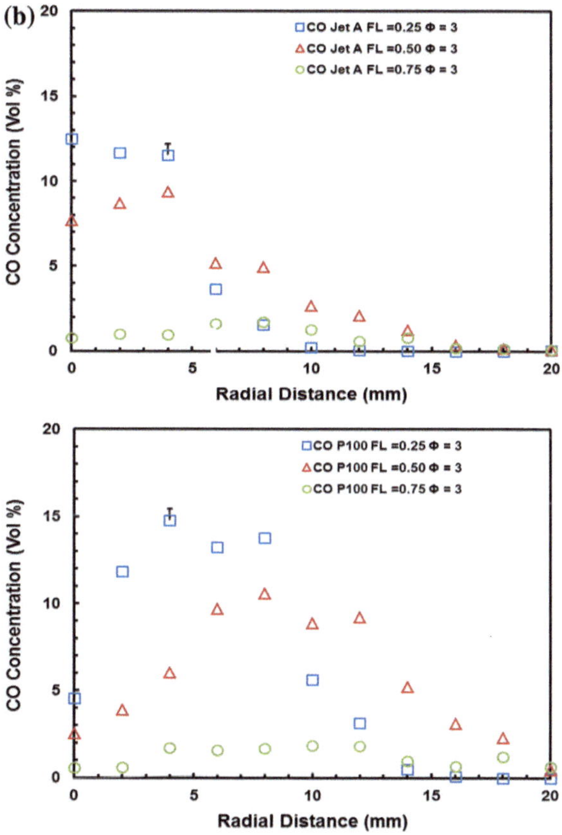

◀**Fig. 10** (continued)

(c)

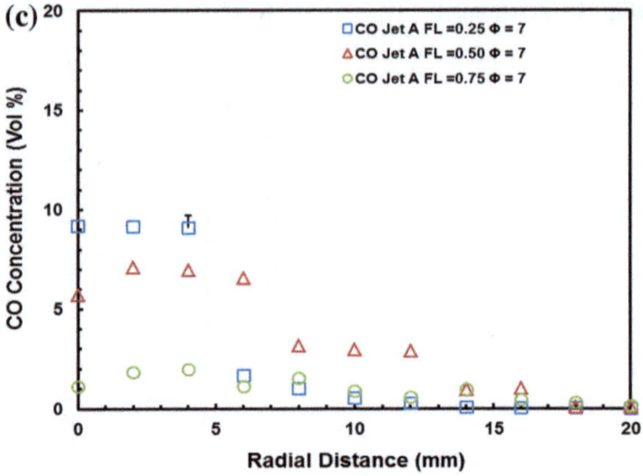

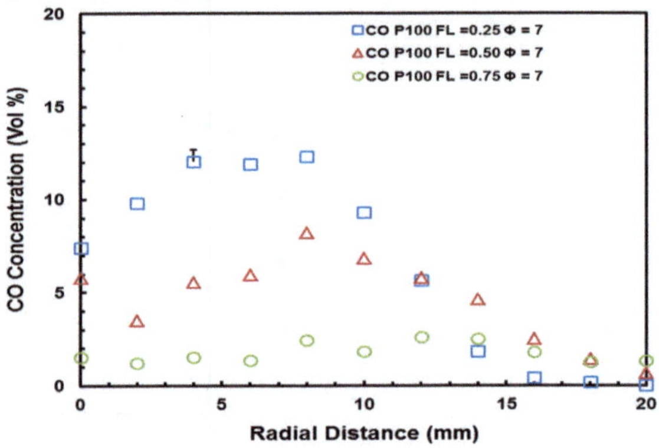

Fig. 11 **a** NO concentration profiles in Jet A and P100 flames, $\Phi = 2$. **b** NO concentration profiles in Jet A and P100 flames, $\Phi = 3$. **c** NO concentration profiles in Jet A and P100 flames, $\Phi = 7$

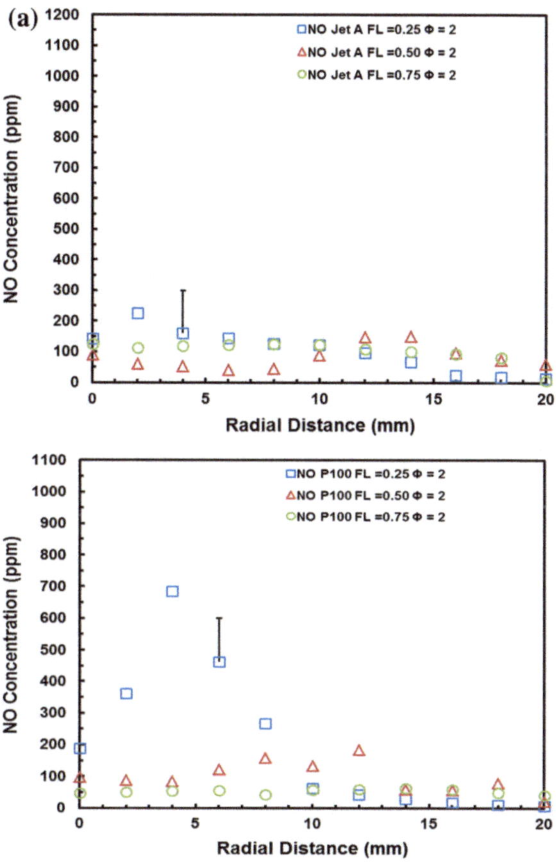

◀**Fig. 11** (continued)

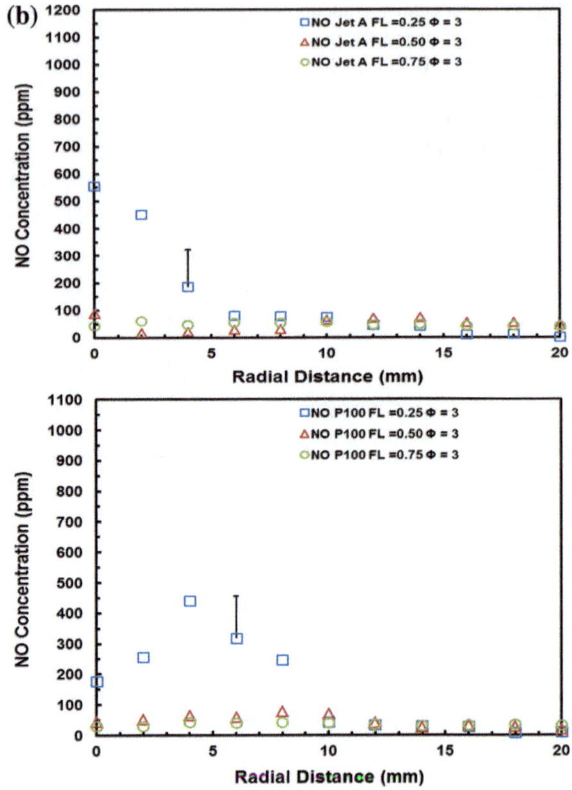

◀**Fig. 11** (continued)

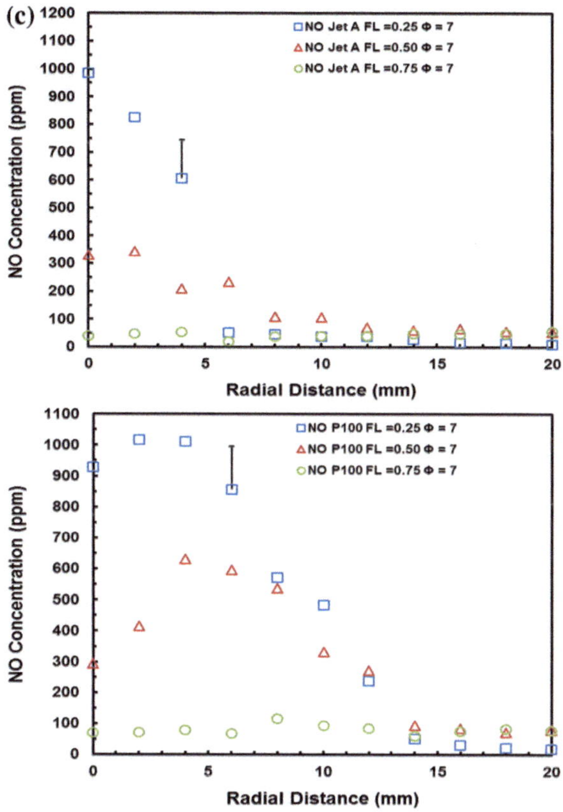

Fig. 12 **a** Radial soot volume concentration in Jet A and P100 flames, $\Phi = 2$. **b** Radial soot volume concentration in Jet A and P100 flames, $\Phi = 3$. **c** Radial soot volume concentration in Jet A and P100 flames, $\Phi = 7$

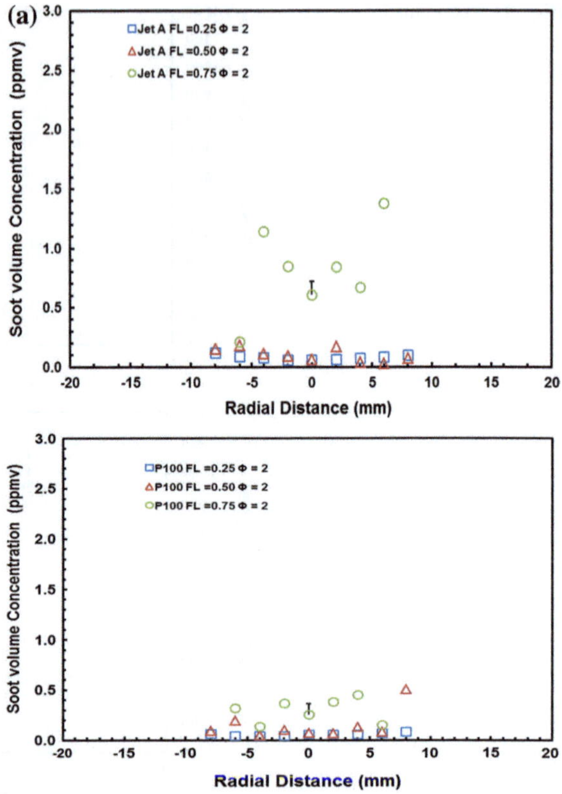

◀**Fig. 12** (continued)

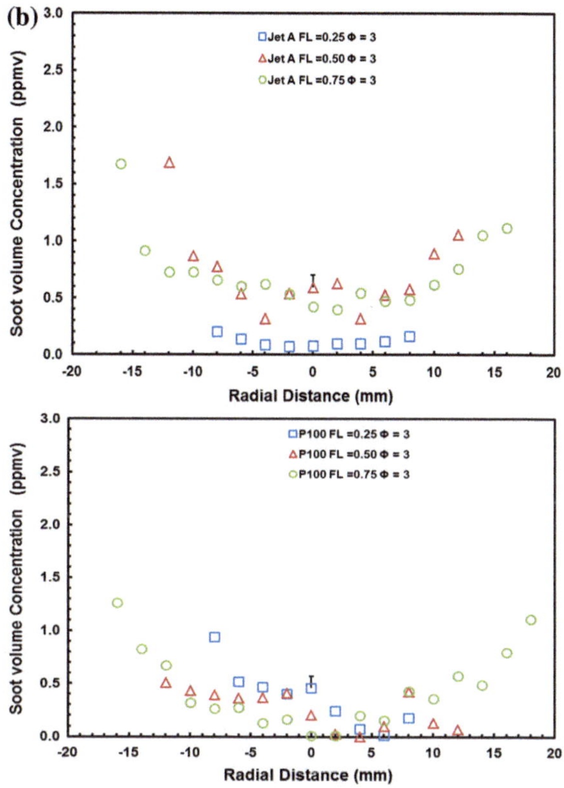

◀**Fig. 12** (continued)

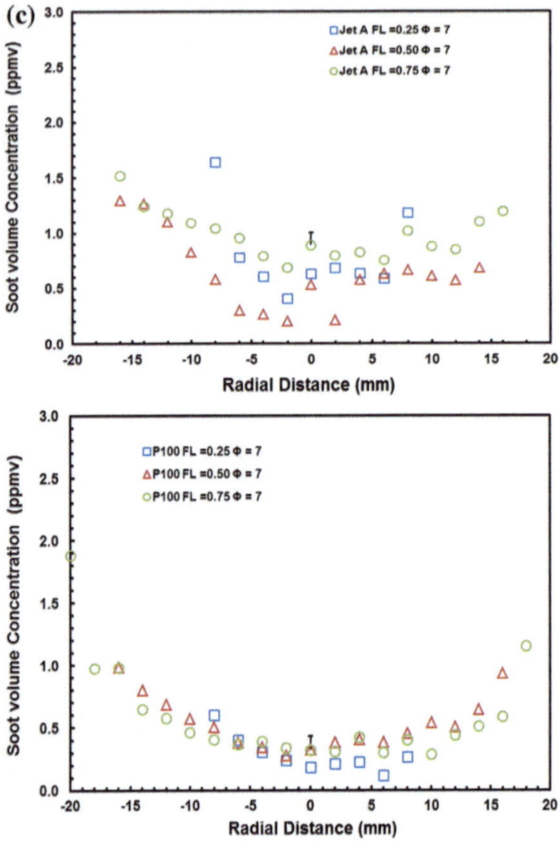

6 Summary and Conclusions

In summary, the global and in-flame combustion characteristics of pre-vaporized and partially premixed flames of Jet A and PME were studied at initial equivalence ratios of 2, 3, and 7. Based on the results obtained from measurements, the following conclusions were drawn:

1. The global CO emission index was lower in the PME flames than in the Jet A flames due to the significant oxygen content in PME. The global NO emission index in the Jet A and P100 flames was comparable at the equivalence ratios of 3 and 7, whereas the global NO emission index of Jet A flame was higher than that of P100 flames at the equivalence ratio of 2.
2. The radiative fraction of heat release of the P100 flame was lower than that of the Jet A flame at the equivalence ratio of 7, whereas the values were comparable at equivalence ratios of 2 and 3. This behavior was due to the increased concentration of radiating gases and the reduction of soot content in the PME flames.

3. Peak temperatures were recorded at Φ = 2 in both Jet A and P100 flames, and the reported values were comparable. As the equivalence ratio was increased, the peak temperatures were significantly reduced. This suggests that a mechanism other than the thermal mechanism played a significant role in NO production under these conditions. Previous studies with other fuels have highlighted the increased contribution of the prompt mechanism of NO_x formation at high equivalence ratios in partially premixed flames (Xue and Aggarwal 2003; Briones et al. 2007; Love et al. 2011).
4. The overall soot content in the PME flames was lower than that of the Jet A flames. The differences in temperature profiles in the two flames suggest differences in the soot particle size and morphology.

Acknowledgments The financial assistance provided by US Department of Energy and NSF EPSCoR is gratefully acknowledged.

References

Barajas P, Parthasarathy RN, Gollahalli SR (2012) Combustion characteristics of biofuels in porous media burners at an equivalence ratio of 0.8. J Energy Res Technol 134:021004-1–021004-5

Briones AM, Som S, Aggarwal S (2007) Effect of multistage combustion on NO_x emissions in methane-air flames. Combust Flame 149:448–462

Bryce D, Ladommatos N, Zhao H (2000) Quantitative investigation of soot distribution by laser-induced incandescence. Appl Opt 39:5012–5022

Canakci M, van Gerpen J (2003) Comparison of engine performance and emissions for petroleum diesel fuel, yellow grease biofuel and soybean oil biofuel. Trans the Am Soc Agric Biol Eng 46:937–944

Choi SC (2009) Measurement and analysis of the dimensionless extinction constant for diesel and biodiesel soot: influence of pressure, wavelength and fuel-type. PhD dissertation, Department of Mechanical Engineering and Mechanics, Drexel University, Philadelphia, Pennsylvania, USA

Dhamale N, Parthasarathy RN, Gollahalli SR (2011) Effects of turbulence on the combustion properties of partially-premixed flames of canola methyl ester and diesel blends. J Combust, article ID 697805. doi: 10.1155/2011/697805

Erazo J, Parthasarathy RN, Gollahalli SR (2010) Atomization and combustion of canola methyl ester biofuel spray. Fuel 42:3735–3742

Grisanti M, Parthasarathy RN, Gollahalli SR (2013) Experimental investigation of flame extinction limits of canola and soy methyl ester. In: 51st AIAA aerospace sciences meeting, AIAA 2013-0737, Grapevine, Texas, USA

Habib Z, Parthasarathy RN, Gollahalli SR (2010) Performance and emission characteristics of biofuel in a small-scale gas turbine. Appl Energy 87:1701–1709

Hashimoto N, Ozawa Y, Mori N, Yuri I, Hisamatsu T (2008) Fundamental combustion characteristics of palm methyl ester (PME) as alternative fuel for gas turbines. Fuel 87:3373–3378

Jha SK, Fernando S, To SD (2008) Flame temperature analysis of biodiesel blends and components. Fuel 87:1982–1988

Karavalakis G, Alvanou F, Stournas S, Bakeas E (2009) Regulated and unregulated emissions of a light duty vehicle operated on diesel/palm-based methyl ester blends over NEDC and a non-legislated driving cycle. Fuel 88:1078–1085

Love ND, Parthasarathy RN, Gollahalli SR (2009a) Effect of iodine number on NO_x formation in laminar flames of oxygenated biofuels. Int J Green Energy 6:323–332

Love ND, Parthasarathy RN, Gollahalli SR (2009b) Rapid characterization of radiation and pollutant emissions of biodiesel and hydrocarbon liquid diesel fuels. J Energy Res Technol 131:012202-1–012202-8

Love N, Parthasarathy RN, Gollahalli SR (2011) Concentration measurements of CH and OH radicals in laminar biofuel flames. Int J Green Energy 8:113–120

Masjuki H, Sapuan S (1995) Palm oil methyl esters as lubricant additive in a small diesel engine. J Am Oil Chem Soc 72:609–615

McCormick R, Graboski M, Alleman T, Herring A (2001) Impact of biodiesel source material and chemical structure on emissions of criteria pollutants from a heavy-duty engine. Environ Sci Technol 35:1742–1747

Merchan-Merchan W, Sanmiguel SG, McCollam S (2012) Analysis of soot particles from biodiesels and diesel fuel air flames. Fuel 102:525–535

Morton CD, Tran VH, Parthasarathy RN, Gollahalli SR (2013) Combustion characteristics of spray flames of canola methyl ester/diesel blends in a furnace. In: International mechanical engineering conference and exposition, IMECE 2013-63004, ASME, New York, USA

Sequera AJ, Parthasarathy RN, Gollahalli SR (2011) Effects of fuel injection timing in the combustion of biofuels in a diesel engine at partial loads. J Energy Res Technol 133:022203–022208

Singh VN, Parthasarathy RN, Gollahalli SR (2013) Radiation and emission characteristics of laminar partially premixed flames of petroleum diesel and canola methyl ester blends. J Petrol Sci Res 2:97–103

Tran V, Morton C, Parthasarathy RN, Gollahalli SR (2014) Pool fires of biofuels and their blends with petroleum diesel. Int J Green Energy 11:595–610

Tsai J, Chen S, Huang K, Lin Y, Lee W, Lin C, Lin W (2010) PM, carbon and PAH emissions from a diesel generator fuelled with soy-biodiesel blends. J Hazard Mater 179:237–243

Turns S (2011) An introduction to combustion, 3rd edn. McGraw Hill, New York

Xue H, Aggarwal SK (2003) NO_x emissions in n-heptane/air partially premixed flames. Combust Flame 132:723–741

Yagi S, Iino H (1962) Radiation from soot particles in luminous flames. In: Eighth international symposium on combustion. The Combustion Institute, pp 288–293

Process and Reactor Level Simulations of Coal-Direct Chemical-looping Combustion

Ramesh K. Agarwal, Subhodeep Banerjee, Xiao Zhang, Zheming Zhang and Ling Zhou

Abstract Reducing carbon emissions from fossil-fueled power plants has been an active area of research in recent years. One technology that appears to be very promising for high-efficiency low-cost carbon capture is chemical-looping combustion (CLC) (Leion et al. 2009a). CLC involves combustion of fuels (either gas or solid) by heterogeneous chemical reactions with an oxygen carrier, usually a particulate metal oxide. Because of the absence of air in the fuel reactor, the combustion products are not diluted by other gases (e.g., N_2), resulting in high purity of CO_2 available at the fuel reactor outlet. Also, the net energy release from a CLC process is theoretically identical to that from conventional combustion of the fuel (Abad et al. 2012; Linderholm et al. 2013; Mattisson et al. 2009a). Research by Lyngfelt et al. (2001) has shown that the energy cost of solid circulation, which is the only energy cost of separation, is a very small percentage (approximately 0.3 %) of the total energy released by the combustion process compared to other pre-combustion technologies such as the oxy-fuel combustion in which the oxygen separation process consumes nearly 15 % of the electricity generation (Hong et al. 2009a, b). Therefore, CLC holds significant promise as a next-generation combustion technology due to its potential to allow zero CO_2 emission with little effect on the efficiency of the power plant.

Keywords CLC process model · CLOU process model · Bituminous coal and anthracitic coal · Multiphase flow · Hydrodynamic characteristics · Cold flow simulations

1 Introduction

In recent years, the use of coal for CLC process has gained increasing interest since coal is likely to maintain its dominant role in the power generation sector in the foreseeable future. Additionally, other solid fuels such as biomass also hold

R.K. Agarwal (✉) · S. Banerjee · X. Zhang · Z. Zhang · L. Zhou
Washington University in St. Louis, 1 Brookings Drive, St. Louis MO 63130, USA
e-mail: rka@wustl.edu

A.K. Agarwal et al. (eds.), *Novel Combustion Concepts for Sustainable Energy Development*, DOI 10.1007/978-81-322-2211-8_14

potential for use in CLC. The utilization of coal in a CLC process can be achieved by two pathways. The coal can be first gasified in a standalone gasifier to subsequently introduce the freshly converted syngas to the fuel reactor. From the perspective of CLC process, such a scenario is essentially identical to the one that uses gaseous fuel. The CLC process utilizing gaseous fuel has been widely studied both numerically and experimentally (Dennis et al. 2006; Lyngfelt et al. 2001; Mahalatkar et al. 2011). On the other hand, if solid coal (pulverized) is directly fed into the CLC system, such a process is known as the coal-direct chemical-looping combustion (CD-CLC). The CD-CLC concept eliminates the necessity of a gasification chamber and therefore reduces the complexity of the entire power generation system. In case of CD-CLC, there have been two proposed options as to how the metal oxide will participate in the coal combustion. One of these options is to make the coal gasification occur in an identical reactor by the oxygen carrier wherein the coal is gasified in situ by H_2O and/or CO_2 supplied as a fluidization agent (Cao et al. 2006). Such a process is known as the in situ gasification chemical-looping combustion (iG-CLC). The other option is to utilize special oxygen carrier which releases gaseous oxygen under reactor conditions to sustain the combustion of solid coal in the fuel reactor. Mattison et al. (2009a) have proposed the so-called chemical-looping combustion with oxygen uncoupling (CLOU) as a variant of CD-CLC process.

There are two major concerns about the CD-CLC process due to the solid/gas mixing in the fuel reactor and the agglomeration between oxygen carrier and the coal ash. It has been identified that char gasification is the bottleneck of the fuel conversion rate in the iG-CLC process (Leion et al. 2009a) and agglomeration-induced loss of reactivity can occur when the ash concentration is high (Rubel et al. 2011). To address both these concerns, the spouted fluidized bed configuration for the fuel reactor has been proposed for CD-CLC where instead of evenly distributing the gas feeding at the inlet, a high-speed center jet is assigned. The presence of the center jet can lead to high circulation rates of solid particles, which enhance the solid–gas mixing for a faster gasification process. The enhanced particle circulation can also help to rub off the deposit of ash on oxygen carrier to reduce agglomeration (Shen et al. 2009).

In this article, two CD-CLC systems—an iG-CLC system and a CLOU system—have been considered for process-level modeling. Process-level simulations are performed for the CLOU process using ASPEN Plus and the results are compared with the experimental data. The validated model is then used to investigate the scale-up for industrial-scale plants. In addition, detailed numerical simulations of multiphase cold flow and reacting flow are performed using the Eulerian/Lagrangian approach by combining the computational fluid dynamics (CFD) modeling for gas flow with Descrete element method (DEM) for particle movement. The commercial software ANSYS Fluent is used in these simulations. The simulations show excellent agreement with the experimental results for cold flow performance of an experimental laboratory-scale iG-CLC apparatus. More importantly, the simulations are used to modify the design configuration for enhanced system performance. These simulations provide valuable insight in the various physical processes critical to the efficient performance of the CLC process. Future areas of promise in CD-CLC research are also discussed including issues related to industrial-scale deployment and retrofitting in existing coal-fired power plants.

2 CLC Process Simulation Using ASPEN Plus

ASPEN Plus is a process simulation software that employs basic engineering relationships such as mass and energy balance, and multiphase and chemical reaction models in modeling a process at system level. It consists of flow sheet simulations that calculate stream flow rates, compositions, properties, and operating conditions. It has been demonstrated that ASPEN Plus can successfully model the CLC processes, and the results of simulations are in excellent agreement with the observed data. ASPEN Plus is employed to model the effect of different rates of coal feed, air flow, and oxygen carrier feed in the CLC process. Based on the results of these simulations, a relationship between the energy output and the amount of coal, air flow, and oxygen carrier is derived and optimized for maximum energy output.

2.1 Validation of the CLC Process Simulation with Experiment

The CLC process simulation in ASPEN Plus was validated against the experimental work of Sahir et al. (2012). The physical and chemical properties of the Colombian coal used as the solid fuel in the experiment are summarized in Table 1.

The schematic of the flow sheet for this simulation is shown in Fig. 1. The coal is first pulverized and dried before it is pressurized and introduced into a shell gasifier to be partially oxidized to form syngas. The oxygen carrier material used is a mixture of 60 wt% Fe_2O_3 and 40 wt% inert Al_2O_3 as support. The molar ratio of steam and carbon is maintained at unity for the process model. The syngas composition at the gasifier outlet is 34.5 % CO, 50.3 % H_2, 12.3 % H_2O, and 2.4 % CO_2. The syngas is converted completely to CO_2 and H_2O in the fuel reactor, while the Fe_2O_3 in the oxygen carrier is reduced to Fe_3O_4. The outflow from the fuel reactor is a concentrated stream of H_2O and CO_2. After condensing the stream, high-purity CO_2 is obtained. The reduced oxygen carrier is fed into the air reactor where the oxidation reaction takes place with an 80 % conversion of Fe_3O_4 to Fe_2O_3.

Table 1 Physical and chemical properties of Colombian coal

Proximate analysis (wt%)				Ultimate analysis (wt%)					Energy
Moisture	Volatile matter	Fixed carbon	Ash	C	H	N	S	O	LHV (MJ/kg)
3.3	37.0	54.5	5.2	80.7	5.5	1.7	0.6	11.5	29.1

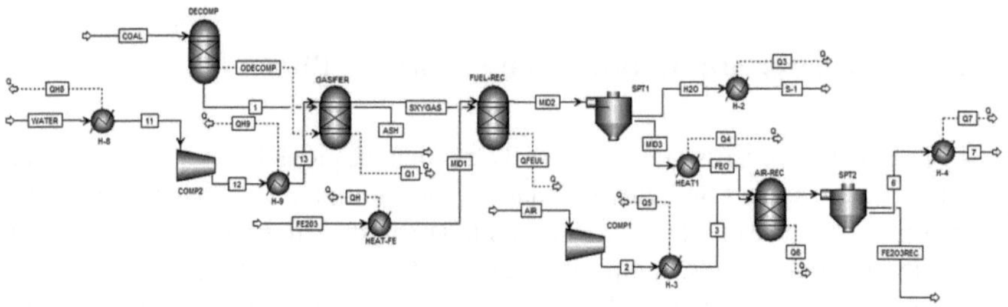

Fig. 1 Overall flow sheet of the CLC model in ASPEN Plus

The various process models used in ASPEN Plus flow sheet in Fig. 1 are summarized in Table 2. The coal devolatilization is defined by the RYIELD reactor, followed by the gasification of coal represented by the RGIBBS reactor. Another RGIBBS reactor defines the actual syngas combustion and the corresponding reduction of the oxygen carrier. These blocks together represent the fuel reactor. The flow sheet within the ASPEN Plus simulation package cannot model this entire reaction with one reactor. As a result, the fuel reactor simulation is broken down into several different reactor simulations. The air reactor is also modeled as an RGIBBS reactor.

The energy balance of the CLC process model was analyzed using the input values from the experiment of Sahir et al. (2012). The input values and the energy requirements for the various units and streams in Fig. 1 are presented in Table 3; this will be referred to as the baseline case in rest of the paper. Energy is consumed mainly in the compressor processes. Compressed air is required in the air reactor to regenerate Fe_2O_3 from Fe_3O_4; the air compressor for the combustor compresses the air to 18 atm. Another compressor is used to compress the steam for the gasifier.

Table 2 Process models used in different parts of the CLC process in ASPEN Plus

Name	Model	Function	Reaction formula
DECOMP	RYIELD	Coal devolatilization	Coal → volatile matter + char
BURN	RGIBBS	Gasification	Char + volatile matter → $CO_2 + H_2O$
FUEL-R	RSTOIC	Carrier reduction reaction	$3Fe_2O_3 + CO \rightarrow 2Fe_3O_4 + CO_2$ $3Fe_2O_3 + H_2 \rightarrow 2Fe_3O_4 + H_2O$
AIR-R	RSTOIC	Carrier oxidation reaction	$4Fe_3O_4 + O_2 \rightarrow 6Fe_2O_3$

There is a large amount of energy produced in the air reactor, but the fuel reactor needs to be supplied with energy. This is because the net heat work in the fuel reactor is the summation of the heat work from the DECOMP, GASIFER, and FUEL-R blocks. Although FUEL-R produces energy because of the combustion of syngas, the combined energy requirement of DECOMP and GASIFIER is more than the energy produced in FUEL-R. Summing the energy requirements of each individual stream, the total energy obtained from the CLC process is 554.2 kW.

The results shown in Table 3 for the baseline case with a coal feed rate of 100 kg/h are in excellent agreement with those reported by Sahir et al. (2012). These calculations validate our CLC model developed in ASPEN Plus.

2.2 Effect of Various Parameters on the Energy Output of the CLC Process Simulation

With the successful validation of the process simulation of the CLC experiment of Sahir et al. (2012) in the previous section, the ASPEN Plus simulation is expanded to consider the effect of varying the air flow rate and the oxygen carrier feeding rate. Additionally, scaled-up simulations are also conducted to determine these effects for an industrial-scale plant.

Table 3 Input and output values for the baseline case corresponding to the work of Sahir et al. (2012)

Input values	Coal (kg/h)	100
	Steam (kg/h)	140
	Air flow rate (kg/h)	71
	Temperature of fuel reactor (°C)	950
	Temperature of air reactor (°C)	935
	Fe_2O_3 flow in the fuel reactor (kg/h)	59,21
	Al_2O_3 in the system (kg/h)	3,951
	Particle density (kg/m^3)	3,200
Energy balance (kW)	Fuel reactor	−161.8
	Air reactor	688.0
	Cool air reactor exhaust	135.4
	Cool flue gas	148.3
	Cool oxygen carrier for air reactor	40.9
	Reheat oxygen carrier for fuel reactor	−42.7
	Heat steam	−69.8
	Heat air	−184.1
	Net	554.2

2.2.1 Effect of Varying the Air Flow Rate on the Energy Output

The recent paper of Mukherjee et al. (2014) suggests that it is favorable to operate the air reactor of the CLC process at higher temperatures with excess air supply in order to achieve higher power efficiency. In order to evaluate the effect of air supply on energy output, the baseline case of Table 3 is considered by varying the air flow rates. The results are presented in Fig. 2. From Fig. 2, it can be seen that with an increase in air flow rate, the net energy output increases and achieves a maximum for a certain air flow rate. If the air flow rate is further increased from its maximum value (i.e., the value corresponding to the maximum energy output), the energy output starts decreasing albeit very slowly. This result implies that there exists a certain rate of air supply around 900 kg/h to obtain the maximum energy output for 100 kg/h of coal supply. At this flow rate in the air reactor, 131.06 kW of additional energy is generated, which is 23.6 % more than the baseline case given in Table 2, indicating that the reaction in the air reactor was not complete for the baseline experimental case. Increasing the air supply ensures the 80 % conversion of Fe_3O_4 to Fe_2O_3.

2.2.2 Effect of Varying the Oxygen Carrier Feed

The oxygen carrier plays a vital role in the CLC process; it reacts with the syngas in the fuel reactor and reacts with the air in the air reactor. Both of these reactions contribute significantly to the net energy output. Figure 3 presents the energy output for different oxygen carrier feeding rates in the system with varying air flow rates. As expected, Fig. 3 shows that for a given air flow rate, a higher oxygen carrier feeding rate yields more energy output. However, when the oxygen carrier feeding

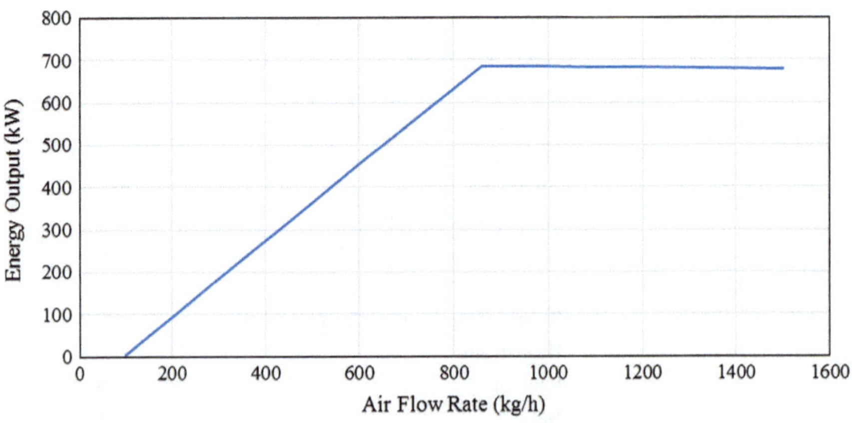

Fig. 2 Energy output for various air flow rates for 100 kg/h of coal feed

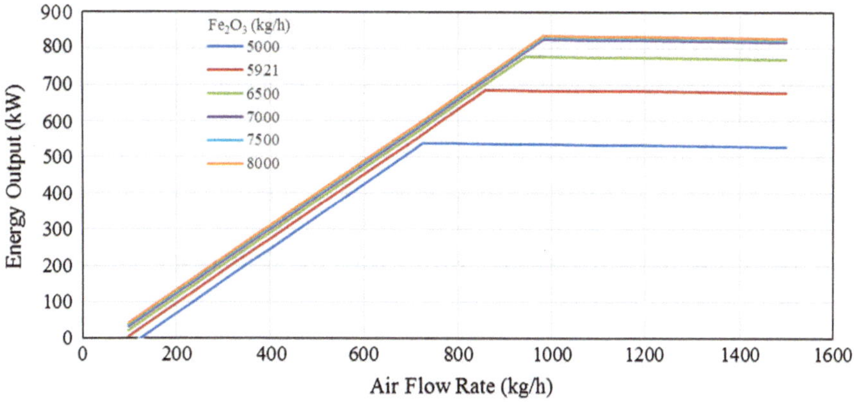

Fig. 3 Energy output for various oxygen carrier feeds and air flow rates for 100 kg/h of coal feed

rate is increased above a certain threshold value, the marginal increase in energy output becomes extremely small. The red line in Fig. 3 represents the baseline experimental case (Fe_2O_3 at 5,921 kg/h) for which the maximum energy output is 685.26 kW with around 900 kg/h air flow rate. For the threshold Fe_2O_3 rate of 7,000 kg/h, the maximum energy output of 824.33 kW occurs at around 1,000 kg/h air flow rate. Thus, 138.97 kW of additional energy output is obtained by increasing the oxygen carrier rate from 5,921 to 7,000 kg/h. Therefore, for maximum energy output with a coal feeding rate of 100 kg/h, the optimum rates of air flow and oxygen carrier feed are 1,000 and 7,000 kg/h, respectively. In other words, the optimum ratio of coal, air, and oxygen carrier is 1:10:70 for maximum energy output.

It should be noted that the above ratio of coal, air, and oxygen carrier feeds is based on using Colombian coal. Compared to the baseline case corresponding to the work of Sahir et al. (2012), a net increase in power of 48 % can be obtained by increasing the air flow rate by 40.25 % and the OC feeding rate by 18.22 % to attain this optimum ratio for the given coal feed. Other parameters that may also influence the energy output include the temperature and pressure of the reactors, types of coal, particle size, etc., which are not investigated in this work.

3 CLOU Process Simulation Using ASPEN Plus

When solid fuel such as coal is used in a CLC system, it is often the case that the reactivity of char is low because of the limited contact of oxygen carrier and gasified coal. An alternative process known as the chemical looping with oxygen uncoupling (CLOU) has been proposed to overcome the low reactivity of the char gasification stage in the CD-CLC (Leion et al. 2009b; Mattisson et al. 2009a, b; Rydén et al. 2011). CLOU is based on the strategy of using materials which release gaseous oxygen in the fuel reactor to serve as the oxygen carrier, thereby allowing the solid fuel to burn with gaseous oxygen. These materials can also be regenerated in high-temperature environment, for example, in the air reactor. Thus, in CLOU, the slow gasification step in the chemical-looping combustion with solid fuels is avoided, resulting in a much faster solid conversion (Mattisson et al. 2009b).

3.1 Experimental Apparatus and Results of Abad et al. (2012) for CLOU Process

A CLOU test apparatus directly using solid coal as the fuel with a 1.5-kWth output was recently built by Abad et al. (2012). The experimental setup basically consists of two interconnected fluidized bed reactors joined by a loop seal, a cyclone for gas–solid separation for transport of only solid from the air reactor to the fuel reactor and a valve to control the circulation of solid flow rate in the system. To the authors' knowledge, this experiment is the first time that the CLOU process has been demonstrated in an experiment utilizing two interconnected fluidized bed reactors using a solid fuel. The solid fuel used in the experiment was a bituminous Colombian coal "El Cerrejon." It should be noted that the coal was subjected to a thermal pre-treatment for pre-oxidation in order to avoid coal swelling and bed agglomeration. Coal was heated at 180 °C in the atmospheric air for 28 h. Proximate and ultimate analyses of the pre-treated coal are given in Table 4. Both the experiments and the ASPEN Plus simulations are based on this pre-treated coal. The coal particle size used in this study is 200–300 μm. Oxygen carrier particles are prepared by spray drying, containing 60 wt% CuO and 40 wt% $MgAl_2O_4$ as supporting material. The inclusion of supporting material is to increase the reactivity, durability, and fluidizability of the oxygen carrier (Hossain and de Lasa 2008). The particle size of the oxygen carrier varies between 100 and 200 μm. The effect of operating conditions on the combustion and CO_2 capture efficiencies is investigated.

The oxygen carrier decomposes in the fuel reactor exhausting gaseous oxygen to the surroundings. The oxygen burns the volatiles and char produced from coal pyrolysis in the fuel reactor. The re-oxidation of the oxygen carrier takes place in the air reactor, consisting of a bubbling fluidized bed followed by a riser. N_2 and unreacted O_2 leave the air reactor passing through a high-efficiency cyclone and a filter before the stack.

3.2 Validation of the CLOU Process Simulation with Experiment

ASPEN Plus can be employed in studying the CLOU process for designing and sizing the reactors, for predicting the reaction conversion efficiency, and for

Table 4 Properties of bituminous Colombian coal "El Cerrejon"

Coal	Proximate analysis (wt%)				Ultimate analysis (wt%)						Energy
	Moisture	Volatile matter	Fixed carbon	Ash	C	H	N	S	O	Ash	LHV (kJ/kg)
Fresh	7.5	34.0	49.9	8.6	70.8	3.9	1.7	0.5	7.20	15.9	25,880
Pre-treated	2.3	33.0	55.9	8.8	65.8	3.3	1.6	0.6	17.6	11.1	21,899

understanding the reaction equilibrium behavior. For Abad et al.'s experimental setup (2012), we established an ASPEN Plus model as shown in Fig. 4.

As summarized in Table 5, coal devolatilization is defined by the RYIELD reactor, followed by the gasification of coal represented by the RGIBBS reactor. The RSTOIC reactor defines the actual fuel combustion. It should be noted here that these three reactor blocks together in Fig. 4 represent the fuel reactor in Abad et al.'s experiments (2012). The flow sheet within the ASPEN Plus simulation package cannot model this entire reaction with one reactor. As a result, the fuel reactor is broken down into several different reactor simulations. The air reactor is modeled as an RSTOIC reactor. The molar flow rate of CuO exiting and Cu_2O

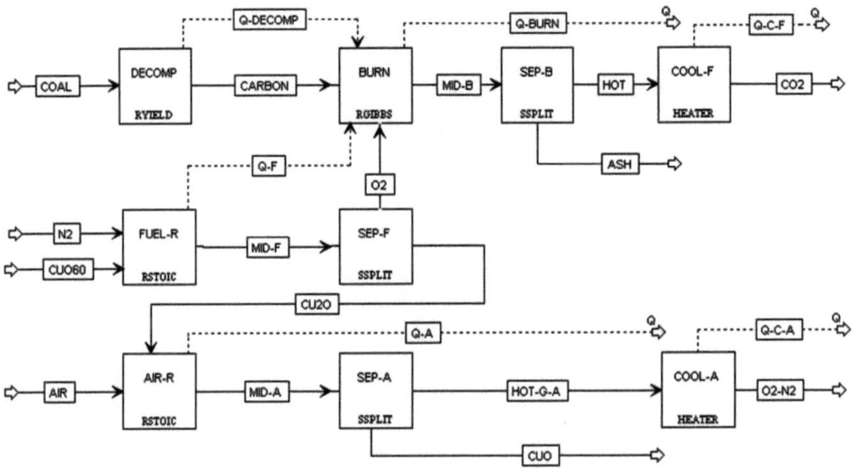

Fig. 4 Overall flow sheet of CLOU process in ASPEN Plus

Table 5 Process models used in different parts of CLOU process in ASPEN Plus

Name	Model	Function	Reaction formula
DECOMP	RYIELD	Coal devolatilization and gasification	Coal → volatile matter + char
BURN	RGIBBS	Syngas and char burn with O_2	Char + volatile matter + O_2 → CO_2 + H_2O
FUEL-R	RSTOIC	Carrier reduction reaction	$4CuO \rightarrow 2Cu_2O + O_2$
AIR-R	RSTOIC	Carrier oxidation reaction	$2Cu_2O + O_2 \rightarrow 4CuO$
SEP-F	SSPLIT	O_2 and Cu_2O Separation	~
SEP-A	SSPLIT	CuO and air Separation	~
SEP-B	SSPLIT	Separation—ash and flue gas	~
COOL-F	HEATER	Flue gas cooler—fuel reactor	H_2O (gas) → H_2O (liquid)
COOL-A	HEATER	Flue gas cooler—air reactor	~

feeding in two separate blocks is defined to be identical to represent the circulation of oxygen carrier within the system; such circulation cannot be defined explicitly in the ASPEN Plus modeling. More detailed descriptions of the model setup can be found in the authors' other work (Zhou et al. 2013). As outputs of the process simulations, gas concentrations at the fuel reactor outlet, oxygen carrier efficiency, energy analysis, system scale-up, and effects of different compositions of the coal are investigated.

Concentrations of CO_2, H_2O, O_2, O_2S, NO, NO_2, and N_2 are monitored at the fuel reactor flue stream of CO_2, and concentration of O_2 is measured at the air reactor flue stream of N_2 and O_2. Dry CO_2 and H_2O concentrations are evaluated by Eq. (1) below.

$$CO_{2,\mathrm{Dry}} = \frac{CO_{2,\mathrm{Flue}}}{(1 - H_2O_{\mathrm{Flue}})} \quad O_{2,\mathrm{Dry}} = \frac{O_{2,\mathrm{Flue}}}{(1 - H_2O_{\mathrm{Flue}})} \tag{1}$$

In Eq. (1), $CO_{2,\mathrm{Flue}}$ and H_2O_{Flue} are the fractional concentrations of the respective gases at the outlet flue stream of the reactors. The dry concentration is used to obtain a purer CO_2 stream without water since impurities in the CO_2 streams could have an effect on their sequestration behavior. Power output is monitored at Q-Burn, Q-A, Q-C-A, and Q-C–F at locations marked by "Q" in the ASPEN Plus flow sheet in Fig. 4.

3.2.1 Oxygen Carrier Efficiency

One of the most important indicators of the performance of a CLOU system is the oxygen carrier efficiency, ΔOC, which is defined as the fraction of oxygen carrier conversion of CuO into Cu_2O and O_2. It is the ratio of transferable oxygen in the current oxygen carrier to that in the fully oxidized oxygen carrier as given by Eq. (2).

$$\Delta OC = \frac{\left(F_{CO_2,\mathrm{out,FR}} + F_{O_2,\mathrm{out,FR}} + 0.5\left(F_{CO,\mathrm{out,FR}} + F_{H_2O,\mathrm{out,FR}}\right) - 0.5\,\mathrm{m_{coal}}\left(\frac{f_{H_2O}}{M_{H_2O}} + \frac{f_O}{M_O}\right)\right) M_{O_2}}{0.25 F_{CuO}} \tag{2}$$

In Eq. (2), F_x is the concentration of dry gas x in the flue stream of the fuel reactor, m_{coal} is the coal feeding rate, f_i is the mass fraction of coal in the compound i, and M_x is the molecular weight of compound x.

As shown in Fig. 5, ΔOC varies linearly with the coal feeding rate. This implies the abundance of oxygen carrier in the system. However, the air reactor efficiency, defined as the fractional oxygen carrier conversion of Cu_2O and O_2 into CuO, does not have a linear relation with respect to coal feeding rate due to variation in air flow within the air reactor. It is also important to obtain the conversion characteristics of the CLOU system. Hence, the CO_2 and O_2 concentrations in the fuel

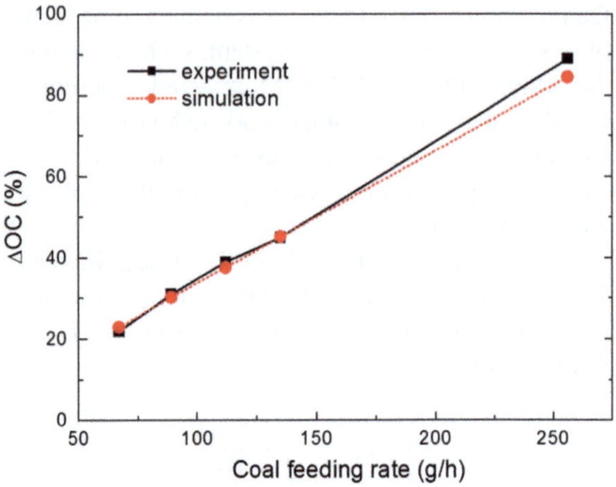

Fig. 5 Comparison of oxygen carrier efficiency between the simulation and the experiment

reactor flue stream and the O_2 concentration in the air reactor flue stream are examined.

3.2.2 Gas Concentration

Figure 6a compares the CO_2 and O_2 concentration in the fuel reactor flue stream obtained from both the experiment and the simulation for different coal feeding rates. The concentration of CO_2 increases and concentration of O_2 decreases almost linearly with an increase in the coal feeding rate. The deviation from linearity at low coal feeding rates is likely due to the incomplete mixing of oxygen carrier and coal, which is significantly mitigated at higher coal feeding rates. The relatively low concentration of O_2 in the flue stream, as shown in Fig. 6a, suggests that the decoupling of oxygen from the oxygen carrier occurs in an on-demand manner, i.e., it is determined by the presence of coal in the system. This is supported by Fig. 6b, which shows that the depletion of O_2 in the air reactor increases with an increase in the coal feeding rate. Overall, the CO_2 and O_2 concentration levels predicted by the ASPEN Plus are in good agreement with the experimental data. Possible improvements in the simulation results can be obtained by calculating a more accurate value for char combustion and mass flow rate of solids.

3.2.3 Energy Balance

Figure 7 compares the thermal power output from the experiment and the ASPEN Plus simulation for the entire CLOU system for different coal feeding rates. It can be seen that the estimated power output obtained from the ASPEN Plus simulation is in reasonably good agreement with the experimental values. The slight

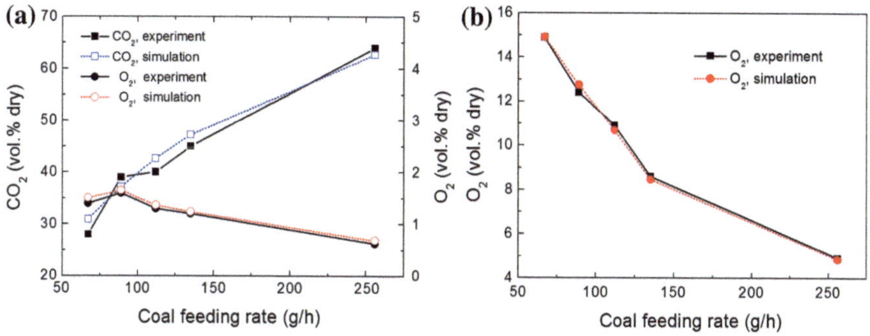

Fig. 6 **a** CO_2 and O_2 concentration at fuel reactor outlet and **b** O_2 concentration at air reactor outlet

Fig. 7 Comparison of overall power between the simulation and the experiment

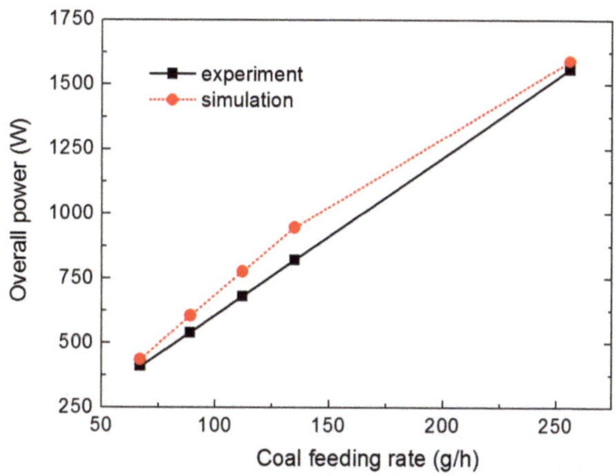

discrepancy between the simulation and the experiment could be attributed to different working conditions in the experimental apparatus that are slightly different from the corresponding block parameters in ASPEN Plus. In addition, losses at each location in the experimental apparatus are inevitable, but these cannot be accurately captured in the simulation.

3.3 *Scaled-up Simulations of CLOU Process*

Scaling up is an essential step toward the realization and optimization of commercial power plants. Four scaled-up simulations were conducted by increasing the coal feeding rate. The oxygen carrier circulation rate and air supply rate were also scaled up accordingly to meet the demand of the increased coal feed. Other modeling parameters such as the reactor efficiency and coal decomposition rate remain

Table 6 Scaled-up simulations based on increasing the coal feeding rate

Test no.	Input				Output			
	m_s (kg/h)	m_{coal} (g/h)	v_{n2} (L_N/h)	v_{air} (L_N/h)	Power (W)	$CO_{2,FR}$ (vol. %)	$O_{2,FR}$ (vol. %)	$O_{2,AR}$ (vol. %)
Base case	9	256	186	1980	1591.4	62.67	0.69	4.95
Scale-up 1	18	512	372	3960	3182.8	62.67	0.69	4.95
Scale-up 2	27	768	558	5940	4774.2	62.67	0.69	4.95
Scale-up 3	36	1024	744	7920	6365.6	62.67	0.69	4.95
Scale-up 4	45	1280	930	9900	7957	62.67	0.69	4.95

unchanged for the scaled-up simulations. The total thermal power output and main gas compositions for the flue stream from the air reactor and the fuel reactor are summarized in Table 6. It can be seen that the total power output increases linearly with increase in coal feeding; however, the flue stream composition remains steady for all the scaled-up simulations. Considering the principles of energy and mass balance on which ASPEN Plus modeling is based, linearity in the scaled-up results is expected since nonlinear effects such as the energy losses at multiple locations are inherently omitted in the modeling process.

3.4 Simulations of CLOU Process Using Different Types of Coal

It is also important to test the performance of a CLOU system for various types of coal. Three different types of coal are used with the $Cu_{60}AlMg$ oxygen carrier: bituminous Colombian coal "El Cerrejon," anthracite coal from El Bierzo, and lignite coal from Teruel basin (Abad et al. 2012). The selections of these coal samples are based on the consideration of diversity in their volatile matter content and carbon content since it is of great interest to investigate the performance of CLOU process over a wide range of coal properties. Therefore, samples from three different coal ranks, bituminous, anthracitic, and lignite are considered in this modeling. The detailed properties of these coals are summarized in Table 7.

Process simulations for various types of coal are conducted for different coal feeding rates. The CO_2 concentration in the flue stream of the fuel reactor outlet and the overall energy released are compared in Fig. 8a, b, respectively. For all three types of coal, the flue stream CO_2 concentrations and the power outputs increase with the higher coal feeding rate as well as with the higher carbon content in the coal. Since the carbon content of bituminous coal and anthracite coal is similar, the results for these two types of coal show similar characteristics for the CO_2 concentration in the flue stream and in the overall power output. In contrast, lowest CO_2 concentration in the flue stream and least power output are obtained for the lignite coal due to its low carbon content. Overall, characteristics of the CO_2

Table 7 Properties of three types of coals considered for CLOU process modeling

Coal type	Proximate analysis (wt%)				Ultimate analysis (wt%)						Energy
	Moisture	Volatile matter	Fixed carbon	Ash	C	H	N	S	O	Ash	LHV (kJ/kg)
Bituminous	2.3	33.0	55.9	8.8	65.8	3.3	1.6	0.6	17.6	11.1	21899
Anthracite	1.0	7.5	59.9	31.6	60.7	2.1	0.9	1.3	2.4	32.6	21900
Lignite	12.6	28.6	33.6	25.2	45.4	2.5	0.6	5.2	8.5	37.8	16250

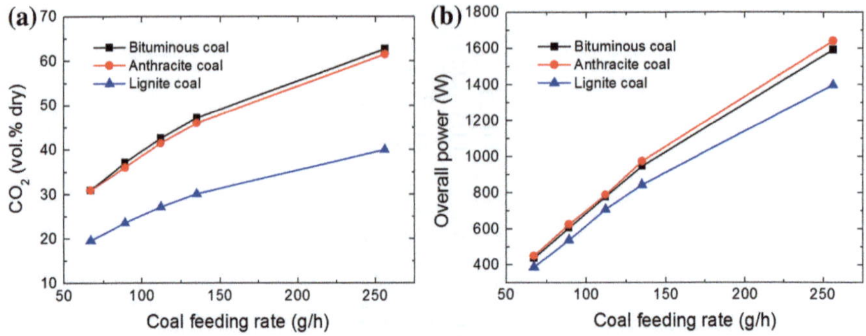

Fig. 8 **a** CO_2 concentration in flue stream at fuel reactor outlet and **b** overall energy output

concentration in the flue stream and the power output for different types of coal have similar increasing trend with higher coal feeding rate. However, the performance of the CLOU process depends greatly on the coal reactivity, which is mainly related to the coal rank. The high-rank coal can produce CO_2 of almost twice the purity compared to the low-rank one, which suggests that a more active oxygen carrier is needed when low-rank coal is utilized as fuel in a CLOU process.

4 CFD/DEM Simulations Using ANSYS Fluent

The APSEN Plus simulations discussed above are rich in information on the overall validity and performance of a conceptual system-level design. However, they lack the capability of capturing the equally critical detailed physics expected in a CLC process, such as solid/gas mixing, separation, and erosion. The successful design and optimization of a CD-CLC system also requires the development of credible simulation models for multiphase fluid flow which include solid particle dynamics and chemical kinetics.

The introduction of spouted fluidized bed in the context of CD-CLC has generated a great deal of interest in recent years because of its several technical benefits. First, the introduction of high-velocity jet is likely to create strong mixing of solids and gas avoiding loss of reactivity due to the ash agglomeration with the oxygen carrier (Rubel et al. 2011). Second, the oxygen carrier particles need to be of a relatively large diameter to be spoutable (Geldart 1973) and the relatively larger particles are likely to lead to easier separation of ash from the recirculating oxygen carrier. A significant differentiation of a spouted fluidized bed from others (e.g., a bubbling fluidized bed) is the intense particle–particle and particle–wall collision. The research of Gryczka et al. (2009) on spouted beds with 2-mm-diameter particles has suggested that accurate numerical representation of particle dynamics is not likely to be achieved using the multiphase granular solid-phase approximation due to "the inadequacies of the continuum model." Such inaccuracy has been identified to originate from the non-physical closure for the continuum model such as the frictional solids viscosity or the solids pressure. To address these inaccuracies, simulations using Eulerian/Lagrangian modeling approach, which has a stronger physical foundation, has been introduced in recent literature. In this approach, the particulate phase is modeled at the individual particle level tracking their movement using the Lagrangian framework and is coupled with the fluid field for interphase mass, momentum, and energy exchange. It should be noted that the inter-particle collision is described using a deterministic model. Because particles are treated as discrete elements, this approach is also known as the discrete element method (DEM). Therefore, a series of transient simulations of quasi-3D spouted fluidized bed apparatus for CD-CLC process are performed using the CFD/DEM coupled model. Unsteady tracking of individual solid particles and their interactions with one another as well as the ambient fluid is enabled to provide the most accurate and realistic representation of the multiphase flow field. The CFD/DEM simulation approach has been used in various applications which require the modeling of particle/fluid interaction (Chu 2010; Feng and Yu 2009; Kloss et al. 2009; Yang and Duan 2013; Zhu et al. 2008); however, its application in the context of close-loop spouted fluidized bed system has been quite limited and preliminary. By establishing the accurate models of CD-CLC configurations validated against laboratory-scale cold flow experimental data, it is expected that this work should help in gaining deeper insights as well as provide best practices for modeling and designing an industrial-scale CD-CLC process.

In the coupled CFD/DEM approach to multiphase simulation of CD-CLC process, the fluid motion is obtained by solving the incompressible continuity, Navier–Stokes, and energy equations. The motion of particulate solids is modeled by the Newtonian equation of motion. By introducing source terms related to movement and interaction of solid particles in the fluid dynamics equations and terms due to forces experienced by the solid particles in the Newtonian equations for the motion of the solid particles and the mass and heat transfer from the particles to the fluid, the coupling of solid–fluid interaction is achieved. On the other hand, the inter-particle momentum transport is described using soft ball collision model. Detailed mathematical description of the modeling process can be found in the authors' other work (Zhang et al. 2014). These simulations employ the commercial simulation package ANSYS Fluent, release version 14.5 (ANSYS 2012a, b). All particles are tracked individually as parcels with the actual particle diameter.

4.1 Cold Flow Simulation of Spouted Fluidized Bed Experiment at TU-Darmstadt

Simulations employing the discrete element method (DEM) coupled with the hydrodynamics solver are first conducted for validating the model. The modeling and simulation is conducted for the laboratory-scale quasi-3D experimental CLC fuel reactor at Darmstadt University of Technology (TU-Darmstadt) (Alobaid et al. 2013). For the cold flow simulations, only the hydrodynamics and particle motion are considered, without chemical kinetics and heat transfer. The TU-Darmstadt experiments were performed with a spouted fluidized bed similar to that in the test rig used by Link (1975). It is a Plexiglas model 100 cm in height, 15 cm in width, and 2 cm in depth. The apparatus was loaded with 36,500 glass beads with an average diameter of 2.5 mm and a density of 2,500 kg/m^3. A high-speed air jet was supplied through a centrally placed nozzle with dimension of 1 cm (width) by 2 cm (depth), while low-speed background air flow was introduced through the two side panels. The glass beads used in the experiment fall into the category of Group D particles based on their size and density according to Geldart's powder classification (1973). Such particles are generally very hard to get fluidized because once the bubbles are formed, they tend to rise slower than the rest of the gas percolating through the emulsion and coalesce rapidly as a result. According to Link (1975), the background flow is necessary to pre-fluidize the bed so that prominent gas bubbles can form when high-velocity central jet is supplied. The top of the domain in the apparatus was open to the atmosphere. A high-speed camera was used to record the distribution of the particles with a 1-ms resolution in time. The static pressure was monitored at four different height locations of z = 2, 12, 22, and 40 cm by pressure sensors. The experiment was initially loaded with 36,500 glass beads with an average diameter of 2.5 mm and a density of 2,500 kg/m^3. The TU-Darmstadt experimental apparatus and the corresponding computational mesh are shown in Fig. 9.

The computational domain shown in Fig. 9 consists of 11,000 hexahedral cells. The multiphase simulation is set up with the continuum fluid (air) as the primary phase and the discrete particulate solid (glass beads) as the secondary phase. The interaction between particles and fluid, that is the two-way exchange of momentum, is defined by the Syamlal-O'Brien drag law (1989). In the simulation, a total of 36,500 particles are released with a small offset in z-direction and start to experience free fall due to gravity. When the aggregate kinetic energy of the particles becomes sufficiently small, the bed is considered settled and the simulation time is set as zero. Such an initialization is important since it ensures the random packing of the particles in the static bed. For the continuous gas phase, the velocity inlet boundary condition is applied at the bottom panels of the domain, with 22.69 m/s gas velocity at the center panel and 0.85 m/s gas velocity at the side panels corresponding to the experimental conditions (Alobaid et al. 2013). Pressure outlet boundary condition with zero gauge pressure is applied to the top surface of the domain. Standard non-slip boundary condition is applied to the lateral walls. For the discrete phase, all walls are treated as reflecting surfaces with coefficient of restitution of 0.97. To ensure numerical accuracy, second-order discretization in both space and time is applied. Due to large value of the particle spring constant, the particle-tracking time step is set sufficiently small to capture the key particle collisions. The key numerical parameters are summarized in Table 8.

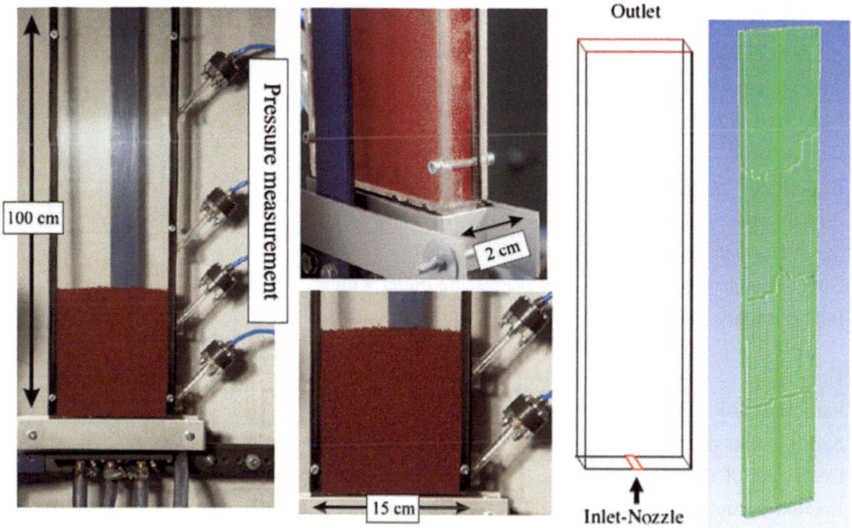

Fig. 9 Spouted fluidized bed apparatus at TU-Darmstadt (*left*) and CFD model (*right*)

The simulation is carried out on a Dell workstation with quad-core Intel Xeon CPU. Each run of 0- to 500-ms simulation time requires about 48 h of CPU time. The simulation results for the particle distribution are presented in Fig. 10 alongside the experimental observation for the first 300 ms of the jet injection.

The simulation results of the average bed height and its comparison with the experiment are presented in Fig. 11a. The transient static pressure response at the four monitoring locations of $z = 2$, 12, 22, and 40 cm is shown in Fig. 11b.

As high-velocity gas enters the static bed from the central nozzle, it suddenly experiences great resistance due to the presence of the particles. Due to the momentum conservation, the static gas pressure near the central nozzle suddenly builds up and is immediately felt by the lowest pressure transducer as shown in Fig. 11b. This pressure buildup can be explained by considering the solid pressure, which is the additional pressure felt by a surface within the bed due to the presence of the solid column above it. Because of both viscous friction and the pressure gradient, part of the kinetic energy of the jet is transferred to the ambient particles, forming a notable gas bubble originating from the central nozzle. The remaining particles start to move upward making room for the growth of the gas bubble as seen in Fig. 10. The gas bubble keeps a steady growth rate for about 450 ms as can be seen from Figs. 10 and 11a during which the particles above the gas bubble get accelerated, while the initial pressure buildup is gradually relieved. Around 450 ms, the spouted bed height reaches its maximum value and the gas bubble begins to collapse.

Table 8 Key modeling parameters for simulation of the TU-Darmstadt experiment

Average diameter of particles	2.5 mm
Average density of particles	2,500 kg/m^3
Mass load of particle in bed	0.73 kg
Inlet velocity	22.69 m/s
Background gas velocity	0.85 m/s
Primary phase	Air
Secondary phase	Glass beads
Outlet boundary condition	Pressure outlet with $P_{out,gage} = 0$ Pa
Inlet boundary condition	Velocity inlet with central jet velocity of 22.69 m/s, background flow velocity of 0.85 m/s
Drag law	Syamlal-O'Brien
Particle collision law	Spring-dashpot
Spring constant	410 kN/m
Coefficient of restitution	0.97
Friction coefficient	0.1–0.5 based on relative velocity
Numerical schemes	Phase coupled SIMPLE, second-order upwind for momentum eq., QUICK for volume fraction, second-order implicit in time
Time step size	Particle: 5×10^{-5} s, Fluid: 5×10^{-4} s

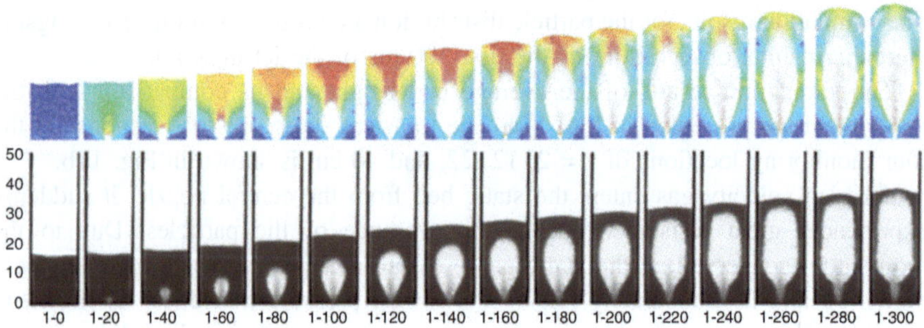

Fig. 10 Particle distribution in the reactor for the first 300 ms of jet injection (*top* Fluent CFD/DEM simulation, *bottom* TU-Darmstadt experiment)

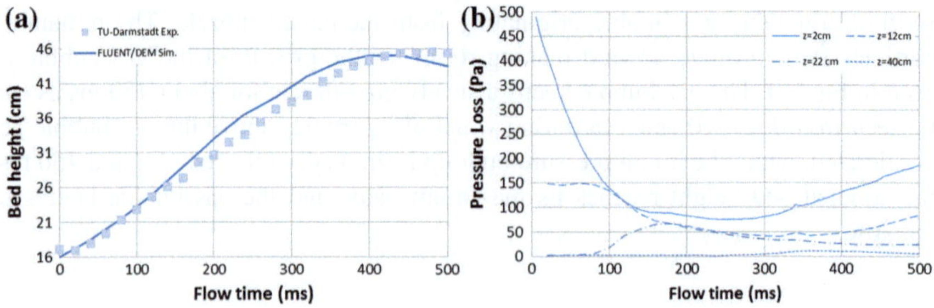

Fig. 11 **a** Time variation of bed height and **b** static pressure at various heights

In Fig. 11b, it is seen that the initial pressure buildup is immediately felt at both locations z = 2 and 12 cm, suggesting the rapid propagation of a pressure wave. However, due to the momentum exchange from gas to solids, such pressure buildup is not felt at z = 22 and 40 cm until about 80 and 300 ms, respectively. Figures 10 and 11 clearly demonstrate that the Fluent CFD/DEM solver is capable of accurately capturing the dynamics of the solid–gas interaction in the fuel reactor of a CD-CLC configuration. There is excellent agreement between the computed expansion of bed height with the experimental result including the expansion rate and also for the prediction of bubble burst. These results show the technical advantage of the Eulerian–Lagrangian modeling approach in modeling the spouted fluidized bed. It should, however, be noted that the computed transient pressure response shown in Fig. 11b is likely to be underestimated because of the relatively coarse mesh employed in the flow field; the volume fraction calculation in the computational cells for the CFD/DEM approach is a technical limitation that requires a minimum cell volume and precludes further mesh refinement.

4.2 Cold Flow Simulation of Complete CD-CLC System Using Spouted Fluidized Bed Fuel Reactor

With the successful numerical simulation of the TU-Darmstadt experiment in the previous section, we consider the simulation of bed expansion, particle separation, and oxygen carrier recirculation in a complete three-dimensional CD-CLC model. Necessary peripherals such as a cyclone separator and loop seal are added to the Plexiglas test rig of the TU-Darmstadt experiment to form a complete recirculating CD-CLC configuration. The geometry and computational model of such a configuration are shown in Fig. 12.

In the 3D CD-CLC configuration shown in Fig. 12, the fuel reactor is where the coal particles get combusted by the oxygen carrier. The reduced oxygen carrier is then transported to the cyclone for separation from the flue gas and then transferred into the downcomer. Additional gas supply is introduced from the bottom of the loop seal to aerate the particle deposition and enable their recirculation back into the fuel reactor. To ensure the adequacy of particle deposition in the system for its recirculation, an additional 11,000 particles are deposited into the downcomer and the loop seal during the initialization stage of simulation. The velocity of the central jet is increased to 40 m/s so that sufficient gas momentum can be transferred to the particles for them to reach the top of the reactor. The aeration velocity in the loop seal is set at 1 m/s. All other parameters in the simulation remain unchanged from the quasi-3D study of the TU-Darmstadt experiment reported in the previous section. With the same computer hardware platform, each run requires about 24 h of CPU time per 200 ms of simulation time. The particles' velocities and their distributions in this CD-CLC configuration for the first 800 ms are examined and are presented in Fig. 13.

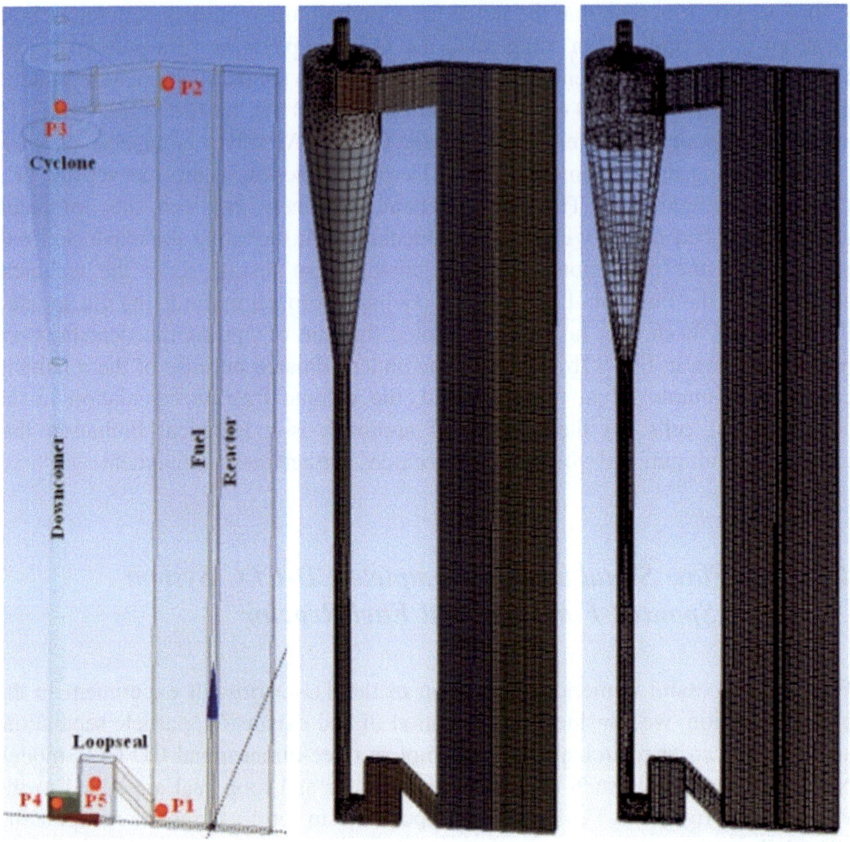

Fig. 12 Geometry outline with five pressure taps (*left*), mesh (*middle*), and wireframe (*right*) of the complete CD-CLC configuration

Based on the particle tracks in Fig. 13, a prominent gas bubble can be observed to form during the first 480 ms of the high-velocity central jet injection. Due to the presence of the particle recirculation duct on one side, the bubble becomes slightly asymmetric. The leading front of the spouted bed reaches the ceiling of the fuel reactor around 480 ms, and a large number of particles are slammed into the cyclone through the connecting duct, as can be seen between 520 and 600 ms. Between 480 and 800 ms, the pressure buildup in the fuel reactor vanishes and the remaining particles in the fuel reactor fall back into the fluidized bed. In the meantime, particles in the cyclone get separated from the flue stream and begin to deposit in the downcomer. In addition, recirculation of small number of particles from the loop seal back into the fuel reactor is observed from about 60–360 ms. From around 400 ms onward, there is a large number of stagnant particles at the bottom of the fuel reactor. At the same time, the high-velocity jet forms an unsteady pathway for the air to rapidly bypass the dense bed regime in the fuel reactor. Such a pathway is very undesirable since no new gas bubbles can get initiated due to its

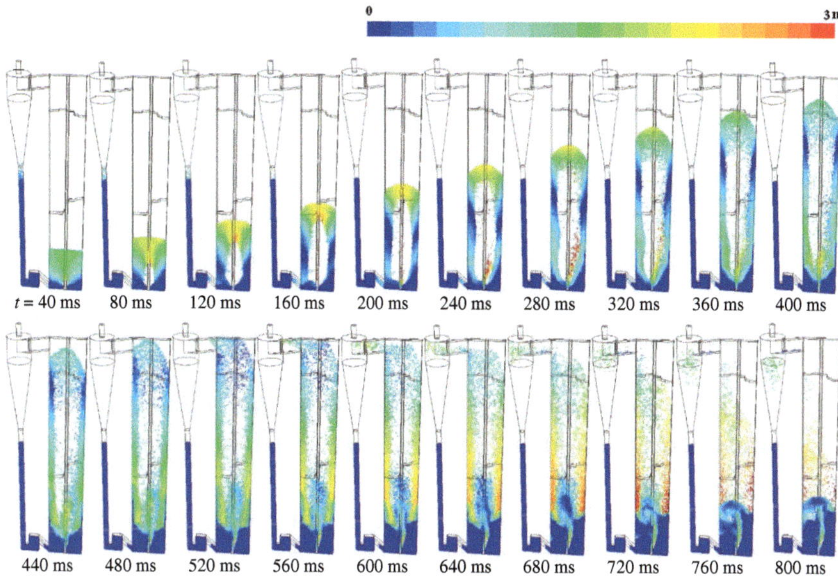

Fig. 13 Particle tracks colored by velocity magnitude in the complete CD-CLC configuration

presence. This is because the critical pressure buildup required to form a gas bubble arises when the injected air experiences the isotropic resistance in a static bed. Due to the unsteady pathway, the energy transferred to the particles becomes minimal since most of the air bypasses the fluidized bed through the pathway and it becomes impossible for the particles to reach the top and participate in the recirculation. Additionally, the recirculation of the solid particles appears to stagnate once the bubble in the fuel reactor is formed. Because the formation of the gas bubble and solids' recirculation is primarily driven by the pressure at various locations in the system, the static pressure readings at various pressure tap locations are investigated to better understand the behavior observed from the particle tracks in Fig. 13; the pressure tap data at t = 400 and 800 ms are shown in Fig. 14.

In Fig. 14, a large drop in pressure can be noted as the particles move from the fluidized bed ($P1$) to the top of the fuel reactor ($P2$) and then to the cyclone ($P3$). As the particles deposit in the downcomer, the pressure builds up ($P4$) and exceeds that in the loop seal ($P5$) and fluidized bed ($P1$), providing the necessary gas flow for particle recirculation. At both times t = 400 and 800 ms, there is a small positive pressure difference between the downcomer and the fluidized bed of about 100 Pa. With such a small pressure difference, the gas flow is not likely to provide enough momentum to the particles to trigger the solids recirculation, which implies poor recirculation of solid particles from the loop seal to the fuel reactor; this result quantitatively confirms the qualitative observations made from Fig. 13.

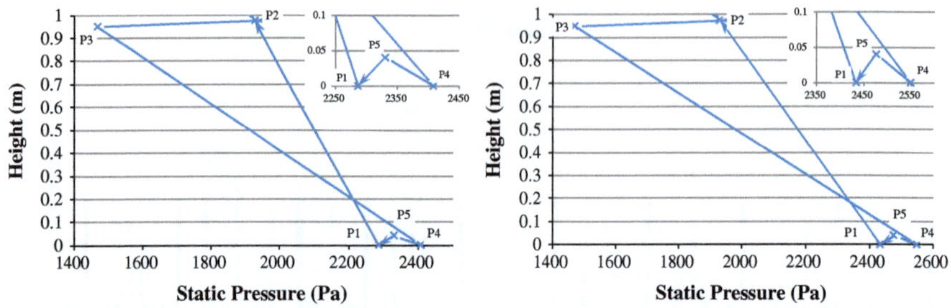

Fig. 14 Static pressure at pressure taps *P*1–*P*5 at t = 400 ms (*left*) and 800 ms (*right*) in the complete CD-CLC configuration

4.3 Cold Flow Simulation of Modified Complete CD-CLC System Using Spouted Fluidized Bed Fuel Reactor

The formation of new gas bubbles and continuous solid recirculation from the loop seal to the fuel reactor are the two key factors for successful CD-CLC operation, which were not satisfactorily achieved in the simulation results for the CD-CLC configuration considered in the previous section; modifications to the configuration are necessary to address these two factors for improved performance. We introduce three modifications to the CD-CLC configuration as shown in Fig. 15. First, a chute is added to the bottom of the fuel reactor. The dimensions of the central jet injection nozzle remain unchanged, while the two side panels of the chute are at about 60° from the horizontal. Second, the loop seal and solid recirculation duct are elevated by 20 mm. Finally, the height of the fuel reactor and downcomer is reduced by 250 mm. The rationale behind each of these modifications is discussed below.

The addition of the chute at the bottom is to induce gravity-driven particle circulation into the central jet. Such a consideration originates from the critical angle of repose (θ) for particulate materials (Lee and Herrmann 1993; Miller and Byrne 1966; Samadani and Kudrolli 2001; van Burkalow 1945; Zhou et al. 2001), which is defined as the steepest angle of descent of the slope relative to the horizontal plane when the particulate material on the slope face is on the verge of sliding. Due to the existence of θ, a dead zone in which particles are stagnant can be identified in a spouted fluidized bed as discussed by Takeuchi et al. (2004, 2005) and as evident from Fig. 13. The poor recirculation of solid particles observed in Fig. 13 is thus also partially explained by the fact that the recirculation duct was connected to the bottom of the fuel reactor, i.e., to the dead zone. Since the particles in the dead zone are stationary regardless of ambient gas flow, high pressure is always present at the bottom of the fuel reactor, impeding the transport of the particles pneumatically from the loop seal. Introducing the bottom chute eliminates the dead zone, relieving the pressure in the fuel reactor for better recirculation of particles. Similarly, elevating the loop seal and recirculation duct serves to reduce the pressure at the connecting point of the recirculation duct into fuel reactor for further improved particle recirculation. Lastly, the reduction in height of the fuel

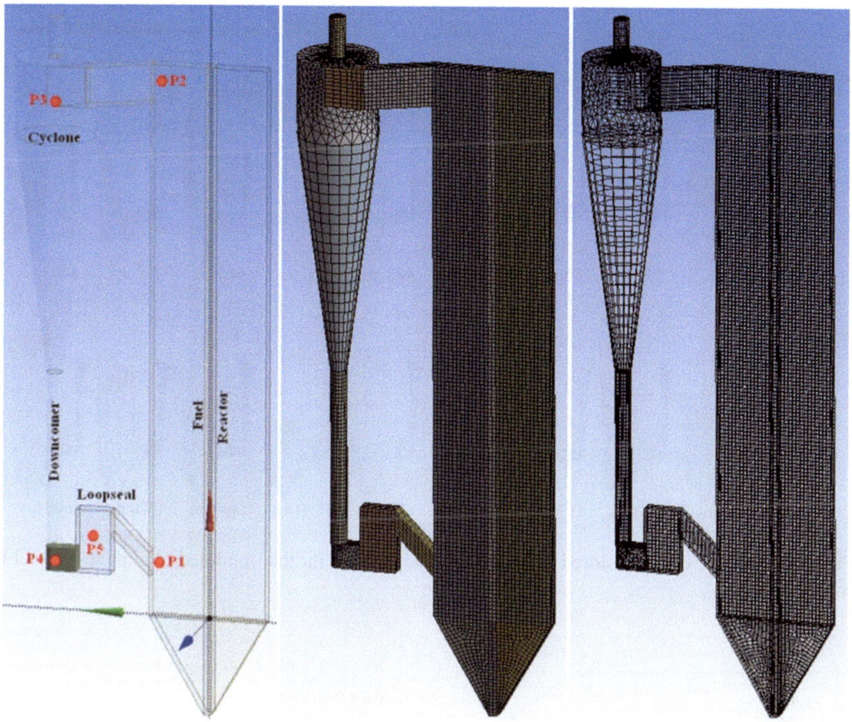

Fig. 15 Geometry outline with five pressure taps (*left*), mesh (*middle*), and wireframe (*right*) of the modified complete CD-CLC configuration

reactor and downcomer is to compensate for the height added by the bottom chute so that sufficiently large number of particles can still reach the top of the fuel reactor with unaltered fluidization conditions.

The computational cost of simulation for the modified CD-CLC configuration shown in Fig. 15 is similar to that for the original case of Fig. 12. The simulation results of particles distributions for the first 1,600 ms of jet injection are shown in Fig. 16.

With the proposed design modifications, the improvement in the CD-CLC system performance is very significant. As seen in Fig. 16, a gas bubble is formed in the first 480-ms shooting a large portion of solid particles into the cyclone. From 520 to 1,340 ms, the remaining particles fall back into the bed, while the particles in the cyclone get separated from the flue stream and drop into the downcomer. Encouragingly, the formation of a second gas bubble is observed from around 1,380 ms onward once a large number of falling particles reach the bed. In addition, continuous solid recirculation from the loop seal is observed in the process along with the elimination of the dead zone in the fuel reactor. The recirculation of solid particles can also be assured quantitatively by examining the static pressures as shown in Fig. 17 at the five pressure tap locations ($P1$–$P5$) shown in Fig. 15 at t = 400, 800, 1,200, and 1,600 ms.

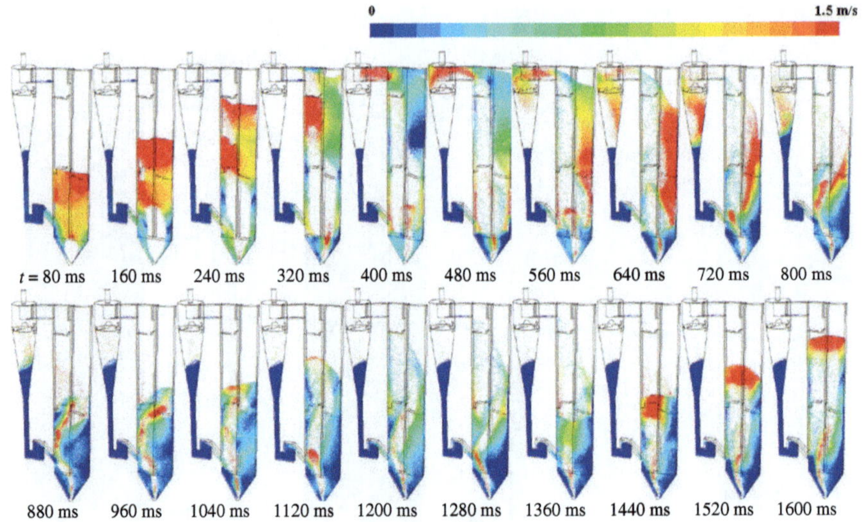

Fig. 16 Particle tracks colored by velocity magnitude in the modified complete CD-CLC configuration

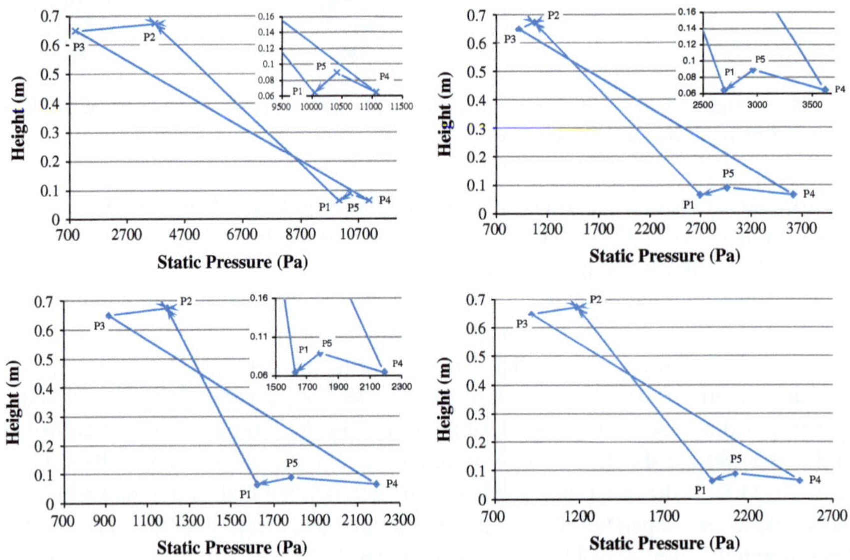

Fig. 17 Static pressure at pressure taps *P*1–*P*5 at t = 400 ms (*upper left*), 800 ms (*upper right*), 1200 ms (*lower left*), and 1600 ms (*lower right*) in the modified complete CD-CLC configuration

Comparing Fig. 17 to Fig. 14, i.e., the static pressures at *P*1–*P*5 in the modified and original CD-CLC configurations, respectively, it is clear that the pressure difference between the downcomer and the fluidized bed (*P*4–*P*1) has increased by almost an order of magnitude, which confirms the improved solid particle recirculation in the modified configuration of Fig. 15. In addition, it is believed that the elimination of the dead zone and enhanced solid recirculation from the loop seal makes the central jet experience a greater solid load in the fuel reactor in the modified fluidized bed configuration, which in turn prevents the formation of the unsteady pathway seen in the original configuration and triggers the formation of the gas bubble. However, more in-depth study of this aspect is needed in future work to gain better understanding of this key mechanism. Nevertheless, the simulation results shown in Figs. 16 and 17 clearly demonstrate that the two major concerns in the original CD-CLC configuration of Fig. 12, i.e., the failure to form new gas bubbles and poor recirculation of solid particles have been eliminated in the modified configuration of Fig. 15.

4.4 Reacting Flow Simulation of Modified CD-CLC System Using Spouted Fluidized Bed Fuel Reactor

The next step in developing a complete CD-CLC model for design and optimization is to incorporate chemical reactions into the CFD/DEM approach. This is achieved by considering the simple reaction described in Eq. (3) and applying the surface combustion model in ANSYS Fluent (ANSYS 2012a, b).

$$12Fe_2O_3(s) + CH_4 \rightarrow 8Fe_3O_4(s) + CO_2 + 2H_2O \quad (3)$$

Iron (Fe) being one of the cheapest and most abundant metals available on Earth, Fe-based metal oxides such as Fe_2O_3 are well-suited for use as the oxygen carrier in CD-CLC operation. However, given the density of Fe_2O_3 being 5,240 kg/m^3, experience has shown that when using particles of relatively large diameter as required for a spouted bed, the large mass compared to the glass beads prevents successful fluidization. There are also issues with particle agglomeration with Fe_2O_3 at high temperatures. To mitigate these drawbacks, it has been proposed to consider iron dispersed on various support materials (Hossain and de Lasa 2008). Johansson et al. (2004) showed that 60 % Fe_2O_3 supported on $MgAl_2O_4$ spinel sintered at 1,100 °C provided excellent reactivity and sufficient hardness, and its apparent density of 2,225 kg/m^3 makes it an ideal candidate for spouted fluidized bed operation. Therefore, this material is used as the oxygen carrier in the reacting flow simulation.

The $MgAl_2O_4$ is non-reacting, so the only reaction that takes place in the fuel reactor in the modified CD-CLC configuration of Fig. 15 is that described in Eq. (3) whereby the Fe_2O_3 is reduced to Fe_3O_4. The components of the inlet flow and the reaction rates are taken from the experimental work of Son and Kim (2006). The flow injection in the simulation comprises 10 % CH_4 and 10 % H_2O by mass fraction; the remaining 80 % is N_2 unlike the experiment where 75 % N_2 and 5 % CO_2 were used by Son and Kim (2006). This change is necessary to isolate the CO_2

formed by the reaction in order to track the reaction progress without the small mass fractions of generated CO_2 being overshadowed by the injected CO_2. The aeration in the downcomer and the loop seal is non-reacting and comprises solely of N_2. This helps to identify the recirculation of particles from the loop seal to the fuel reactor; the particles that originate in the loop seal will have a smaller mass fraction of Fe_3O_4 since they were initially exposed to inert N_2. Based on preliminary trials to investigate the momentum transfer to the particles, the velocity of the central jet and the aeration velocity in the loop seal are set at 60 and 2 m/s, respectively. All other parameters in the simulation remain unchanged from the cold flow simulations reported in the previous sections. The particle distributions and velocities are inspected at 80-ms intervals and presented in Fig. 18 for the first 1,600 ms of flow injection.

The particle tracks in Fig. 18 show the same initial behavior as in the cold flow case (see Fig. 16); the first gas bubble reaches the top of the reactor around 400 ms and deposits a large number of solid particles into the cyclone. Between 480 and 880 ms, the pressure buildup in the fuel reactor vanishes and the remaining particles in the fuel reactor fall back into the fluidized bed, while the particles in the cyclone fall into the downcomer. Continuous recirculation of solid particles from the loop seal back into the fuel reactor is also evident. Once the particles start to settle back into the fluidized bed and aided by the recirculation of particles from the loop seal, the pressure buildup due to the jet injection is partially restored and subsequent gas bubbles are formed around 960 ms and 1,440 ms. However, in these cases, the kinetic energy transferred to the particles is insufficient to carry them to the top of the reactor and the bubbles collapse prematurely. This can be explained by the

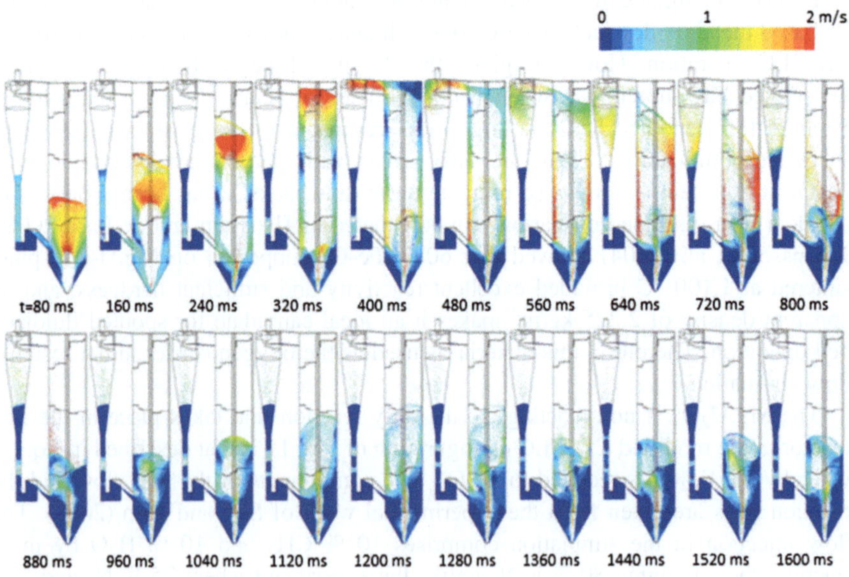

Fig. 18 Particle tracks colored by velocity magnitude in reacting flow

unsteady pathway that forms around 480 ms by the high-velocity jet in the absence of the initial packed bed, which allows the energy in the jet to bypass the fluidized bed and prevents the critical pressure buildup as discussed in Sect. 4.2. Since the gas bubble formation and the solid particle recirculation are both driven by the pressures at various locations in the system, the static pressure readings at pressure taps *P*1–*P*5 are presented in Fig. 19 to assess the qualitative behavior observed from the particle tracks in Fig. 18.

The subplot at 400 ms in Fig. 19 shows a large pressure buildup of around 1,900 Pa at *P*1 (fluidized bed). At 800 ms, when the bubble has collapsed, the initial pressure buildup is lost and the pressure at *P*1 drops to around 1,250 Pa. Subsequently, at 1,200 and 1,600 ms, the pressure at P1 increases slightly to around 1,400 and 1,350 Pa, respectively. This renewed increase in pressure is in line with the observation of the second and third gas bubbles formations; the slight increase also explains why the subsequent bubbles did not carry sufficient kinetic energy to reach the top of the reactor. Figure 19 also shows a consistent positive pressure differential between the loop seal and the fluidized bed (*P*4–*P*1), which corroborates the continuous recirculation of particles from the loop seal to the fuel reactor observed from the particle tracks.

In the absence of experimental results of CD-CLC operation with Group D-type particles from Geldart's powder classification (1973), the successful incorporation of chemical reactions into the multiphase flow simulation is judged by inspecting the formation of Fe_3O_4 and CO_2 as a result of the reaction described in Eq. (3). These results are presented in Figs. 20 and 21, respectively.

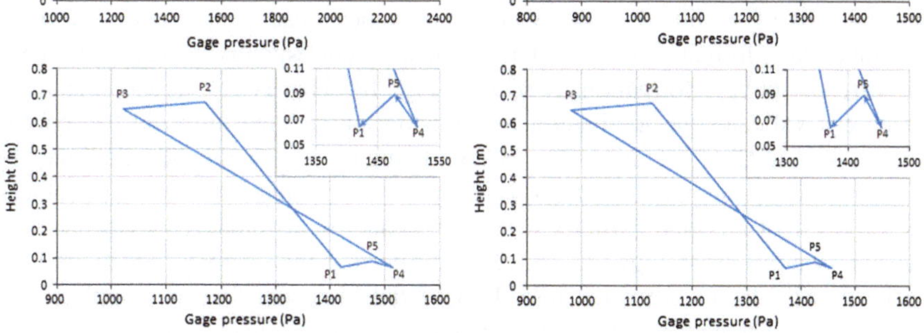

Fig. 19 Static pressure at pressure taps *P*1–*P*5 at t = 400 ms (*upper left*), 800 ms (*upper right*), 1,200 ms (*lower left*), and 1,600 ms (*lower right*) in reacting flow

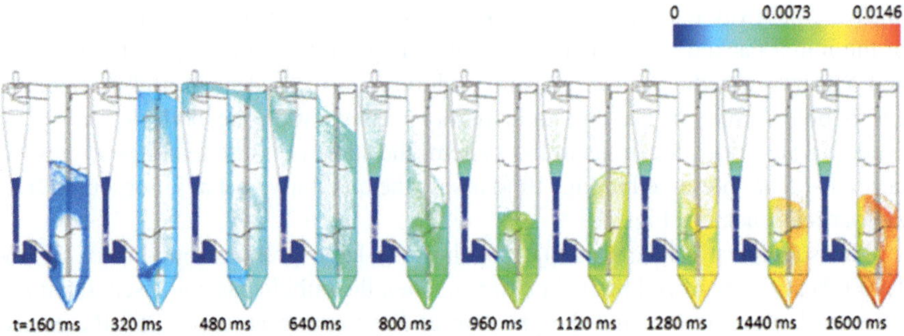

Fig. 20 Particle tracks colored by mass fraction of Fe_3O_4 relative to original mass of Fe_2O_3

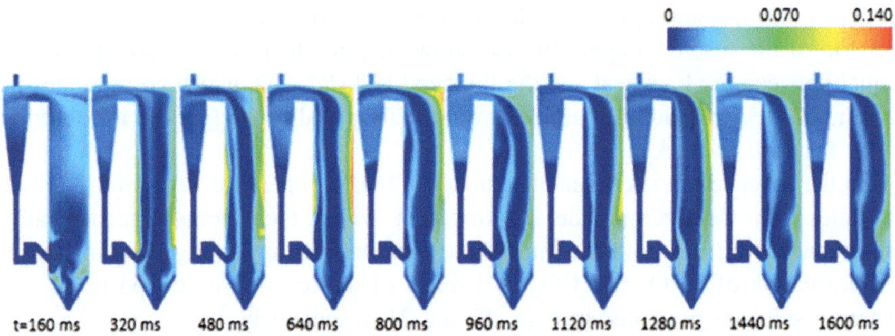

Fig. 21 Contours of CO_2 mass fraction produced by reaction of Fe_2O_3 with CH_4

It should be noted that the mass fraction of Fe_3O_4 in Fig. 20 is in relation to the mass of Fe_2O_3; the total mass of each particle also includes 40 % inert $MgAl_2O_4$ by mass; thus, the mass fraction of Fe_3O_4 as a function of total particle mass would be proportionately smaller. Figure 20 indicates that the mass fraction of Fe_3O_4 progressively increases for the particles inside the fuel reactor as the simulation advances as expected. From Fig. 21, the mass fraction of CO_2 rises quickly for the first 640 ms of simulation, after which it becomes steady, representing nearly a constant rate of formation of CO_2. The initial spike corresponds to the time frame of the first gas bubble formation as seen in Fig. 18. This is an encouraging result because it validates the motivation for utilizing the spouted fluidized bed configuration for CD-CLC operation. The heavy mixing of gases and solids facilitated by the bubble or spout creates vortical pathways, which increase the residual time and the exposure of the solid particles with the fuel and improve the reactivity. Overall, the observations from Figs. 20 and 21 indicate that the chemical reactions have been successfully incorporated into the CFD/DEM model for the complete CD-CLC reactor. The particles with a lower mass fraction of Fe_3O_4 near the left wall of the fuel reactor indicate that these particles originate in the loop seal where the flow is non-reacting; the constant stream of such particles provides further evidence of continuous recirculation from the loop seal into the fuel reactor.

The key shortcoming from the results of the simulation of the complete CD-CLC system using 60 wt% Fe_2O_3 particles as the oxygen carrier is that the high jet velocities required to fluidize these relatively heavy particles lead to the formation of an unsteady pathway through the bed after the first bubble collapses that hinders the pressure buildup required for subsequent bubbles to reach the top of the reactor and continuously recirculate. Since this material is already among the lightest Fe-based oxygen carrier studied by Johansson et al. (2004) and lighter alternatives likely to be expensive, one alternate way to mitigate this problem is to reduce the height of the fuel reactor. The central jet velocity is specified such that the first bubble has enough kinetic energy to reach the top of the reactor; with a smaller reactor, this velocity can be reduced which in turn reduces the likelihood of the unsteady pathway to form. However, this also reduces the residence time of the fuel in the reactor, which can greatly affect the rate of fuel conversion, particularly when using solid coal that has to go through a slow gasification process before the reaction with the metal oxide can occur. Since the unsteady pathway only forms after the first bubble collapses, another alternative is to use a cyclic flow injection whereby the jet is turned off intermittently to allow the bed particles to re-settle down into the original packed bed configuration, which resets the fluidization to the initial bubble formation stage once the jet is turned back on. Cyclic injections are already used in the laboratory-scale CLC experiments to switch between N_2 and CH_4 in lieu of separate fuel and air reactors as considered in the work of Son and Kim (2006); their operational feasibility on an Industrial scale can also be readily studied.

5 Conclusions

In this paper, ASPEN Plus is first employed to model and study the CLC and CLOU processes at the system level. The CLC process model is validated against previous work and shows good agreement, following which further studies are conducted to investigate the effect of varying the air flow rate and oxygen carrier feeding rate. It is found that the maximization of energy output from CLC can be accomplished by using the optimum ratio of coal, air, and oxygen carrier in the system equal to 1:10:70, generating a 48 % net increase in power over the baseline experimental case of Sahir et al. (2012). In the CLOU process modeling also using ASPEN Plus, excellent agreement is obtained between the experimental results and the simulation for various quantities such as the oxygen carrier conversion kinetics, flue stream O_2 and CO_2 concentrations, and power output. It is clearly demonstrated that the ASPEN Plus can provide a creditable process simulation platform for the study of CLOU process. Scaled-up simulations of CLOU process were also conducted using different types of coal and coal feeding rates. The results show that the total power output is nearly linear with the increase in coal feeding rate and the carrier circulation. Such linearity in general will not be expected for actual scale-up, since the ASPEN Plus system modeling software neglects miscellaneous energy losses in the system. Furthermore, the coal rank appears to have significant impact on the flue stream concentration of CO_2 and overall energy output; bituminous coal and anthracitic coal show significantly improved CLOU performance over lignite coal. The similarity in the CLOU performance of bituminous coal and anthracitic coal can be explained by the fact that they have similar carbon content, which indicates that char gasification is no longer a performance factor of high relevance in CLOU since the presence of oxygen enables the solid–gas combustion to take place without gasification.

To account for changes in the hydrodynamic characteristics, detailed multiphase flow simulations are conducted using the CFD/DEM coupled approach in ANSYS Fluent. The initial reactor-level simulation shows excellent agreement with the experimental results obtained by the research group at Darmstadt University of Technology. Based on the insights from reactor-level modeling, simulations for a complete three-dimensional CD-CLC configuration are performed to determine the cold flow performance of the configuration. It is observed from the simulations' results that the continuous formation of new bubbles and solid recirculation from the loop seal to the fuel reactor—two key factors important for successful CD-CLC operation—are not satisfactorily achieved for the configuration used in the experiment at TU-Darmstadt. Subsequently, a new configuration was designed to redress these issues, i.e., to enhance the particle recirculation in the entire system and to increase the possibility of continuous bubble formation by eliminating the dead zone in the spouted bed and the resulting pressure imbalance. The cold flow simulations results for the modified CD-CLC configuration confirmed that significant improvements on the performance can be achieved with the introduction of proper design modifications.

Finally, the CFD/DEM cold flow model for the modified CD-CLC configuration was expanded to consider a chemically reacting flow. The oxygen carrier selected was 60 wt% Fe_2O_3 supported by 40 wt% $MgAl_2O_4$ due to its favorable properties for CD-CLC operation. The system behavior was investigated in the presence of chemical reactions between the Fe_2O_3 and gaseous CH_4. It was found that the initial bubble formation and continuous recirculation of solid particles from the loop seal to the fuel reactor were adequate and that the chemical reactions were successfully incorporated. One shortcoming observed in the simulation of the complete CD-CLC system with reactive flow was that subsequent bubbles formed in the fuel reactor lacked sufficient kinetic energy required to reach the top of the reactor and into the cyclone. By investigating the pressures in the system, it was found that the high-velocity inlet jet formed a straight pathway that bypassed much of the particle bed after the first bubble collapsed, which in turn prevented the critical pressure buildup required for subsequent bubbles to develop. Based on this insight, a few alternatives are proposed to circumvent this issue, which will be investigated in the future work. More importantly, the system performance of the modified CD-CLC configuration demonstrates the necessity of developing a credible CFD/DEM model with high fidelity for future optimization designs of industrial-scale CD-CLC systems.

Acknowledgements The authors gratefully acknowledge the financial support for this work from the Consortium for Clean Coal Utilization (CCCU) at Washington University in St. Louis.

References

Abad A, Adánez-Rubio I, Gayán P, García-Labiano F, de Diego LF, Adánez J (2012) Demonstration of chemical-looping with oxygen uncoupling (CLOU) process in a 1.5 kWth continuously operating unit using a Cu-based oxygen-carrier. Int J Greenhouse Gas Control 6:189–200

Alobaid F, Ströhle J, Epple B (2013) Extended CFD/DEM model for the simulation of circulating fluidized bed. Adv Powder Technol 24:403–415

ANSYS Inc. (2012a) ANSYS fluent theory guide. Canonsburg, PA

ANSYS Inc. (2012b) ANSYS fluent user's guide. Canonsburg, PA

Cao Y, Pan W, Green B, July RV, Re V, Recei M, December V (2006) Investigation of chemical-looping combustion by solid fuels 1. Proc Anal Energy Fuels 20:1836–1844

Chu K (2010) CFD-DEM simulation of complex particle-fluid flows. PhD dissertation, University of New South Wales, Sydney, Australia

Dennis JS, Scott SA, Hayhurst AN (2006) In-situ gasification of coal using steam with chemical looping: a technique for isolating CO_2 from burning a solid fuel. J Energy Inst 79(3):187–190

Feng Y, Yu AB (2009) CFD-DEM modeling of gas fluidization of particle mixtures with size and density differences. In: Proceeding of 7th international conference of CFD in the minerals and process industries, Melbourne, Australia

Geldart D (1973) Types of gas fluidization. Powder Technol 7(5):285–292

Gryczka O, Heinrich S, Deen NG, van Sint Annaland M, Kuipers JAM, Jacob M, Mörl L (2009) Characterization and CFD-modeling of the hydrodynamics of a prismatic spouted bed apparatus. Chem Eng Sci 64(14):3352–3375

Hong J, Chaudhry G, Brisson JG, Field R, Gazzino M, Ghoniem AF (2009a) Analysis of oxy-fuel combustion power cycle utilizing a pressurized coal combustor. Energy 34(9):1332–1340

Hong J, Chaudhry G, Brisson JG, Field R, Gazzino M, Ghoneim AF (2009b) Performance of the pressurized oxy-fuel combustion power cycle with increasing operating pressure. In: Proceedings of 34th international technical conference clean coal and fuel systems, Clearwater, FL

Hossain MM, de Lasa HI (2008) Chemical-looping combustion (CLC) for inherent separations—a review. Chem Eng Sci 63(18):4433–4451

Johansson M, Mattisson T, Lyngfelt A (2004) Investigation of Fe_2O_3 with $MgAl_2O_4$ for chemical-looping combustion. Ind Eng Chem Res 43:6978–6987

Kloss C, Goniva C, Aichinger G, Pirker S (2009) Comprehensive DEM-DPM-CFD simulations—model synthesis, experimental validation and scalability. In: Proceedings of 7th international conference CFD in the minerals and process industries, Melbourne, Australia

Lee J, Herrmann HJ (1993) Angle of repose and angle of marginal stability: molecular dynamics of granular particles. J Phys A: Math Gen 26(2):373–383

Leion H, Lyngfelt A, Mattisson T (2009a) Solid fuels in chemical-looping combustion using a NiO-based oxygen carrier. Chem Eng Res Des 87(11):1543–1550

Leion H, Mattisson T, Lyngfelt A (2009b) Using chemical-looping with oxygen uncoupling (CLOU) for combustion of six different solid fuels. Energy Procedia 1(1):447–453

Linderholm C, Lyngfelt A, Dueso C (2013) Chemical-looping combustion of solid fuels in a 10 kW reactor system using natural minerals as oxygen carrier. Energy Procedia 37:598–607

Link JM (1975) Development and validation of a discrete particle model of a spout-fluid bed granulator. PhD dissertation, University of Twente, Enschede, The Netherlands

Lyngfelt A, Leckner B, Mattisson T (2001) A fluidized-bed combustion process with inherent CO_2 separation; application of chemical-looping combustion. Chem Eng Sci 56(10):3101–3113

Mahalatkar K, Kuhlman J, Huckaby ED, O'Brien T (2011) Computational fluid dynamic simulations of chemical-looping fuel reactors utilizing gaseous fuels. Chem Eng Sci 66 (3):469–479

Mattisson T, Lyngfelt A, Leion H (2009a) Chemical-looping with oxygen uncoupling for combustion of solid fuels. Int J Greenhouse Gas Control 3(1):11–19

Mattisson T, Lyngfelt A, Leion H (2009b) Chemical-looping with oxygen uncoupling using CuO/ZrO_2 with petroleum coke. Fuel 88:683–690

Miller RL, Byrne RJ (1966) The angle of repose for a single grain on a fixed rough bed. Sedimentology 6(4):303–314

Mukherjee S, Kumar P, Hosseini A (2014) Comparative assessment of gasification based coal power plants with various CO_2 capture technologies producing electricity and hydrogen. Energy Fuels 28(2):1028–1040

Rubel A, Zhang Y, Liu K, Neathery J (2011) Effect of ash on oxygen carriers for the application of chemical-looping combustion to a high carbon char. Oil Gas Sci Technol Rev d'IFP Energies Nouv 66(2):291–300

Rydén M, Lyngfelt A, Mattisson T (2011) Combined manganese/iron oxides as oxygen carrier for chemical-looping combustion with oxygen uncoupling (CLOU) in a circulating fluidized bed reactor system. Energy Procedia 4:341–348

Sahir AH, Cadore AL, Dansie JK (2012) Process analysis of chemical looping with oxygen uncoupling (CLOU) and chemical looping combustion (CLC) for solid fuels. In: Proceedings of 2nd international conference chemical looping, Darmstadt, Germany

Samadani A, Kudrolli A (2001) Angle of repose and segregation in cohesive granular matter. Phys Rev E 64(5):051301

Shen L, Wu J, Gao Z, Xiao J (2009) Reactivity deterioration of NiO/Al_2O_3 oxygen carrier for chemical looping combustion of coal in a 10 kWth reactor. Combust Flame 156(7):1377–1385

Son SR, Kim SD (2006) Chemical-looping combustion with NiO and Fe_2O_3 in a thermo-balance and circulating fluidized bed reactor with double loops. Ind Eng Chem Res 45:2689–2696

Syamlal M, O'Brien TJ (1989) Computer simulation of bubbles in a fluidized bed. AIChE Symp Ser 85:22–31

Takeuchi S, Wang XS, Rhodes MJ (2004) Discrete element simulation of a flat-bottomed spouted bed in the 3-D cylindrical coordinate system. Chem Eng Sci 59(17):3495–3504

Takeuchi S, Wang XS, Rhodes MJ (2005) Discrete element study of particle circulation in a 3-D spouted bed. Chem Eng Sci 60(5):1267–1276

van Burkalow A (1945) Angle of repose and angle of sliding friction: an experimental study. Geol Soc Am Bull 56(6):669–707

Yang C, Duan Y (2013) CFD-DEM model for simulating solid exchange in a dual-leg fluidized bed. Chem Eng Technol 36(11):1907–1914

Zhang Z, Zhou L, Agarwal R (2014) Transient simulations of spouted fluidized bed for coal-direct chemical looping combustion. Energy Fuels 28(2):1548–1560

Zhou L, Zhang Z, Chivetta C, Agarwal R (2013) Process simulation and validation of chemical-looping with oxygen uncoupling (CLOU) process using Cu-based oxygen carrier. Energy Fuels 27(11):6906–6912

Zhou YC, Xu BH, Yu AB, Zulli P (2001) Numerical investigation of the angle of repose of mono-sized spheres. Phys Rev E 64(2):021301

Zhu HP, Zhou ZY, Yang RY, Yu AB (2008) Discrete particle simulation of particulate systems: a review of major applications and findings. Chem Eng Sci 63(23):5728–5770

Acoustic Reynolds Stress: The Source of Coherent Structures During Combustion Instability

Chellappan Balaji, Ashwin Kannan, Ramgopal Sampath and Satyanarayanan Chakravarthy

Abstract Combustion was 'taken for granted' from prehistoric times until the last century. Concerns about environmental pollution from combustion emissions dominated the end of that century. Global warming due to greenhouse gas emissions has predominated pollutant emissions as a major cause of concern against concern by the end of the first decade of the present century. Yet, a majority of the world's energy and transportation needs are met from combustion of coal, oil and biomass. This trend is likely to continue well into the foreseeable future. Thus, a 'business as usual' approach seems to be adopted globally, particularly during the years of economic downturn

Keywords Flow and acoustics · Flame-acoustic coupled simulations · Combustion instability · Acoustic Reynolds stress (ARS) · Combustion-acoustic lock-on

1 Introduction

Combustion was 'taken for granted' from prehistoric times until the last century. Concerns about environmental pollution from combustion emissions dominated the end of that century. Global warming due to greenhouse gas emissions has predominated pollutant emissions as a major cause of concern against concern by the end of the first decade of the present century. Yet, a majority of the world's energy and transportation needs are met from combustion of coal, oil and biomass. This trend is likely to continue well into the foreseeable future. Thus, a 'business as

C. Balaji
Metacomp Technologies Pvt. Ltd, Chennai 600020, India

A. Kannan · R. Sampath · S. Chakravarthy (✉)
National Centre for Combustion Research and Development, Department of Aerospace Engineering, Indian Institute of Technology Madras, Chennai 600036, India
e-mail: src@ae.iitm.ac.in

A.K. Agarwal et al. (eds.), *Novel Combustion Concepts for Sustainable Energy Development*, DOI 10.1007/978-81-322-2211-8_15

usual' approach seems to be adopted globally, particularly during the years of economic downturn.

The present chapter is on combustion instability, which has been a problem plaguing engine design, development and operation for several decades now. It is a phenomenon wherein the interaction between the fluid flow processes such as unsteady heat release, shedding of vortical structures, etc. and acoustics leads to production, growth and sustenance of high amplitude sound. Some of the significant mechanisms of combustion instability involve vortex shedding (Schadow and Gutmark 1992) and equivalence ratio fluctuations (Lieuwen and Zinn 1998), both leading to heat release rate fluctuations that act as a source to acoustic field.

Combustion instability has been a problem of the proverbial 'six blind men and an elephant'. The acousticians look at the combustion process merely as a source of unsteady heat release. The combustion scientists investigate unsteady flames without regard to the excitation of acoustics in the engine subject to certain boundary conditions. The unsteady fluid mechanics that is coupled with combustion is usually left uncovered in this process.

The vortex shedding mechanism, in particular, for instance, is prevalent because combustors usually adopt geometric constructs such as sudden expansions, bluff-bodies or a strong swirl flow to create regions of flow recirculation that would facilitate flame holding. However, such recirculation zones involve inherently unsteady shear layers that oscillate and in turn, lead to fluctuating heat release. The important aspect to consider here is the dynamics of combusting shear layers which are likely to be very different from shear layer dynamics observed usually under cold flow conditions. For instance, it is well known that in the wake flow of a bluff-body flame holder, a Karman vortex street of alternating pairs of vortices is usually observed under isothermal flow conditions. These are the dominant large-scale flow structures in an otherwise turbulent wake. However, under statistically steady combustion conditions, the Karman vortex street is replaced by extended shear layers from the separation points due to the dilatation caused by the heat release in the flame held by the bluff body. These extended shear layers exhibit Kelvin–Helmholtz instability waves that result in relatively small-scale vortical roll-ups as the most dominant flow structures under such reacting flow conditions. On the contrary, large-scale coherent structures (of the order of the Karman vortex street, but symmetrical, in the case of bluff-body flame holders) have been reported during the advent of combustion instability in burners. The present work attempts to focus on the origin or reappearance of such flow structures under combustion conditions when intense acoustic excitation and feedback as during instability ensues.

The acoustic-flame-flow interaction involved in combustion instability is governed by the compressible Navier–Stokes (NS) equations. Many investigators have solved the compressible NS equations directly to study combustion instability in practical combustors, e.g., (Poinsot and Candel 1988; Menon and Jou 1991). Further, approaches such as URANS (Brookes et al. 2001)/LES (Huang and Yang 2004) are also performed to compute flows that are prone to combustion instability. A hybrid approach involving compressible LES and a Helmholtz acoustic solver is also reported (Roux et al. 2005; Selle et al. 2006).

These computations based on compressible NS equations can provide a reasonably accurate description of the underlying flow physics. However, the computational resource and time required to compute practical turbulent flow fields are enormous. This motivates the development of reduced order modelling approaches to compute these flows. In a majority of the reduced order models, the linear evolution equations for the acoustic field are considered, and the unsteady heat release rate from the flame is modelled to act as a source of the acoustic field, in a somewhat ad hoc manner. In these approaches, the acoustic variable, usually the acoustic velocity, is related to the heat release rate to close the acoustic equation.

The acoustic equations are generally solved in the frequency domain and the unsteadiness of the flame (or its heat release rate) in response to the acoustic velocity at the inlet in the frequency range of interest is computed or measured first and used in the acoustic evolution equation subsequently. This is known as the flame transfer function (FTF). Recently, Boudy et al. (2011) adopted the flame-describing function (FDF) obtained at different amplitudes. Non-linear effects such as transient frequency shifts are triggered when the FDF is used. Vortex-based flame models are used when the dominant mechanism for heat-release rate fluctuations that drives duct acoustics are due to large-scale vortical structures that are formed in the flame stabilising regions (Culick 1988; Sterling 1993; Matveev and Culick 2003).

Most of these models presume the characteristic time scale of the flame as the same as that of the duct acoustic modes. During the operation of combustors, however, it is observed that the flame dynamically evolves as it interacts with the acoustic field (Searby 1992; Balachandran et al. 2008; Chakravarthy et al. 2007). The time scales of the flow and acoustics are observed to approach each other (Chakravarthy et al. 2007) as the system enters into instability mode.

Wu et al. (2003) considered the dynamic evolution of the flame as it got coupled with the duct acoustics. The flame was shown to drive the acoustics by inducing a jump in the longitudinal velocity and the acoustics affected the flame through the global acceleration term. Tyagi et al. (2007) also considered the simultaneous evolution of a non-premixed flame in the Burke–Schumann geometry along with the duct acoustics.

The present work adopts a framework of simultaneous multiple length and time scales, wherein the flow/flame evolves at long time and short length scales, whereas the acoustics evolves at short time and long length scales. This approach results in work done due to the fluid expansion as the source to the acoustic field and the stress due to acoustic velocity (acoustic Reynolds stress, ARS) as the contribution from the acoustic field to the flow.

Acoustics interacts with the base flow in the interior region through the ARS (Nyborg 1965; Lighthill 1978). The mixing or 'stirring' characteristics of the ARS is highlighted by (1965), who also points out the wide application of the ARS in areas of heat transfer, surface reaction, biological cell changes etc., in enhancing the rate at which these processes occur. Tanabe et al. (2000) investigated the importance of the ARS in enhancing the evaporation rate of droplets that are present in an acoustic standing wave field. In this paper, evidence of the ARS promoting mixing and combustion by churning out large-scale vortical structures is presented. These events eventually result in intermittent combustion and drive the combustor to an unstable mode of operation.

2 Mathematical Formulation

The schematic of the problem with specific context to a dump combustor is shown in Fig. 1. The flow/combustion field is modelled in a shorter length scale h and longer (convective) time scale $t = h/u$, whereas the acoustics is modelled in a longer length scale L (for longitudinal acoustic modes) and shorter time scale $\tau = L/c$. h is characteristic of the lateral dimensions of the combustor, such as the step height, and L is the length of the combustor. u and c are reference flow and sound speeds. The fluid is considered to be a calorically perfect gas and all the diffusion coefficients are assumed equal and constant. The conservation equations are non-dimensionalised with shorter length and time scales, h and τ, respectively (Chakravarthy et al. 2007). Apart from other non-dimensional quantities, the resulting set of equations have $M = u/c$ and h/L, the Mach number and the ratio of length scales, respectively. Taken together, these constitute the simultaneous multiple length and time scales. Both of these quantities are taken to be unequally small in general, here. To handle two small parameters simultaneously, each of them is, in turn, represented as a function of a single small number ε through the relations $M = \varepsilon^m$ and $h/L = \varepsilon^n$. Here, we choose $m = 1$, and let n evolve in the solution as a marker of the compactness of flame due to its coupling with the acoustics.

Any dependent variable φ is expanded according to the ansatz . For $m = 1$, this is the conventional Mach number expansion, with M as the small parameter. Splitting the spatial and temporal derivatives for the two scales considered, expanding the variables as powers of ε and comparing equi-order terms in ε, we decompose the conservation equations for the flow and acoustics in their respective length and time scales. In the resulting set of equations, it is observed that both the flow and acoustics affect the fluid velocity to leading order. Temporally averaging the base-flow equations and spatially averaging the acoustic equations results in the

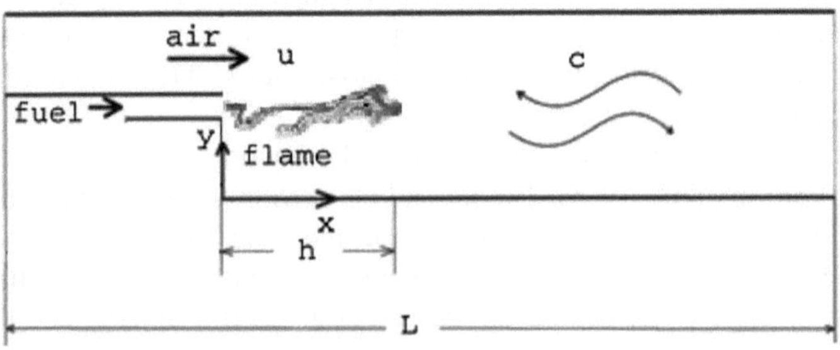

Fig. 1 Schematic of the backward facing step combustor

definition of new velocity variables that can be identified distinctly with the base flow and the acoustics. The time-averaging over the slower of the two time scales formally ensures that the slower process would pose to be rate-limiting in the coupling between the two, and that the evolution of the coupled system would occur at the slower time scale only. The averaging also results in the appearance of explicit interaction terms between the base-flow and acoustics. The final sets of conservation equations for the flow and acoustics are given below (Chakravarthy et al. 2007).

Flow:

$$\frac{\partial \rho_0}{\partial t} + \frac{\partial}{\partial x_j}\left(\rho_0 \overline{u_{0j}}^{\tau}\right) = 0 \tag{1}$$

$$\begin{aligned} &\frac{\partial}{\partial t}\left(\rho_0 \overline{u_{0i}}^{\tau}\right) + \frac{\partial}{\partial x_j}\left(\rho_0 \overline{u_{0i}}^{\tau}\overline{u_{0j}}^{\tau}\right) + \frac{1}{\gamma}\frac{\partial \overline{\rho_2^{\tau}}}{\partial x_i} \\ &\qquad - \frac{1}{Re}\left(\frac{\partial^2 \overline{u_{0i}}^{\tau}}{\partial x_j^2} + \frac{1}{3}\frac{\partial}{\partial x_i}\left(\frac{\partial \overline{u_{0k}}^{\tau}}{\partial x_k}\right)\right) = -\frac{\partial}{\partial x_j}\left(\rho_0 \overline{u_{0i}'^{\tau} u_{0j}'^{\tau}}^{\tau}\right) \end{aligned} \tag{2}$$

$$\frac{\mathrm{d}\rho_0}{\mathrm{d}t} + \gamma \rho_0 \frac{\partial \overline{u_{0j}}^{\tau}}{\partial x_j} = (\gamma - 1) Da\, Q_{R0} + \frac{\gamma}{Pe}\frac{\partial^2 T_0}{\partial x_j^2} \tag{3}$$

$$\frac{\partial \rho_{K0}}{\partial t} + \frac{\partial}{\partial x_j}\left(\rho_{K0}\overline{u_{0j}}^{\tau}\right) - \frac{1}{ReSc}\frac{\partial^2 \rho_{K0}}{\partial x_i^2} - Da_K \omega_K = 0 \tag{4}$$

$$p_0 = \rho_0 T_0 \tag{5}$$

Acoustics:

$$\frac{\partial \rho_1}{\partial \tau} + \frac{\partial}{\partial \xi_j}\left(\overline{\rho_0}^{x} u_{0j}'^{\tau}\right) = -\frac{\partial}{\partial \xi_j}\left(\overline{\rho_0}^{x}\overline{\overline{u_{0j}}^{\tau}}^{x} + \overline{\rho_0'^{x}\overline{u_{0j}}^{\tau' x}}^{x}\right) \tag{6}$$

$$\overline{\rho_0}^{x}\frac{\partial u_{0i}'^{\tau}}{\partial \tau} + \frac{1}{\gamma}\frac{\partial p_1}{\partial \xi_i} = 0 \tag{7}$$

$$\frac{\partial p_1}{\partial \tau} + \gamma p_0 \frac{\partial u_{0j}'^{\tau}}{\partial \xi_i} = -\gamma p_0 \frac{\partial \overline{\overline{u_{0j}}^{\tau}}^{x}}{\partial \xi_i} \tag{8}$$

$$\begin{aligned} &\frac{\partial \overline{\rho_{K1}}^{x}}{\partial \tau} + \frac{\partial}{\partial \xi_i}\left(\overline{\rho_{K0}}^{x} u_{0j}'^{\tau}\right) \\ &\qquad = -\frac{\partial}{\partial \xi_i}\left(\overline{\rho_{K0}}^{x}\overline{\overline{u_{0j}}^{\tau}}^{x} + \rho_{K0}'^{x}\overline{\overline{u_{0j}}^{\tau' x}}^{x}\right) \end{aligned} \tag{9}$$

$$p_1 = \rho_1 \overline{T_0}^{x} + \overline{\rho_0}^{x} T_1 \tag{10}$$

In these equations, ρ_0, $\overline{u_{0j}}^{\tau}$, $\overline{p_2}^{\tau}$, ρ_{K0} and T_0 are base-flow mixture density, velocity, pressure (hydrodynamic), partial density of species K and temperature, respectively, and ρ_1, p_1, ρ_{K1} and T_1 are the corresponding acoustic quantities. *Re, Pe, Sc,* Da and Da_K, are the Reynolds, Peclet, Schmidt, Damköhler and Damköhler for species *K*, respectively. With these variables and non-dimensional quantities, Eqs. (1)–(5) can be identified as the NS equations for incompressible flow with temperature-dependent density governing the combustion zone, and Eqs. (6)–(10) as the linearized Euler equations governing the acoustic zone. Explicit coupling terms, viz., the flow divergence over acoustic length scale (RHS of Eq. (8)) and the acoustic Reynolds stress (ARS, RHS of Eq. (2)) are brought out naturally without any ad hoc treatment.

In this formulation, the acoustic damping does not show up. This is because, the length scale associated with the acoustic boundary layer where most of the damping occurs is not considered. Besides, bulk of the acoustic losses could occur at the boundaries, which are not governed by the above equations. Accounting for acoustic damping is important to predict the long-time behaviour of the system and eventually predict the limit cycle amplitude of the system. Towards this end, the above formulation is extended to include acoustic damping by considering a visco-thermal friction term (Chakravarthy et al. 2007) in the acoustic momentum balance equation Eq. (7); the resultant equation is given below,

$$\overline{\rho_0}^{x}\frac{\partial u'^{\tau}_{0,i}}{\partial \tau}+\frac{1}{\gamma}\frac{\partial p_1}{\partial \xi_i}=-2\alpha\overline{\rho_o}^{x}u'^{\tau}_{0,i}, \tag{11}$$

where α is the nondimensional Helmholtz–Kirchhoff wall-attenuation coefficient (Chakravarthy et al. 2007), non-dimensionalised by the duct length *L*. In combustion flow fields, acoustic attenuation is increased due to the presence of water vapour (Chakravarthy et al. 2007). This is approximately taken into account by increasing α up to 15 times the calculated value.

When the flow is turbulent, the range of flow scales span from the integral time/length to the Kolmogorov time/length scales. There are three possibilities that can arise in the position of the acoustic time scale relative to the turbulent spectrum. The acoustic time scale could be shorter than the Kolmogorov time scale, or it can be longer than the integral time scale, or it can lie in between these two scales. The first two possibilities are relatively straightforward to handle but rare in occurrence; the most common candidate is the last possibility, wherein the acoustic time scale falls in between the integral and Kolmogorov time scales.

With LES, the temporal average is taken over the acoustic time scale *t*, which implicitly imposes a filter width D_a; the spatial grid should be fine enough to resolve this. The temporal average of the convective term in the flow momentum equation results in the cross-correlation of the temporal fluctuations, which can further be decomposed spatially as

$$\overline{u_{0i}^{\prime\tau} u_{0j}^{\prime\tau}}^{\tau} = \overline{\overline{u_{0i}^{\prime\tau}}^{x} \overline{u_{0j}^{\prime\tau}}^{x}}^{\tau} + \overline{u_{0i}^{\prime\tau\prime x} u_{0j}^{\prime\tau\prime x}}^{\tau} + \overline{\overline{u_{0j}^{\prime\tau}}^{x} u_{0j}^{\prime\tau\prime x}}^{\tau} + \overline{u_{0i}^{\prime\tau\prime x} \overline{u_{0j}^{\prime\tau}}^{x}}^{\tau} \tag{12}$$

The first term is the ARS, but the other terms are sub-grid scale stresses that are to be modelled. The second term refers primarily to turbulent fluctuations, but the last two terms indicate turbulence-acoustic interaction (TAI) terms.

3 Solution Methodology

Simulations reported in this work are performed using modified versions of open source codes FASTEST (http://www.fnb.tu-darmstadt.de/forschung_fnb/software_fnb/software_fnb.en.jsp, 2002) and CLAWPACK (http://depts.washington.edu/clawpack/, 2013), which are employed to solve the base flow (Eqs. (1)–(5)) and acoustic sets of equations (Eqs. 8 and 11), respectively. Only the momentum and energy balance equations are solved for acoustic field, since its eigenvalues are only two ($+c$ and $-c$, where c is the speed of sound (Chakravarthy et al. 2007)).

In FASTEST, the spatial discretisation is second order accurate and the deferred correction procedure (Ferziger and Peric 2002) is followed to discretise the convective terms. Flux limiters that blend the first order upwind and the second order central differencing schemes depending upon the local distribution of a scalar are also employed. Temporal discretization is performed with second order accurate Crank–Nicolson technique. The matrix inversions are performed by the strongly implicit procedure (SIP) (Ferziger and Peric 2002).

In all the acoustic computations reported in the present work, the first order Godunov-split method is employed. This method in combination with high-resolution methods to compute wave propagation yields results that are very close to methods that employ the second order accurate Strang splitting (Leveque 2002).

Turbulence in the flow is handled by means of LES, and the sub-grid scale (SGS) terms (including the TAI terms) are simulated by adopting the monotonically implicit LES (MILES) approach (Chakravarthy et al. 2007). A single-step global chemical reaction is adopted with laminar finite-rate Arrhenius kinetics to compute the production rate of species.

In-house benchmarks are performed using FASTEST to compute flow through backward facing step (Armaly et al. 1983; Friedrich and Arnal 1990) and channel flows (Hussain and Reynolds 1975). Its capability to compute incompressible variable density flow is validated following (Nicoud 2000). CLAWPACK is tested to reproduce the duct natural frequencies in the presence of mean temperature gradient (Sujith et al. 1995).

The acoustic damping is the only aspect that is modelled in an ad hoc fashion, and together with this, the strength of the dilatation source term driving the acoustics is artificially tuned to calibrate for acoustic amplitudes against the experimental data.

In general, the domain, grid and time step size are different for the flow and acoustic codes, and are chosen in order to compute the underlying physics accurately (discussed next). Both the flow and acoustic computations are performed in MPI-based parallel environment. MPI-based specialised modules are added that can perform spatial/temporal averaging and interpolation of flow quantities from the flow grid to the acoustic grid and vice versa. Simple arithmetic average is followed to perform spatial average in view of the fineness of the grids used. Quadratic Shepard scheme is used to perform data interpolation in 3D (Renka 1988).

The flow computations and acoustic computations are performed using 16 and 8 parallel processors, respectively. Additional 6 and 4 processors are used to handle the forward and backward data transfer. A total of 34 processors spread across two nodes (24 processors per node) are used to perform a coupled simulation.

4 Results and Discussion

Flame-acoustic coupled simulations are performed for the laboratory combustor (Chakravarthy et al. 2007) (schematic as shown in Fig. 1) containing a backward facing step for flame stabilisation with fuel injection at the step corner. The length of the combustor is 1.4 m and the step plane is located 0.4 m from the inlet. The step height $h = 0.03$ m and the expansion ratio across the step plane is 2. The span wise extent of the duct is 0.06 m. Coupled 3D simulations are performed for *Re* in the range of 9,000–59,000, at a fixed fuel flow rate of 142 mg/s, for which experimental data on acoustic pressure measurements are available for direct comparison (Chakravarthy et al. 2007).

A reduced domain of 0.15 and 0.4 m (5 and 13 h) upstream and downstream of the step plane, respectively, is considered for the turbulent combustion computations. The range of scales expected in terms of non-dimensional wave number k for $Re = 18{,}000$ and 33,000 are $1.0\text{–}1.5 \times 10^3$ and $1.0\text{–}2.5 \times 10^3$, respectively. Multiblock structured mesh with a total size of 1.04 million nodes is used. Besides resolving the near-wall regions down to a wall y^+ of 2–3, the grid employed resolves scales up to $k = 15.2$ in the regions where the mesh is coarsest.

Random inlet velocity perturbations with amplitude of 20 % of the mean inlet velocity are superposed on the mean velocity and are prescribed at the inlet plane. No-slip boundary condition is used at the walls. Outflow boundary condition is used at the exit plane of the computational domain. At the exit plane, the velocity is corrected such that the instantaneous balance in volume flow rate is satisfied taking into account the dilatation that occurs in the computational domain. This improves the global convergence and also ensures the advection of vortical structures from the exit plane. The walls of the combustor are maintained at 700 K as the approximation of hot walls witnessed in the experiments.

The dynamic viscosity, thermal conductivity, species diffusivity, specific heat capacity and gas constant for all the species are kept constant and are taken as 1.84×10^{-5} Ns/m^2, 0.0425 W/mK, 2.67×10^{-5} m^2/s, 1,354 J/kg K and 296.9 J/kg K, respectively. The inlet temperature at both the air and fuel ducts is specified as 298 K corresponding to experiments. The thermodynamic pressure is taken as 101,325 Pa. A simplified reaction of the form F + 3O + 12N → 4P + 12N is considered. The molecular weight of all the species is taken equal as 28 kg/kmol. The reaction rate is taken to be of Arrhenius type and the pre-exponential factor and activation temperature are taken as $3.3 \times 10^{12}(\mathrm{gmol/cm})^{(1-\upsilon_1-\upsilon_2)}/\mathrm{s}$ and 2,300 K; υ_1 and υ_2 are the exponents in the rate equation for the fuel and oxidizer (air) species, respectively, and are taken as 0.15 and 1.15, respectively. The calorific value of the fuel is taken as 50 MJ/kg. Most of the parameters are chosen such that the process closely resembles methane-air combustion overall.

For the acoustic computations, the entire length of the duct as in the experiment is considered as the domain, and an uniform grid with a spatial resolution of ~8 mm is employed. An open-open boundary condition is prescribed at the inlet and exit planes of the duct; other surfaces are treated as walls. The flow properties and the acoustic source are spatially averaged in 3D over half the step height as the reference flow length scale and communicated to the acoustic solver. Upstream of the embedded domain, they are taken as constant corresponding to the experimental conditions at 298 K.

In the embedded region, they are communicated at the start of every flow time step. For the downstream part, the flow domain's exit plane average properties are computed and communicated to the acoustic solver at the start of every flow time step.

All the computations reported in this work are calibrated against experimental data (Chakravarthy et al. 2007) by introducing factors that modify the acoustic source and damping (rhs of Eqs. (8) and (11), respectively). These are kept as 0.3 and 15, respectively.

Figure 2a shows the evolution of pressure monitored at the top wall just downstream of the step plane for *Re* = 18,000 and 33,000. The pressure amplitude is observed to vary with time for *Re* = 18,000 at a time scale larger than the acoustic time scale, whereas for *Re* = 33,000, the time scale over which the pressure amplitude varies becomes close to the acoustic time scale, and therefore, no distinct long time scale variation in its amplitude is seen. Figure 2b shows the spectra of these time series. A shift in the dominant frequency of the coupled system is seen for *Re* = 33,000. This corresponds to lock-on of the duct acoustics to the flow, and is observed to occur in experiments (Chakravarthy et al. 2007) at the same *Re*.

Figure 3 shows the computational and experimental comparisons of Helmholtz number *He*, Strouhal number *St*, and pressure amplitude, corresponding to the dominant frequency, along the top wall downstream of the step. The shift from the constant *He* to a linearly increasing trend in the $33{,}000 < Re < 45{,}000$ range (Fig. 3a) correspond to the lock-on of the duct acoustic frequency to the vortex shedding mode (*St* = 0.2 (Chakravarthy et al. 2007)). The coupled computation using the multiple time and length scales approach is able to reproduce this behaviour. Good comparison is obtained for variation of *St* with *Re* as well (Fig. 3b). The combustion-acoustic lock-on is characterised by non-linear variation

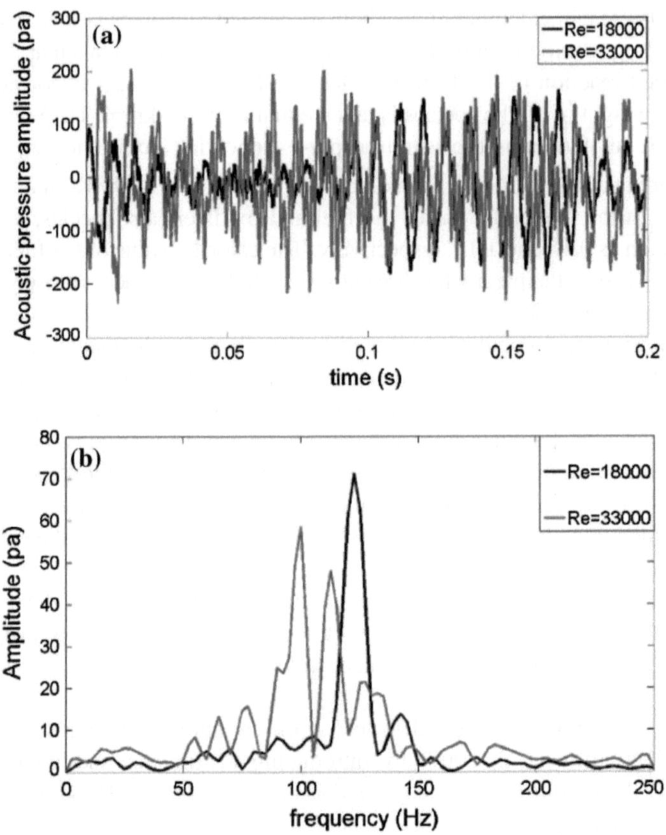

Fig. 2 **a** Evolution of pressure for Re = 18,000 and 33,000 and **b** its frequency spectra, monitored near the top wall just downstream of the step plane

in the pressure amplitude, a trend that the coupled computations are able to predict quite well (Fig. 3c). Exact match in the amplitudes is a matter of fine-tuning the calibration on damping, but the key features are the frequency shift and nonlinear amplitude rise corresponding to lock-on.

Combustion-acoustic lock-on is also characterised by evolution coherent large-scale vortical structures (Chakravarthy et al. 2007). Figure 4 compares the heat release rate contours obtained at mid-span plane for Re = 18,000 (no lock-on) and 33,000 (lock-on). The vortical structures rolled up due to mean shear in the velocity field for Re = 18,000 (Fig. 4a) are distinctly different from the evolution of large-scale vortex roll-up that is clearly evident for Re = 33,000 (Fig. 4b) with more contribution from the ARS.

Figure 5 compares the magnitude of the ARS field in the mid-span location for Re = 18,000 and 33,000. At Re = 18,000, the combustion predominantly occurs in the shear layer and in the vortical structures that are rolled up at far downstream

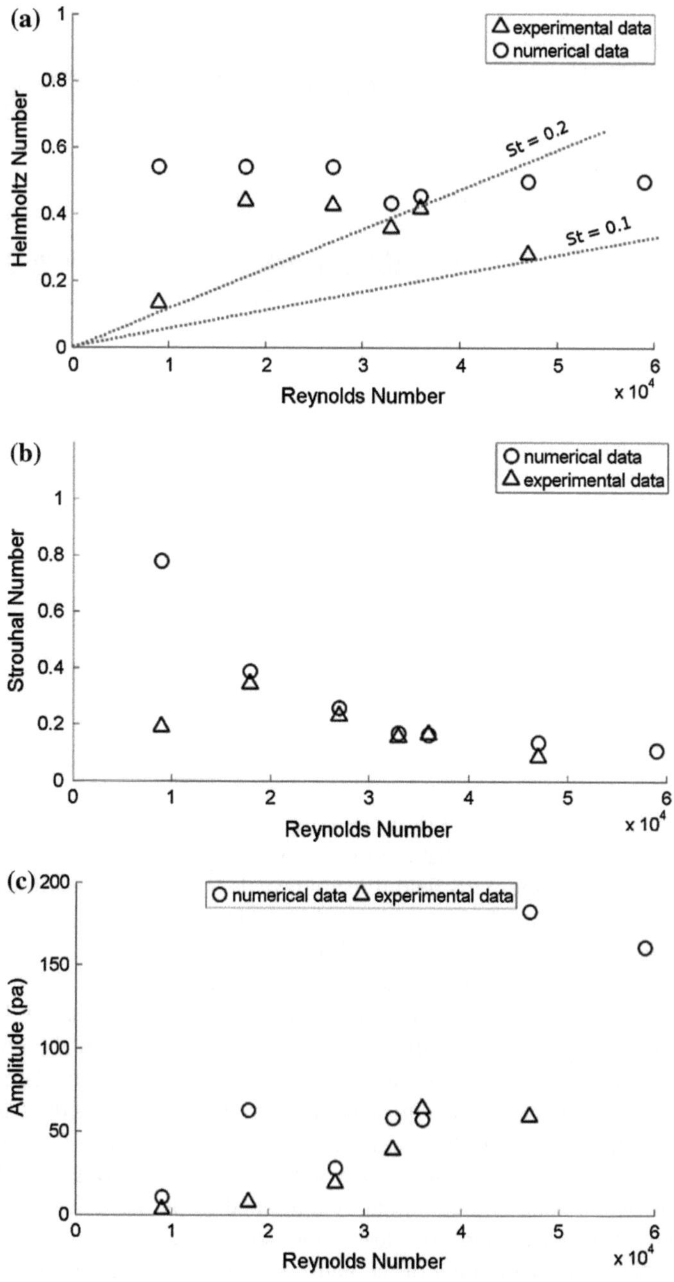

Fig. 3 Variation of Helmholtz number with *Re*, showing **a** lock-on and **b** the corresponding variation in Strouhal number, and **c** the non-linear variation in pressure amplitude with *Re*. Results from coupled computations are compared with experimental data

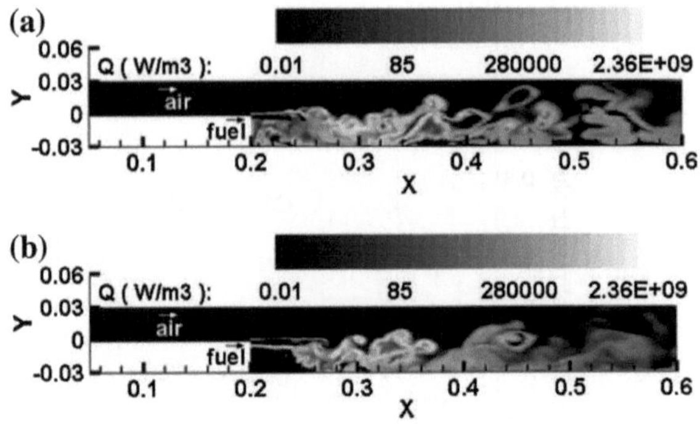

Fig. 4 Heat release rate contour in the mid-span plane for **a** *Re* = 18,000 and **b** *Re* = 33,000

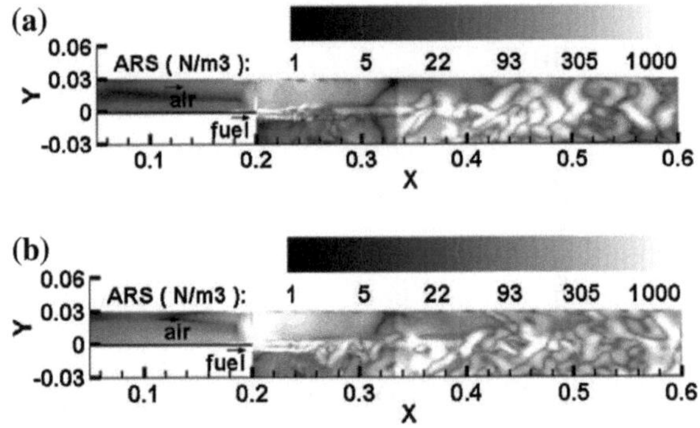

Fig. 5 Contours of ARS in mid-span plane for **a** *Re* = 18,000 and **b** *Re* = 33,000

(Fig. 5a), whereas for *Re* = 33,000, the combustion occurs predominantly in the large-scale vortical structures that are rolled up relatively near to the step (Fig. 5b). Close examination of the ARS field directly reveals that the regions of active ARS are adjoined by regions of active combustion. This is because the ARS promotes mixing, which aids in the combustion of the reactants.

Another interesting feature of a combustor in instability mode is that the flame length reduces dramatically when compared to its stable mode operation. In order to quantify this change from the computations, the flame length in the combusting field is computed as the axial distance over which 5–95 % of the total heat release rate occurs. Figure 6 shows the variation of the average flame length as a function of *Re*. Dramatic decrease in the flame length is observed as the *Re* is increased, moving the combustor from no lock-on conditions to lock-on mode. The shortening of the combustion zone is predominantly attributed to the large-scale mixing and subsequently efficient combustion due to the action ARS. In fact, the lock-on is made possible by the shortening of the flame, such that the characteristic flow time scale evolves to approach the acoustic time scale to enable resonance.

In order to further substantiate the role of the ARS in churning large-scale vortical structures, the flow components of velocity are monitored at $x/h = 4$ in the mid-span plane at the centerline of the duct for *Re* = 18,000 and 33,000. Figure 7

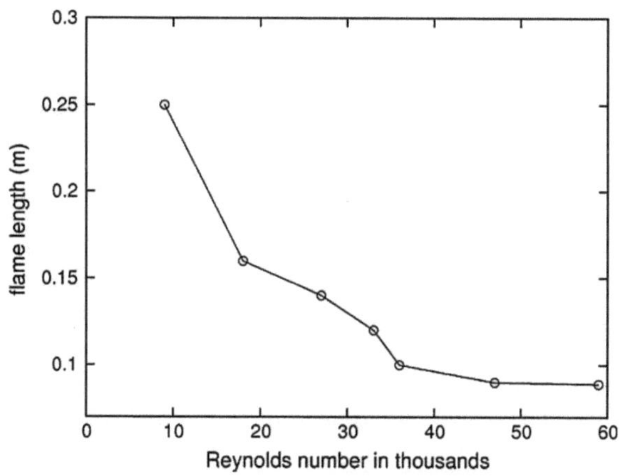

Fig. 6 Variation of average flame length with increase in Reynolds number

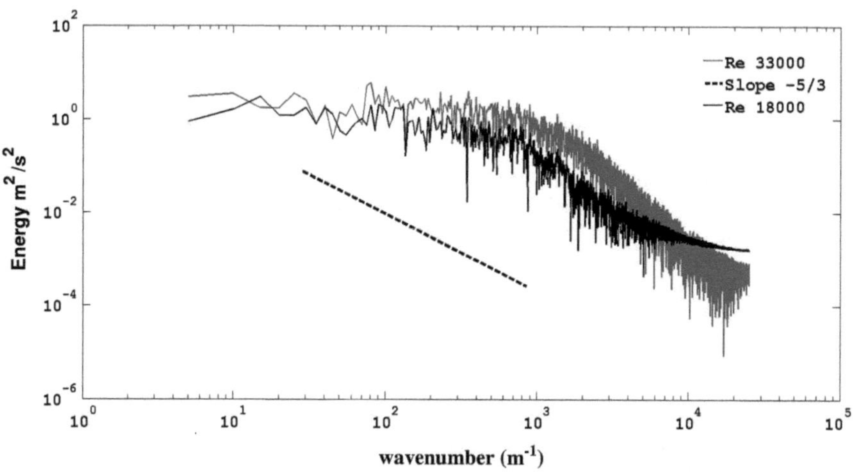

Fig. 7 Comparison of turbulent kinetic energy spectra at $x/h = 4$ for Re = 18,000 and 33,000

shows the turbulent kinetic energy (TKE) spectra for these cases. It is clearly seen that TKE is higher in large scales for Re = 33,000 when compared to 18000. Also, there is a sharp decrease in TKE in small scales for Re = 33,000 relative to 18,000. This reveals that the large scales are being formed at the expense of small scales at Re = 33,000, bringing out evidence for the ARS to churn these large-scale structures. The ARS reorganises the flow field into large-scale coherent structures and prevents their decay into small scales, as observed in experiments.

5 Conclusion

The present work adopts an approach of simultaneous multiple time and length scales of flow and acoustics to demonstrate that the time scales actually hasten to approach each other in a lock-on, leading to combustion instability, rather than presume equal time scales of the two processes à priori in an ad hoc manner. The results match with the experimental observations of shift in the dominant frequency of pressure oscillations that accompanies a rise in their amplitude, as *Re* is varied.

The formulation formally brings out the coupling terms between the flow and the acoustics. The dilatation due to heat release is the acoustic energy source, but more importantly, the acoustic Reynolds stress (ARS) is the source of the momentum to the base flow. The ARS churns the flow to produce large-scale coherent vortices at the expense of the small-scale structures as observed in experiments, which promotes fuel-air mixing and aids combustion, leading to intense heat release until the reactants are quickly consumed. Effectively, it causes high amplitudes of heat release rate fluctuations that drive the acoustics, in turn. The simulations are able to show the evolution of the flame into a compact source of heat release rate fluctuations, rather than necessitating such an assumption that most analysis adopt.

Besides being based on a rigorous framework, the present approach brings out rich physics of the problem, and is a computationally affordable alternative to predict combustion instability in turbulent reacting flows in ducted geometries.

Acknowledgments This work was partly supported by the Deutsche Forschungsgameinschaft. The National Centre for Combustion Research and Development is supported by the Department of Science and Technology, Government of India.

References

Armaly BF, Durst F, Pereira JCF, Schonung B (1983) Experimental and theoretical investigation of backward-facing step flow. J Fluid Mech 123:473–496

Balachandran R, Chakravarthy SR, Sujith RI (2008) Characterization of an acoustically self-excited combustor for spray evaporation. J Propul Power 24:1382–1389

Boudy F, Durox D, Schuller T, Candel SM (2011) Nonlinear mode triggering in a multiple flame combustor. Proc Combust Inst 33:1121–1128

Brookes SJ, Cant RS, Dupere IDJ, Dowling AP (2001) Computational modeling of self-excited combustion instabilities. J Eng Gas Turbines Power 123:322–326

Chakravarthy SR, Shreenivasan OJ, Böhm B, Dreizler A, Janicka J (2007) Experimental characterization of onset of acoustic instability in a nonpremixed half-dump combustor. J Acou Soc Am 122:120–127

Culick FEC (1988) Combustion instabilities in liquid-fuelled propulsion system–an overview. AGARD-CP-450

Ferziger JH, Peric M (2002) Computational methods for fluid dynamics. Springer, Berlin

Friedrich R, Arnal M (1990) Analysing turbulent backward-facing step flow with the lowpass-filtered Navier-Stokes equations. J Wind Eng Ind Aerodyn 35:101–128

Huang Y, Yang V (2004) Bifurcation of flame structure in a lean-premixed swirl-stabilized combustor: transition from stable to unstable flame. Combust Flame 136:383–389

Hussain AKMF, Reynolds WC (1975) Measurements in fully developed turbulent channel flow. J Fluid Eng 97:568–580
Leveque RJ (2002) Finite volume methods for hyperbolic problems. Cambridge university press, Cambridge
Lieuwen TC, Zinn BT (1998) The role of equivalence ratio oscillations in driving combustion instabilities in low NO_x gas turbines. Proc Combust Inst 27:1809–1816
Lighthill J (1978) Acoustic streaming. J Sound Vib 61:391–418
Matveev KI, Culick FEC (2003) A model for combustion instability involving vortex shedding. Combust Sci Technol 175:1059–1083
Menon S, Jou WH (1991) Large-eddy simulations of combustion instability in an axisymmetric ramjet combustor. Combust Sci Technol 75:53–72
Nicoud F (2000) Conservative high-order finite-difference schemes for low-Mach number flows. J Comput Phys 159:71–97
Nyborg WL (1965) Acoustic Streaming, Physical Acoustics Part B (ed. WP. Mason, II), Academic Press
Poinsot TJ, Candel SM (1988) A nonlinear model for ducted flame combustion instabilities. Combust Sci Technol 61:121–153
Renka RJ (1988) Multivariate interpolation of large sets of scattered data. ACM Trans Math Softw 14:151–152
Roux S, Lartigue G, Poinsot TJ, Meier U, Bérat C (2005) Studies of mean and unsteady flow in a swirled combustor using experiments, acoustic analysis, and large eddy simulations. Combust Flame 141:40–54
Schadow SC, Gutmark EJ (1992) Combustion instability related to vortex shedding in dump combustors and their passive control. Prog Energy Combust Sci 18:117–132
Searby G (1992) Acoustic instability in premixed flames. Combust Sci Technol 81:221–231
Selle L, Benoit L, Poinsot T, Nicoud F, Krebs W (2006) Joint use of compressible large-eddy simulation and Helmholtz solvers for the analysis of rotating modes in an industrial swirled burner. Combust Flame 145:194–205
Sterling JD (1993) Nonlinear analysis and modelling of combustion instabilities in a laboratory combustor. Combust Sci Technol 89:167–179
Sujith RI, Waldherr GA, Zinn BT (1995) An exact solution for one-dimensional acoustic fields in ducts with an axial temperature gradient. J Sound Vib 184(3):389–402
Tanabe M, Morita T, Aoki K, Satoh K, Fujimori T, Sato J (2000) Influence of standing sound waves on droplet combustion. Proc Combust Inst 28:1007–1013
Tyagi M, Chakravarthy SR, Sujith RI (2007) Unsteady combustion response of a ducted non-premixed flame and acoustic coupling. Combust Theory Model 11:205–226
Wu X, Wang M, Moin P, Peters N (2003) Combustion instability due to the nonlinear interaction between sound and flame. J Fluid Mech 497:22–53

Developing Surrogates for Liquid Transportation Fuels: The Role of Spherically Symmetric Droplet Combustion

C. Thomas Avedisian

Abstract The finite supply of transportation fuels has generated renewed interest to improve their performance in power and propulsion devices. However, the complexity of real fuels prohibits developing their combustion chemistries and property databases needed for simulating performance in engines to identify operational regimes for improving fuel efficiency. Surrogates offer the means to address these concerns if they can be shown to replicate certain combustion targets of the real fuel, and to result in combustion properties similar to real fuels when burned in a suitable configuration that is amenable to detailed numerical modeling. This paper examines the role which droplet combustion can play in the development of surrogates for complex transportation fuels. Recognizing that spray combustion is far too difficult to model and that droplets represent the fine-grid structure of sprays, the combustion dynamics of fuel droplets are examined in an environment that seeks to remove external convective influences to simplify the transport field and produce spherical symmetry in the droplet burning process that can be modeled using a detailed numerical simulation approach. The one-dimensional flames and transport dynamics that result are shown to be well positioned to evaluate the efficacy of surrogate fuel performance. Recent efforts are summarized that have used the spherical droplet flame configuration to evaluate the performance of surrogate fuel blends. Experiments are discussed for promoting spherical symmetry, and results are presented to show the efficacy of some surrogate blends to replicate the performance of gasoline and jet fuels using the spherical droplet configuration. Some results are also included from detailed numerical modeling of biodiesel droplets that incorporate complex combustion chemistry, and unsteady transport, vaporization, and variable property effects that illustrate the potential for high fidelity predictions needed for developing surrogates using the spherically symmetric droplet flame configuration.

This paper is based on a presentation given at the Indo-US International Workshop on Novel Combustion for Sustainable Energy Development, Kanpur, India, January 2–4, 2014.

C.T. Avedisian (✉)
Sibley School of Mechanical and Aerospace Engineering, Cornell University,
Ithaca, NY 14853-7501, USA
e-mail: cta2@cornell.edu

A.K. Agarwal et al. (eds.), *Novel Combustion Concepts for Sustainable Energy Development*, DOI 10.1007/978-81-322-2211-8_16

Keywords In-cylinder combustion · Surrogate fuels · Low-dimensional combustion · Fuel spray · Multiphase processes

1 Introduction

In 2013, almost 90 million barrels of liquid fuels were consumed in power and propulsion devices worldwide. The United States was the largest consumer at 20 % (USEIA 2014). The transportation sector accounted for approximately 70 % of this consumption. The emissions produced, and the fact that much of the oil used is imported, has motivated investments in alternative propulsion and energy technologies (e.g., electric, wind, solar, geothermal) that would also reduce reliance on foreign supplies. However, these alternatives are not yet at the stage to significantly impact petroleum use or fully meet the transportation and energy needs of modern societies (NSF 2011). And none of them currently matches the cost and energy density of liquid fuels. Biofuels derived from nonfood feedstocks have generated significant interest to complement, and eventually replace, petroleum-based fuels. There are, however, issues that need to be resolved for them to realize their full potential as an alternative to petroleum fuels (Dirks et al. 2012). Nonetheless, the need for liquid fuels is expected to remain even as the supply of petroleum-based fuels is limited (NSF 2011).

Internal combustion and aircraft engines are the primary consumers of liquid fuels. A further stimulus for improving engine efficiency comes from new regulations (i.e., in the U.S.) for the ground transportation sector that mandate fuel economy levels of 54.5 miles per gallon by 2025 for new model cars as shown in Fig. 1 (FR 2012). As such, development of technologies that improve the efficiency

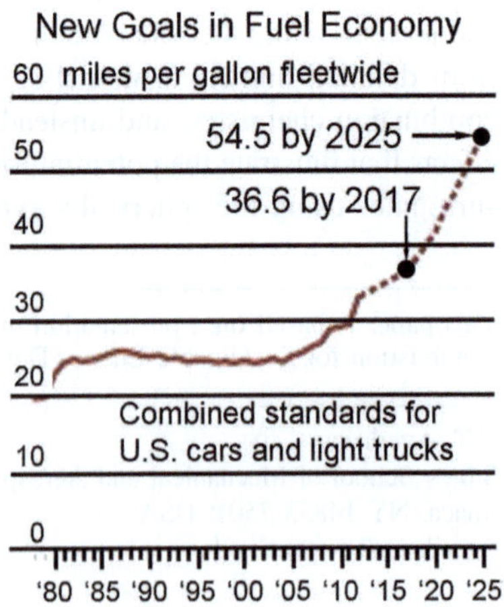

Fig. 1 New fuel economy targets for 2025 (FR 2012)

of combustion engines stands to have an immediate impact on petroleum fuel supplies and their clean combustion.

Considering the design process for bringing new engine technologies to the marketplace, experimental testing and prototype evaluation have been the standard approach of the transportation sector. The emergence of powerful computational capabilities is beginning to change this paradigm. For example, the Cummins ISB 6.7 l diesel was developed entirely by computer simulation (PRECISE 2011), with the only testing coming from performance evaluation after prototype fabrication and operation. The new engine also achieved mileage targets and emission constraints of the design.

The prediction of in-cylinder processes is a challenging task owing to the difficulty of incorporating all of the relevant physics for engine processes in a first-principles solver. The presence of liquid droplets associated with spray injection and the combustion chemistry that provides the chemical pathways for oxidation of the fuel are particularly problematic. Sub-model inputs and approximations are typically needed to enable predictions of performance. For example, the often used KIVA code (Amsden 1999; Lee and Reitz 2013) requires sub-models for soot and the real fuel combustion chemistry, the phase equilibrium (distillation) characteristics of the fuel, adjustable parameters to make the spray liquid and vapor penetrations match experimental data and values of the crank angle where 10 % of the fuel is burned (Wang et al. 2013), thermofluid properties, turbulence, droplet collision dynamics, and evaporation for multicomponent droplets.

The context of the challenges involved with predicting the performance of engines powered by real transportation fuels is illustrated by considering their compositions. Real petroleum-based fuels—gasoline, jet, diesel—consist of hundreds of miscible constituents with a wide range of volatilities, sooting propensities, and thermo-physical properties. Table 1 lists broad chemical classes present in a typical jet fuel (i.e., Jet A is basically JP8 minus additives for lubricity, de-icing, and antistatic effects (Edwards and Maurice 2001), and Table 2 lists the major components for diesel.

Figure 2 illustrates the mole percents of the broad chemical classes' hydrocarbons in a typical gasoline blend (Tsang 2003). The highly multicomponent nature of real fuels is further complicated by the variability of manufacturing, with

Table 1 Constituents of a typical Jet A fuel (Colket et al. 2007)

Component	(v %)
Paraffins	55.2
Monoclycloparaffins	17.2
Dicycloparaffins	7.8
Tricycloparaffins	0.6
Alkyl benzenes	12.7
Indans + tetralins	4.9
Naphthalene	<0.2
Substituted naphthalenes	1.3

Table 2 Typical diesel fuel components (Farrell et al. 2007)

Component	(v %)
Normal and isoparaffins	25–50
Cyclo-paraffins	20–40
Aromatics	15–40

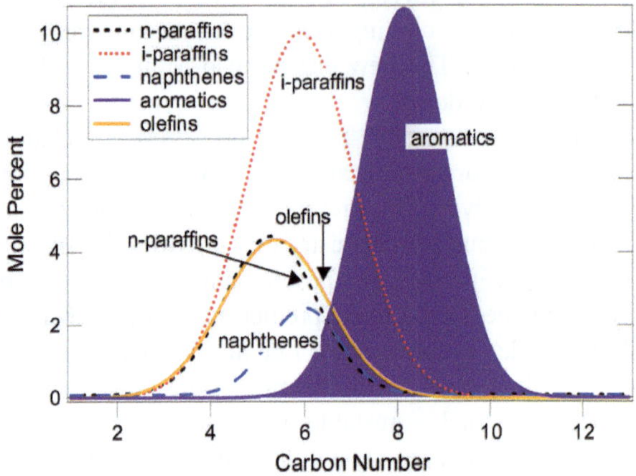

Fig. 2 Typical gasoline composition showing concentration by chemical classes (adapted from Tsang 2003)

compositions that can vary depending on the nature of the crude, time of year, and the refinery (Farrell et al. 2007). The result is that fuels produced from different production runs or refineries may not have precisely the same properties.

The multicomponent characteristic of real fuels makes it a significant challenge to develop the required inputs for simulation, including predicting the phase equilibrium behavior, properties, and combustion kinetics as noted above. The problem is further exacerbated by the new generation of biofuels about which much effort has been devoted to production but little to evaluating combustion performance.

The only viable approach to simulating combustion performance of real fuels in engines is to develop a "surrogate" for them. A surrogate is defined as a blend of relatively fewer compounds of known chemical species and mixture fractions which are selected to match certain thermochemical aspects ("targets")[1] of the real fuel (Dooley et al. 2010, 2012; Liu and Avedisian 2012; Zhang et al. 2007; Mueller et al. 2012; Narayanaswamy et al. 2013; Pitz and Mueller 2011). With appropriate blending and mixing strategies, the surrogate may replicate most (but often not all) of the combustion targets around which it is developed. It is advantageous that the number of constituents of the surrogate blend should be as low as possible in order to make tractable the prediction of properties and phase equilibrium behavior, as well as to develop combustion chemistry. The simplest surrogate would be a single

[1] A combustion "target" is defined here as a variable used for matching a surrogate with a real fuel (e.g., molecular weight, derived cetane number, liquid density). A combustion "property" is defined as a variable measured in a combustion configuration.

component hydrocarbon (e.g., n-heptane = #2 diesel fuel (Wang et al. 2013) or combining the kinetic mechanism of methyl butanoate with heptane for a soybean biodiesel (Lee et al. 2013).

A step-up from a single component hydrocarbon is a binary blend which have been used as surrogates (i.e., a n-dodecane/iso-octane = a multicomponent paraffinic jet fuel in a specific mixture (mole) fraction of [i.e., 0.519/0.481 (Dooley et al. 2012)]). Three and four component blends have been successfully developed for some transportation fuels (Dooley et al. 2010, 2012). Higher-order surrogates have been formulated with eight components to represent a certification gasoline and reference diesel fuel (Mueller et al. 2012). It is clearly advantageous to have a surrogate with as few components as possible to enable predictions of physical and thermodynamic properties with minimal effort.

This paper reviews the development of surrogates from the perspective of liquid fuels and the role that droplet combustion dynamics can play in this process. Experimental results are presented that show the efficacy of a simplified droplet burning configuration for evaluating the performance of gasoline and jet fuel. Some results related to detailed numerical simulation of droplet combustion of biofuel surrogates are also included.

2 The Canonical Configuration for Liquid Fuel Combustion

Developing a surrogate involves several tasks: (1) determining the broad chemical classes of the real fuel; (2) selecting the surrogate components; (3) calculating the fractional amounts of the surrogate components through a process that minimizes the difference between selected combustion targets of the surrogate and real fuel (e.g., by various optimization methods (Narayanaswamy et al. 2013; Dooley et al. 2010, 2012)); (4) comparing combustion properties of the surrogate to the real fuel as measured in a suitable combustion configuration; and (5) assessing the efficacy of the surrogate to produce high fidelity predictions of combustion properties measured in a suitable combustion configuration.

Considering combustion properties pertinent to liquid fuels, at one extreme is the stochastic environment of a piston engine (Fig. 3), with its swirling and turbulent motion, spray dynamics, and radiation.

Ab initio models for engines do not exist. The KIVA code (Amsden 1999) is one of the more advanced solvers of the transport equations for in-cylinder combustion dynamics that includes models for spray injection. Nonetheless, detailed numerical modeling of spray combustion is also beyond current capabilities (i.e., when including, simultaneously, detailed combustion kinetics, radiative and unsteady effects, droplet formation, evaporation). This leaves processes at the level of individual droplets to consider. Here too, there are significant challenges.

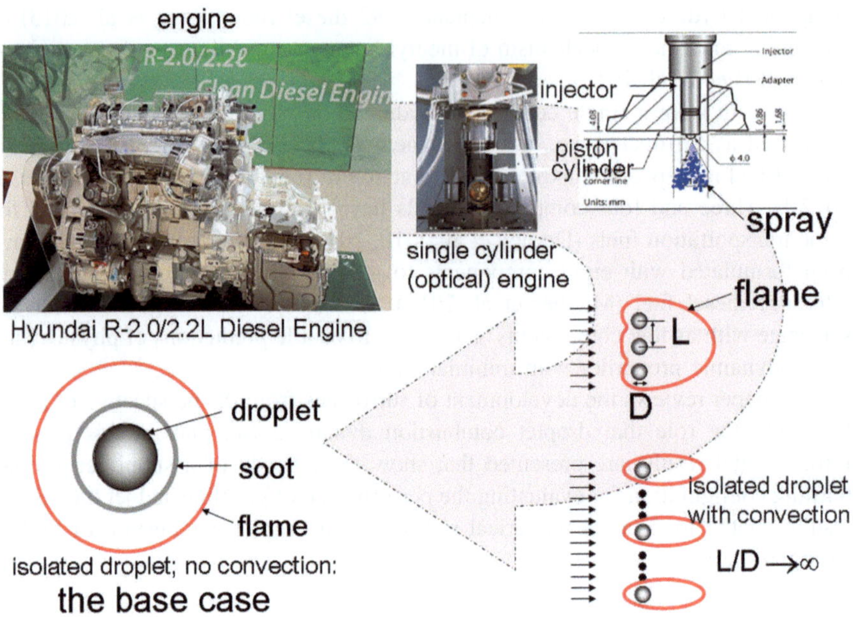

Fig. 3 Context of the spherically symmetric droplet burning configuration to the stochastic environment of an engine

A hierarchy for combustion of a liquid fuel is envisioned as depicted in Fig. 3 that includes a full spray, 3-D multi-droplet cloud, and 2-D array with inter-droplet spacing "L". For finite L, none of these configurations is yet amenable to ab initio modeling that includes simultaneously the effects of time-dependent gas and liquid transport, moving boundary of the liquid/gas interface as evaporation occurs, variable properties, detailed combustion chemistry, and radiation. Even if L is arbitrarily "large" such that a single isolated drop is realized but convective effects are still present, detailed numerical simulation has not yet been accomplished, though approaches have included some (but not all) of these elements (Sirignano 1999).

Currently, it appears that only for the single stationary isolated droplet ($L \rightarrow \infty$ in Fig. 3), burning in a quiescent ambience with no convective effects (forced or buoyancy-induced) can detail numerical simulation based on a first-principles solution be carried out that incorporates all the physical processes mentioned above, except for soot formation (Farouk et al. 2013, Liu et al. 2013a; Cuoci et al. 2005). Figure 4a is a schematic of this configuration.

The gas flow will be one-dimensional and only in the radial direction, and the flow is created entirely by the evaporation process. If soot forms [Fig. 4b, c (Liu et al. 2014a)], the soot aggregates will be trapped between the droplet and flame by the forces acting on the particles (Avedisian 1997). The configuration of Fig. 4 represents the "base case" of liquid fuel combustion as a building block for complex droplet configurations that lead to spray dynamics.

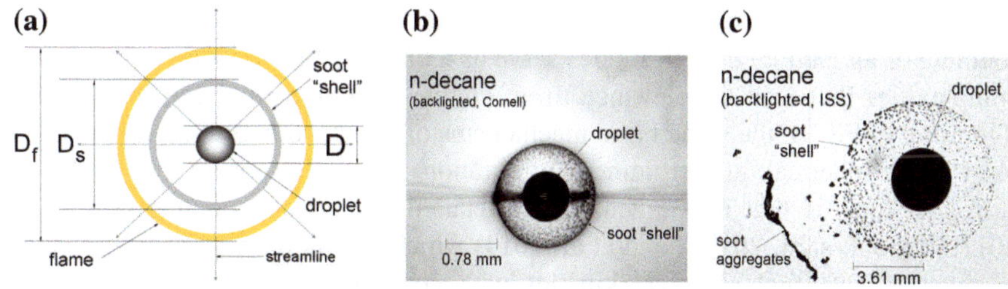

Fig. 4 **a** Schematic representation of an isolated droplet burning with spherical symmetry. **b** Backlighted photograph of a "small" fiber-supported n-decane droplet burning in the standard atmosphere at normal atmospheric pressure and showing the soot shell; **c** "large" free or unsupported n-decane droplet burning in the MDCA of the International Space Station in the standard atmosphere

The attractiveness of a low-dimensional transport configuration for developing surrogates has been recognized. The principle advantage is that detailed numerical modeling can be applied, thereby allowing a rigorous test of the combustion kinetics. For example, configurations have included shock tubes and rapid compression machines (Gauthier et al. 2004; Yahyaoui et al. 2007; Werler et al. 2014), counter flow flames and jet-stirred reactors (Choi et al. 2011; Dagaut et al. 2008), and premixed and non-premixed flow reactors (Chaos et al. 2007; Bieleveld et al. 2009), among others. They require that the fuel be pre-vaporized before entering the combustion zone. Pre-vaporization eliminates the intimate relationship between fuel vaporization and multicomponent effects for sustaining the flame that is intrinsic to the combustion of sprays and droplets in engines.

With liquid fuel injection comes evaporation and moving boundary effects; the configuration of Fig. 4 incorporates these processes. Phase equilibrium of the fuel determines the evaporation rate of liquid in a spray; and the temperature of an isolated droplet is also controlled by this process. Unsteady gas and liquid transport, soot formation, and radiation are intrinsic to in-cylinder processes in which the configuration of Fig. 4 also includes. The isolated droplet case incorporates these aspects as a natural consequence of the burning process. There is also carryover of such aspects as soot morphology and combustion chemistry of the base case of Fig. 4 to the more complex conditions of an engine. As such, the combustion symmetry of an isolated droplet burning with no convective effects is considered to be an appropriate platform for developing combustion properties for developing surrogate fuels.

It is also worth-mentioning that developing surrogates from the perspective of individual droplets is attractive for the very small liquid volumes needed for droplet combustion experimentation. In spray or engine tests, tens of gallons are typically required to evaluate performance. New fuel blends can be in short supply (e.g., algal biodiesel) with the availability of a sufficient fuel supply limiting the range of conditions that can be evaluated in experiments. On the other hand, droplet combustion experiments utilize very small volumes, on the order of 10^{-8} gallons per burning event, and will not be restricted by this concern.

The configuration of Fig. 4 is obviously not the same as the stochastic environment of an engine, though it does serve as a building block for a progression of complexities that lead to relevance to an engine environment. Even for the configuration of Fig. 4, the same basic mechanisms of combustion present in an engine exist in the spherical droplet flame configuration. Its greatest virtue for developing surrogate fuels is that it is amenable to detailed numerical modeling (Liu et al. 2013a; Farouk et al. 2013; Fahd et al. 2014) which allows for a more rigorous test of the surrogate fuel chemical kinetics with combustion properties that include vaporization and liquid-phase multicomponent effects. For these reasons, the isolated droplet burning geometry can be viewed as a canonical configuration for liquid fuel combustion.

3 Developing a Surrogate for Liquid Transportation Fuels

Surrogate components are selected with reference to broad chemical classes of the real fuel (Dooley et al. 2010, 2012; Huber et al. 2010; Mueller et al. 2012; Narayanaswamy et al. 2013). Once identified, their fractional amounts need to be determined. A simple approach involves sweeping through compositions to identify ones where the real and surrogate fuel targets match. This primitive approach looses its appeal for more than one target or binary system (Avedisian 2008).

For a general multicomponent surrogate, constrained optimization or regression techniques are used to identify the fractional amounts that reduce the difference between the real and surrogate fuel targets (Dooley et al. 2010, 2012; Narayanaswamy et al. 2013; Mueller et al. 2012; Huber et al. 2010). There is no universal set of targets for developing a surrogate. They are linked to the application (Pitz et al. 2007). Targets that have been considered for transportation fuels include various combinations of the hydrogen-to-carbon (H/C) ratio, derived cetane number (DCN), threshold sooting index (TSI), average molecular weight (MW_{ave}), liquid density, and advanced distillation curve to characterize the phase behavior of the fuel. Typically, four targets are involved in the matching process.

In the next sections, we discuss the performance of several surrogates for liquid transportation fuels from the perspective of the spherical droplet flame configuration. We begin with a brief discussion of the experimental designs for creating near-spherically symmetric droplet burning conditions. We then present some experimental results and include discussion of the performance of a surrogate developed using the optimization process mentioned above. Some results are also discussed for detailed numerical modeling of the spherical droplet flame configuration for a biodiesel surrogate.

4 Creating Spherically Symmetric Droplet Burning Conditions

The computational simplicity of the one-dimensional droplet flame in Fig. 4 belies the experimental difficulty of creating this configuration; it is arguably the most difficult combustion configuration to create in all of combustion science. To create spherically symmetric droplet burning conditions, an environment is required where the dynamic parameters that control convection are "small": The Rayleigh number $\left(Ra = \frac{g\beta\Delta TD^3}{\alpha\nu}\right)$ and Reynolds number $\left(\mathrm{Re} = \frac{UD}{\nu}\right)$ are most relevant, where g, β, ΔT, D, α U$_\infty$, and ν are gravity, thermal expansion coefficient, characteristic temperature difference (e.g., flame and droplet temperatures), droplet diameter, gas thermal diffusivity, relative velocity between the droplet and surrounding gas, and kinematic viscosity, respectively (D_f may also be used as the characteristic dimension though it is related to the droplet diameter). It is seen that there are a number of ways that convective influences on the burning process can be reduced. The gravitational level is typically reduced to promote a small Ra, while various experimental configurations are designed to form, deploy, and ignite the droplet with minimal disturbances and residual droplet motion to make Re suitable small.

Reduced gravity is created by performing the experiments in a free fall facility, either on Earth over a distance that determines the experimental time [i.e., a drop tower (Avedisian et al. 1988)] or in an orbiting craft [e.g., the International Space Station (ISS) (Dietrich et al. 2014)] which is essentially in free fall as it orbits the Earth.

The experimental design must form, deploy, and ignite the test droplet without any significant accompanying motion to these operations. The droplet can either be physically restricted by some sort of a support fiber attached to the droplet (Avedisian and Jackson 2000) or free of any support structures that do not impart any residual motion to the droplet during the deployment process (Avedisian et al. 1988; Okajima and Kumagai 1975; Hara and Kumagai 1990; Liu et al. 2014b; Avedisian and Callahan 2000; Jackson et al. 1991).

The experimental designs are broadly tailored to "small" droplets corresponding to initial diameters (D_o) of D_o < 1 mm and "large" droplets corresponding to D_o > 1.5 mm. Small droplet experiments are most suitable for experimental times on the order of several seconds (i.e., as in drop towers). Large droplet experiments will typically require several tens of seconds to observe the complete droplet burning history and are best examined in the environment of the ISS where the experimental time is unlimited.

For small droplets, radiative effects are not important. For large droplets, radiation can be important to promote extinction with the possibility of "cool flames" (Nayagam et al. 2012; Farouk and Dryer 2013; Dietrich et al. 2014) that are characterized by a low-temperature combustion regime of burning. By varying D_o from about 0.5 mm up to about 6 mm, the complete range of physical phenomenon a liquid fuel could experience in a combustion environment can be accessed.

A small droplet experimental configuration is schematically shown in Fig. 5. It involves deploying a test droplet at the vertex of two very small diameter fibers crossed at an angle, essentially by propelling the test droplet onto the vertex with a droplet generator (i.e., it is very difficult to deploy a droplet on the order of 0.5 mm onto a 14-μm-diameter support structure by other methods). At that point, and after the period of free fall begins, the droplet is then ignited by two sparks positioned on opposite sides of the droplet to provide some degree of symmetry to the ignition process. Digital video cameras record the burning history.

An unsupported droplet can be created using this same basic approach by essentially removing the crossed fibers but still retaining the droplet trajectory. At the apex of the trajectory, cameras positioned there, along with the sealed gas in which the deployment process is carried out, are then literally released into free fall (Okajima and Kumagai 1975; Avedisian et al. 1988). The downward motion of the droplet falling at the same rate as the imaging system then creates the illusion of a levitated droplet. Though creating unsupported droplets is very difficult, if successful images like those shown in Fig. 4b and c will result.

For large droplets, the multiuser droplet combustion apparatus (MDCA) is designed to form and deploy unsupported or free-floating droplets using the approach schematically illustrated in Fig. 6 (Banu 2008; Robbins and Shinn 2010, and Dietrich et al. 2014).

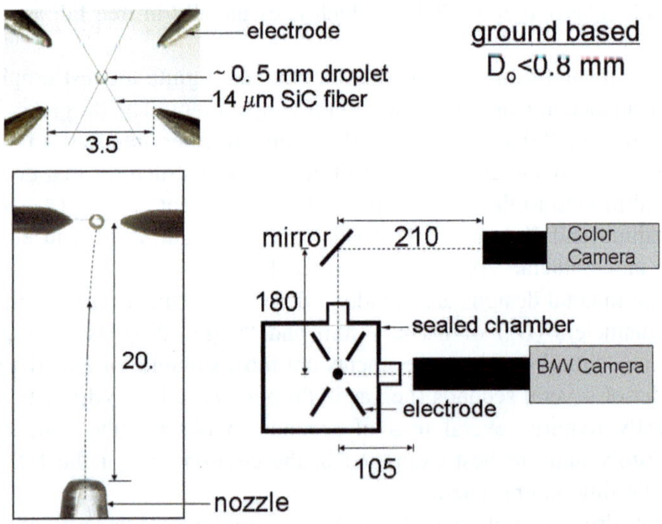

Fig. 5 Droplet deployment design for ground-based experiments (Liu et al. 2012). Numbers are in millimeters. Camera arrangement for recording the droplet burning history is shown. To create an unsupported droplet, the 14-μm fiber is removed and the sealed chamber is released into free fall when the droplet reaches the apex of its approximately 20 mm deployment trajectory

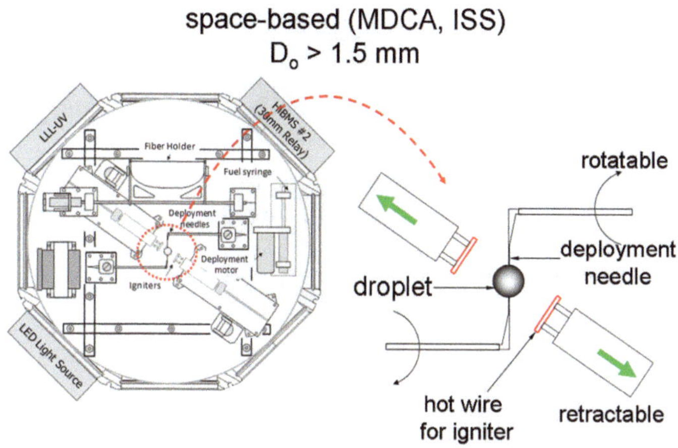

Fig. 6 Droplet deployment design for the multi-user droplet combustion facility (MDCA) (Dietrich et al. 2014; Robbins and Shinn 2010; Banu 2008). Droplet deployment is in microgravity by rapid retraction of two deployment needles to separate the droplet from the needles

Test droplets are ignited by hot wires on support tubes that are linearly retracted to provide an unobstructed ambience for burning. Hardware on the MDCA is adjusted through communication links from the ISS to a control center at NASA Glenn and through video and audio feeds to the author's laboratory at Cornell (Fig. 7). The MDCA also has the capability for supporting a test droplet onto an 80-μm-diameter fiber.

For a fiber-supported droplet, there is always a concern that the fiber will influence the droplet burning history. The effects involved can be distortion of the droplet shape (e.g., by surface tension effects), an influence on heat transfer to the droplet through the fiber, and/or inducing an asymmetric flow field in the surrounding gas. It was shown (Liu et al. 2014b) that for the design of Fig. 5 which incorporates a fiber of approximately 14 μm diameter and $D_o < 0.6$ mm, the influence of a fiber on burning is minimal and the droplet behaves as essentially as if it were free-floating when burning in the standard atmosphere, which is the environment in which this particular method has been used. The influence of a fiber on the gas phase flow field is revealed by using soot aggregates formed during the burning process as essentially particles that seed the flow. For droplets supported by a single fiber, rather unusual flow patterns can occur that are significantly different from the idealization of Fig. 4b and c (Avedisian and Jackson 2000).

For large droplets, Fig. 8 shows the gas motion that can be induced by the 80-μm fiber employed on the MDCA design (Liu et al. 2014b).

Interestingly, it was found that the onset of the motions shown in Fig. 8 coincided with a significant increase in the droplet burning rate as shown in Fig. 9. Nonetheless, regardless of the droplet and fiber diameters for any supported droplet design, an influence of the fiber will always be expected as the droplet evaporates because the droplet diameter decreases while the fiber diameter is fixed.

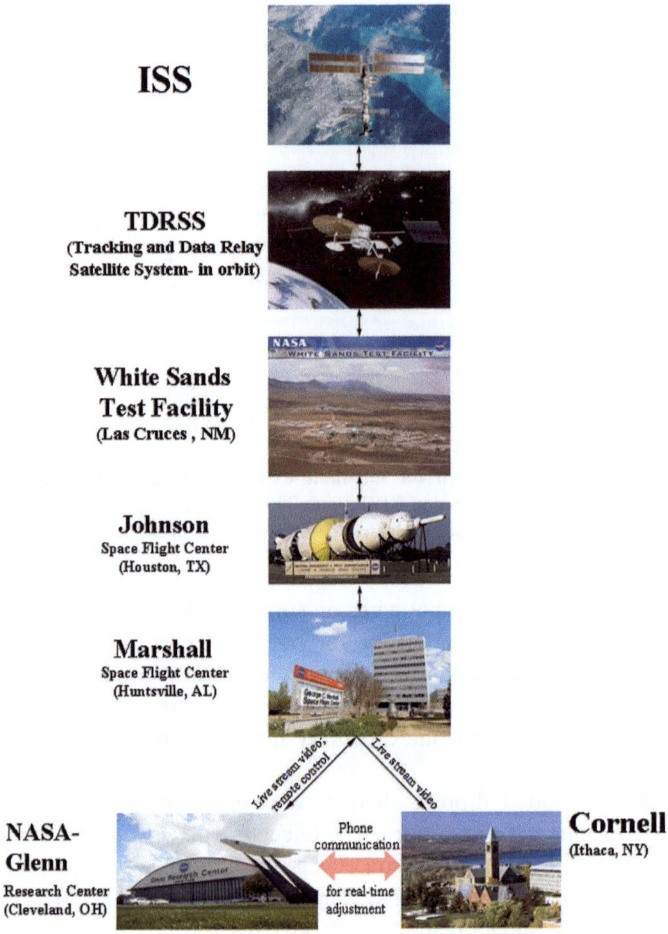

Fig. 7 Communication network for controlling operations on the MDCA. Hardware controls originate from NASA Glenn

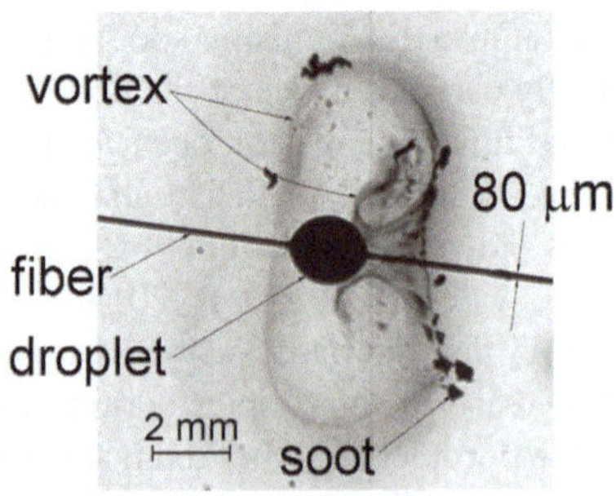

Fig. 8 Photograph of a fiber-supported n-decane droplet an n-decane droplet with D_o = 1.82 mm at $t/D_o^2 = 0.4$ s/mm^2 after ignition. Soot aggregates reveal a vortex-like motion adjacent to the fiber (Liu et al. 2014a)

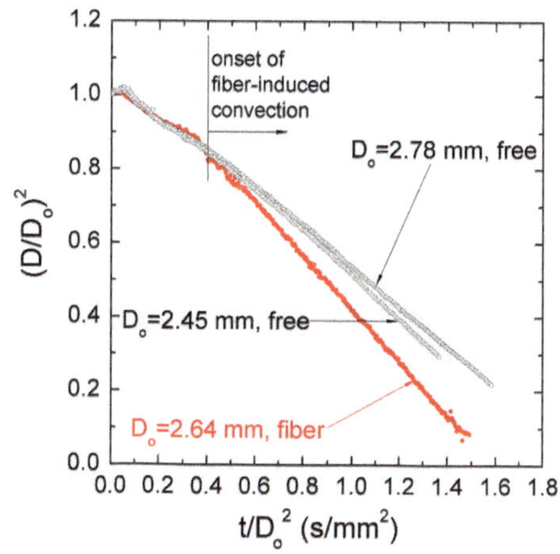

Fig. 9 Evolution of "large" droplet diameters of the indicated initial values showing the effect of an 80-μm support fiber using the experimental design of Fig. 6 on the ISS (Liu et al. 2014b). The ratio of droplet diameter of the fiber is large enough for no influence of a fiber to have been anticipated. The clear deviation of the burning rate of the fiber-supported droplet from the free droplet is attributed to the onset of convection and the *vortical motions* shown in Fig. 8

5 Visualization of the Droplet Burning Process

The main diagnostic for the droplet burning experiments discussed here is photographic. Figure 10 shows the burning history of a backlighted unsupported n-heptane droplet that illustrates development of the soot shell for D_o = 0.76 mm (Jackson and Avedisian 1998). The flame is not visible in this sequence as is typical

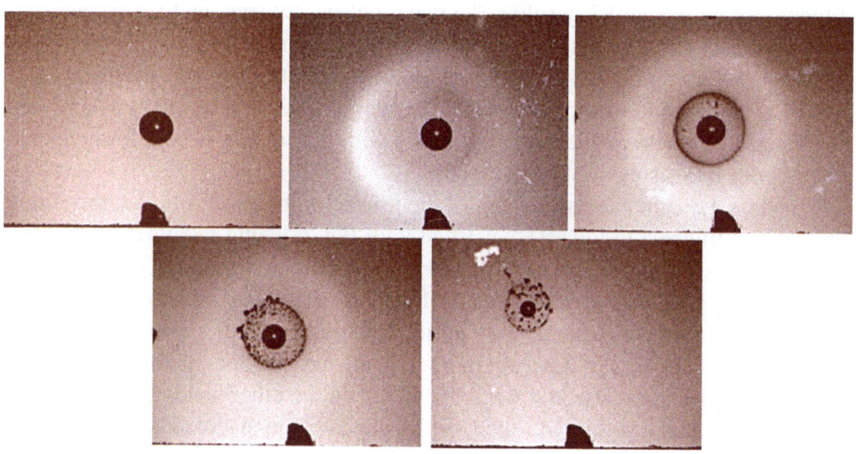

Fig. 10 Selected images from an unsupported droplet burning sequence for n-heptane (D_o = 0.76 mm) showing development of the soot shell (Jackson and Avedisian 1998). Flame is also visible in this sequence. Camera marker on the film is at the bottom of the photographs

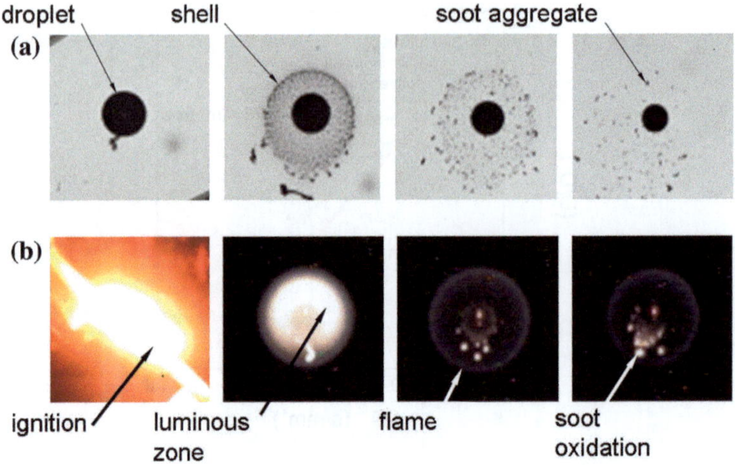

Fig. 11 Selected photographs from the burning history of a D_o = 2.78 mm n-octane droplet showing backlighted images (**a**) and development of the sooting dynamics, and flame-illuminated images (**b**) showing development of the flame (Liu et al. 2014b)

of backlighting. Figure 11 displays backlighted images of an unsupported large n-octane droplet (D_o = 2.78 mm Liu et al. 2014a) from ISS experimentation, as well as a flame-illuminated sequence (Fig. 11b) of the same droplet (i.e., no backlighting is used). The soot shell is again clearly visible as are individual aggregates that formed.

When developing a surrogate, qualitative comparisons of the flame luminosity can provide a clue about the efficacy of the surrogate. Figure 12 compares the flame luminosity for several small hydrocarbon droplets considered as surrogate constituents for an octane 87 gasoline (heptane, octane, toluene, and iso-octane).

The images are arranged according to their 'brightness' (i.e., the sensitivity of the eye to wavelengths in the visible region of the electromagnetic spectrum due to oxidation and incandescence of soot aggregates that are transported to the flame). The image brightness qualitatively correlates with the amount of soot formed, being greater with increasing soot formation.

Based on a qualitative assessment of the image brightness in Fig. 12, the sooting propensities (highest to lowest) consistent with Fig. 6 would be toluene > heptane/toluene > gasoline > iso-octane > heptane/iso-octane > heptane. The backlighted images show the intensity of the soot shell and are consistent with this approximate ordering. It could be concluded that if matching sooting propensity were an important consideration in developing a surrogate for gasoline, neither heptane, iso-octane, or their mixtures would work. Toluene, on the other hand, has a noticeably brighter flame than gasoline. As such, one could envision that blending toluene with heptane or iso-octane would produce a feasible surrogate for gasoline, at least in terms of sooting propensities.

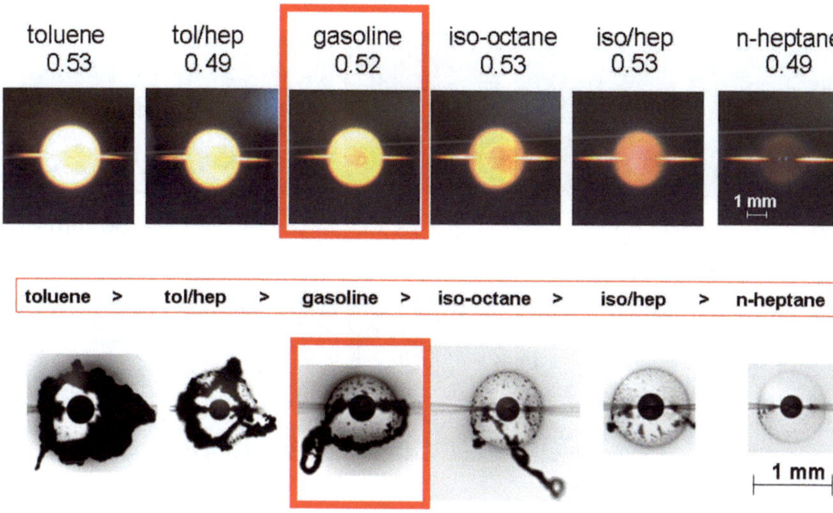

Fig. 12 *Upper row* of images are taken from self-illuminated color videos, arranged from highest to lowest intensity. Numbers denote the initial droplet diameters in mm. The images were selected from each sequence for maximum qualitative luminosity. *Lower row* of images are backlighted video that show soot dynamics; the images were selected for the thickest soot cloud in the burning sequence. The indicated sooting trend is based on a qualitative assessment of the flame image intensity (Liu and Avedisian 2012). The *red box* is a reference for gasoline

For sufficiently large D_o, the flame will extinguish owing to radiation (Chao et al. 1990). The flame diameter becomes so large that the flame temperature drops below that needed to sustain the combustion process. Figure 13 shows an example of the dynamics of this process for an n-decane droplet studied in the MDCA where an oscillatory behavior is noted. The mechanism for this process is not fully understood, though it may be related to droplet size and motion.

As burning progresses, the droplet diameter decreases and the flame is reduced perhaps for chemical kinetic reasons. Continued reduction of temperature will extinguish the flame. When the droplet moves (even slightly), it enters a fresh oxidizing ambience on the front side, while on the back side, the gas is depleted of oxygen. The flame is then hottest on the front, while on the opposite side, it is colder and extinction can occur locally as manifested by a hole appearing in the flame that propagates around the droplet as oxygen is used up. As the extinction boundary moves to the front of the droplet, it enters a region of gradually enriched oxygen concentration and higher temperature due to the flow and the flame recedes to the back side as combustion is re-established, thereby completing one cycle of an oscillation. This sort of oscillatory phenomenon has often been associated with "cool flame" phenomena (Nayagam et al. 2012; Farouk and Dryer 2013; Dietrich et al. 2014).

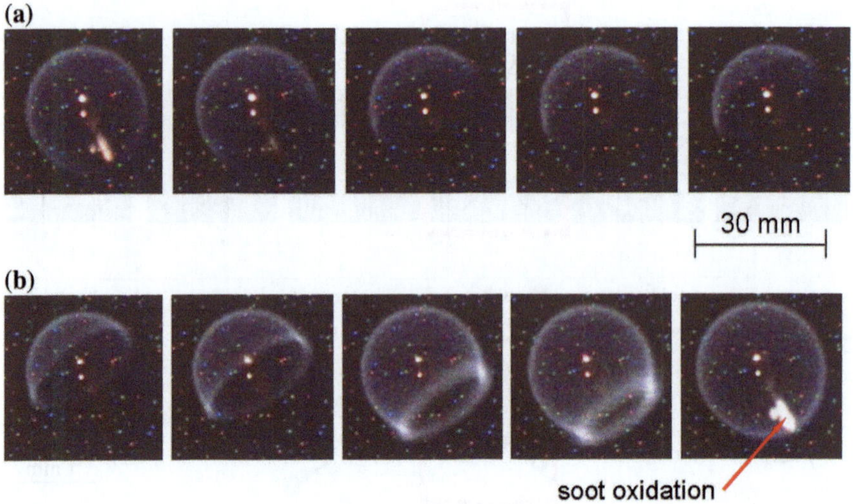

Fig. 13 Self-illuminated images showing oscillations of a n-decane droplet flame ($D_o = 5.1$ mm) as it radiatively extinguishes (Liu et al. 2014a). Time between images is 0.17 s. Flame peels away from the droplet (the droplet is not visible in these flame images) and then closes around the droplet as the flame is re-established. Specks in background are from video sensor imperfections. (Also see this website, http://www.smithsonianmag.com/science-nature/zero-g-fire-pulses-jellyfish-space-station-180952454/). **a** Flame opens-up. **b** Flame closes-back

6 Combustion Properties of the Spherical Droplet Flame Configuration

As remarked previously and depicted in Fig. 4, the combustion properties for the spherical droplet flame are the droplet diameter (D), the flame diameter (D_f), and the soot shell diameter (D_s). These dimensions are obtained by a frame-by-frame analysis of the digital video record of the droplet burning history. Computer-based algorithms have been developed for this purpose (Dembia et al. 2012). The droplet diameter is relatively easy to measure while the flame, and especially the soot shell diameter, can be challenging and time-consuming to extract from images like those shown in Fig. 11 because the flame and soot shell boundaries are not especially distinct. As such, the effort to extract droplet dimensions can be quite arduous, especially if soot obscures part of the droplet boundary which can create gaps in the measurements. The following illustrates results of using combustion properties obtained from the experimental designs of Figs. 5 and 6 to assess performance of surrogates.

To be effective, a surrogate should replicate the combustion properties of the real fuel in a suitable combustion configuration. Arguments have been previously presented for the suitability of the spherical droplet flame configuration to serve that purpose. Using the evolution of droplet diameter as a combustion property for the spherical flame configuration (Fig. 4), Fig. 14 compares an 87 octane gasoline (no ethanol) with toluene and mixtures of heptane and toluene. A binary mixture is the

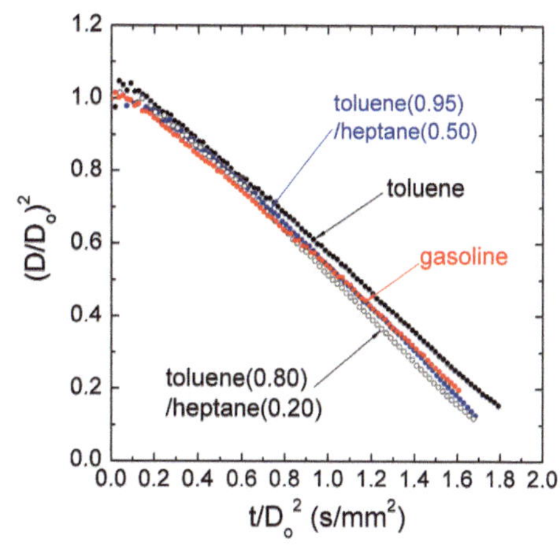

Fig. 14 Comparison of the evolution of droplet diameter of gasoline with toluene and mixtures of toluene with heptane (Liu et al. 2012). The closest match with gasoline is for 5 % heptane in toluene

simplest blend for a surrogate. Varying the composition is a relatively simple task for a binary system, and optimization techniques are not needed to identify the composition that produces the best match of combustion properties between the surrogate and real fuel.

Heptane and iso-octane were found to have essentially the same droplet burning characteristics for the configuration of Fig. 4. Figure 14 shows that neither heptane (nor iso-octane since heptane and iso-octane burn in an almost identical manner (Liu et al. 2012) can replicate the evolution of droplet diameter, which is consistent with the less luminous flames of heptane and iso-octane compared to gasoline.

Toluene droplet flames are, however, more luminous than gasoline (Fig. 12). As such, from a mixing perspective, blending toluene with heptane (or iso-octane, for that matter) can result in sooting propensities that are closer to gasoline. Furthermore, because toluene burns slower than gasoline while heptane burns faster as shown in Fig. 14, it will be possible to identify a heptane mixture fraction of heptane and toluene that closely matches the evolution of droplet diameter of gasoline. Figure 14 shows that a toluene/heptane mixture containing 5 % heptane and 95 % toluene accomplishes this quite well.

It is important to note that a surrogate may not be able to replicate all combustion properties of a real fuel. Considering the data in Fig. 15 which show the relative position of the flame to the droplet, gasoline droplet flames are much closer to the droplet surface than are toluene or heptane flames. A binary mixture evidently does not have enough flexibility to match more than a few combustion properties (one in this case). Including more constituents in a surrogate provides more flexibility to match combustion properties of a surrogate with a real fuel. Jet fuel surrogates (Dooley et al. 2010, 2012) are a case in point.

Two surrogates were formulated for jet-A: one containing three components (3CS) and the other four components (4CS). The surrogates were developed around four targets (average molecular weight (MW), DCN, TSI, and hydrocarbon-to-carbon ratio (H/C)) and the mixture fraction of the three or four components determined by

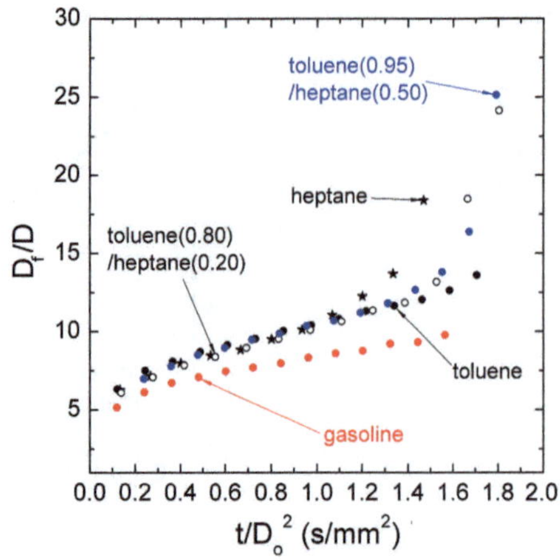

Fig. 15 Comparison of flame standoff ratio of gasoline with toluene and mixtures of toluene and heptane (Liu et al. 2012). The gasoline droplet flame is consistently closer to the droplet than for toluene or heptane/toluene mixture droplets, as contrasted with the closer match of the evolution of droplet diameter for a toluene/heptane mixture with gasoline shown in Fig. 14

an optimization algorithm. Table 3 lists the properties of the surrogates and the mixture fractions of the components. The 3CS surrogate matched only the DCN and H/C ratio, but not the TSI or MW. The 4CS surrogate was able to match the four targets. A test of these surrogates for the configuration of Fig. 4 was reported by Liu et al. (2013b). Figure 16 shows the results.

It is evident from Fig. 16 that the 3CS and 4CS blends represent jet-A quite well, with the 4CS being slightly better. This better agreement may be attributed to the fact that the 4CS matched four targets, while the 3CS, only two. Moreover, the relative position of the flame to the droplet for the 4CS was in slightly better agreement with jet-A than the 3CS as shown in Fig. 17, though as for that both surrogates were reasonably well correlated with jet-A.

As remarked previously, one of the more attractive features of the spherical droplet flame configuration of Fig. 4a is its one-dimensional gas transport that facilitates developing a numerical simulation of the burning process that can include detailed combustion chemistry, unsteady effects, and spectral radiative sub-models. Armed with this capability, it is possible to use measured combustion properties to evaluate some of the numerical inputs required for predictions [e.g., modifying the combustion chemistry to provide a better match between predicted and measured combustion properties (e.g., Werler et al. 2014)]. To illustrate the sort of comparative process that lies at the heart of using a detailed numerical simulation to predict combustion properties for the spherically symmetric droplet flame configuration, we consider methyl decanoate (MD) droplets burning under conditions that promote spherical droplet flames. MD is sometimes considered a surrogate for biodiesel derived from rapeseed oil as the feedstock.

Table 3 Representative properties and compositions of the three (3CS) and four (4CS) component surrogates of Dooley et al. (2010, 2012)

Fuel	Pure component	Formula	Density[a] (g/cm^3)	MW (g/mole)	BP (K)	Molar ratio[c]	Volume fraction (%)
3CS	n-decane	$C_{10}H_{22}$	0.726	142.28	443.7	0.427	50.86
	iso-octane	C_8H_{18}	0.688	144.23	372.4	0.33	33.34
	toluene	C_7H_8	0.862	92.14	383.8	0.243	15.80
4CS	n-dodecane	$C_{12}H_{26}$	0.745	170.33	489.5	0.40	50.36
	iso-octane	C_8H_{18}	0.688	144.23	372.4	0.29	26.67
	1,3,5-trimethylbenzene	C_9H_{12}	0.861	120.19	437.9	0.07	5.54
	Propylbenzene	C_9H_{12}	0.858	120.19	432.4	0.23	17.43
Jet-A	–	$C_{10.17}H_{19.91}$[c]	0.800	142.01[c]	478–573[b]	–	–

[a] Measured using a digital density meter (Mettler Todelo DA-100M) at 24.6 °C
[b] Colket et al. (2007)
[c] Dooley et al. (2010, 2012)

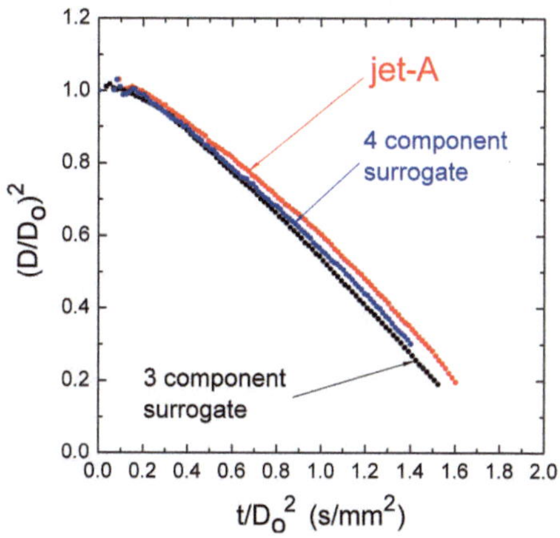

Fig. 16 Comparison of the evolution of droplet diameter for a 3CS and 4CS (Table 3) with jet-A (Liu et al. 2013b)

A recent study (Liu et al. 2013a) showed the efficacy of a detailed numerical simulation of the configuration of Fig. 4a to predict combustion properties of MD using the kinetic scheme of Divart et al. (2012). The reaction kinetics incorporated 238 species and 1,244 reaction mechanisms. As such, the simulation solved 238 individual species conservation equations simultaneously along with detailed formulations for radiative transfer, physical property correlations, and unsteady heat transfer in the droplet and surrounding gas.

Figures 18 and 19 compare measured droplet diameters and relative positions of the droplet to the flame with predicted values. The experimental method (cf. Fig. 5) employed droplets suspended from SiC fibers. As the fiber thermal conductivity was not known, values of 5.2, and 0 W/mK (a free droplet) were employed in the computation.

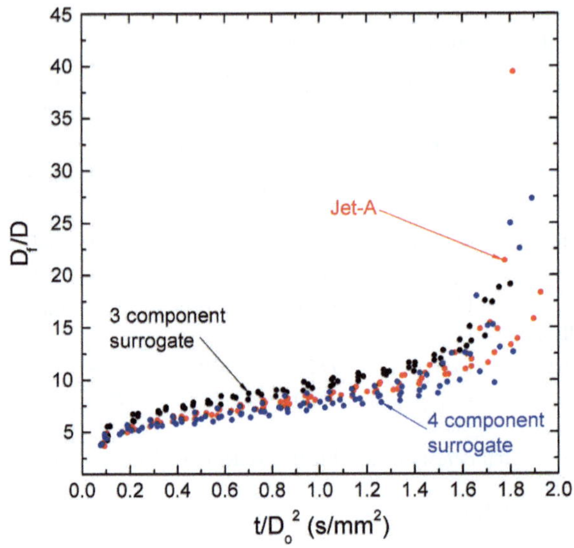

Fig. 17 Evolution of the measured relative position of the flame to the droplet for the 3CS and 4CS (Liu et al. 2013b in comparison with the flame position of jet-A

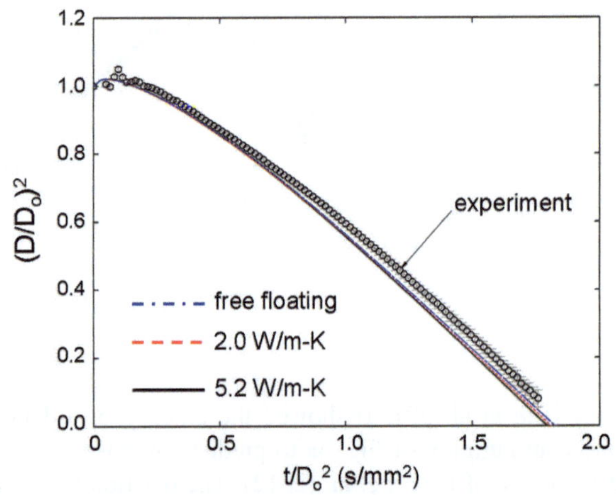

Fig. 18 Comparison of the measured evolution of droplet diameter for a methyl decanoate droplet with a detailed numerical simulation (*solid line*) of the spherically symmetric droplet burning process. The droplets were supported by a SiC fiber (Fig. 5), and predictions for various fiber thermal conductivities are shown (from Liu et al. 2013a)

The agreement between simulated and measured combustion properties for MD is very good as shown in Figs. 18 and 19. This agreement is testimony to the relevance of the Dievart et al. (2012) kinetic mechanism for MD. It is this sort of comparative analysis that lies at the heart of evaluating the inputs needed to

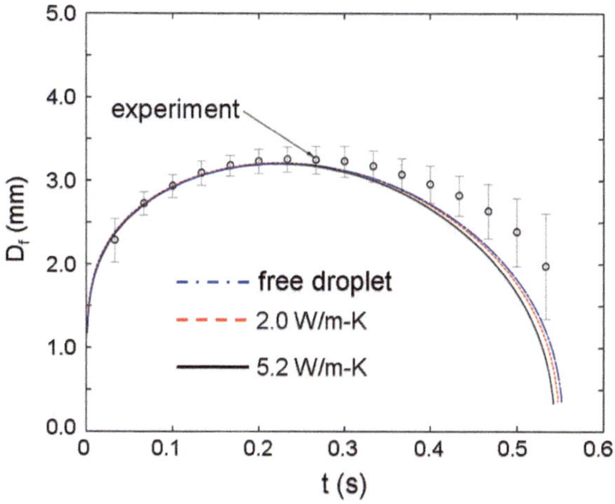

Fig. 19 Comparison of the measured evolution of flame (or outer luminous zone) diameter for a methyl decanoate droplet with a detailed numerical simulation (*solid lines*). Data are for the same droplet as Fig. 18. The droplets were supported by a SiC fiber (Fig. 5), and predictions for various fiber thermal conductivities are shown (from Liu et al. 2013a)

simulate more complex combustion configurations. Armed with this information, detailed modeling can then be used to determine with some confidence the operational conditions under which the formation of gas products responsible for degrading fuel efficiency can be minimized, thus increasing fuel efficiency of combustion engines.

7 Summary and Conclusions

The development of surrogates for liquid transportation fuels is one of the most important problems that will impact the ability to predict in-cylinder processes of combustion engines using detailed numerical simulation. Real fuels are so complex that it is not currently possible to develop kinetic and property database inputs needed to simulate engine performance. Surrogates are the only viable way to address this problem.

The traditional approach to assessing the performance of surrogate fuels has been to use low-dimensional combustion configurations that require the liquid to be pre-vaporized. This restriction eliminates from consideration the performance of a surrogate when multiphase processes are important such as spray injection which sets the initial conditions for in-cylinder combustion processes. The combustion of an isolated droplet burning with spherically symmetric gas phase transport provides a platform that folds into the assessment of a liquid surrogate some of the basic evaporation and phase equilibrium dynamics of sprays, while at the same time providing a transport process that is amenable to detailed numerical modeling.

Results were summarized that illustrate the efficacy of the spherical flame configuration to provide combustion properties useful for assessing the performance of surrogate fuels. Though it will not generally be possible to match all combustion properties of a combustion configuration used in the development of a surrogate fuel, the current capabilities show that complex transportation fuels can benefit from including combustion properties obtained from the spherical flame configuration in assessing the performance of surrogate fuels.

Acknowledgments Preparation of this paper was supported in part by the National Aeronautics and Space Administration grant no. NNX08AI51G with Mr. Michael Hicks as the project monitor. The author also benefitted from discussions with Dr. Y.C Liu of the University of Michigan-Flint and Mr. Yuhao Xu of Cornell University. The author is pleased to acknowledge this essential help in the course of his work on droplet combustion processes.

References

Amsden AA (1999) KIVA-3V, Release 2, Improvements to KIVA-3V, Los Alamos National Laboratory, report LA-UR-99-915

Avedisian CT (1997) Soot formation in spherically symmetric droplet combustion. In: Physical and chemical aspects of combustion, Chap. 6. Gordon and Breach, New York pp 135–160

Avedisian CT (2008) Surrogate fuel development: the role of droplet burning. Paper no. IMECE2008-68748, proceedings of IMECE2008, ASME international mechanical engineering congress and exposition, Boston, Massachusetts, Oct 31–Nov 6 2008

Avedisian CT, Callahan BJ (2000) Combustion of nonane/hexanol mixture droplets in microgravity. Proc Combust Inst 28:991–997

Avedisian CT, Jackson GS (2000) Soot patterns around suspended n-heptane droplet flames in a convection-free environment. J Propul Power 16:974–979

Avedisian CT, Yang JC, Wang CH (1988) On low gravity droplet combustion. Proc Roy Soc Lond A420:183–200

Banu B (2008) Fluids and combustion facility (FCF) and combustion integrated rack (CIR). Payload Accommodations Handbook CIR-DOC-4064, NASA John H, Glenn Research Center, Cleveland, Ohio

Bieleveld T, Frassoldati A, Cuoci A, Faravelli T, Ranzi E, Niemann U (2009) Experimental and kinetic modeling study of combustion of gasoline, its surrogates and components in laminar non-premixed flows. Proc Combust Inst 32:493–500

Chao BH, Law CK, T'ien JS (1990) Structure and extinction of diffusion flames with flame radiation, extinction of diffusion flames with flame radiation. Proc Combust Inst 23:523–531

Chaos M, Zhao Z, Kazakov A, Gokulakrishnan P, Angioletti M, Dryer FL (2007) A PRF+ toluene surrogate fuel model for simulating gasoline kinetics. Paper no. E26, 5th US National combustion meeting, Western States Section, Combustion Institute, 25–28 Mar 2007

Choi BC, Choi SK, Chung SH (2011) Soot formation characteristics of gasoline surrogate fuels in counterflow diffusion flames. Proc Combust Inst 33:609–616

Colket M, Edwards T, Williams S, Cernansky NP, Miller DL, Egolfopoulos F, Linstedt P, Seshadri K, Dryer FL, Law CK, Friend D, Lenhert DB, Pitsch H, Sarofim A, Smooke M, Tsang W (2007) Development of an experimental database and kinetic models for surrogate jet fuels. Paper no. AIAA2007-770, 45th aerospace sciences meeting, Reno, 8–11 Jan 2007

Cuoci A, Mehl M, Buzzi-Ferraris G, Faravelli T, Manca D, Ranzi E (2005) Autoignition and burning rates of fuel droplets under microgravity. Combust Flame 143(2005):221–226

Dagaut P, Togbe C (2008) Oxidation kinetics of butanol–gasoline surrogate mixtures in a jet-stirred reactor: Experimental and modeling study. Fuel 87:3313–3321

Dembia CL, Liu YC, Avedisian CT (2012) Automated data analysis of consecutive digital images from droplet combustion experiments by a MATAB-based algorithm. Image Anal Stereology 31:137–148

Dietrich DL, Nayagam V, Hicks MC, Ferkul PV, Dryer FL, Farouk T, Shaw BD, Suh HK, Choi MY, Liu YC, Avedisian CT, Williams FA (2014) Droplet combustion experiments aboard the International Space Station. Microgravity Sci Technol. doi:10.1007/s12217-014-9372-2 (open access)

Dievart P, Won SH, Dooley S, Dryler FL, Ju Y (2012) A kinetic model for methyl decanoate combustion. Combust Flame 159:1793–1805

Dirks LC, Dirks GW, Wu J (2012) Evolving perspectives on biofuels in the United States. Front Energy 6:379–393

Dooley S, Won SH, Chaos M, Heyne J, Ju Y, Dryer FL, Kumar K, Sung CJ, Wang H, Oehlschlaeger MA, Santoro RJ, Litzinger TA (2010) A jet fuel surrogate formulated by real fuel properties. Combust Flame 157:2333–2339

Dooley S, Won SH, Jahangirian SJUY, JU Y, Dryer FL, Wang H, Oehlschlaeger MA (2012) The combustion kinetics of a synthetic paraffinic jet aviation fuel and a fundamentally formulated experimentally validated surrogate fuel. Combust Flame 159:3014–3020

Edwards T, Maurice LQ (2001) Surrogate mixtures to represent complex aviation and rocket fuels. J Propul Power 17:461–466

Fahd EA, Liu YC, Avedisian CT, Dryer FL, Farouk TI (2014) A detailed numerical simulation of spherically symmetric n-butanol droplet combustion and comparisons with experimental data. Proceedings of the Combustion Institute, vol 35 (in press)

Farouk T, Dryer FL (2013) Isolated alkane droplet combustion in microgravity: cool flames. Paper no.: 1G17, 8th U.S. National combustion meeting, Combustion Institute, Park City, Utah, 19–22 May 2013

Farouk TI, Liu YC, Avedisian CT, Dryer FL (2013) Sub-millimeter sized methyl butanoate droplet combustion: Microgravity experiments and detailed numerical modeling. Proc Combust Inst 34:1609–1616

Farrell JT, Cernansky NP, Dryer FL, Friend DG, Hergart CA, Law CK, McDavid RM, Mueller CJ, Patel AK, Pitsch H (2007) Development of an experimental database and kinetic models for surrogate diesel fuels. SAE paper no. 2007-01-0201

FR (2012) Federal Register, National Archives and Records Administration, vol 77(199), pp 62623–63200, 15 Oct 2012

Gauthier BM, Davidson DF, Hanson RK (2004) Shock tube determination of ignition delay times in full-blend surrogate fuel mixtures. Combust Flame 139:300–311

Hara H, Kumagai S (1990) Experimental investigation of free droplet combustion under microgravity. Proc Combust Inst 23:1605–1610

Huber ML, Lemmon EW, Bruno TJ (2010) Surrogate mixture models for the thermophysical properties of aviation fuel Jet-A. Energy Fuels 24:3565–3571

Jackson GS, Avedisian CT (1998) Combustion of unsupported water-in-heptane emulsion droplets in a convection-free environment. Int J Heat Mass Transf 41:2503–2515

Jackson GS, Avedisian CT, Yang JC (1991) Soot formation during combustion of unsupported methanol/toluene mixture droplets in microgravity. Proc Roy Soc Lond A435:359–368

Lee CH, Reitz RD (2013) A comparative study on CFD simulation of spray penetration between gas jet and standard KIVA-3V spray model over a wide range of ambient gas densities. J Mech Sci Technol 26:4014–4025

Lee Y, Jang K, Han K, Huh KY (2013) Simulation of a heavy duty diesel engine fueled with soybean biodiesel blends in low temperature combustion. SAE paper no. 2013-01-1100

Liu YC, Avedisian CT (2012) A comparison of the burning characteristics of sub-millimeter droplets of binary mixtures of iso-octane, n-heptane and toluene with a commercial unleaded gasoline. Combust Flame 159:770–783

Liu YC, Farouk TI, Savas AJ, Dryer FL (2013a) On the spherically symmetrical combustion of methyl decanoate droplets and comparisons with detailed numerical modeling. Combust Flame 160:641–655

Liu YC, Savas AJ, Avedisian CT (2013b) Spherically symmetric droplet combustion of three and four component miscible mixtures as surrogates for Jet-A. Proc Combust Inst 34:1569–1576

Liu YC, Rah JK, Trenou KN, Hicks MC, Avedisian CT (2014a) Experimental study of initial diameter effects on convection-free droplet combustion in the standard atmosphere for n-heptane, n-octane and n-decane: International Space Station and ground-based experiments. Paper no. AIAA-2014-1019, 52nd aerospace sciences meeting, National Harbor, Md., 13–17 Jan 2014

Liu YC, Xu Y, Hicks MC (2014b) The role of micro-convection induced by support fiber in droplet Combustion processes. Proceedings of the Combustion Institute, vol 35 (in press)

Mueller CJ, Cannella WJ, Bruno TJ, Bunting B, Dettman HD, Franz JA, Huber ML, Natarajan M, Pitz WJ, Ratcliff MA (2012) Methodology for formulating diesel surrogate fuels with accurate compositional, ignition-quality, and volatility characteristics. Energy Fuels 26:3284–3303

Narayanaswamy K, Pepiot P, Pitsch H (2013) Jet fuels and Fischer-Tropsch fuels: surrogate definition and chemical kinetic modeling. Paper # 070RK-0273, Western States section meeting, Combustion Institute, 19–22 May 2013

Nayagam V, Dietrich DL, Ferkul PV, Hicks MC, Williams FA (2012) Can cool flames support quasi-steady alkane droplet burning? Combust Flame 159:3583–3588

NSF (2011) Transforming combustion research through cyberinfrastructure. Committee on Building Cyberinfrastructure for Combustion Research, National Research Council, The National Academies Press. ISBN-13: 978-0-309-16387-3, April (http://www.nap.edu/catalog.php?record_id=13049)

Okajima S, Kumagai S (1975) Further investigations of combustion of free droplets in a freely falling chamber including moving droplets. Proc Combust Inst 15:401–407

Pitz WJ, Mueller CJ (2011) Recent progress in the development of diesel surrogate fuels. Prog Energy Combust Sci 37:330–350

Pitz JW, Cernansky NP, Dryer FL, Egolfopoulos FN, Farrell JT, Friend DG, Pitsch H (2007) Development of an experimental database and chemical kinetic models for surrogate gasoline fuels. SAE paper no. 2007-01-0175

PRECISE (2011) U.S. Department of Energy, Office of Energy Efficiency and Renewable Energy and Science, March 3. http://www1.eere.energy.gov/vehiclesandfuels/pdfs/presice_rpt.pdf

Robbins J, Shinn C (2010) Multi-user droplet combustion apparatus FLEX-2. Reflight Safety Data Package MDC-DOC-1790A, NASA John H. Glenn Research Center, Cleveland, Ohio

Sirignano WA (1999) Fluid dynamics and transport of droplets and sprays. Cambridge University Press, Cambridge

Tsang W (2003) Workshop on combustion simulation databases for real transportation fuels, report no. NISTIR 7155, National Institute of Standard and Technology, Gaithersburg, MD, 4–5 Sept 2003

USEIA (2014) U.S. Energy Information Administration, Independent Statistics and Analysis. March. http://www.eia.gov/forecasts/steo/pdf/uncertainty.pdf

Wang H, Reitz JD, Yao M, Yang B, Jiao Q, Qiu L (2013) Development of an n-heptane-n-butanol PAH mechanism and its application for combustion and soot production. Combust Flame 160:504–519

Werler M, Cancino LR, Schiessl R, Mass U, Schulz C, Fikri M (2014) Ignition delay times of diethyl ether measured in a high-pressure shock tube and a rapid compression machine. Proceedings of the Combustion Institute, vol 35 (in press)

Yahyaoui M, Djeballi-Chaumeix N, Dagaut P, Paillard CE, Gail S (2007) Experimental and modeling study of gasoline surrogate mixtures oxidation in jet stirred reactor and shock tube. Proc Combust Inst 31:385–391

Zhang HR, Eddings EG, Sarofim AD (2007) Criteria for selection of components for surrogates of natural gas and transportation fuels. Proc Combust Inst 31:401–409

Part IV
Emissions

Urban Traffic Emissions and Associated Environmental Impacts in India

Ajay S. Nagpure and Bhola R. Gurjar

Abstract Rapid increase in urban population has been observed in Indian subcontinent during recent decades. The recent data shows that level of urbanization in India has increased from 27.81 % in 2001 to 31.16 % in 2011. About 34 % of urban growth has been observed only between 2001 (286 million) and 2011 (377 million). Substantial increases in urban population throughout the country have placed heavy demands on urban transport systems and personal vehicles. With this increment, road transport became a major source of air pollution in Indian urban agglomeration. For better understanding of current status of urban infrastructures and its associated impacts, there is always need to update and review of each sector for policy implementation. Current chapter attempts to highlight the recent scenario of Indian road transportation sectors and its impact on local and regional air quality and public health.

Keywords Urban traffic · Transport systems · Vehicular emissions · Air pollution · Public health

1 Introduction

Surface transportation in the Indian subcontinent is a huge source of local air pollution and greenhouse gas emissions and therefore contributes to local level air quality deterioration, human health risks, and global scale climate change (TERI 2001a). Traffic statistics in India show that the number of motor vehicles has increased from 0.3 million in 1951 to 142 million in 2011. The number of registered

A.S. Nagpure
Center for Science, Technology and Public Policy,
Hubert H. Humphrey School of Public Affairs,
University of Minnesota, Minnesota, USA

B.R. Gurjar (✉)
Department of Civil Engineering, Indian Institute of Technology Roorkee,
Roorkee 247 667, India
e-mail: brgurjar@gmail.com

A.K. Agarwal et al. (eds.), *Novel Combustion Concepts for Sustainable Energy Development*, DOI 10.1007/978-81-322-2211-8_17

vehicles in the country grew at a compound annual growth rate (CAGR) of 9.9 % between 2001 and 2011 (MoS&PI 2014). Of these, 32 % are concentrated in 23 metropolitan cities of India. Delhi itself accounts for about 8 % of the total registered vehicles and has more registered vehicles than those in Mumbai, Kolkata, and Chennai taken together (TERI 2001b). The rapid increase in the number of vehicles has significantly raised the emission loads of various air pollutants. Pachauri and Sridharan (1998) found that total pollution load from transportation sector in India has increased 68 times from 1947 (150 Tg) to 1997 (10,300 Tg). During this period, CO emission had the largest share (43 %) of the total pollutants from transport sector, followed by NO_x (30 %), HC (20 %), SPM (5 %), and SO_2 (2 %). According to CPCB (2000), twelve major metropolitan cities in India produce 0.35 Gg of NO_x, 1.91 Gg of CO, and 0.67 Gg of VOC annually from vehicular emissions alone. The quantum of vehicular pollutants emitted has been found highest in Delhi followed by Mumbai, Bangalore, Kolkata, and Ahmadabad (TERI 2001a).

Several studies have shown that vehicular emissions are a major culprit for the air quality degradation in Indian cities (Gurjar et al. 2004, 2010a; Nagpure et al. 2011, 2014; Kumar et al. 2011; Nagpure and Gurjar, 2012; Nagpure et al. 2013; Babaee et al. 2014). Urban population, number of cities, and urban area are growing at faster rate in the country resulting in the increasing demand of energy, goods, and services. With these demands, both private and public sector transportation systems are expanding rapidly in the country, even the current improved technologies in transportation sector are insufficient to counteract this growth in sustainable way. The future forecast of increasing demands therefore consistently shows the deteriorating air quality in the urban areas. This congruence has contributed to urban air pollution problem directly related to exhaust and non-exhaust motor vehicle emissions of particulate matter and other hazards pollutants. Because of sources of these emissions, the public health implications are substantial. Therefore, better understanding of traffic emissions and associated environmental impacts in India is needed. Current article presents an overview of urban traffic emissions and associated impacts in the Indian context (Gurjar et al. 2004, 2010a; Nagpure et al. 2011, 2014).

2 Vehicular Emission Sources and Pollutants

Figure 1 shows the types of emission source from vehicles. Burning fuel in engine provides energy to vehicle to move on the road. GHG emissions (e.g., CO_2, NO_x) and most of the pollutants (e.g., CO, HC, PM, benzene, 1–3 butadiene, formaldehyde, acetaldehyde, total aldehyde, and total PAH) from vehicle are the by-product and this process is called exhaust emission. In addition to exhaust, vehicle also emits pollutants (e.g., VOCs) from other activities such as fuel evaporation during rifling and vehicle operation, and non-exhaust emission (e.g., PM 10, PM 2.5) from break wear, tire, and road surface wear (Gurjar et al. 2004, 2010a; Nagpure et al. 2011, 2014).

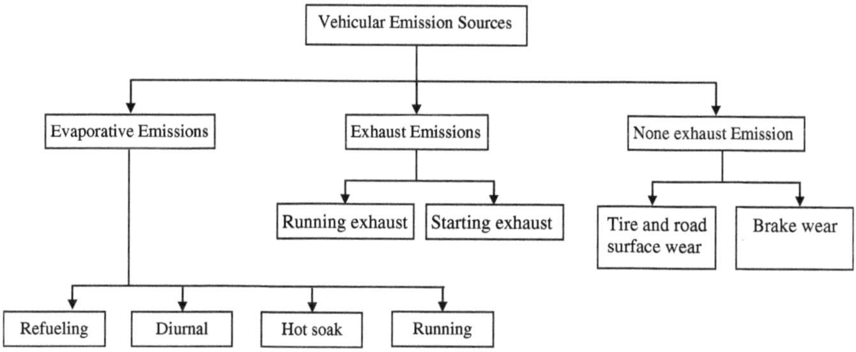

Fig. 1 Emission sources from individual vehicle

3 Contribution of Vehicular Emissions in Air Quality of Indian Megacities

Ambient air quality in Indian cities frequently reaches levels high enough to cause substantial impacts on human health and urban environmental quality (Singh et al. 1997; Sharma and Khare 2001; Kumar et al. 2004) which is similar for most of the world cities (Molina et al. 2007). Over the past decades, the urban road transportation has increased, which in turn has significantly increased vehicular emissions (Gurjar et al. 2004). As a result, in recent decades, urban traffic emissions and their impacts on air quality at the local, regional, and even worldwide level have received great attention from researchers, policy makers, and the general public. Several studies have been conducted by various researchers to understand the contribution of transportation sector in derationing air quality of Indian urban agglomeration. Gurjar et al. (2004) mentioned in their study that thermal power plants were the major sources for SO_2 and TSP from 1991 to 2000 in Delhi, whereas the on-road vehicles accounted most to NO_X, CO and non-methane volatile organic compound (NMVOC) emissions (80 %). The contribution of CO_2 emission from transport sector has significantly increased (27 % in 1990 and 39 % in 2000) during same duration. Auto Fuel Policy Report (CRRI 2002) states that the emission range of NO_X from transport sector is 66 to 74 % in megacity Delhi. CPCB (1995) data show that almost 50 % of the emission is from vehicular activities, followed by domestic, industrial, and power plants. It was found that from 1990–1991 to 1995–1996, annual NO_X emission from gasoline consumption increased from 3.5 to 4.5 Gg, respectively, whereas it was 8 Gg in 1990–1991 and 12.8 Gg in 1995–1996 from diesel (Sharma et al. 2002a). According to Xie and Shah (2002), diesel-driven vehicles were the major source of NO_X emission in Delhi (Fig. 2), whereas least contribution was from two and three-wheelers. Figure 2 shows the contribution of various pollutants from different sector in Delhi. As shown in Fig. 2, road dust is a major contributor (59 %) for particulate matter (PM) followed by power plants and vehicular emissions in national capital of Delhi (Government of Delhi 2010; Guttikunda et al. 2011).

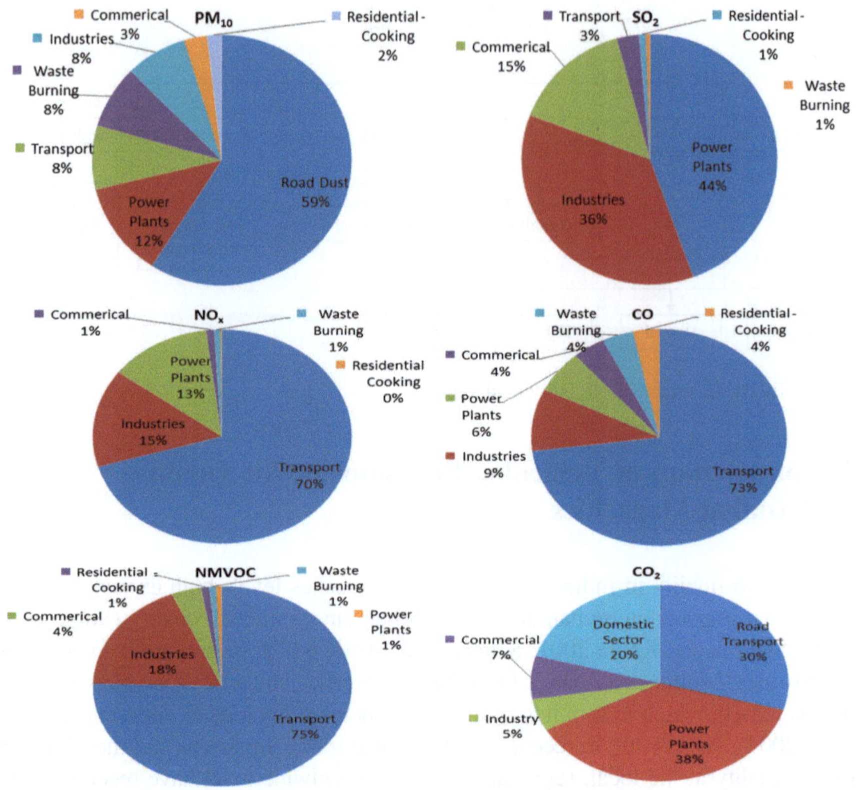

Fig. 2 Contribution of different sector for emissions of various pollutants in megacity Delhi. *Source* Government of Delhi 2010; Guttikunda et al. 2011

Also the vehicles are the major source of NO*x*, CO, and NMVOC (Fig. 2). Ramachandra and Shwetmala (2009) have stated in their study that one square kilometer area of Delhi emits about 221 Gg, 16 Mg, and 130 Mg of CO_2, CH_4 and NO_X from surface transportation respectively, in the area of one km^2 are only coming from surface transport in Delhi. Recently, Central Pollution Control Board (CPCB 2010a) analysis shows the predominance of private vehicles, especially cars and two-wheelers for GHG emissions in Delhi.

Mumbai is the financial and commercial capital of India. The rapid increases in vehicle population have been observed in megacity Mumbai during last few decades. It not only leads traffic congestion but also responsible for severe air pollution in megacity. During 1994, the air pollutants emissions from Mumbai on-road vehicles were about 52 %; however, the share of automobile sector for local air pollutants has increased tremendously during recent years (URBAIR 1997). The recent study conducted by Central Pollution Control Board mentioned that PM is mainly contributed by vehicles in Mumbai (35 %) and also accountability of surface transportation for CO, SO_2, NO_X, and HC is about 26, 1, 12, and 26 %, respectively, in Mumbai (CPCB 2010b). Figures 3 and 4 show the clear picture of source apportionment study of

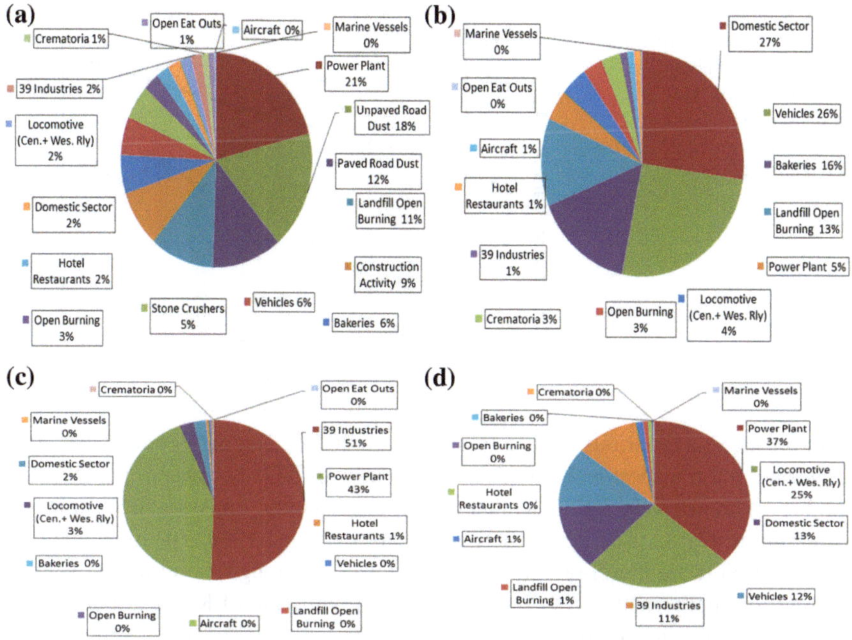

Fig. 3 Contribution of different sector for emissions: **a** PM, **b** CO, **c** SO_2, **d** NO_x in megacity Mumbai. *Source* CPCB, 2010b

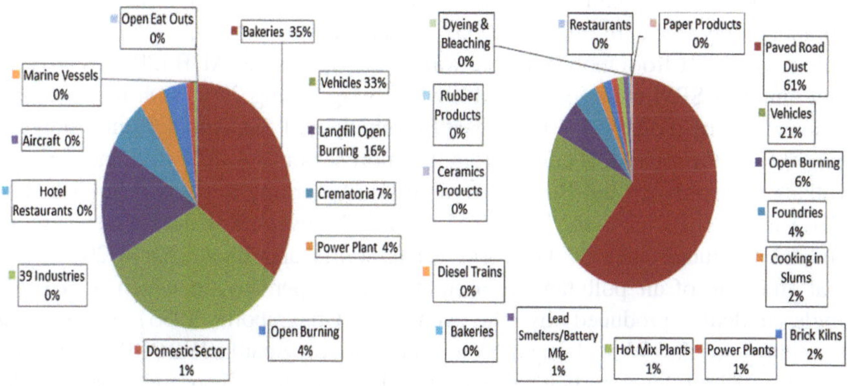

Fig. 4 HC emissions from different sector from various sector in megacity Mumbai (*Left Side*) and RPM emissions from various sectors in megacity Kolkata (*Right Side*) [1]. *Source* CPCB 2010b

megacity Mumbai. Based on TERI (2002), I have drawn estimated emissions from transportation sector in Mumbai against vehicle population and ambient concentration between year 1996 and 2005 (Fig. 5). Estimation done by TERI (2002) stated that annual emissions from transportation sector in 1993 were 16.33 Gg, which increased to 27.72 Gg in 2005, and it is projected that by 2010, it will be 30 Gg.

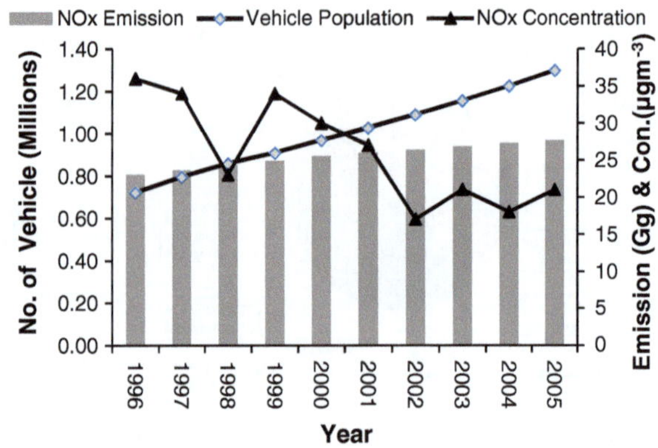

Fig. 5 Comparison of vehicle population growth, emission, and ambient concentration of NO*x* in Mumbai during 1996–2005. *Source* based on TERI (2002), CPCB (2008)

In recent years, various studies (Gurjar et al. 2010a; Nagpure et al. 2013, 2014) suggested that transport is the predominant source of NO*x* because of the increased number of motor vehicles in Kolkata. Estimated annual emission from transport increased from 1.82 to 25.55 Gg between years 1970 and 1990. The main vehicular sources of NO*x* are diesel-driven trucks and buses. Although diesel-driven vehicles account for approximately 10 % motor vehicle population of Kolkata, but they are responsible for almost 90 % of NO*x* emissions caused by motor vehicle (ESS 2002). ADB (2005) concluded that within the vehicle fleet, buses (53.7 %) are the principal source of NO*x*. Sharma et al. (2002b) estimated the NO*x* emissions from gasoline consuming vehicles were 0.3 Gg in 1990–1991 and increased to 0.4 Gg in 1995–1996, while the emission from diesel was 2.6 and 3.2 Gg for the same period. Another study done by ADB (2005) estimated that in 2003, the total annual emission of NO*x* from mobile source was 93.85 Gg. As per ADB (2005) study, the total annual RSPM emission in KMA was 76 Gg during 2003 with major contributions from road dust (61 %) followed by vehicles (21 %). Another study suggests that 50 % of the total SPM comes from transport sector and 48 % from industries in Kolkata (Chakraborty and Bhattacharya 2004). According to a report published by WHO and UNEP, due to the high SPM and RSPM levels, Kolkata is placed among the most polluted cities of the world (Chakraborty and Bhattacharya 2004). The total emission of air pollutants is around 0.71 Gg per day, in which 0.24 Gg is single-handedly produced by the vehicles (Chakraborti 1993) in Kolkata. According to the West Bengal Pollution Control Board (WBPCB 2006), the presence of high-level particulate matter in the ambient air is a major problem for the megacity Kolkata. The major sources of air pollution in Kolkata are the industrial emissions and automobile emissions. While Asian Development Bank (ADB 2006) estimation suggests that for respirable particulate matter (RPM), paved road dust is the major contributor in megacity Kolkata followed by vehicles and open burning (Fig. 4). As per ADB, in case of NO_x, vehicles (74 %) are the major source of NO_x followed by industrial sources (26 %). Ghose et al. (2004) stated that approximately 70 % of the total pollution load of the city is contributed by vehicles in Kolkata.

4 Traffic Emissions and Its Impacts on Human Health

According recent studies, the traffic emission effect on human health is the one of the major concern in Indian cities. Several researchers indicated the adverse effect of outdoor air pollution on human health in Indian cities. As discussed in above section, that traffic emission is an important contributor for local air pollution in Indian cities which directly reflects its association with leading health effects due to urban traffic. Primary emissions of NO_2, CO, and VOC from vehicles are also responsible for ground level O_3 formation by photochemical reaction which is responsible for allergic asthmatics by augmenting allergic responses (Steinberg et al. 1991). The emission of SO_2, PM, NO, and acid aerosols also causes inflammation of bronchial mucous (Karen and Michak 1991; Giuseppe et al. 1993). The exposure of urban traffic emissions is most vulnerable to sick and old people. According to Guttikunda and Jawahar (2012), the health impacts of air pollution from the transport sector are not insignificant and the nature of the issue is that those areas with the most population density are most affected. The health impact analysis estimates up to 49,500 deaths in 2010 and 158,500 in 2030 due to road transport in India. The results of health risk analyses study conducted by Guttikunda and Jawahar (2012) is given in Fig. 6. However, Guttikunda and Jawahar (2012) only estimated the direct impacts of urban traffic emissions, but they mentioned emission indirectly responsible for increase in inflammation, cardiac conditions, decrease in fertility, cancer, and premature birth.

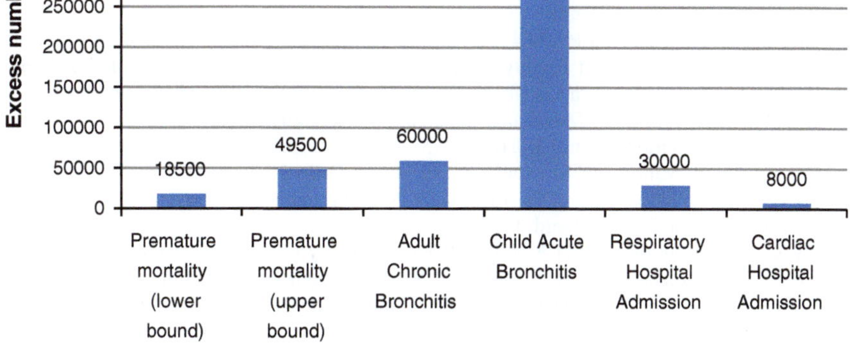

Fig. 6 Traffic emission and associated health effects estimation in India. *Source* Guttikunda and Jawahar (2012)

5 Conclusions

Emissions from on-road traffic are a major contributor to urban air pollution in Indian urban agglomerations. They are also important for their accountability for global GHG emissions. Because of high population and economic growth, the number of vehicles has increased significantly in India which is responsible for rising contribution of Indian transport sector for local air pollution and global climate change. However, significant progress has been made by India for reducing the amount of pollutants from motor vehicles despite substantial growth in vehicle numbers and vehicle kilometer traveled by introducing emission norms and clean fuel. Moreover, the effects of these measures have only been seen in megacities such as Delhi and Mumbai, and smaller cities still need attention of regulating authorities. With rising vehicle population in megacities, these measures are not enough to control the worsening air pollution in megacities. It is cleared from the studies that motor vehicles emissions have playing significant impact on human health, and it is anticipated that it will continue to play a central role for human health risk in Indian cities. Transport sector in India still needs more attention for reducing its effects on urban air quality and human health.

References

ADB (Asian Development Bank) (2005) Strengthening Environmental Management at the State level Cluster) Component E-Strengthening Environmental Management at Waste Bengal Pollution Control Board. Final Report, Volume V-Air Quality Management, November 2005

ADB (Asian Development Bank) (2006) Final Report Volume V: Air Quality Management. Published 2006. Accessed March 10, 2012. www.wbpcb.gov.in/html/downloads/adb_ta_3423_air_fin_rep.pdf

Babaee S, Nagpure AS, DeCarolis JF (2014) How much do electric drive vehicles matter to future U.S. emissions? Environ Sci Technol 48(3):1382–1390

Chakraborti D (1993) Calcutta environmental profile: air pollution. In: Sivaramakrishnan KC (ed) In managing urban environment in india: towards an agenda for action, vol 3. The Times Research Foundation, Calcutta

Chakraborty D, Bhattacharya P (2004) Air quality management strategy for Kolkata City under India Canada Environment Facility Project—A Case Study. West Bengal Pollution Control Board. http://www.cleanairnet.org/baq2004/1527/articles-59332_bhattacharya.pdf

CPCB (2008) Environmental Data Bank. http://cpcbedb.nic.in/ngtpmain.asp

CPCB (Central Pollution Control Board) (1995) Annual Reports 1993–1994. Central Pollution Control Board, Ministry of Environment and Forest, Delhi-32, www.cpcb.nic.in

CPCB (Central Pollution Control Board) (2000) Polluting industries. Parivesh newsletter 2000. Central Pollution Control Board, Delhi

CPCB (Central Pollution Control Board) (2010a) Air quality assessment, emissions inventory & source apportionment studies: Mumbai, National Environmental Engineering Research Institute, November, 2010. www.cpcb.nic.in/Mumbai-report.pdf. Accessed 28 Apr 2012

CPCB (Central Pollution Control Board) (2010b) Status of the Vehicular Pollution Control Programme in India. Published 2010a. Accessed April 28, 2012. www.cpcb.nic.in/upload/NewItems/NewItem_157_VPC_REPORT.pdf

CRRI (Centre Road Research Institute) (2002) Auto fuel policy report—Chapter 1. http://www.petroleumbazaar.com/Library/auto%20fuel%20policy/ch_1.pdf. Chapter 5. http://www.petroleumbazaar.com/Library/auto%20fuel%20policy/ch5.pdf, Chapter 15. http://www.petroleumbazaar.com/Library/auto%20fuel%20policy/ch15.pdf

ESS (2002) (Environmental Software and Services GmbH AUSTRIA) Air pollution a detail report, Delhi, Mumbai, Kolkata. www.ess.co.at/GAIA/CASES/IND/DEL/DELpollution.html. http://www.ess.co.at/GAIA/CASES/IND/BOM/BOMpollution.html

Ghose MK, Paul R, Banerjee SK (2004) Assessment of the impacts of vehicular emissions on urban air quality and its management in Indian context: the case of Kolkata (Calcutta). Environ Sci Policy 7(4):345–351

Giuseppe MC, Francesco F et al (1993) Effect of environment on atopic status and respiratory disorders in children. J Allergy Clin Immunol 92:616–645

Government of Delhi (2010) Department of environment, executive summary of inventorization of green house gases—Sources and Sinks in Delhi, 2010. http://www.delhi.gov.in/wps/wcm/connect/5fa00f80410c7d319880fa579a6b604f/Executive+Summary.pdf?MOD=AJPERES&CACHEID=5fa00f80410c7d319880fa579a6b604f. Accessed 12 Feb 2011

Gurjar BR, van Aardenne JA, Lelieveld J, Mohan M (2004) Emission estimates and trends (1990–2000) for megacity Delhi and implications. Atmos Environ 38:5663–5681

Gurjar BR, Nagpure AS, Kumar P, Sahni N (2010a) Pollutant emissions from road vehicles in Mega-City Kolkata, India: Past and Present Trends. Indian Journal of Air Pollution Control, Vol. X No.2 September 2010 pp 18–30. www.mypages.surrey.ac.uk/m01455/Gurjar-Kumar-Kolkata(2011).pdf

Gurjar BR, Jain A, Sharma A, Agarwal A, Gupta P, Nagpure AS, Lelieveld J (2010b) Human health risks in megacities due to air pollution. Atmos Environ 44(2010):4606–4613. doi:10.1016/j.atmosenv.2010.08.011

Guttikunda S, Jawahar P (2012) Road transport in India 2010–30, emissions, pollution & health impacts, simple interactive models for better air quality, Synopsis Report, Urban Emission info

Guttikunda S, Calori G, Velay-Lasry F, Ngo R (2011) Air quality forecasting system for cities: modeling architecture for Delhi. Simple interactive models for better air quality, SIM-air Working Paper Series: 36, 2011

Karen E, Michak FW (1991) Smoking cessation and acute airway response to ozone. Arch Environ Health 46:288–295

Kumar VA, Patil RS, Nambi KSV (2004) A composite receptor and dispersion model approach for estimation of effective emission factors for vehicles. Atmos Environ 38:7065–7067

Kumar P, Gurjar BR, Nagpure A, Harrison RM (2011) Preliminary estimates of particle number emissions from road vehicles in megacity Delhi and associated health impacts. Environ Sci Technol 45:5514–5521. doi:10.1021/es2003183

Molina LT, Kolb CE, de Foy B, Lamb BK, Brune WH, Jimenez JL, Molina MJ (2007) Air quality in North America's most populous city overview of MCMA-2003 Campaign. Atmos Chem Phys Discuss 7:3113–3177

MoS&PI (Ministry of Statistics and Programme Implementation) (2014) Statistical Year Book, India 2014. http://mospi.nic.in/Mospi_New/upload/SYB2014/ch20.html. Accessed 13 July 2014

Nagpure AS, Gurjar BR (2012) Development and evaluation of vehicular air pollution inventory model. Atmos Environ 59:160–169

Nagpure AS, Gurjar BR, Kumar P (2011) Impact of altitude on emission rates of ozone precursors from gasoline-driven light-duty commercial vehicles. Atmos Environ 45:1413–1417

Nagpure AS, Sharma K, Gurjar BR (2013) Traffic induced emission estimates and trends (2000–2005) in Megacity Delhi. Urban Climate 4:61–73. http://www.sciencedirect.com/science/article/pii/S2212095513000114

Nagpure AS, Gurjar BR, Martel JC (2014) Human health risks in national capital territory of Delhi due to air pollution. Atmos Pollut Res 5(2014):371–380

Pachauri RK, Sridharan PV (1998) Looking back to think ahead: growth with resource enhancement of environment and nature. Tata Energy Research Institute, New Delhi, p 346

Ramachandra TV, Shwetmala (20009) Emissions from India's transport sector: state wise synthesis. Atmos Environ 43:1–8 doi:10.1016/j.atmosenv.2009.07.015

Sharma P, Khare M (2001) Modelling of vehicular exhausts—a review. Transp Res Part D 6 (3):179–198

Sharma C, Dasgupta A, Mitra AP (2002a) Future scenarios of inventories of GHGs and urban pollutants from Delhi and Calcutta. Proceedings of IGES/APN Mega-City Project, (Rihga Royal Hotel Kokura, Kitakyushu Japan), 23–25 Jan 2002

Sharma C, Dasgupta A, Mitra AP (2002b) Inventory of GHGs and other urban pollutants from transport sector in Delhi and Calcutta. Proceedings of IGES/APN Mega-City Project

Singh A, Sarin SM, Shanmugam P, Sharma N, Attri AK, Jain VK (1997) Ozone distribution in the urban environment of Delhi during winter months. Atmos Environ 31(20):3421–3427

Steinberg JJ, Janet L, Gilson DG (1991) The pathobiology of ozone-including damage. Arch Environ Health 46:80–87

TERI (The Energy and Resources Institute) (2001a) State of environment report for Delhi 2001. Report No. 2000EE65, supported by the department of environment. Government of National Capital Territory, Delhi, TERI, New Delhi

TERI (The Energy and Resources Institute) (2001b) India state of the environment, Sponsor: United Nations Environment Programme. http://www.teriin.org/projects/ES/ES1999EE45.pdf

TERI (The Energy and Resources Institute) (2002) Pricing and infrastructure costing for supply and distribution of CNG and ULSD to the transport sector. Mumbai, India, Tata Energy Research Institute, New Delhi August 2002

URBAIR (1977) In: Shah Jitendra J,Nagpal T (eds) Urban air quality management in Asia. Greater Mumbai Report. World Bank Technical Paper No. 381, Washington, DC

WBPCB (West Bengal Pollution Control Board) (2006) Green Governance Newsletter. Published 2006. Accessed May 3, 2012. http://www.wbpcb.gov.in/html/newslet

Xie J, Shah J (2002) Reducing transport air pollution: the case of two-stroke engine vehicles in Asian Cities. The World Bank, International Conference on Sustainable Transport and Clean Air, Jakarta, 29–31 May 2000

Comparison of Primary and Secondary Emissions from an Internal Combustion Engine

Tarun Gupta, Avinash K. Agarwal and Pravesh Chandra Shukla

Abstract Diesel engines are among the most efficient power sources. Diesel engine emit relatively lower amounts of CO and HC emissions as compared to the gasoline engines but higher amounts of oxides of nitrogen (NO_X) and particulate matter (PM). NO_X and PM are both associated with deleterious effects on human health. Polycyclic aromatic hydrocarbons (PAHs) and trace metals are two most toxic and harmful class of chemical species present in the engine exhaust. Diesel emission is composed of a complex mixture of many organic compounds (OC) or soluble organic fraction (SOF), nitrates, sulfate, metals, and irritants (such as acrolein, ammonia, PAHs) which are typically adsorbed over elemental carbon (EC) core.

Keywords Primary emissions · Secondary emissions · Particle bound PAHs emissions · Elemental and organic carbon emissions · Particle number-size distribution · Biodiesel · Toxicity

1 Introduction

Excessive utilization of diesel engines given their fuel economy, durability and power advantage has resulted in exponentially high levels of diesel exhaust emissions. This calls for a thorough understanding of the nature and fate of diesel emissions in the ambient environment. Diesel emissions are mainly classified as regulated emissions (oxides of nitrogen, unburnt hydrocarbons, CO and particu-

T. Gupta (✉) · A.K. Agarwal · P.C. Shukla
Department of Mechanical Engineering, Indian Institute of Technology Kanpur, Kanpur 208016, India
e-mail: tarun@iitk.ac.in

A.K. Agarwal et al. (eds.), *Novel Combustion Concepts for Sustainable Energy Development*, DOI 10.1007/978-81-322-2211-8_18

lates) and unregulated emissions (aldehydes, benzene toulene xylene (BTX), PAHs, etc.) (Di et al. 2009). Current emission norms look only at the mass concentration of tail pipe exhaust. However, recent studies have linked the deleterious human health effects to the overwhelmingly large number of tiny diesel exhaust particles (Diaz et al. 2012). Efforts to reduce diesel emissions have traditionally focused on advancements in combustion, engine technology, improved fuel injection systems, and employment of after-treatment technologies (Khair et al. 2006). Although mass emissions have lowered significantly from the modern diesel engines, particle number emissions continue to rise up (Kittelson 1998). Bergmann et al. (2009) showed that after-treatment devices namely diesel particulate filters (DPF) significantly decrease the overall mass of tail-pipe emissions. Nevertheless, it leads to an increase in the number concentration of nuclei mode particles downstream of DPF and enhances the formation of secondary organic aerosol (SOA) as it provides required surface area for condensation of gas phase organics as SOF (Bergmann et al. 2009).

Formation of particles in the exhaust is a very complex process. Various volatile organic species condensed over tiny particles emanate out of the Tailpipe. It gets diluted and cools down when emitted in the atmosphere. Gas-to-particle phase transition of low-volatility compounds is enhanced by the free-radicals mediated photo-oxidation reactions as well as aqueous oxidation mechanisms active during this period (Kaul et al. 2011). The tail-pipe emission dilutes and cools down when it comes into the atmosphere and thus secondary emissions like SOA are formed. Aerosol properties are significantly affected by the level of dilution and the temperature of the ambient environment (Robinson et al. 2007). SOA has adverse effects on human health, climate, and environment (Baltensperger et al. 2008).

Inhalation exposure to diesel exhaust has shown to cause cough, bronchial irritation, nausea, lightheadedness and phlegm. Nasal intake of diesel exhaust aggravates the immune response to the antigens (Diaz-Sanchez et al. 1994, 1997). A study by Tsien et al. 1997 suggested that organic fraction of the diesel engine emissions cause most harmful health effects. The inflammatory potential of the exhaust particles in the lungs of rats aggravates with ozone (Madden et al. 2000). The current emission standards focus solely on the primary emissions. Therefore, it is essential to investigate the secondary particles formed via atmospheric photochemical aging of primary diesel emissions.

Biodiesel has emerged as an important alternative fuel for diesel engines. It is important to study the toxicity of biodiesel exhaust particulates in primary and secondary emissions which in turn largely depend upon the overall SOF content.

2 Primary Particulate Composition and Its Formation

2.1 Particulate Composition

Engine exhaust particles are composed of volatile (soluble) and non-volatile (insoluble) fractions. Soluble part consist organic carbon (OC), sulphate and nitrate compounds whereas insoluble part comprises of carbonaceous (soot) fraction and ash content (Eastwood 2008). The level of particle toxicity depends on its chemical composition and size. Figure 1 (Kittelson 1998) shows a typical chemical composition of diesel particulates. The soluble fraction is the most toxic component of the particulates. This fraction can also be present in the gas phase and it can eventually get condensed over the surface of solid particle. Such primary particles (15–40 nm in diameter) formed can further agglomerate and form bigger particles (Eastwood 2008). The size of these particles is expressed as electrical mobility diameter. Electrical mobility diameter refers to the particle velocity at which it migrates in an applied electrical field.

2.2 Particulate Formation

Many past studies have explored the soot formation process in diesel exhaust. Figure 2 explains the critical steps of the particle formation. Diesel is injected in the combustion chamber at high pressure and temperature. Spray penetration and fuel atomization depends on the actual combustion chamber pressure and temperature. Higher degree of fuel penetration and atomization results in improved air and fuel mixing. Injected fuel at high pressure forms tiny droplets that get dispersed in the chamber with the incoming air flow. Excess air ratio (γ) is not uniform within the combustion chamber. 'γ' value is very low near the droplet surface and attains

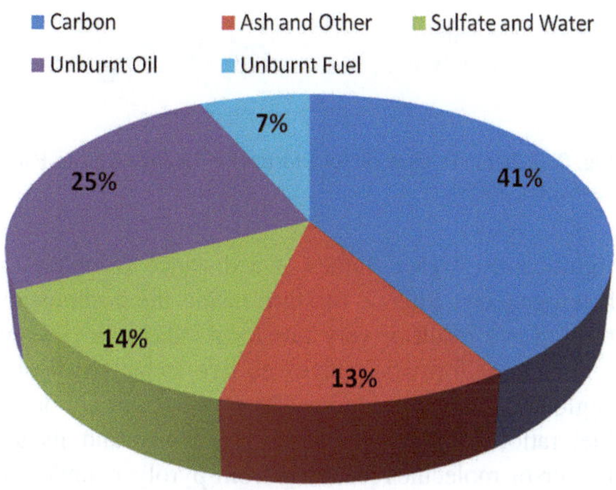

Fig. 1 Typical particle composition for a heavy-duty diesel engine tested in a heavy-duty transient cycle (adapted from Kittleson 1998)

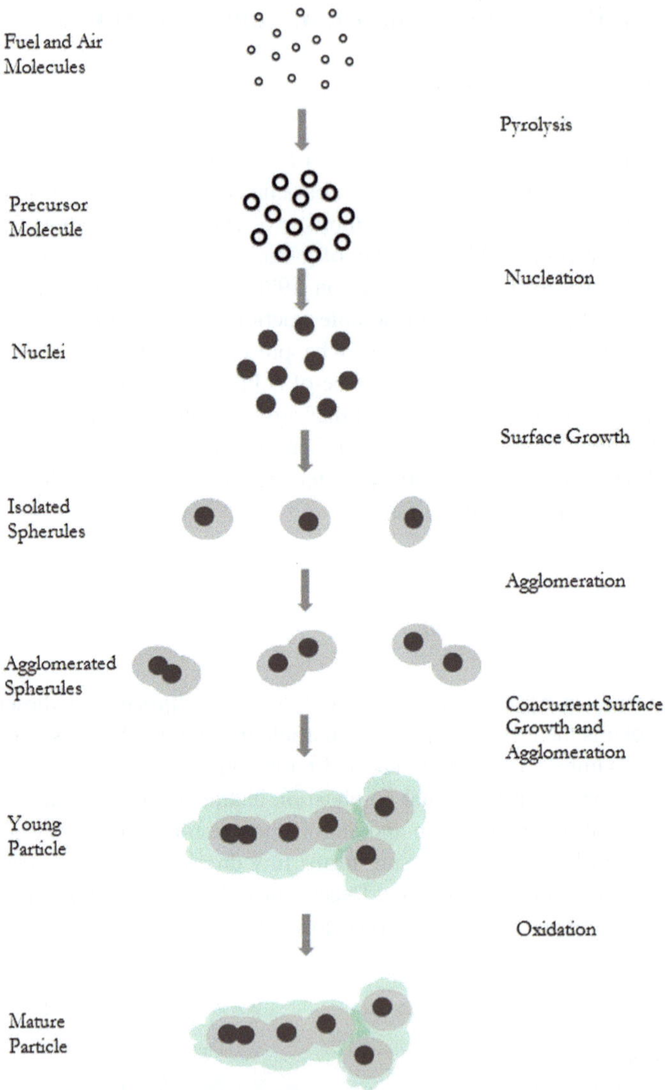

Fig. 2 Schematic view of soot formation (Adopted from Particulate emissions from vehicles by Eastwood, 2008)

significantly higher values at a distance from the droplet surface. This leads to heterogeneous fuel-air mixture inside the combustion chamber. Incomplete combustion is a result of very low local excess air (less oxygen) ratio near the droplet.

The soot formation involves five steps: pyrolysis, nucleation, surface growth, agglomerates and oxidation. Pyrolysis is largely dependent upon three factors: air-fuel ratio, combustion chamber pressure and its temperature (Eastwood 2008). Precursor molecules resulting from pyrolysis undergo nucleation process and form nuclei (<3 nm in size). Nuclei formation is followed by surface growth under which H molecules is striped off and results in higher C/H ratio. The size of the spherule is $\sim$20 nm. These spherules are the building blocks for agglomeration and form agglomerated particles. Spherules form chain and eventually agglomerated particles. The fifth step is surface oxidation. In the past, several researchers (Li and Wallace 1995) have come up with semi-empirical models for studying the soot formation mechanism.

3 Secondary Organic Aerosol (SOA)

Organic compounds are a dominant portion of the ambient aerosols (Seinfeld and Pandis 2012). Current analytical methods have enabled us to quantitatively evaluate these organic fractions. The oxidation of volatile organic carbon (VOC) results in the formation of SOA. The composition and formation mechanism for SOA needs to be carefully studied to effectively control it. Risk assessment of SOA desires availability of such information.

3.1 SOA Formation

Oxidation of the reactive organic gases (ROG) results in the SOA formation. Both natural as well as anthropogenic sources like solvents, vegetation and ocean emissions, combustion of wood and biomass, wood and fossil fuels etc. are responsible for ROG formation (Seinfeld and Pandis 2012; Jacobson et al. 2000). The most common ROG found in the atmosphere are alkanes, alkenes, aromatics and phenols.

Oxidation of precursor ROGs results in formation of semi-volatile organic carbon (SVOC). The species having their saturation vapor pressure below a certain pressure are known as semi-volatile (Turpin et al. 2000). It is required that for SOA formation, the rate of oxidation of precursor ROGs should be high. Next comes the partitioning of the SVOCs into the aerosol phase. There should be sufficient amount of SVOC having low saturation vapor pressure. Thus, SOA formation inherently depends on reactivity or volatility (Grosjean and Seinfeld 1967). Figure 3 shows a schematic view of SOA formation in the atmosphere.

3.2 Oxidation of Reactive Organic Gases (ROGs)

(a) **Gas Phase Oxidation**—O_3, OH radical and NO_3 radical are the 3 major ambient oxidizing. Photolysis of ozone in presence of moisture produces hydroxyl radical (OH). There are compounds (oxygenated aliphatics, saturated

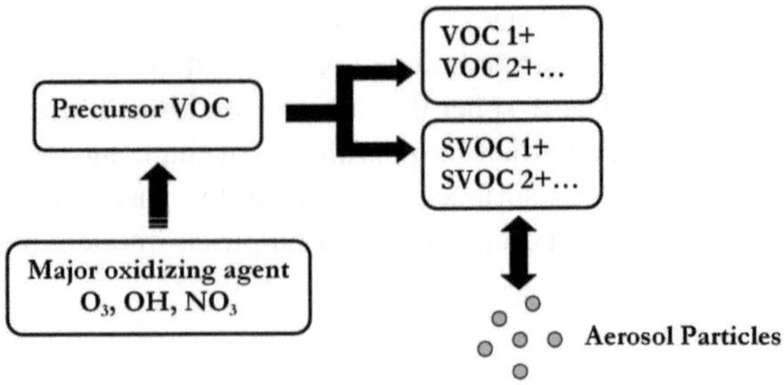

Fig. 3 Schematic overview of SOA formation (adapted from Dusek et al. 2000)

aliphatics, aromatics) which are not as reactive and show limited activity in presence of OH radicals. ROGs that have a double bond can be oxidized with O_3 and NO_3. Table 1 depicts the reactivity of ROG groups with these oxidizing agents.

A few of these oxidation products are seen in particle phase due to their lower vapor pressure such as polyols, dicarboxylic acids, amino acids and other such multifunctional compounds (Saxena and Hildemann 1996; Holes et al. 1997; Forstner et al. 1997a, b; Limbeck and Puxbaum 1999; Blando et al. 1998; Rogge et al. 1993).

(b) **Liquid Phase Oxidation**—Most of the precursor gases emitted by anthropogenic sources are water insoluble but their hygroscopicity increases with aging. Fogs and clouds being rich in oxidizing agents leads to adsorption of hygroscopic VOCs. Carboxylic acids, glyoxal, esters, and organo sulfur compounds are the usual products of fog and cloud processing (Blando and Turpin 2000). Studies have revealed several mono and di-carboxylic acids found in rain water and fog (Saxena and Hildemann 1996; Blando and Turpin 2000; Facchini et al. 1999). Formaldehyde can also get collected in the fog droplets (Facchini et al. 1992). A recent study (Aumont et al. 2000) suggested that carboxylic acid formation is most dominated by VOC oxidation in aqueous phase.

Table 1 Reactivity of ROGs towards oxidizing species in urban air

Organic species	Ozone	OH radical	NO_3 radical
Alkanes, Cycloalkanes	$\leq 10^{-23}$	$(0.3–8) \times 10^{-11}$	$\leq 10^{-17}$
Oxygenated aliphatics	$\leq(2.2 \times 10^{-21})$	$(0.2–6) \times 10^{-11}$	$\leq(1.4 \times 10^{-16})$
Aromatics	$\leq(6 \times 10^{-21})$	$(0.1–6) \times 10^{-11}$	$\leq 10^{-17}$
Alkanes, Cycloalkenes and other olefins	2×10^{-18} to 1.5×10^{-15}	$(0.8–12) \times 10^{-11}$	6×10^{-17} to 3×10^{-11}

Units cm^3 $molecule^{-1}$ s^{-1} at 289 K
Adapted from Grosjean and Seinfeld 1967

4 Objective

Particulate matter in diesel engine is generated due to incomplete combustion in the nanometer (D50 < 50 nm) and ultrafine (D50 < 100 nm) size ranges (Donaldson et al. 1996). Ultrafine and nano particles contribute 1–20 % of total mass but comprise more than 90 % of the total particle number and surface area. Particle size influences the atmospheric residence time of the particles, the optical properties of the particles, and the particle surface area and ability to participate in photochemical oxidation/aging and finally health effects of the particles.

This study compares diesel exhaust (DE) and biodiesel exhaust (BDE) for their toxicity with a strong emphasis on following parameters:

(1) Total particle number and mass concentration from diesel and biodiesel exhaust.
(2) EC and OC content.
(3) PAHs in Diesel and B20 exhaust.
(4) Particle morphology.

5 Experimental Setup

For both primary and secondary particles, number, mass, and surface area distributions with respect to particle size were measured at various engine loads (0, 12.5, 25, 37.5, 50, 62.5, 75, 87.5 and 100 %) for rated engine speed (1,800 rpm). Chemical characterization of primary and secondary emissions was carried out using real-time EC, OC and total particle-bound PAHs measurements. Particulate laden filters were observed by scanning electron microscopy (SEM) at 1,000× magnification.

A modern CRDI diesel engine (Tata; Safari DICOR 3.0 L) with an eddy current dynamometer (Dynomerk; ECB 300) were used for the comparative study of mineral diesel and B20. The experimental setup for this study (Fig. 4) comprises of an artificial photochemical chamber.

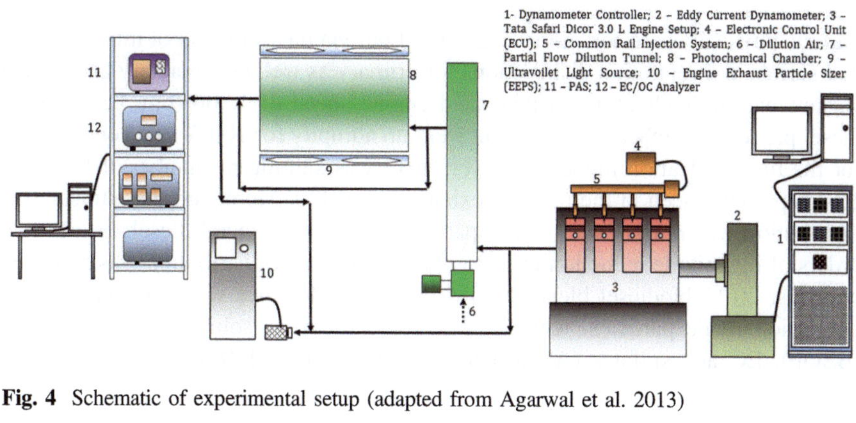

Fig. 4 Schematic of experimental setup (adapted from Agarwal et al. 2013)

The chamber walls were constructed with fluorinated ethylene propylene (FEP) membrane to avoid contamination and minimize wall losses. The chamber walls were completely transparent to the UV spectrum. A 40 μm thick sheet of cellulose acetate was used to filter out UV light with wavelengths below 300 nm (UV-C spectrum). It was kept in between the UV lamp banks and the photochemical chamber. It was designed to keep the artificial light spectrum as close as possible to the sunlight spectrum. A humidity measurement sensor (Testo; 605H1) was kept at the outlet of the photochemical chamber to monitor temperature and humidity of the chamber. Chamber leak tests were performed to avoid any interference from the ambient aerosol. FEP membrane was cleaned with ethanol and demineralized water. Zero air supply was used to flush the photochemical chamber before starting an experiment. Zero air supply setup comprised of an external compressor, pressure regulators, chemical scrubbers and a temperature controller, all contained in one enclosed unit. Air supply was used to flush the photochemical chamber to remove trace pollutants remaining from the previous experiment. Experimental setup was equipped with EC/ OC analyzer (Sunset Laboratory; Semi-continuous field v.4) (Gupta et al. 2011); online PAHs analyzer (EcoChem Labs; PAS 2000) (Bae et al. 2004; Ott and Siegmann 2006) to measure total particle bound PAHs; engine exhaust particle sizer spectrometer (TSI; 3090) to measure the number-size distribution of nanoparticles (5.6–560 nm electrical mobility diameters). In order to ensure complete particle formation mimicking atmospheric conditions, a partial flow dilution tunnel was employed to dilute the engine exhaust (Dwivedi et al. 2006). Approximately 36 % wall losses were measured using the penetration of EC through the photochemical chamber in the absence of UV light. All measurements reported here for EC, OC and total particle bound PAHs in the secondary emissions were carefully corrected for chamber wall losses.

6 Results and Discussion

6.1 Physical Characterization of Particulates

Particle physical characterization was carried out via measurement of number, mass and surface concentration distributions for primary as well as secondary particles at various engine loads at a constant rated engine speed of 1,800 rpm. For primary diesel particles, number concentration went up abruptly for 100 % rated engine load for the tiny nucleation mode. However, there was significant number of less than 10 nm particles seen in the primary exhaust from B20. Moreover, the number-size distribution was much flatter as compared to that from mineral diesel. Nearly two orders of magnitude reduction in particulate number-size distribution was achieved with just 20 % biodiesel blend. Majority of this difference was explained by the nucleation mode particles, which comprises of benzene soluble organic fraction (BSOF), EC and sulfates (Abdul-Khalek and Kittelson 1995). At higher engine loads, a relatively higher cylinder temperature leads to lower amount of unburnt organic species. This led to lower BSOF in the emitted particles. With increasing engine load the total number of particles emitted has lower BSOF which is critical for growth of soot nuclei. Sulfur content of the fuel also decides the number of sulfate nuclei formed in the cylinder. As biodiesel is essentially sulfur free, it further explains lower number concentration of primary particles from B20. Biodiesel contains approximately 10 % (w/w) oxygen this may also led to the sharp reduction in particle number-size distribution for the primary particles.

In comparison to the primary emissions, we see a striking decrease in particle number concentration in secondary aerosols (Fig. 5). This can be attributed to gravitational settling, wall losses, sampling losses and various other chemical reactions occurring inside the photo-chemical chamber (Gupta et al. 2011; Ruiz et al. 2007). Lower particle number concentration in secondary emissions was noticed for no load conditions for both the fuels. It is interesting to note that tiny particles (510 nm) emitted by B20 were prevailing as secondary aerosols. Particles emitted (in primary emissions) at higher engine loads were dominated by relatively larger particles, which somehow showed lower tendency to undergo agglomeration.

At 25 % engine load for both the fuels, a sudden increase in number concentration was observed. It is possible that the conditions inside the photochemical chamber (37.5 °C and 50 % RH) were the most favorable for SOA formation for this case. In line with the observations of Robinson et al. 2007, a sign of fresh nuclei formation in the photochemical chamber was observed. It seems that the particles were in a heterogeneous transient phase within the chamber. This further explains high variability in the secondary aerosol size distribution.

For both fuels as the engine load increased, the peak of particle number concentration shifted toward larger diameters for primary as well as secondary aerosols. The particle number concentration for both primary and secondary diesel exhaust was an order of magnitude higher than B20 (Agarwal et al. 2011). At a constant rated engine speed of 1,800 rpm, the total PM mass in primary emissions from diesel increases with engine load (Fig. 6). Particle number concentrations at lower engine loads were higher, but particle diameters were smaller. Hence, they cannot contribute significantly to the PM mass. For B20, at lower engine loads, PM mass in the primary emissions increased with engine load, with a maxima at 50 % engine load. The PM mass in primary and secondary emissions from diesel was higher than B20 by an order of magnitude. Numerous studies in the literature have indicated similar trends (Agarwal et al. 2011; Cowley et al. 1993; Den Ouden et al. 1994; Kalligeros et al. 2003; Lange 1991). Oxygen content of B20 favors efficient fuel combustion resulting in the lower PM mass (Akasaka et al. 1997; Owen et al. 1995).

Total surface area of the particles is directly proportional to the particle's number and is inversely proportional to the particle size. Reduction in primary emission particle number concentration and increase in the particle size were observed with increasing engine load for B20. Higher number concentration at lower engine loads (Fig. 5) for primary diesel emissions leads to lower PM mass and higher particulate surface area (Fig. 6). For B20, the trend of particulate surface area curve is very well correlated to the PM mass curve, with a peak at 50 % engine load. At 100 % engine

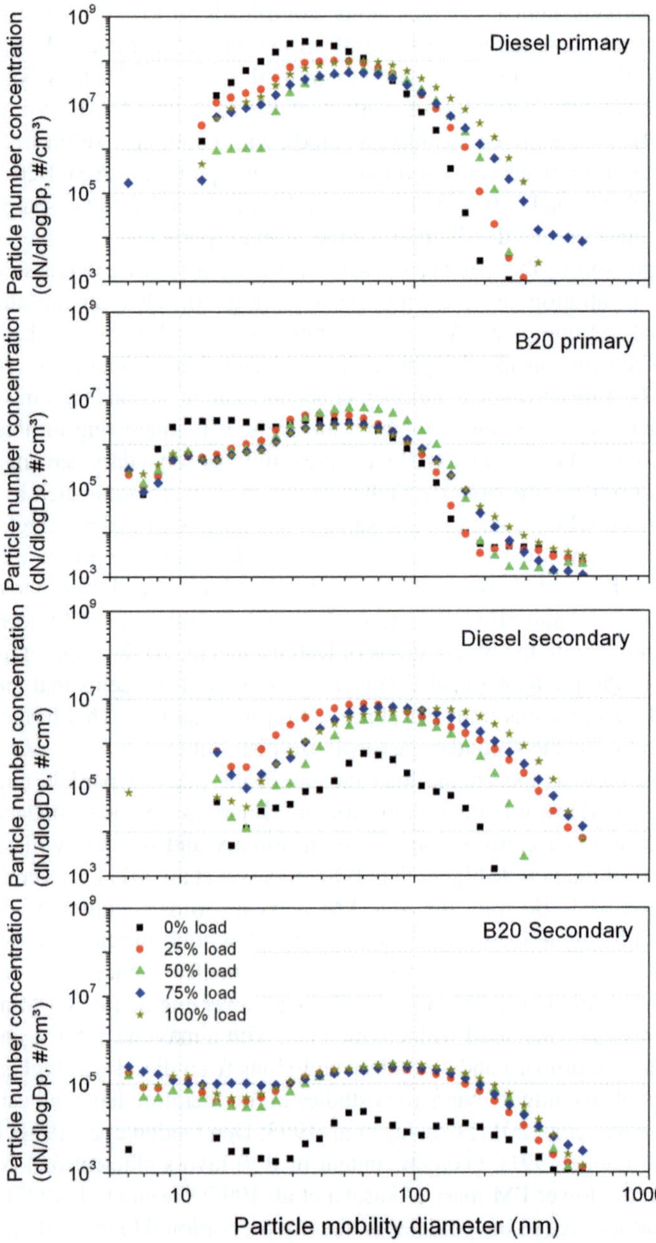

Fig. 5 Particle number concentration versus mobility diameter for primary and secondary exhaust fuelled with diesel and B20 (adapted from Agarwal et al. 2013)

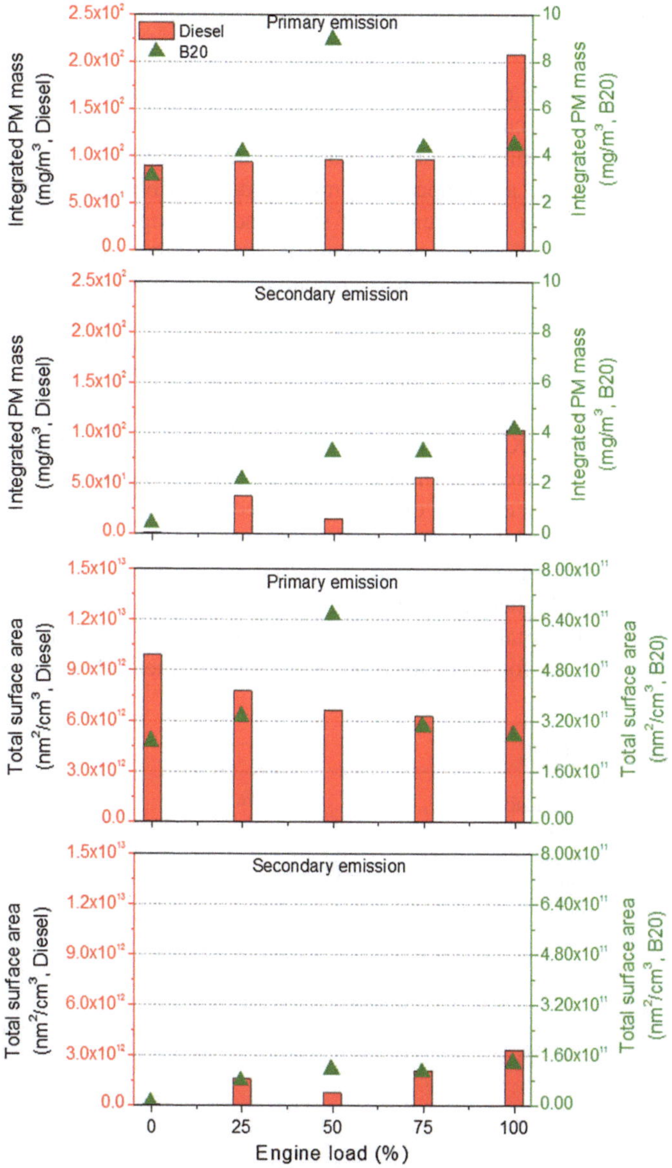

Fig. 6 Integrated particulate matter (*PM*) mass and surface area concentrations versus engine load for primary and secondary exhaust fuelled with diesel and B20 (adapted from Agarwal et al. 2013)

load, the mode of number-size distribution (Fig. 2) shifts toward right and the area under the curve is maximum, which results in the maximum particle surface area for primary diesel emissions (Fig. 6). At all engine loads, the surface area distribution in primary emissions from diesel is an order of magnitude higher than B20.

6.2 Chemical Characterization of Particulates

Higher particle number concentration for secondary emissions (Fig. 5) and correspondingly higher particulate surface area (Fig. 6) was seen with increasing engine load. The surface area distribution in secondary emissions from B20 was lower by an order of magnitude as compared to mineral diesel. Enhanced surface area of particles corresponds to more number of sites readily being available for the condensation of toxic organic species (Brüske et al. 2010). Hence, secondary emissions from diesel have relatively larger toxic potential as compared to B20.

Increase in EC upon increasing engine load has also been recorded in past studies (Gupta et al. 2011). Fuel-to-air ratios were relatively higher at higher engine loads leading to higher EC formation. Lubricating oil pyrolysis at relatively higher temperatures actually results in even higher EC measured at higher engine loads (Sharma et al. 2005). OC increased with an increase in engine load and attained a maxima and then it decreased with further increase in the engine load (Gupta et al. 2011). Higher oxygen content of B20 facilitated more efficient combustion as compared to the mineral diesel (Fig. 7). EC/ OC ratio was observed to be higher for diesel at higher engine loads. Whereas, for B20 it was relatively lower as compared to mineral diesel at higher engine loads.

At 1,800 rpm, the total particle bound PAHs were found to increase with higher engine loads (Fig. 8) for both the fuels. For both the fuels, PAHs in secondary

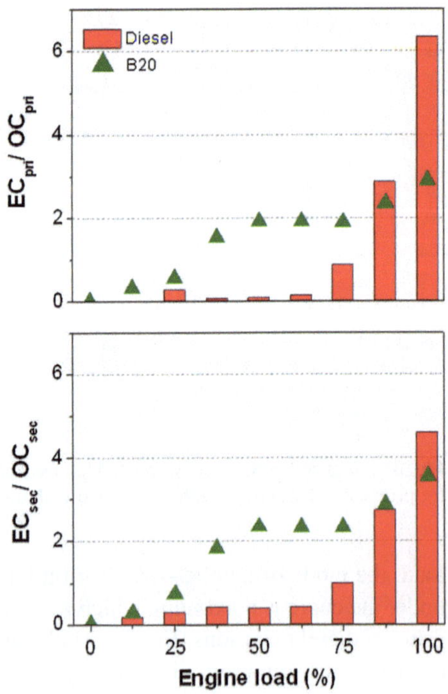

Fig. 7 Variation of EC/OC ratio for primary and secondary versus engine load for primary and secondary exhaust fuelled with diesel and B20 (adapted from Agarwal et al. 2013)

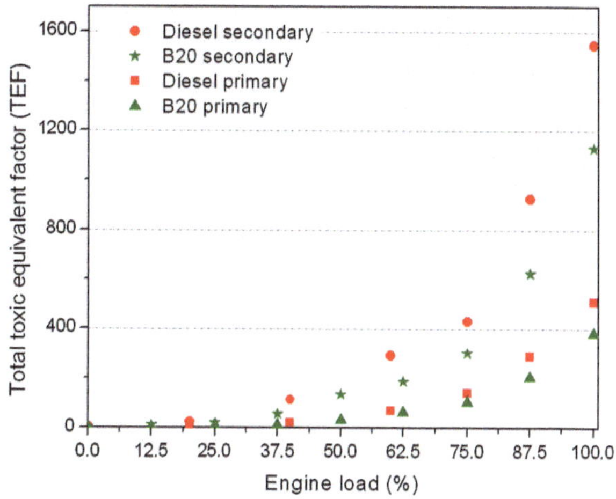

Fig. 8 Variation of total particle-bound PAHs versus engine load for primary and secondary exhaust fuelled with diesel and B20 (adapted from Agarwal et al. 2013)

emissions were higher as compared with primary emissions, at all engine loads. High in-cylinder temperature facilitates cyclization of unburned fuel and lubricating oil, leading to formation of PAHs at higher engine loads (Ravindra et al. 2008). A large amount of gaseous PAHs undergo phase transition and condense as particles as a result of cooling and aging. PAHs adsorb on particles and persist during incomplete combustion inside the engine (Atal et al. 1997). Even though there was a major reduction in PM mass for B20 as compared to diesel, however the particle bound PAHs were nearly comparable for both the fuels. This suggests that the emitted particles from B20 were relatively heavily laden with PAHs.

6.3 Toxic Equivalent Potential Evaluation

To fully understand the overall toxicity of PAHs, toxic potential of individual PAHs can be evaluated and a summation can be carried out. Integration of the individual toxicity of PAHs can be carried out by multiplying the toxic equivalent factors (TEFs) of the individual PAH species with their respective concentrations. TEFs for PAHs were obtained from Nisbet and Lagoy 1992 and CAEPA 1993. A total of 8 PAHs and 2 Nitro-PAHs were considered. Pan et al. 2000 showed the nitro-PAHs and PAHs emissions from diesel and B100 (soy methyl ester), and B20. They utilized a Cummins B5.9 engine. PAHs and nitro-PAHs concentrations were obtained by using gas chromatography and mass spectrometry. For our calculations, percentage distribution of PAHs species from the study of Pan et al. (2000) have been employed. Overall PAHs concentration in our study is attributed to individual PAHs based on the Pan et al. 2000 study.

The trend of total toxic equivalent potential (Fig. 9) was similar to particle-bound PAHs emissions (Fig. 8). Diesel showed slightly higher toxic potential

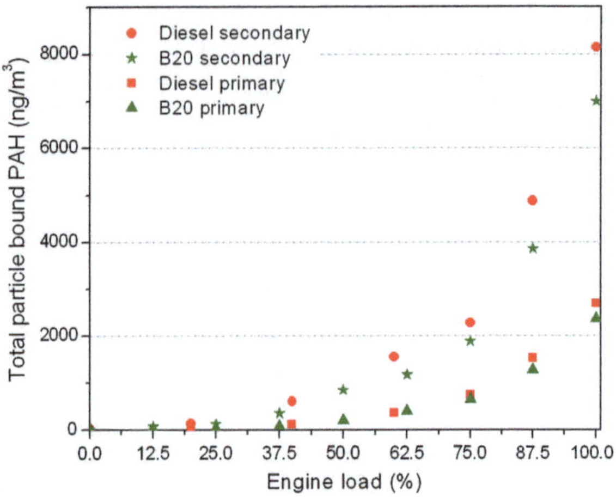

Fig. 9 Values of total toxic equivalent potentials with respect to engine load for primary and secondary exhaust fuelled with diesel and B20 (adapted from Agarwal et al. 2013)

compared to B20 for both primary and secondary particles. Primary and secondary particles TEFs for diesel was higher at 100 % engine load. Overall TEFs were 509 and 381 for diesel and B20 (primary particles), respectively, at 100 % engine load. TEFs were 1,547 and 1,130 for diesel and B20 (secondary particles), respectively, at 100 % engine load. B20 predominantly has individual PAHs with relatively lower toxicity. B20 particulates had significantly enhanced PAHs loading while their contribution to overall toxicity was far lower as compared to those from mineral diesel.

6.4 Primary and Secondary Particulates Emission

Primary emission particles were carefully collected over 47 mm diameter quartz filter papers after diluting the exhaust with HEPA filtered clean air. No significant collection of particles was noticed at the no load condition (Fig. 10). This maybe due to the fact that at no load conditions, most of the nuclei mode particulates are formed and nearly no contribution from EC (Zielinska et al. 2003). Due to the light colored OC dominantly present over the filter, they may have completed masked the tiny EC present in the sample. Also, the white filter background did not helped in providing the right contrast needed for such images. Aged particulate collection was wet in nature when compared to primary exhaust particulates measured over filter paper (Fig. 10). For low and intermediate engine loads (25 and 50 % load) a significant amount of particles got collected. Aged particles were wet, denser and bigger (coagulation and aggregation) than the primary particles (Agarwal et al. 2013).

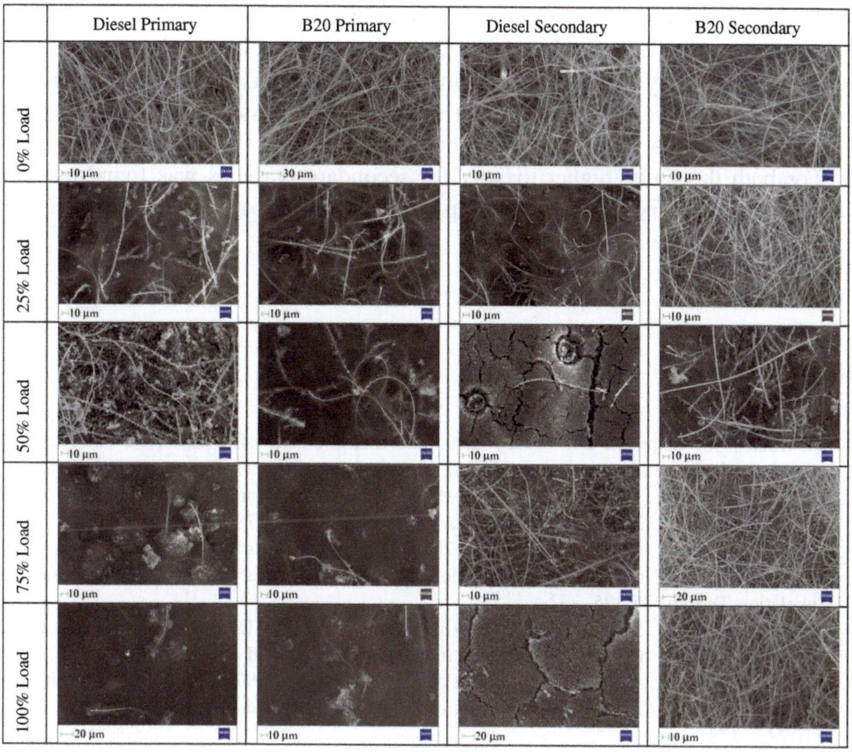

Fig. 10 SEM images for mineral diesel and B20 for primary particulates (1,000× magnification) and aged particulates (1,000× magnification) (adapted from Shukla et al. 2014)

7 Conclusions

Particle number emissions, PM mass emissions, EC/ OC and PAHs emissions were compared for diesel and B20. At higher engine loads, lower amounts of BSOF were measured. BSOF content decreased as the engine load increased and an excess phase transition to gas phase was noticed.

- Secondary emissions for both the fuels showed that the particle number-size distribution peak was shifted towards the right. Number concentration was recorded to be two orders of magnitude lower for B20 as compared to mineral diesel, for both primary as well as secondary emissions.
- Lower number of primary particles in B20 exhaust can be explained by the relatively lower amount of sulphur being present in the B20 as compared to the mineral diesel.
- For primary as well as secondary emissions, overall PM mass was an order of magnitude lower for B20 exhaust as compared to the mineral diesel.

- It was seen that particle collected on the filter substrates was quite low for aged particulates for B20 at 75 and 100 % engine loads, respectively.
- PAHs were very high for both primary and secondary particulates in case of diesel. PAHs were extremely high in secondary emissions for both the fuels. B20 particulates were loaded to a higher degree with PAHs.
- For both the fuels, higher toxicity for secondary particles was found as compared to primary particles. B20 particles showed overall lower toxicity as compared to the mineral diesel.

References

Abdul-Khalek IS, Kittelson DB (1995) Real time measurement of volatile and solid exhaust particles using a catalytic stripper. In: SAE technical paper

Agarwal AK, Gupta T, Kothari A (2011) Particulate emissions from biodiesel vs diesel fuelled compression ignition engine. Renew Sustain Energy Rev 15:3278–3300

Agarwal AK, Gupta T, Dixit N, Shukla PC (2013) Assessment of toxic potential of primary and secondary particulates/aerosols from biodiesel vis-a-vis mineral diesel fuelled engine. Inhalation Toxicol 25:325–332

Akasaka Y, Suzuki T, Sakurai Y (1997) Exhaust emissions of a DI diesel engine fueled with blends of biodiesel and low sulfur diesel fuel. In: SAE technical paper

Atal A, Levendis YA, Carlson J, Dunayevskiy Y, Vouros P (1997) On the survivability and pyrosynthesis of PAH during combustion of pulverized coal and tire crumb. Combust Flame 110:462–478

Aumont B, Madronich S, Bey I, Tyndall GS (2000) Contribution of secondary VOC to the composition of aqueous atmospheric particles: a modeling approach. J Atmos Chem 35:59–75

Bae M-S, Schauer JJ, DeMinter JT, Turner JR, Smith D, Cary RA (2004) Validation of a semi-continuous instrument for elemental carbon and organic carbon using a thermal-optical method. Atmos Environ 38:2885–2893

Baltensperger U, Dommen J, Alfarra MR, Duplissy J, Gaeggeler K, Metzger A, Facchini MC, Decesari S, Finessi E, Reinnig C (2008) Combined determination of the chemical composition and of health effects of secondary organic aerosols: the POLYSOA project. J Aerosol Med Pulm Drug Delivery 21:145–154

Bergmann M, Kirchner U, Vogt R, Benter T (2009) On-road and laboratory investigation of low-level PM emissions of a modern diesel particulate filter equipped diesel passenger car. Atmos Environ 43:1908–1916

Blando JD, Turpin BJ (2000) Secondary organic aerosol formation in cloud and fog droplets: a literature evaluation of plausibility. Atmos Environ 34:1623–1632

Blando JD, Porcja RJ, Li T-H, Bowman D, Lioy PJ, Turpin BJ (1998) Secondary formation and the Smoky Mountain organic aerosol: an examination of aerosol polarity and functional group composition during SEAVS. Environ Sci Technol 32:604–613

Brüske I, Hampel R, Socher MM, Rückerl R, Schneider A, Heinrich J, Oberdörster G, Wichmann H-E, Peters A (2010) Impact of ambient air pollution on the differential white blood cell count in patients with chronic pulmonary disease. Inhalation Toxicol 22:245–252

CAEPA (1993) Benzo[a]Pyrene as a toxic contaminant. Part B: health assessment

Cowley L, Stradling R, Doyon J (1993) The influence of composition and properties of diesel fuel on particulate emissions from heavy-duty engines. In: SAE technical paper

Den Ouden C, Clark R, Cowley L, Stradling R, Lange W, Maillard C (1994) Fuel quality effects on particulate matter emissions from light-and heavy-duty diesel engines. In: SAE technical paper
Di Y, Cheung C, Huang Z (2009) Experimental investigation on regulated and unregulated emissions of a diesel engine fueled with ultra-low sulfur diesel fuel blended with biodiesel from waste cooking oil. Sci Total Environ 407:835–846
Diaz EA, Chung Y, Papapostolou V, Lawrence J, Long MS, Hatakeyama V, Gomes B, Calil Y, Sato R, Koutrakis P (2012) Effects of fresh and aged vehicular exhaust emissions on breathing pattern and cellular responses-pilot single vehicle study. Inhalation Toxicol 24:288–295
Diaz-Sanchez D, Dotson AR, Takenaka H, Saxon A (1994) Diesel exhaust particles induce local IgE production in vivo and alter the pattern of IgE messenger RNA isoforms. J Clin Invest 94:1417
Diaz-Sanchez D, Tsien A, Fleming J, Saxon A (1997) Combined diesel exhaust particulate and ragweed allergen challenge markedly enhances human in vivo nasal ragweed-specific IgE and skews cytokine production to a T helper cell 2-type pattern. J Immunol 158:2406–2413
Donaldson K, Beswick PH, Gilmour PS (1996) Free radical activity associated with the surface of particles: a unifying factor in determining biological activity? Toxicol Lett 88:293–298
Dusek U, Amann M (2000) Secondary organic aerosol–formation mechanisms and source contributions in Europe. IIASA interim report IR-00-066
Dwivedi D, Agarwal AK, Sharma M (2006) Particulate emission characterization of a biodiesel vs diesel-fuelled compression ignition transport engine: a comparative study. Atmos Environ 40:5586–5595
Eastwood P (2008) Particulate emissions from vehicles. Wiley, New York
Facchini M, Fuzzi S, Lind J, Fierlinger-Oberlinninger H, Kalina M, Puxbaum H, Winiwarter W, Arends B, Wobrock W, Jaeschke W (1992) Phase-partitioning and chemical reactions of low molecular weight organic compounds in fog. Tellus B 44:533–544
Facchini MC, Fuzzi S, Zappoli S, Andracchio A, Gelencsér A, Kiss G, Krivácsy Z, Mészáros E, Hansson HC, Alsberg T (1999) Partitioning of the organic aerosol component between fog droplets and interstitial air. J Geophys Res: Atmos (1984–2012) 104:26821–26832
Forstner HJ, Flagan RC, Seinfeld JH (1997a) Secondary organic aerosol from the photooxidation of aromatic hydrocarbons: molecular composition. Environ Sci Technol 31:1345–1358
Forstner HJ, Flagan RC, Seinfeld JH (1997b) Molecular speciation of secondary organic aerosol from photooxidation of the higher alkenes: 1-octene and 1-decene. Atmos Environ 31:1953–1964
Grosjean D, Seinfeld JH (1967) Parameterization of the formation potential of secondary organic aerosols. Atmos Environ 23(1989):1733–1747
Gupta T, Kothari A, Srivastava DK, Agarwal AK (2010) Measurement of number and size distribution of particles emitted from a mid-sized transportation multipoint port fuel injection gasoline engine. Fuel 89:2230–2233
Gupta T, Dixit N, Agarwal AK, Gupta S (2011) The secondary organic carbon (SOC) formation from a CRDI automotive diesel engine exhaust. In: SAE technical paper
Holes A, Eusebi A, Grosjean D, Allen DT (1997) FTIR analysis of aerosol formed in the photooxidation of 1, 3, 5-trimethylbenzene. Aerosol Sci Technol 26:516–526
Jacobson M, Hansson HC, Noone K, Charlson R (2000) Organic atmospheric aerosols: review and state of the science. Rev Geophys 38:267–294
Kalligeros S, Zannikos F, Stournas S, Lois E, Anastopoulos G, Teas C, Sakellaropoulos F (2003) An investigation of using biodiesel/marine diesel blends on the performance of a stationary diesel engine. Biomass Bioenergy 24:141–149
Kaul D, Gupta T, Tripathi S, Tare V, Collett J Jr (2011) Secondary organic aerosol: a comparison between foggy and nonfoggy days. Environ Sci Technol 45:7307–7313
Khair MK, Majewski WA (2006) Diesel emissions and their control. In: SAE technical paper
Kittelson DB (1998) Engines and nanoparticles: a review. J Aerosol Sci 29:575–588
Lange W (1991) The effect of fuel properties on particulates emissions in heavy-duty truck engines under transient operating conditions. In: SAE technical paper

Li X, Wallace JS (1995) A phenomenological model for soot formation and oxidation in direct-injection diesel engines. In: SAE technical paper

Limbeck A, Puxbaum H (1999) Organic acids in continental background aerosols. Atmos Environ 33:1847–1852

Madden MC, Richards JH, Dailey LA, Hatch GE, Ghio AJ (2000) Effect of ozone on diesel exhaust particle toxicity in rat lung. Toxicol Appl Pharmacol 168:140–148

Nisbet IC, LaGoy PK (1992) Toxic equivalency factors (TEFs) for polycyclic aromatic hydrocarbons (PAHs). Regul Toxicol Pharmacol 16:290–300

Ott WR, Siegmann HC (2006) Using multiple continuous fine particle monitors to characterize tobacco, incense, candle, cooking, wood burning, and vehicular sources in indoor, outdoor, and in-transit settings. Atmos Environ 40:821–843

Owen K, Coley T (1995) Automotive fuels reference book

Pan J, Quarderer S, Smeal T, Sharp C (2000) Comparison of PAH and nitro-PAH emissions among standard diesel fuel, biodiesel fuel, and their blend on diesel engines. In: Proceedings of the 48 th ASMS conference on mass spectrometry and allied topics, Long Beach, CA

Ravindra K, Sokhi R, Van Grieken R (2008) Atmospheric polycyclic aromatic hydrocarbons: source attribution, emission factors and regulation. Atmos Environ 42:2895–2921

Robinson AL, Donahue NM, Shrivastava MK, Weitkamp EA, Sage AM, Grieshop AP, Lane TE, Pierce JR, Pandis SN (2007) Rethinking organic aerosols: semivolatile emissions and photochemical aging. Science 315:1259–1262

Rogge WF, Mazurek MA, Hildemann LM, Cass GR, Simoneit BR (1993) Quantification of urban organic aerosols at a molecular level: identification, abundance and seasonal variation. Atmos Environ. Part A. General Top 27:1309–1330

Ruiz PA, Lawrence JE, Wolfson JM, Ferguson ST, Gupta T, Kang C-M, Koutrakis P (2007) Development and evaluation of a photochemical chamber to examine the toxicity of coal-fired power plant emissions. Inhalation Toxicol 19:597–606

Saxena P, Hildemann LM (1996) Water-soluble organics in atmospheric particles: a critical review of the literature and application of thermodynamics to identify candidate compounds. J Atmos Chem 24:57–109

Seinfeld JH, Pandis SN (2012) Atmospheric chemistry and physics: from air pollution to climate change. Wiley, New York

Sharma M, Agarwal AK, Bharathi K (2005) Characterization of exhaust particulates from diesel engine. Atmos Environ 39:3023–3028

Shukla PC, Gupta T, Agarwal AK (2014) A Comparative morphological study of primary and aged particles emitted from a biodiesel (B20) vis-à-vis diesel fuelled CRDI engine. Aerosol Air Qual Res 14:934–942

Tsien A, Diaz-Sanchez D, Ma J, Saxon A (1997) The organic component of diesel exhaust particles and phenanthrene, a major polyaromatic hydrocarbon constituent, enhances IgE production by IgE-secreting EBV-transformed human B cells in vitro. Toxicol Appl Pharmacol 142:256–263

Turpin BJ, Saxena P, Andrews E (2000) Measuring and simulating particulate organics in the atmosphere: problems and prospects. Atmos Environ 34:2983–3013

Zielinska B, Goliff W, McDaniel M, Cahill T, Kittelson D, Watts W (2003) Chemical analyses of collected diesel particulate matter samples in the E-43 Project. National Renewable Energy Laboratory, Ann Arbor, MI

Emissions and Soot in Partially Premixed Combustion

Surbhi Kohli and Abhijit Kushari

Abstract Practical combustion systems typically undergo non-premixed mode of combustion, in which the fuel is injected directly into the combustion chamber, which already has compressed air. The emissions tend to be high in this diffusion mode of combustion as compared to premixing mode. But a switch to premixing is not desirable for practical systems as the premixed air-fuel mixtures are more accident-prone. Partial premixing implies that the flame is only partially premixed and in parts, it is also a diffusion flame. This refers to the condition when a small amount of air or oxygen is mixed with the fuel (rich mixture), before it enters the reaction zone and then further at the reaction zone, more quantities of oxidizer are added, so as to make the overall mixture lean and to complete the combustion. Experimental study of partially premixed, co-flow, LPG-air flames are reported in this chapter, which show substantial improvement in emission and soot characteristics in comparison to a pure diffusion flame for a swirl stabilized combustor when the premixing equivalent ratio was kept between 1.5 to 2.0.

Keywords Partially premixed combustion (PPC) · Liquid petroleum gas (LPG) · Equivalence ratio · Diffusion flame · Health hazards · CO and NO emissions

1 Introduction

Humans have unscrupulously used available resources on earth and have extracted the maximum out of them in order to ease their life on this planet. But in doing so, they have also disturbed the ecological balance of nature. Thus, with the excessive exploitation of resources; the entire flora and fauna, including mankind itself is in danger. Omens for this are already being seen—the erratic climatic patterns all over

S. Kohli · A. Kushari (✉)
Department of Aerospace Engineering, Indian Institute of Technology Kanpur, Kanpur, Uttar Pradesh 208016, India
e-mail: akushari@iitk.ac.in

A.K. Agarwal et al. (eds.), *Novel Combustion Concepts for Sustainable Energy Development*, DOI 10.1007/978-81-322-2211-8_19

the world, global warming, extinction of various species, health problems, advent of new diseases, loss of fertile land which ultimately is used to feed us humans and animals, loss of aesthetic beauty of nature and so on. Thus, such alarming happenings are indeed a sign for humans to try and undo the harm, which has already been done to the environment.

It is also a fact, though, that through technological advancements, the quality of life has improved. Talking from the perspective of Aerospace Engineering, travelling time has been greatly reduced and areas presumed to inaccessible at one point of time are now fully connected by air. But the increasing air traffic across the globe has its own set of concerns, and one of the primary issues to be addressed is that of air pollution. While considering the process of gas turbine combustion (aircraft engines), the emissions are already quite low, as the air-fuel mixture is extremely diluted. So, the emissions per unit volume of exhaust are much reduced. Despite this fact, since the airplanes travel at a high altitude, these incredibly small emissions can also affect the atmospheric cycles, climatic patterns and cloud formation. Recently, many erratic weather patterns are being experienced all over the world, part of which has been attributed to the increased air traffic all across the globe. For example, the frequent occurrence of hurricanes in the United States in the recent years has also been correlated with the increased number of flights taking off and flying over the States. Thus, there is an urgent need to further reduce the aircraft emissions, and still more stringent norms are being applied. Another area of concern is that the environment in and around airports has been found to be extremely toxic, mainly because of the pollutants emitted while the aircrafts take off and land there. That poses direct hazard to the passengers and the employees present in the vicinity. Apart from these issues, the more complete the combustion attained in a gas turbine engine, lesser is the fuel consumption, which translates into saving of energy, reduction in operation cost and also corresponds to a reduction in the weight of the aircraft.

Conventional combustion systems use either premixed or non-premixed combustion, which have their own pros and cons. Premixed flames occur when the fuel and air are premixed *before* entering the reaction zone. Hence, the fuel–air mixture itself enters the reaction zone and ignites when subjected to a source of ignition, say a spark plug. On the other hand, in the case of non-premixed or diffusion flames, first the air enters the combustion chamber, and whenever combustion is desired, fuel is injected or sprayed into the air in the combustion chamber and ignited. Relative merits and demerits of premixed and non-premixed combustion are listed in Table 1.

It can be seen that both kinds of combustion phenomena have their own pros and cons. So, an alternative was sought after and partially premixed flames were deemed to be a good compromise between completely premixed and completely diffusion flames. It incorporates the advantages of both and overcomes both their drawbacks. As the name suggests, *partial premixing* implies that the flame is only partially premixed and in parts, it is also a diffusion flame. The term '*partial premixing*' hence, basically refers to the condition when a small amount of air or oxygen is mixed with the fuel (rich mixture), before it enters the reaction zone and then further at the reaction zone, more quantities of oxidizer are added, so as to

Table 1 Differences between premixed and non-premixed combustion

Premixed combustion	Non-premixed combustion
Fuel and air are premixed before entering the combustion chamber	Air enters the combustion chamber first, and fuel is injected into it separately
More complete combustion is expected due to better mixing	Combustion is often incomplete due to insufficient time for mixing
Flame front is thinner, broader and more contained	Flames are less localized as combustion occurs only at the air-fuel interface
Flames are broader and smaller in height	Longer, narrower flames
Flames are bluish in colour	Flames are mostly incandescent yellow
Faster burning, as fuel and air are already mixed	Burning rate is slower as the combustion depends on the contact of the fuel with air, i.e. mixing time is a limiting factor
Dangerous as the premixed mixture is within the flammability limits before the desired ignition, accident prone	Safer to use, as the fuel is injected only when ignition is desired
Emissions of NO_x can be higher, because of higher temperatures; CO, and UHC emissions are minimal	Generally, NO_x emissions are lesser, but CO and unburnt hydrocarbon emissions are too high, indicating incomplete combustion
Examples: spark ignition (SI) engines use premixing	Examples: Includes most of the practical combustion systems; compression ignition (CI) engines, candle flame

make the overall mixture lean and to complete the combustion. The air, which is premixed with the fuel, is termed as *primary air* or the *premixed air*, and the air that is introduced at the reaction zone is the *secondary air* or the *main combustion air* or the *diffusion air*. The idea is to have a little bit of premixing, which would keep the mixture below the rich flammability limit, while the secondary or diffusion air would make the mixture globally extremely lean. This can help prevent accidental explosions of the flammable premixed mixtures.

Partially premixed flames can also occur either by design or under conditions arising due to different phenomena, like poor mixing, spray vaporization, flame lift-off and local extinction, followed by re-ignition occurring in turbulent flames (Xue and Aggarwal 2003). Local variations in the fuel–air ratio or the equivalence ratio can occur due to imperfect and/or progressive mixing of the fuel and the air streams, which can also be termed as partially premixed combustion (PPC) (Mitrovic and Lee 1998). Partially premixed flames produce much less emissions, mainly as a result of much better and improved combustion. They are also deemed to be safer and more stable as compared to pure diffusion or pure premixed flames (Jeong et al. 2006). Various studies have already been conducted on partially premixed flames, both in co-flow (the premixed fuel–air mixture is flowing in the same direction in the combustion chamber as the main combustion air outside, which gives a co-annular configuration), and counter-flow configuration, where the fuel–air premixed mixture is in the opposite direction to the oxidizer flow direction.

Liquefied petroleum gas (LPG) can be chosen as a fuel, mainly because of its easy availability and widespread use in domestic, industrial as well as transport sectors all across India. LPG is a non-toxic, non-corrosive fuel and is free of tetraethyl lead. It also has a higher calorific value as compared to natural gas and other commonly used fuels. The octane rating is high too. It has a broad range of flammability limits and hence can also be used in very lean burning conditions. It has a lower carbon content, which can reduce carbon footprints. An engine working with LPG as fuel produces much less engine noise. Even in case of spills, it vaporizes quickly and is not water-soluble; so it does not harm the soil and underground water sources. But when compared with gasoline and other fuel oils, it has lesser energy density, which increases the equivalent LPG consumption. Another problem is that of cold start for use in extremely cold places. Thus overall, it is not much economical and can also lead to higher emissions (Sreenivasan et al. 2012). Therefore, there needs to be an optimum level of partial premixing, otherwise the emissions might increase.

Ideally, it is desirable to have complete combustion in a combustion system in which the fuel should burn in the presence of oxygen (air) to given CO_2 (for hydrocarbon fuels) and H_2O. But, in actual practice, due to incomplete combustion, and/or due to the presence of very high temperatures, some other reaction also tend to occur given rise to the formation of CO, unburnt hydrocarbons (VHC), volatile organic compounds (VOC), various radicals, NO_x, SO_2, etc., which are detrimental to the human health and environment.

Carbon monoxide is a tasteless, colourless and odourless gas that is slightly lighter than air. The natural atmospheric level is 0.1 ppmv (parts per million by volume), and on an average, exposures at 100 ppm or greater are considered dangerous to human health. When inhaled in large concentrations, it can be toxic to humans and animals. At lower levels of exposure, CO can cause mild effects that can be easily mistaken for flu, like headaches, dizziness, disorientation, nausea and fatigue. At moderate concentrations, angina, impaired vision and reduced brain function may result. In higher concentrations and on prolonged exposure, CO forms carboxy-haemoglobin which reduces the oxygen carrying capacity of blood, which can ultimately lead to death. CO is mostly formed when the fuel–air mixture is rich, and there is insufficient oxygen to oxidize all the carbon in the fuel to CO_2 (incomplete combustion). Also, if the combustion is occurring at stoichiometric or moderately fuel-lean conditions, a significant amount of CO is produced from the dissociation of CO_2. CO formation can also be attributed to the following reasons:

- Inadequate burning rates due to extremely lean fuel–air mixtures, or insufficient residence times: CO is formed in large amounts as the temperature might not be high enough due to the presence of excess of air, which precludes the conversion of CO to CO_2. Similarly, low residence time implies that there is not enough time for the CO to be converted to CO_2 despite the presence of excess of oxygen.
- If the mixing of the air and the fuel is improper or inadequate: This might produce some local regions where the mixture strength is too low to support combustion and some locally over-rich combustion zones; both of which might yield higher concentrations of CO.

- In the post-flame region, there might be a quenching of the chemical reactions, due to the entrainment of the combustor liner cooling air, especially in the primary or premixed zone in a gas turbine engine.

The main factors which influence the combustion efficiency and hence CO emissions are the engine and the combustor inlet temperatures, combustion pressure, primary-zone equivalence ratio; and for liquid fuels, the mean droplet size of the spray is also significant.

The unburned hydrocarbon is a collective term coined for the hydrocarbon species, which are not completely burnt or oxidized to CO_2 and are produced during the course of combustion reaction, and includes unburnt fuel molecules. They mainly occur in the combustor in the form of drops or vapour, as well as the products of thermal degradation of the parent fuel into species of lower molecular weight. They might be formed due to poor atomization (in case of liquid fuels), inadequate burning rates, chilling effects of film-cooling air, etc. The reaction kinetics of UHCs are much more complex than that for CO formation, though more or less the same factors which affect the CO formation would affect the formation of UHCs as well. Compounds like Poly Aromatic Hydrocarbons (PAHs), Benzene, Toluene, Xylene (BTX) and Aldehydes like methanal (HCHO) can also be grouped in this category. Control of these unburned hydrocarbons is vital, as these compounds are extremely hazardous to health, in that they have both short-term as well as long-term health effects. For example, HCHO can cause irritation in eyes over a short term of exposure, and BTX is supposed to be a carcinogen.

'Particulate matter', also known as particle pollution or PM, suspended particulate matter (SPM), and soot, is a complex mixture of extremely small particles and liquid droplets. It basically implies extremely small solid particles suspended in a gaseous or liquid medium. Particle pollution is made up of a number of components, including acids (such as nitrates and sulphates), organic chemicals, metals and soil or dust particles (USEPA 2013). The size of particles is directly linked to their potential for causing health problems. Environmental Protection Agency (EPA) is concerned about particles that are 10 μm in diameter or smaller, because those are the particles that generally pass through the throat and nose and enter the lungs. Once inhaled, these particles can affect the heart and lungs and cause serious health effects. EPA groups particle pollution into two categories (USEPA 2013):

- 'Inhalable coarse particles or PM 10', such as those found near roadways and dusty industries, are larger than 2.5 μm and smaller than 10 μm in diameter.
- 'Fine particles or PM 2.5', like those found in smoke and haze, are 2.5 μm in diameter and smaller. These particles are emitted directly from sources such as forest fires, or they can form when gases emitted from power plants, industries and automobiles, react in the air.

PAHs are clusters of aromatic rings, up to 4 rings in size, which are formed under fuel-rich combustion conditions. Soot is formed when under the appropriate conditions; these PAHs grow in size up to 20–50 nm. Soot is formed from combustion processes in locally fuel-rich zones at elevated temperatures (generally >1,300 K), and is primarily composed of carbon (can be even up to 96 %), up to 10 % hydrogen on a molar basis, and even more in young soot, oxygen and other elements. Formation of soot proceeds in three main chemical and physical steps:

1. Formation and growth of large aromatic hydrocarbons and their transition to particles (inception of soot particles)
2. Growth of these solid particles by the addition of components from the gas phase (surface growth)
3. Coagulation of the primary particles to large aggregates (particle agglomeration).

The net emission of soot depends on the competition between soot formation and oxidation (Hadef et al. 2010). The basic mechanism for the formation of PAHs and soot is that, first a precursor species, like acetylene (ethyne) is formed under fuel-rich conditions. This ethyne and other hydrocarbon fragments combine through pyrolytic reactions (in the absence of oxygen), to form the larger ringed structures, which are known as PAHs. These clusters grow in size, in the absence of oxidation reactions, which break the ring structures. Small particles of a certain critical size are formed by the growth of such aromatic clusters by both physical and chemical means, i.e. coagulation. At this point, these particles are termed as primary soot particles. Many of these soot particles, now grow by particle agglomeration (physical growth) and surface growth (chemical reactions), while remaining suspended with the precursor species, clusters and adolescent soot particles.

Thus, while low temperatures generally inhibit combustion and lead to the formation of more unburnt hydrocarbons, even very high temperatures would hinder combustion. At extremely high temperatures, the hydrocarbons in the fuel are pyrolysed and stripped of the hydrogen atoms. The remaining carbon atoms form soot at these temperatures.

As such particulate matter is known to cause irritation in eyes, coughing and the blackening of face. Particulate matter also increases the risk of coronary heart disease. Soot particles smaller than 300 nm in size are known to penetrate deep into the lungs and its alveoli, and hence can have a negative impact on the health. Soot also contributes to visible air pollution and haze.

Particulate matter and soot formed in gas turbine engines are present at high altitudes, which affect cloud formation and weather patterns. Since these particles are present above the cloud formation zones, they do not come down with rains and hence remain suspended in the stratosphere. Also, PAHs are oxidized more rapidly in the urban atmosphere and may get stabilized if they get adsorbed on the surface of particulates like soot. Then they become quite long-lived and capable of penetrating into the lungs as well. PAHs are also known mutagens, co-carcinogens and carcinogens.

Exhaust smoke is caused by the production of finely divided soot particles in fuel-rich regions of the flame, which in conventional combustors are always close to the fuel spray. But most of the soot produced near to the fuel spray, is consumed in the high-temperature regions downstream. Therefore, in a gas turbine engine, the primary-zone dictates the rate of soot formation, while the intermediate zone determines the rate of soot consumption. The soot concentration is mainly given by the net effect of these two effects. Soot concentration mainly increases with an increase of pressure, which results in chemical changes, extension of flammability limits, etc.

Carbon dioxide is a trace gas present in the earth's atmosphere. As of November 2011, the reported concentration of CO_2 has been estimated to be 390 ppmv or 591 parts per million by mass. It is an important gas, in that it is a prerequisite for the process of photosynthesis in plants. It is deemed as the product of complete combustion of a hydrocarbon fuel in oxygen, along with water. Hence, more complete the combustion; more is the amount of CO_2 produced. Therefore, for the automotive industry, CO_2 is not considered as a pollutant. But since it is a greenhouse gas; the alarming rise in the atmospheric CO_2 concentration has caused environmental concerns over the regulation of CO_2 levels coming out from vehicles, industries as well as aircrafts. Thus, to reduce the CO_2 levels, the aim must be to reduce the fuel consumption for the same efficiency and power.

Nitrogen is the most abundant gas contained in the atmosphere, contributing to about 79 % of the atmosphere's composition by volume. Hence, during combustion in air, nitrogen also takes part provided the temperatures are sufficiently high to produce such oxides of nitrogen. These oxides are highly dangerous pollutants; both for the environment as well as for the human health. Nitric Oxide (NO), Nitrogen dioxide (NO_2), Nitrous Oxide (N_2O) are collectively termed as oxides of Nitrogen or NO_x in short. NO_x is one of the main concerns when it comes to the control of emissions.

Nitric Oxide or Nitrogen Monoxide (NO) is a gas that has a significant role in the biological processes in human beings. The presence of a lone electron on the Nitrogen atom makes it highly reactive. Therefore, as soon as it is exposed to air, it rapidly oxidizes to NO_2. The conversion of NO to NO_2 is favoured at low temperatures. NO might also subsequently convert to nitric acid (HNO_3), which can fall as acid rain on earth. Apart from this, it also participates in ozone layer depletion and is the major precursor of photochemical smog.

Nitrogen dioxide (NO_2) is reddish brown, toxic gas, with a characteristic sharp odour. It is a toxic gas when inhaled, but its acrid smell helps its detection even at low concentrations. Hence, exposure to inhalation can generally be avoided. Current scientific evidence links short-term exposures to NO_2, ranging from 30 min to 24 h, can cause adverse respiratory effects, including irritation in the respiratory tract, airway inflammation in healthy people and increased respiratory symptoms in people with asthma. Long-term exposures at concentrations above 40–100 $\mu g/m^3$, may decrease lung function and increase the risk of respiratory syndromes. Also, symptoms of poisoning (lung oedema) can appear several hours after inhalation of a low, but a potentially fatal dose.

Nitrous Oxide (N_2O) is a colourless, non-flammable gas with a slightly sweet odour and taste. It is also known as 'laughing gas' or 'sweet air'. It is mainly used as an anaesthetic and analgesic in surgery and dentistry. It behaves as a strong oxidizer, similar to molecular oxygen; hence, it is often employed in rocketry and motor racing as an oxidizer to enhance the power output of engines.

On the other hand, nitrous oxide liberates NO, which reacts with ozone. It is the main, naturally occurring regulator of stratospheric ozone. It is also a major greenhouse gas, and has almost 300 times greater Global Warming Potential than CO_2. Also, its long lifetime allows its transportation into the stratosphere, where it can participate in ozone depletion. It is often associated with asphyxiation hazards. Exposure to N_2O might also cause short-term decrease in mental performance, audiovisual ability and manual dexterity. Long-term exposures can cause Vitamin B_{12} deficiency, numbness, reproductive side effects, neurological problems, etc.

In general, NO_x reacts with ammonia, moisture and other compounds to form small particles, which penetrate deeply into the sensitive parts of the lungs like the alveoli. This can cause or aggravate respiratory diseases like emphysema and bronchitis. It is also dangerous for heart patients, and can even cause premature deaths. Also, Ozone is formed when NO_x and VOC react in the presence of heat and sunlight. Ozone might prove hazardous, especially for children and for elderly people, with lung diseases such as asthma.

There are basically four different mechanisms of NO_x formation (Lefebvre 1999): Thermal NO, Nitrous Oxide Mechanism, Prompt NO and Fuel NO. Thermal NO Mechanism is also popularly known as the Zeldovich Mechanism. According to this mechanism, NO is formed by the oxidation of atmospheric nitrogen in high temperature regions of the flame and in the post-flame gases. The thermal NO production declines rapidly as the temperature is reduced. All in all, we can say that the high flame temperatures (>1,800 K), large residence time of the burnt products in the combustor and availability of an excess of oxygen, all would promote the formation of thermal NO_x.

Prompt NO mechanism is also known as the Fenimore NO_x mechanism. The NO_x formed here is in the flame region, very close to the burner and hence is termed as prompt NO_x. Prompt NO_x can be significant in quantity; under low temperature, fuel-rich conditions, where the residence times are short. Such conditions can be created in surface burners, staged combustion systems and gas turbines. This mechanism is generally not prevalent in extremely lean conditions.

Nitrous oxide (N_2O) mechanism is not a major source of NO production and occurs basically under fuel-lean, low-temperature conditions. Based on the origin of the fuel, nitrogen can be present in a bound form in the fuel itself. Plants contain nitrogen in the form of proteins, amino acids, alkaloids, chlorophyll and porphyrins, etc. These were transformed during the coalification process into polycyclic aromatic compounds with pyridinic, pyrrolic or other functional groups, and hence the nitrogen is fixed in the coal, which can get converted to NO_x during combustion. Such a path of NO_x formation is called fuel-bound NO_x. Light distillate fuels contain less than 0.06 % of organically bonded nitrogen or fuel-bound nitrogen, but the heavy distillates may contain as much as 1.8 %. This nitrogen from the fuel might react during combustion to form the fuel NO. Fuel-bound NO is the main source of NO in rocket engines, as no atmospheric nitrogen is available there. The fraction of nitrogen undergoing oxidation to NO, changes slowly with the increasing flame temperature. Gaseous fuels like LPG, which contain little or no fuel-bound nitrogen would have an insignificant amount of NO production through this mechanism.

As has been discussed so far, most of the emissions from practical combustion systems are due to either very high flame temperature or incomplete combustion. Therefore, reducing the combustion temperature can lead to a decrease in NO_x emissions and soot but that will slow down the reaction leading to the formation CO and UHC. PPC addresses both these issues and hence is a good candidate for 'greener' energy production. The scientific literature was explored and it was found that various studies have been done on partially premixed flames, purely premixed and purely diffusion flames, which focus on either or both experimental measurements and computational derivations of various pollutants as well as other parameters like pressure and temperature variations. Some of the literature reviewed has been enlisted here.

Xue and Aggarwal (2003) studied NO_x emissions in *n*-heptane/air partially premixed flames in the counter-flow configuration using Held's mechanism for *n*-heptane and Li and William's mechanism for NO_x. They observed that the rate of NO formation is generally higher in the non-premixed zone, and a NO_x destruction region is present in between the rich premixed and the non-premixed zone. This can be explained by the transport of acetylene from the rich premixed to the non-premixed zone, which forms CH and CH_2 radicals. These radicals enhance prompt NO formation in the non-premixed zone, as well as act to consume NO in the destruction region. The higher concentrations of O and OH radicals in the non-premixed zone also enhance both thermal and prompt NO formation there. In the rich premixed zone, the contribution of thermal NO is higher than prompt NO, and vice versa in the non-premixed zone. They concluded that for prompt NO to form, the presence of O and OH radicals is also necessary in addition to CH radicals. And that there existed an optimum value of Φ, for which $EINO_x$ could be minimized for moderate and high strain rate flames.

Mitrovic and Lee (1998) used a laser induced incandescence (LII) setup to measure the soot volume fraction in ethylene partially premixed flames. The fuel flow rate was maintained constant and the equivalence ratios (Φ) were varied from a pure diffusion flame (∞) to about 2.5. The idea behind keeping the fuel constant was to fix the carbon content, so that any changes in the soot concentration can only be attributed to changes in the efficiency of combustion or the chemistry, rather than due to any change in the fuel content. They observed that initially, on increasing the premixing air, the soot emissions increase till an equivalence ratio of 20. This is explained on the basis that the extra oxygen molecules serve to increase the local radical pool in the soot pyrolysis region, and these radicals are in turn precursors to soot formation. Partial premixing significantly alters the chemistry of intermediate hydrocarbons, which can account for the initial increase in soot production. But eventually, at higher levels of premixing at an equivalence ratio of 10, the soot concentration reduces as expected, because the oxidation of soot dominates over the soot production. Also, the soot is first formed in the annular region, while at downstream locations; the soot is detected in the central region of the flames too.

Sreenivasan et al. (2012) studied partial premixing in co-flow configuration using LPG as the fuel. They observed that for zero and low levels of premixing, the flame height reduces and the soot inception point from the burner tip increases with increase in the secondary air flow rate. Also, with increase in the premixing air flow rate, CO emissions decrease initially and later attain an almost constant value, at around 40 % of premixing air. For lower premixing levels, increase in percentage of secondary air flow rate leads to steep reduction in CO concentrations. This may be because of the increased availability of air and hence oxygen, which promotes the oxidation of CO to CO_2. The CO_2 concentration initially decreases with increasing percentage of premixing air, for a particular secondary air flow rate, till about 40 % of premixing air. On further addition of oxygen/air here, the CO_2 concentration increases, perhaps due to the oxidation of CO to CO_2. At a particular premixing level, the CO_2 concentration increases with an increase in the percentage of secondary air flow rate, till 40 % of premixing air, and for higher levels of premixing, the reverse trend is followed. The NO concentration is quite low, and a minimum is observed around 40–50 % of premixing air. For 50–86 % of premixing air, NO is seen to increase as the secondary air increases from 100 to 150 %, and then decreases at 200 % of secondary air. These effects can be attributed to the combined effects of increased oxygen content and the temperature in the combustor. They concluded that for such a configuration, a premixing level of 45 % is optimum, as it gives minimum emissions and higher flame stability for LPG-air flames.

Mishra et al. (2005) measured the concentrations of hydrocarbons like C_2H_2 (ethyne) in methane-air partially premixed flames using gas chromatography. With the addition of primary air, the flame colour is seen to change from orange of the non-premixed flame to blue. A double flame structure occurs at a premixed equivalence ratio (Φp) of about 2.13. The inner flame height reduces with the increase in the premixing air flow rate and its colour changes from blue to greenish at higher premixing levels. The blue flame indicates the presence of CH radicals, while the green colour indicates a rise in C_2 radical concentration. Initially, the hydrocarbons are located within the inner flame and with the increase in premixing; the non-fuel hydrocarbon concentration shows a decrease for most of the locations within the inner flame. The peak centreline concentrations of ethyne show a monotonic decrease with the increase in premixing, indicating a reduction in soot formation; attributed to a combined effect of fluid mechanics and chemical kinetics.

Mungekar and Atreya (2007) studied the formation of NO, soot and OH radicals and temperature distributions in counter-flow, partially premixed methane-air flames. A double flame structure is observed at sufficient levels of partial premixing. As oxygen is progressively added to the fuel side, the measured NO concentration increases significantly and the NO profile broadens, as compared to a non-premixed flame. The effect of radiation from the soot particles also affects the NO formation mechanisms and NO concentration. At higher premixing levels, two NO formation zones and one NO destruction zone have been identified. As the premixing air increases, the thermal NO formation increases with the increasing temperature, and slowly the formation rate of NO overcomes its destruction rate, leading to a net rise in the NO concentration. The EINO also first increases and then decreases with increasing partial premixing.

Alder et al. (1998) studied the trends of $EINO_x$, for laminar ethane and turbulent methane partially premixed flames, with varying premixed (Φp) and overall equivalence ratios (Φt). The prompt NO_x formation mechanism is seen to be active majorly in the inner premixed flame, while the thermal NO_x mechanism is more prevalent in the outer non-premixed flame zone, for the double flame structure. It is the redistribution of and changes in the contributions of these mechanisms, as well as the NO reburn and reduction reactions occurring in these zones, which dictate the net $EINO_x$ values and the $EINO_x$ minima. For a fixed Φt, the $EINO_x$ decreases with decreasing Φp, reaches a minimum and then increases again. For laminar flames, the closely grouped $EINO_x$ values; for $\Phi p < 2.2$, show lesser drop in the peak flame temperature, which agrees with the prompt NO_x mechanism. For $\Phi p > 2.2$, the drop in temperatures is much higher and the $EINO_x$ values are widely spaced, and hence it corresponds to the Zeldovich mechanism. Similarly, for turbulent flames, the $EINO_x$ values are closely spaced and show lesser sensitivity to Φt for $\Phi p < 1.5$, and are widely spaced and show greater sensitivity to Φt variations, for $\Phi p > 1.5$.

Lyle et al. (1999) studied the emission indices for NO_x, CO and HC in a turbulent partially premixed flame, formed by injecting rich methane-air mixtures through a central burner tube, into a co-flow of air. A double flame structure was also observed around $\Phi p = 1.5 - 1.2$. The NO_x emission index with a fixed fuel flow rate and a fixed Φt, is independent of Φp, the flame height and the global residence time for low premixing levels ($5 \leq \Phi p \leq \infty$). For $1.5 \leq \Phi p \leq 5$, the $EINO_x$ decreases with decreasing flame height and decreasing residence time and reaches a minimum at around $\Phi p = 1.5$. With further premixing till $\Phi p = 1$, the $EINO_x$ shows a rapid increase, with decreasing flame height and global residence time, to a maximum for the completely premixed case. This is due to the formation of thermal NO_x. Similar trends are observed for EINO. In the rich premixed flame, NO is generally formed by the prompt NO_x mechanism, and in the diffusion flame, NO is formed generally by the thermal NO_x mechanism. The temperature increases monotonically, while the flame height and the residence time monotonically decrease with increasing premixing. Below $\Phi p < 1.5$, NO production rate is faster than the reduction in the residence time, so the NO_x formation can be attributed to the wider high temperature zones (thermal NO_x mechanism). The NO_x increase from $\Phi p = \infty$ to $\Phi p = 2$ is also probably due to the slowly increasing temperature. The decrease in $EINO_x$ and EINO from $\Phi p = 2$ to $\Phi p = 1.5$, despite the increase in the mean temperature, can be due to the changes in the contributions of the prompt and re-burn mechanisms. Sufficient partial premixing also results in a 25–35 % reduction in the experimentally observed $EINO_x$. The EICO and EIHC are negligibly small, and of the order of ambient air measurements, indicating complete combustion. Thus, there is no trade-off between reduced NO_x and increased CO and HC concentrations.

Mishra et al. (2006) compared the structures of methane-air and propane-air partially premixed flames and saw that both these alkanes exhibit a double flame structure for a certain range of equivalence ratios. The inner flame for propane is smaller than that of methane. The acetylene concentration is higher in the propane flame, and since acetylene is a precursor of soot, it explains why propane flames are sootier than methane flames. Also, the concentration of CO is higher in the inner flame, while the CO_2 concentration is higher in the outer flame, for both the fuels.

This shows that initially, the fuel is consumed in the inner, rich, premixed flame, forming intermediates and CO, which undergo further complete combustion to form CO_2. The difference in the concentrations is due to the fact that propane has C–C bonds in addition to C–H bonds. The C–C bonds being weaker, initiate the chain reactions sooner for propane, and thus, the auto-ignition temperature of propane is lower than that of methane.

Gore and Zhan (1996) measured the flame heights, mole fractions of pollutants and other major stable gas species like CO_2, CO, H_2, O_2, N_2 and C_2H_2, in a fuel-rich mixture of methane and air, in an over-ventilated co-flow of secondary air. The overall equivalence ratio was fixed at 0.5. They observed that the visible flame length and luminosity decrease with increasing amounts of premixing air. The $EINO_x$ emission index has a value of 1.1 g/kg of fuel for a pure diffusion flame, which remains almost constant near this value till the premixing air reaches $\Phi p = 4$. It is because the decreasing flame height accounts for the decrease in the residence time and thus a decrease in the rate of production of thermal NO_x. An increase in temperature as indicated by the peak in CO_2 and H_2O mole fractions (more complete combustion) and a broadening of the high temperature region occurs with the advent of double flames, which increases the rate of thermal NO_x production. This increase in thermal NO_x might be possibly countered by the decrease in prompt NO_x production, and thus almost no net change in NO_x concentration is evident. EICO remains constant for $\Phi p > 10$, but for $\Phi p \leq 10$, CO emissions decreased monotonically as expected. This supported more complete combustion, even though the Φp for minimum NO_x does not give the minimum CO. The HC emissions also showed a trend similar to EICO, but the Φp for minimum NO_x showed minimum HC emissions as well.

Semerjian et al. (1979) studied emissions of NO_x, CO and Total Hydrocarbons (THC) with respect to the local equivalence ratio for a 'partially' mixed flame. For $\Phi = 0.4$, the THC and CO are maximum at the centre of the burner and decrease towards the edges. This indicates that the fuel is rapidly oxidized at the burner peripheries, since the temperatures are higher in these richer zones, while leaner mixtures and lower temperatures at the centre of the burner favour an increase in CO and THC production. The NO_x concentration shows a reverse trend, being minimum at the centre and higher at the peripheries, as due to lower temperatures, lesser thermal NO_x is formed at the centre and vice versa. For $\Phi = 0.78$, no unburned hydrocarbons are detected, and CO is minimum at the centre, indicating complete combustion and leanest fuel–air mixture. For flames between $0.6 \leq \Phi \leq 1.0$, the NO_x levels are below the corresponding premixed levels while the CO levels are correspondingly higher. The low NO_x levels might be due to convective and radiative heat losses near the walls.

Kim et al. (2009) studied the mixing effects of fuel and air in a lean premixing mode (partial premixing but with a lean premixed mixture) and compared the emissions of NO_x and CO in each case. At higher equivalence ratios, since the oxygen supply is reduced, the temperature of the reaction zone increases, which in turn decreases the CO emissions. Also, the temperature rise favours thermal NO_x formation. Both NO_x and CO emissions are observed to be much lower with a lean premixed combustor, than a diffusion flame combustor. With a decrease in the equivalence ratio, CO emissions increase and the NO_x emissions decrease, as the combustor is cooled with an excess of air. So, the lean premixed combustor is shown to be highly efficient when it comes to NO_x reduction.

Rink and Lefebvre (1987) studied the formation of different pollutants for liquid fuel combustion with changes in spray characteristics, like the mean spray diameter, equivalence ratio. The air supplied was atomized with airblast atomizers, and the fuel used for this study was light diesel oil (DF-2). The UHCs decrease as Φ increases, reaches a minima at around stoichiometric, then increases again. The initial high levels of UHCs at low Φ imply incomplete combustion of the fuel, due to low reaction temperatures, while higher levels of UHCs at higher values of $\Phi > 1$ are due to incomplete combustion of the richer mixture, due to a paucity of O_2. Hence, CO_2 concentration showed a trend opposite to that of UHCs. NO increases as Φ increases, reaches a maximum slightly on the lean side of the stoichiometric conditions, and then reduce. The initial rise in NO is due to an improvement in combustion, leading to higher temperatures, while the subsequent reduction in NO can be due to reduced flame height and lowered residence times. As the atomization quality improves and the mean drop size decreases, the combustion quality also improves, because smaller fuel particles can vaporize much faster and hence combust more quickly. Hence, though the concentration of UHCs is reduced, that of CO_2 and NO increases due to increased residence times, as the combustion now occurs sooner. The particulate emissions show an increase with an improvement in the atomization quality, as the effective fuel–air ratio, i.e. the ratio of the mass of fuel vapours to the mass of air, increases. For lean zones, reduced drop size leads to reduced particulate emissions, as it helped enhance the evaporation rates, and so, most of the fuel could burn in the premixed mode. For richer zones, as the atomization improves, the enhanced evaporation leads to higher effective equivalence ratios, which increases soot formation.

Gülder (1995) studied the effects of oxygen addition on the soot formation in overventilated, co-flow, axisymmetric, laminar diffusion flames of methane, propane and *n*-butane. The addition of oxygen on the fuel side had three main effects: one was the dilution effect (changing carbon content per unit mass of the fuel gas mixture), another was the thermal effect, occurring due to a change in the flame temperature field upon diluent addition, and lastly, a direct chemical interaction (excluding the changes in the rates of chemical reactions, due to changing temperatures), due to changes in the species concentration. Whenever oxygen is added to the fuel side, the soot formation is chemically suppressed. The addition of oxygen on the fuel side promotes the pyrolysis of fuel, and hence the production of hydrocarbon radicals and H atoms which enhance soot formation. Simultaneously aromatic radicals and some important aliphatic radicals like C_2H_3 and C_4H_3 are removed by reactions with molecular oxygen and oxygen atoms. These two opposing effects result in a net suppression in soot formation. Acetylene, formed from vinyl radicals and consumed by the OH, H and O atoms, is known to be a soot precursor. And the higher concentration of H atoms explains the reduced concentration of acetylene and hence of soot. For higher alkanes like propane, *n*-butane, oxygen addition enhances the formation of soot by direct chemical interaction, provided compensation is made for the dilution effects. This enhancement increases with the increase in the oxygen concentration.

Yetter et al. (2000) experimented with asymmetric whirl combustion, in which the fuel was injected off-axis or asymmetrically into the combustor, in an axial direction, while the air is injected separately with purely tangential velocity. The combustor exhibits unusual stability even for equivalence ratios (Φ) less than 0.1. NO levels increase with increasing Φ, primarily because of the temperature rise. But at low NO levels, CO concentration is higher due to lower temperatures at such lean conditions, and so a trade-off exists. But all in all, the recorded CO and NO levels are extremely low, CO being less than 25 ppm, and NO levels less than 15 ppm. It is observed that between temperatures 1,100 and 1,700 K, CO can be reduced without increasing NO_x, as thermal NO_x formation is insignificant at temperatures below 1,800 K.

Cheng et al. (2001) studied the effects of partial premixing on flame characteristics and pollutant emissions in swirling methane flames. The flame length decreases continuously with increasing partial premixing and the flame colour also changes from yellow to blue. This indicates an improvement in combustion with increased partial premixing, and the reduced flame lengths imply reduced residence times. The $EINO_x$ and EICO decrease with increasing premixing, reach a minima at $\Phi p = 3$, simultaneously, and then again start increasing as the premixing air is further increased. Tae et al. (1996) measured NO emissions in laminar partially premixed ethane-air flames, as a function of the co-flow air flow rate. NO production has been found to occur mainly in the annular space, between the inner premixed and the outer non-premixed flame fronts, in the dual flame structure. The inner premixed region is dominated by prompt NO mechanism, and the outer non-premixed region by the thermal NO mechanism. A minimum $EINO_x$ is seen around $\Phi p \approx 2$, and it represents a compromise between thermal and prompt NO_x mechanisms. The visible flame height also decreases with decreasing Φp. The flame resembles a premixed flame at $\Phi p < 2.2$ and a non-premixed flame at $\Phi p > 2.2$. The formation of soot was also initiated around $\Phi p = 2.2$, which could be related to the minima in $EINO_x$ around this point. At lower Φp, the inner premixed zone produces most of the NO_x, while at higher Φp, most of the NO_x is produced by the outer non-premixed zone.

The aim of this research was to obtain an optimum combustion process, which might be a small effort in reaching towards a greener environment. A specially designed burner was designed for this purpose which has been configured in such a way, that the resulting flames are partially premixed, co-flow flames. The objective has been to optimize the level of partial premixing, so as to minimize the emissions for a given fuel flow rate. The idea is to keep the premixed mixture always rich, such that the premixed equivalence ratio, $\Phi p > 1$. Part of the fuel is combusted in this premixing or primary air, and the remaining fuel is combusted in the presence of secondary or diffusion air. The diffusion air helps make the combusting mixture overall very lean.

Emissions like Carbon monoxide (CO), Carbon dioxide (CO_2), Unburnt Hydrocarbons (UHC), Nitric Oxide (NO) and Soot; as well as temperature, have been measured at a central location in the combustor, for certain sets of predefined conditions. The corresponding flame heights have also been analysed and compared at these conditions. Based on the available data, an optimal level of premixing has also decided upon in this study.

2 Experimental Setup and Procedure

The schematic diagrams of the burner configuration and the experimental setup have been shown in Figs. 1 and 2, and a picture of the burner is shown in Fig. 3. The primary (premixing) and the secondary (diffusion) air are both taken from a storage tank connected to an air compressor through separate lines. Both the airflow rates are measured using a rotameter each. The fuel is taken from a domestic LPG tank and its flow rate is also measured using a rotameter. A pressure gauge is provided to read the gauge pressure at the inlet of each rotameter. It is assumed that there is no drop in the pressure across any of the rotameters, as the inlet valve is completely opened for all the measurements. The flow rate is controlled using a needle valve, one provided at the exit of each rotameter.

The premixing air and the fuel pass through the respective rotameters into a premixing chamber, where they are thoroughly mixed. And from there, this premixed mixture is supplied to the burner. This premixed mixture comes out through the centre of a flame holder, having diameter of the central tube as 10 mm; fixed in the centre of a perforated plate in the burner. There is a provision for tangential swirling, through tangential slots cut at right angles at the bottom of the burner. The premixed fuel–air mixture enters the bottom of the burner, swirls through these

Fig. 1 Schematic of the burner used in the experiment

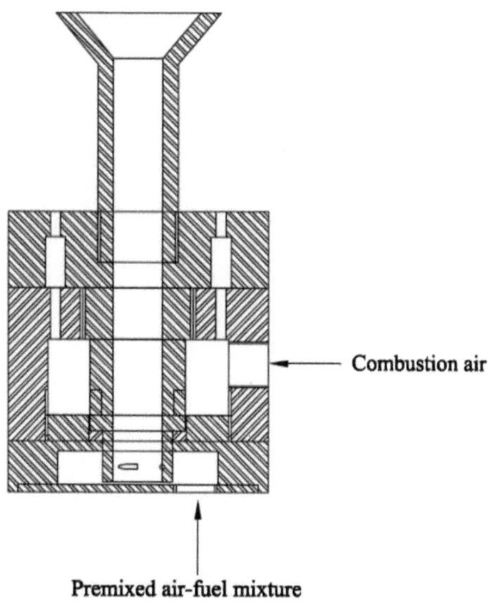

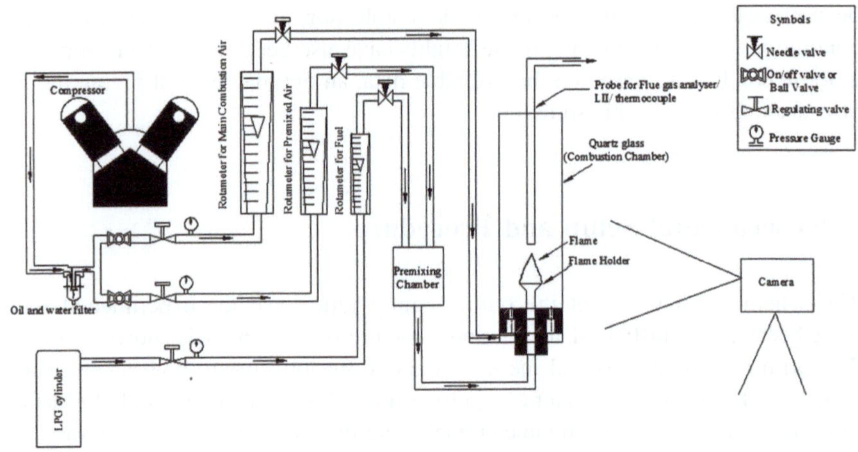

Fig. 2 Schematic of the experimental setup

Fig. 3 Picture of the burner with quartz tube

tangential slots and finally comes out through the centre of the flame holder. The perforated plate used has a circumferential row of 32 holes, with each hole of diameter 2 mm. There is another circumferential row of 24 holes, of 2 mm diameter each, just beneath the perforated plate. This is to facilitate proper ventilation and cooling of the burner, as well as to ensure proper secondary air supply to the reaction zone without any choking.

The diffusion or secondary or the main combustion air passes through a separate supply line to the burner, and comes out of the holes in the perforated plate, on the outside of the flame holder. A cylindrical Quartz glass tube (length 400 mm, inner diameter 60.3 mm, thickness 3 mm) acts as the combustion chamber and is placed around the burner. The chamber is open to the atmosphere at the top. Glass is used as it is needed for optical access during videography and imaging of the flames, and quartz is preferred as it can withstand the high temperatures occurring during combustion without shattering or bending.

For taking measurements, the probes for the emission analyser, the LII as well as the thermocouple were inserted one by one, into the centre of the glass chamber from the top, with the probe tip at a distance of 190, 32 and 305 mm, respectively, as measured from the top of the glass chamber.

Temperature measurements were taken using a K-type thermocouple (with an accuracy of ±0.7 % and 1 °C, with a digital indicator). The thermocouple was placed vertically inside and in the centre quartz tube combustor at a distance of 305 mm, measured from the top of the glass.

Emissions were measured using an AVL Ditest 444 gas analyser set for LPG. The exhaust gases, the concentrations of which were recorded, were Carbon monoxide (CO), Carbon dioxide (CO_2), Oxides of Nitrogen (NO_x) and Unburnt Hydrocarbons (UHCs). The probe was positioned at the centre inside the glass chamber, at a distance of 190 mm from the top of the glass. The position was decided such that the probe was able to suck the flue gases properly. That is why it was decided to keep the probe at the flame tip of the tallest flame.

Measurements of soot primary particle size, soot volume fraction and soot mass concentration were done through the LII 300 instrument manufactured by Artium Technologies. The probe was inserted into the centre of the glass chamber, at a distance of 32 mm from the top of the glass. The probe was connected to an external vacuum pump, which sucked the sample from the probe tip and pumped it to the LII unit.

3 Results and Discussions

For a pure diffusion flame, at the set fuel flow rate of 0.0254 g/s, the flame was deep yellow–orange in colour, with a significant height. The flame was narrow, and had a small bluish component towards the left side. This blue component could be attributed to the effect of swirling, which promoted complete combustion in that region. Also, as seen in the photograph taken of the flame holder (Fig. 4) after

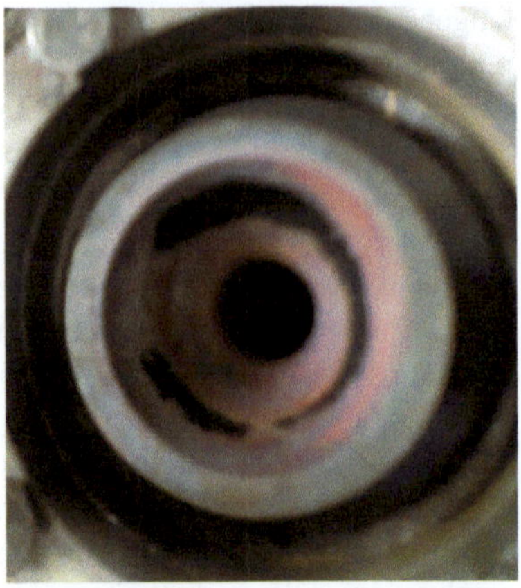

Fig. 4 Flame holder showing soot deposits

prolonged usage, it was observed that there was soot deposition all around the flame holder, except at the left most side. The deposition increased from zero to maximum and then to zero, in the anticlockwise direction, as observed the flame holder's top view. This showed that the swirling effects were more prominent initially towards the left side and this affirmed complete combustion at the left most blue region.

Some images of the ensuing flame, at different values of premixing equivalence ratio (Φp), are presented in Fig. 5. It can be seen that for a fixed value of the diffusion airflow rate, as the premixing air was initially introduced, the flame turned light yellow and more luminous. The flame height was reduced and its width increased considerably. With further premixing, the bluish component, which was

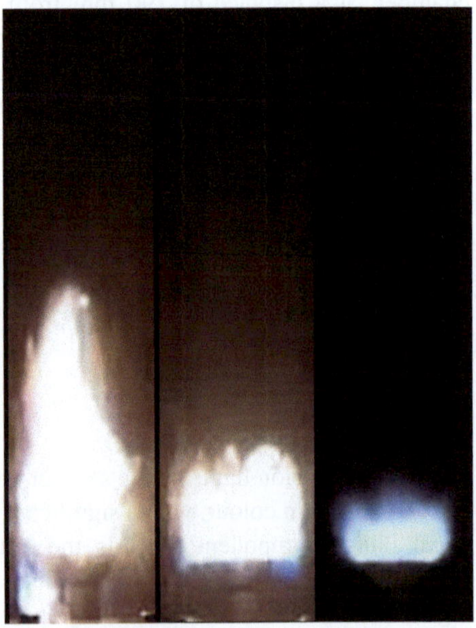

Fig. 5 Flame images, $\Phi p = \infty$, 4.16, 1.38

initially on the left side only, spread all around the base of the flame. This was accompanied with subsequent reduction in the yellowish component and an increase in the bluish component of the flame. The flame height also kept on reducing slightly. Another characteristic feature of the flame noticed was that of the flame being present more on one side of the burner than the other, due to swirling. This is because the swirl is directional, and it makes the air to move in a particular direction (anticlockwise), and hence the flame also has some of these directional characteristics.

Figure 6 represents the variation of temperatures normalized with the temperature of the corresponding diffusion flame. It can be observed that the temperature decreases only slightly initially as the premixing air is introduced and increased. After that there is a steep reduction in temperature. The temperature rises from $\Phi p = 3.32$, till $\Phi p = 2.77$, then reduces slightly at $\Phi p = 2.38$, and then again rises to a higher temperature at $\Phi p = 1.85$. With further addition of premixing air, the temperature falls to its minimum at $\Phi p = 1.11$. The data presented in Fig. 6 show that the temperature for partially premixed flame is lesser than the temperature observed for the corresponding pure diffusion flames. This can be attributed to cooling due to premixing air addition and counters the premise of improvement in combustion in the presence of increased premixing air. At higher primary air flow rates (lower Φp), the swirling efficiency of air increases causing better and thorough mixing of fuel and air resulting in an increase in temperature rather than just increasing the air velocity and causing cooling effect.

Figures 7 and 8 present the variation in CO and UHC concentrations with premixing equivalence ratio. It shows that CO and UHC concentration increases from that for a purely diffusion flame with increasing levels of premixing and peaks at $\Phi p = 3.32$. Figure 6 had shown that the exit temperature decreases in this range and this decrease is especially sharp around this point. As the combustion is hindered due to the decrease in temperature, which in turn is caused by the presence of an excess of primary air, the concentration of CO increases. However, the flame

Fig. 6 Temperature variation with premixing equivalence ratio

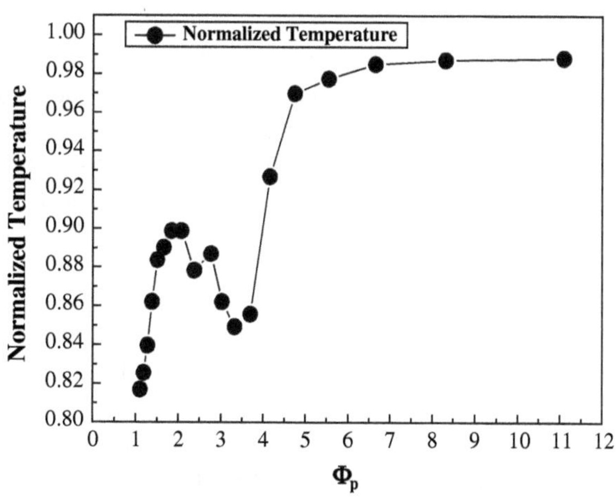

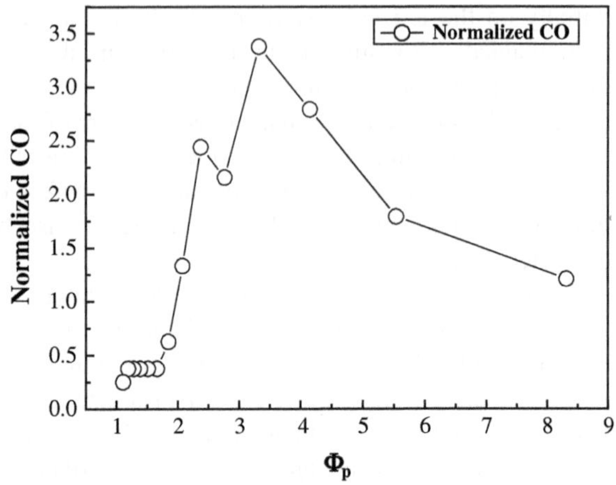

Fig. 7 Variation of CO concentration with premixing equivalence ratio

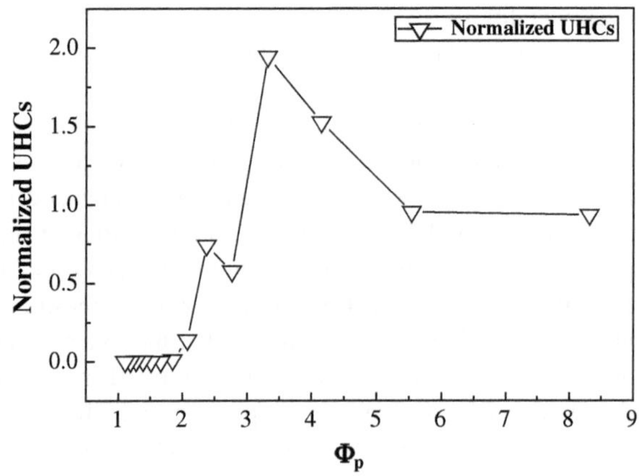

Fig. 8 Variation of UHC concentration with premixing equivalence ratio

height decreases with a decrease in Φp implying better combustion and hence a reduction in CO. It seems that the effect of reduction in flame height that dominates at lower values of Φp causes a drop in CO concentration below $\Phi p = 3.32$.

As the premixing air is increased, leaner conditions prevail in the combustion zone leading to a drop in both the thermal NO_x and prompt NO_x and, hence, the overall NO_x concentration decreases with respect to that produced in a diffusion flame as seen in Fig. 9. Improvement in mixing and combustion leads to higher temperature in the vicinity of $\Phi p = 3.0$ results in an increase in temperature (Fig. 6) and, hence, an increase in NO_x as well. On the other hand, adding premixing air causes an increase in the concentration of soot in the exhaust as compared to that of a pure diffusion flame as seen in Fig. 10. In the presence of premixing air, excess oxygen molecules become available resulting in the increase of local concentration of certain radials, which act as precursors to soot formation. Partial premixing also alters the chemistry of intermediate hydrocarbons, which can account for the

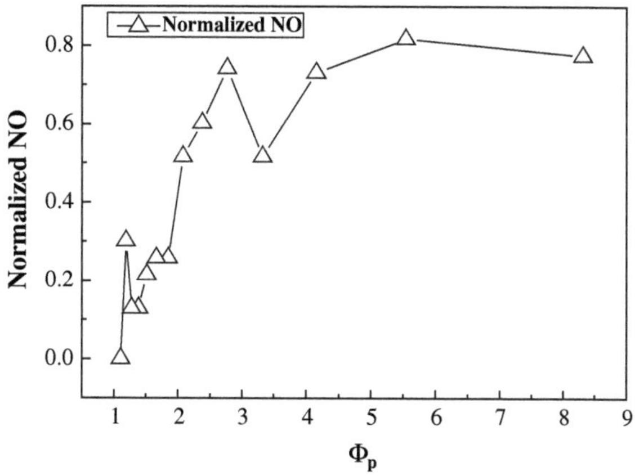

Fig. 9 Variation of NO concentration with premixing equivalence ratio

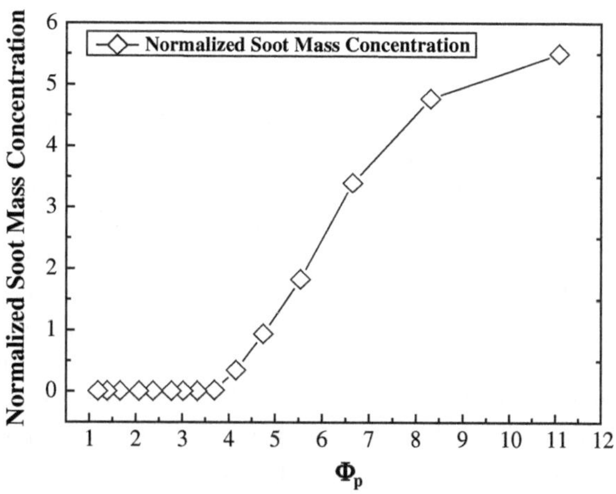

Fig. 10 Variation of soot concentration with premixing equivalence ratio

increase in soot production. However, for higher degree of premixing, soot particles start to get oxidized resulting in a decrease in soot concentration and beyond $\Phi p = 3.77$, the soot is almost completely eliminated.

The primary soot particle size for the set of premixing conditions considered for this study varied between 9 and 28 nm approximately as shown in Fig. 11. Initially, as the premixing air was increased, the particle size remained almost constant up to around $\Phi p = 5$ and beyond that an increase in particle size is observed till $\Phi p = 3.32$. Similar tends in CO and UHC concentrations were observed in Figs. 7 and 8 which implies that during the incomplete and improper combustion, which resulted in higher concentrations of CO and UHC, the soot particles might have agglomerated to form larger size particles.

An interesting area for further study is measurements of pressure oscillations in the combustor that can be correlated with sound vibrations in order to analyse the impact of partial premixing on the acoustic characteristics of the system. Problem areas like

Fig. 11 Variation of average primary soot particle size with premixing air

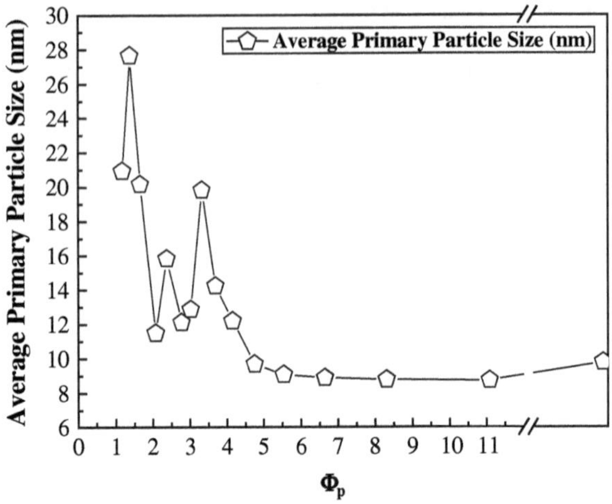

hooting, combustion instabilities can also be addressed. The premixed mixtures were rich for all the cases studied here. A probable area for further study can be lean PPC, where the premixed mixtures employed are already lean. Measurements of radical species can also be done, along with measurement of temperatures at various radial and axial locations along the burner. This can give an idea about the heat release rate, the temperature field, reaction mechanisms and mechanisms of formation of soot, etc. The use of various sophisticated instruments like Particle Image Velocimetry (PIV) and Planar Laser Induced Fluorescence (PLIF) might prove to be helpful in providing information about flow fields, flow velocities, the planar profiles of temperatures, radical concentrations and various gases like NO. Moreover, similar studies can be undertaken at extremely low levels of premixing, wherein the mixture cannot ignite without the addition of secondary air. This would prove to be advantageous when dealing with the safety aspects of premixed and PPC.

4 Conclusions

From the above study conducted on partially premixed, LPG/air flames, it is observed that this optimum premixing equivalence ratio (Φp) is found to be between 1.5 and 2. At such high premixing levels, the emissions of CO, UHCs, NO and soot are seen to be considerably reduced. Trade-off between CO, UHCs and NO was not there in this range of Φp. Another observation is that the soot primary particle size is comparatively higher for partially premixed cases, than for a pure diffusion flame. This also is advantageous, as larger-sized pollutant particles are prevented from entering the human body through its self-defence mechanisms. Hence, the health hazards posed to travellers and people working or staying in the vicinity of airports are reduced. But again, though the particle size of soot is more at moderate to high levels of premixing as compared to that at lower premixing levels;

it is still less than 10 μm, which is the limit imposed by EPA for non-inhalable coarse particles. It is also noted that for $\Phi p \leq 2$, the flame is stable and its height is considerably reduced. This can be an added advantage, as reduced flame heights help to reduce the combustor length, the amount of material used for the combustor, its manufacturing cost, and most importantly, the weight of the combustor. This would prove to be extremely useful when considered from the point of view of aircrafts. The combustor temperatures observed are also not that high for $\Phi p \cong 2$. This implies that a combustor used with such a burner configuration need not be made of a highly heat resistant material, which again implies a reduction in the manufacturing cost for an aircraft engine. The study reported here establishes the fact that the quality of combustion and the emissions can be improved considerable by partially premixing the fuel. However, there is a critical limit of premixing beyond which not much improved is gained in terms of the combustor performance.

References

Alder BJ, Lyle KH, Laurendeau NM, Gore JP (1998) Effect of overall equivalence ratio on minimum nitric oxide emission index in laminar and turbulent partially premixed flame. Combust Sci Technol 140:461–468

Cheng TS, Chao YC, Wu DC, Hsu HW, Yuan T (2001) Effects of partial premixing on pollutant emissions in swirling methane jet flames. Combust Flame 125:865–878

Gore JP, Zhan NJ (1996) NO_x emission and major species concentrations in partially premixed laminar methane/ air co-flow jet flames. Combust Flame 105:414–427

Gülder OL (1995) Effect of oxygen on soot formation in methane, propane, and n-butane diffusion flames. Combust Flame 101:302–310

Hadef R, Geigle KP, Meier W, Aigner M (2010) Soot characterization with laser-induced incandescence applied to a laminar premixed ethylene-air flame. Int J Therm Sci 49:1457–1467

Jeong YK, Jeon CH, Chang YJ (2006) Evaluation of the equivalence ratio of the reacting mixture using intensity ratio of chemiluminiscence in laminar partially premixed CH_4-air flames. Exp Thermal Fluid Sci 30:663–673

Kim HS, Arghode VK, Gupta AK (2009) Combustion characteristics of a lean premixed LPG-air combustor. Int J Hydrogen Energy 34:IO45–IO53

Lefebvre AH (1999) Gas turbine combustion. Taylor and Francis, London

Lyle KH, Tseng LK, Gore JP, Laurendeau NM (1999) A study of pollutant emission characteristics of partially premixed turbulent jet flames. Combust Flame 116:627–639

Mishra TK, Datta A, Mukhopadhyay A (2005) Concentration measurements of selected hydrocarbons in methane/air partially premixed flames using gas chromatography. Int J Therm Sci 44:1078–1089

Mishra TK, Datta A, Mukhopadhyay A (2006) Comparison of the structures of methane-air and propane-air partially premixed flames. Fuel 85:1254–1263

Mitrovic A, Lee TW (1998) Soot formation characteristics of laminar partially premixed flames. Combust Flame 115:437–442

Mungekar H, Atreya A (2007) NO formation in counterflow partially premixed flames. Combust Flame 148:148–157

Rink KK, Lefebvre AH (1987) Pollutant formation in heterogeneous mixtures of fuel drops and air. J Propul Power 3(1):5–10

Semerjian HG, Ball IC, Vranos A (1979) Pollutant emissions from partially mixed turbulent flames. Symposium (International) on combustion, vol 17, pp 679–687

Sreenivasan R, Koli SK, Raghavan V (2012) Experimental study of effects of coflow air and partial premixing on liquid petroleum gas flames. International Scholarly Research Network, ISRN thermodynamics, Article ID 202715. doi:10.5402/2012/202715

Tae KK, Alder BJ, Reisel JR, Laurendeau NM (1996) Exhaust and in-situ measurements of nitric oxide in laminar partially premixed C_2H_6-air flames: effect of premixing level at constant burner tube flow rate. Energy Fuels 10:1060–1066

United States Environmental Protection Agency USEPA (2013) http://www.epa.gov/pm

Xue H, Aggarwal SK (2003) NO_x emissions in n-heptane/ air partially premixed flames. Combust Flame 132:723–741

Yetter RA, Glassman I, Gabler HC (2000) Asymmetric whirl combustion: a new low NO_x approach. Proc Combust Inst 28:1265–1272

Low-Emission, Fuel-Flexible Combustion of Liquid Fuels

Ajay K. Agrawal

Abstract In this chapter, we will show that with proper fuel injector and combustor designs, it is possible to cleanly combust different liquid fuels (diesel, straight vegetable oil, and highly viscous glycerol) in the same combustion system without requiring upstream processing and/or hardware modifications. The results challenge the traditional combustor design approach relying upon strict fuel property specifications. Instead, we illustrate that fuels with widely different physical and chemical properties can be cleanly utilized in a combustion system specifically designed to achieve fuel flexibility. Wide-spread application of this latter combustor design strategy can result in significant economic benefits in the emerging markets for the alternative fuels. The impetus of this study is to engineer innovative liquid combustion systems to handle conventional and alternative fuels while reducing harmful environmental impact from various emissions. We will focus on continuous flow combustion systems, thereby precluding reciprocating systems such as diesel engines from the scope of the present study.

Keywords Liquid fuel combustion · Biofuel combustion · Fuel flexibility · Low-emission combustion · Straight vegetable oil (SVO)

1 Introduction

More than 80 % of the world's energy supply presently requires combustion of solid, liquid, and gaseous fuels. In particular, liquids fuels are used almost exclusively for vehicular transportation by automotive and aircraft engines. Liquid fuels are also used in applications such as boilers, industrial and home furnaces, and power generating gas turbines. Liquid fuel combustion systems of the next generation must provide high system efficiency, low emissions, and fuel flexibility.

A.K. Agrawal (✉)
Department of Mechanical Engineering, University of Alabama, Tuscaloosa, AL, USA
e-mail: aagrawal@eng.ua.edu

A.K. Agarwal et al. (eds.), *Novel Combustion Concepts for Sustainable Energy Development*, DOI 10.1007/978-81-322-2211-8_20

For example, rise in air traffic with global business, military, and private demands has resulted in international recognition of environmental concerns associated with aviation emissions. In response, the governing organizations are adopting increasingly stringent regulations to limit emissions from the next generation of aero-propulsion systems. In particular, nitric oxides (NO_X) emissions remain a major concern owing to the harmful ozone and smog at ground levels and to the formation of ozone, acid rain, and contrails in the troposphere (Porter 2007). In addition to reducing emissions, higher fuel efficiency will require future aviation gas turbines with higher pressure ratios. While higher pressure ratio tends to reduce emissions of soot and greenhouse gas, NO_X emissions tend to increase. Since NO_X production is sensitive to reaction-zone temperature, it is necessary to avoid "hot spots" by burning liquid fuel premixed with air at relatively lean conditions.

In spite of the impressive gains in arenas of efficiency and emissions, present combustion systems still rely upon petroleum-derived fuels—gasoline, diesel, and jet fuels. However, existing and future combustion systems for vehicular transportation, power generation, etc., will increasingly utilize alternative energy sources to achieve fuel flexibility, while minimizing our dependence on petroleum-based fuels. In particular, biofuels produced from renewable biomass such as plant or animal matter offer the possibility of one day becoming carbon neutral, meaning any carbon released would have been absorbed during the fuel's life cycle. Biofuels also offer other benefits: allowing energy independence and providing a renewable energy source that can be around long after the world's fossil fuel reserves have exhausted.

Biodiesels are among the most widely studied liquid biofuels because of their similarity to diesel fuels (Graboski and McCormick 1998; Raghavan et al. 2009; Song et al. 2007; Bolszo and McDonell 2009a, b, Wang et al. 2011; Pan et al. 2009; Li et al. 2011; Park et al. 2011). Biodiesels are produced by transesterification of source oils, such as vegetable oil (VO). The transesterification process is necessary to modify the physical properties of the source oil to produce biodiesel, which can readily replace diesel fuel in an end-use device. However, transesterification requires substantial energy input and results in glycerol by-product. With increasing biodiesel production, glycerol can become an unintended waste stream, with its supply far exceeding the demand in the cosmetics, medical, and chemical industries.

Table 1 lists the key thermo-physical properties No. 2 diesel, biodiesel, straight VO, and glycerol (Panchasara et al. 2009a, b). The energy content of biodiesel is about 12 % less than that of petroleum-based diesel fuel on a mass basis. Biodiesel is an oxygenated fuel, which contains 10–11 % oxygen by weight. The kinematic viscosity of biodiesel is about 1.3 times that of diesel. Still, biodiesel properties are closest to those of the diesel, and thus, biodiesel is the usual replacement for diesel. In contrast, at room temperature, the kinematic viscosity of VO is over 13 times that of diesel. VO also has higher auto-ignition temperature and higher specific heat capacity compared to biodiesel or diesel. Because of these adverse thermo-physical properties, VO cannot be used as a diesel substitute in existing liquid fuel combustion systems, and thus, it is processed to generate the biodiesel fuel.

The situation is less favorable when comparing glycerol with diesel. The low heat value (LHV) of glycerol is only 36 % of that of diesel on mass basis. In

Table 1 Fuel properties (Panchasara et al. 2009b)

Property	Diesel	Biodiesel	VO	Glycerol
Molecular weight (kg/k mol)	142.2	291.5	277.9	92.1
Density at 25 °C (kg/m^3)	834.0	880.0	925.0	1,260.0
Viscosity at 25 °C (mm^2/s)	3.88	5.61	53.74	740.0
Surface tension at 25 °C (m N/m)	28.2	31.1	30.1	62.5
LHV (MJ/m^3)	37,198	33,442	34,225	20,400
Specific heat (J/kg K)	1,750	2,000	2,269	2,430
Latent heat of evaporation (kJ/kg)	250	200	–	662
Boiling point (°C)	195–372	338–380	–	290
Auto-ignition temperature (°C)	315	–	445	370

comparison with diesel, glycerol has kinematic viscosity approximately 200 times greater and over twice the surface tension. Glycerol also has higher enthalpy of vaporization and higher vaporization and auto-ignition temperatures than diesel. Thus, significant thermal energy from the flame must be recirculated to the liquid to ensure proper glycerol vaporization before combustion. In summary, glycerol is extremely difficult to breakup into droplets to form a spray and the resulting glycerol droplets are rather difficult to vaporize and combust. For these reasons, existing combustion systems cannot utilize glycerol as fuel.

In this research, we will show that with proper fuel injector and combustor designs, it is possible to cleanly combust all of the fuels listed in Table 1, without any upstream processing and/or hardware modifications. The study challenges the traditional approach to combustor design, which relies on upstream fuel processing to ensure strict fuel property specifications. Instead, the study illustrates that fuels with widely different physical and chemical properties can be utilized, cleanly, in a combustion system specifically designed to achieve fuel flexibility. Wide-spread application of this latter combustor design strategy can result in significant economic benefits in the emerging markets for the alternative fuels. The impetus of this study is to engineer innovative liquid combustion systems to handle conventional and alternative fuels while reducing harmful environmental impact from various emissions. We will focus on continuous flow combustion systems, thereby precluding reciprocating systems such as diesel engines from the scope of the present study.

2 Liquid Fuel Combustion

Figure 1 illustrates a typical liquid fuel combustion system, denoted as lean direct injection or LDI combustion. The liquid fuel is injected directly into the strongly swirling flow of combustion air. A fuel injector is used to atomize the liquid, i.e., disintegrate the fuel into droplets in a turbulent spray. Numerous fuel injector designs are available with many based on twin-fluid atomization, whereby a gas (usually air, denoted as atomizing air) is used to create the fuel spray by the transfer of either kinetic energy or potential energy, as discussed in Sect. 3. The atomization process generally takes place in two steps: during primary atomization, the fuel is broken into ligaments and blobs at the injector exit. Subsequently, ligaments and blobs interact with the atomizing air and/or combustion air to undergo secondary atomization resulting in a fuel spray containing droplets of a wide range of diameters.

Droplets in the spray vaporize and premix with air, depending upon the flow, thermal, and chemical processes. Fuel vaporization depends upon parameters such as the boiling and/or fuel volatility characteristics, latent heat of vaporization, and initial droplet diameter. Clean combustion requires appropriate fuel droplet size distribution, for a range of operating conditions, such that the droplets are both large enough to penetrate into the combustion chamber, but also small enough to prevaporize within the short residence time of the spray/flame region. If they can be ignited, large droplets in a spray burn in the diffusion flame mode, which results in unacceptably high emissions of NO_X, carbon monoxide (CO), soot, and unburned hydrocarbons (UHCs). The LDI combustion concept in Fig. 1 illustrates excellent fuel prevaporization and fuel–air premixing. The fuel and airflow rates are

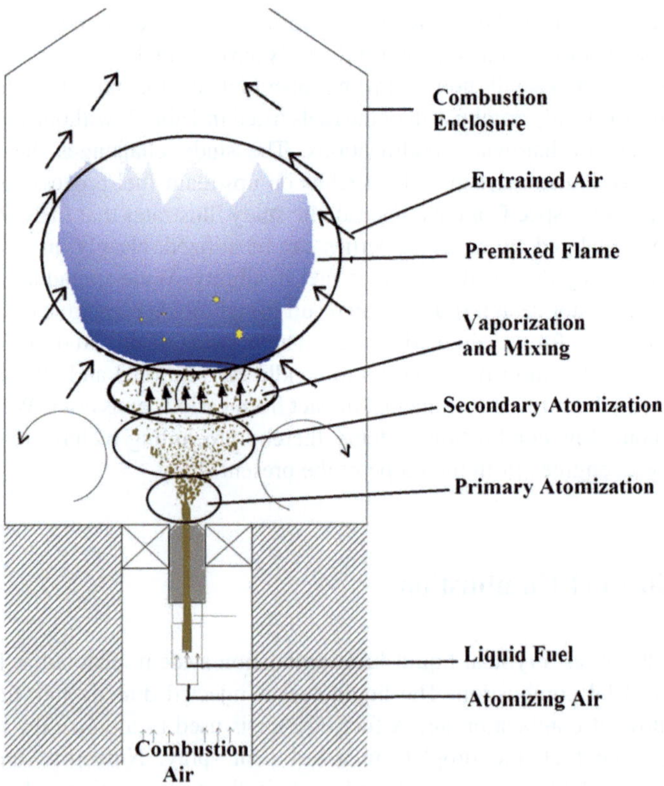

Fig. 1 Illustration of lean direct injection (LDI) combustion

controlled to produce the reactant mixture near the lean blow-off (LBO) limit which favors lean premixed (LPM) combustion producing very low emissions of harmful species.

3 Liquid Fuel Atomization

A spray serves as the heart of almost every type of liquid-fueled combustion system. In what Lefebvre (1989, 1999) has defined as "classical" atomization, a liquid jet or sheet, subjected to destabilizing forces, breaks up after it leaves the injector. In the classical atomization, a liquid jet or sheet initially disintegrates into ligaments which in turn break up into droplets, whose size depends on the characteristic length scale of the liquid jet or sheet. If the Weber number, the ratio of the jet's kinetic energy to its surface tension, is small, surface instabilities control the breakup. The fluid's viscosity and the interaction at the air–liquid interface become influential with larger Weber numbers. Lefebvre (1989, 1999) denotes "prompt" atomization as spray production immediately at the injector exit or even within the injector, depending on the internal structure of the injector. Prompt atomization also occurs with the interaction of the liquid with a source of highly energetic air. Excellent reviews concerning the preparation of fuel sprays optimized for lean combustion include those written by McDonell and Samuelsen (1991), Razdan (1998), Mansour (2005), and Nakamura et al. (2006). Some of the characteristics Lefebvre (1989, 1999) listed for an ideal fuel injector include good atomization over a wide range of fuel flow rates either steady or transient, unaffected by flow instabilities, low power requirements, scalability, resistant to blockages, and delivering a uniform spray.

A high pressure gradient across the injector will ensure a large velocity differential at the jet's liquid–air interface, which, in turn, leads to rapid classical or even prompt breakup in pressure atomizers. This simple arrangement can be problematic if the fuel flow rate is to extend over a large range. The orifice size and pressure gradient must accommodate the largest anticipated fuel flow rate, and the pressure gradient cannot be maintained for low fuel flow rates. A second method transfers the kinetic energy from the surrounding air to the fuel jet through the use of air assist or air blast. The air-assist method employs a low volume flow rate of high-velocity atomizing air to break up the fuel jet. The air-blast atomizer delivers a large volume flow rate of low-velocity atomizing air to both break up the fuel jet or sheet and to deliver the resulting spray to the combustion zone. For the air-assist and air-blast (AB) atomization procedures, the production of either small volumes of high-pressure air or large volumes of low-pressure air can be expensive in terms of power requirements. Prefilming the fuel represents a third type of process in which the liquid fuel flows over a surface in such a manner that it leaves the surface as a sheet. The thin liquid sheet can subsequently be subjected to air blast on both sides to affect fine droplet spray. Prefilming reduces the volume of air needed to atomize the liquid fuel. Combinations of these three processes in a wide variety of geometries, which can sometime become quite complex, are used in a majority of liquid fuel combustion systems.

A fourth process known as effervescent atomization (EA) injects high-pressure gas directly into the liquid in a mixing chamber upstream of the injector discharge orifice (Sovani et al. 2001). The injected gas bubbles into the liquid and produces a two-phase mixture that flows through the orifice, and upon exiting the orifice, the gas bubbles expand rapidly to shatter the surrounding liquid into droplets. The liquid breakdown mechanism with EA atomization relies on potential energy unlike the kinetic energy-driven AB and air-assist atomizers. The EA process requires small atomizing airflow rate to produce a spray with fine droplets, the atomizer performance is relatively insensitive to the liquid's kinematic viscosity, and the larger diameter of the injector orifice alleviates clogging problems and simplifies fabrication. However, in the EA process, the spray angle is small, air must be pressurized to the liquid's pumping level, and the spray produced can exhibit unsteadiness because of the long residence time for the two-phase flow between the mixing chamber and injector discharge orifice.

We have employed a novel process initially presented by Ganan-Calvo (2005), so-called flow blurring (FB), to achieve low-emission, fuel-flexible combustion. As illustrated in Fig. 2, the FB mechanism consists of a liquid supply tube and a circular orifice in a plate oriented perpendicularly to the axis of the tube and situated downstream of the tube exit. The tube outer wall is tapered at the outlet, and the sharpened edge of the orifice in the plate is the same diameter as the inner diameter of the tube (d). When the axial distance between the tube and the orifice (H) is about the size of the tube diameter ($H/d \sim 1$), the liquid and gas interact similar to an AB atomizer. However, for shorter distances ($H/d \sim 0.25$), the atomizing gas enters a short distance into the liquid tube where it bubbles and mixes with the liquid. The resulting bubbly two-phase mixture exits the injector orifice, where the gas phase rapidly expands to create fine liquid droplets. FB atomization is similar to the EA process, although the two-phase mixing in a FB atomizer occurs only at the tip of the liquid tube as dictated by the geometry (H/d). Following the introduction by

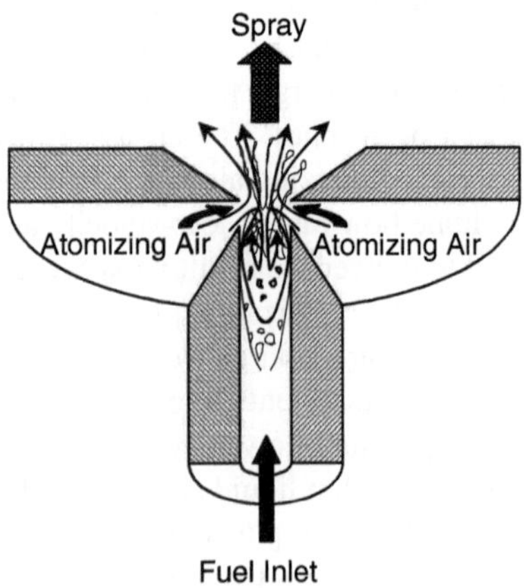

Fig. 2 Working principle of the flow-blurring (FB) injector

Ganan-Calvo (2005), our research group has developed the FB atomization concept to achieve low-emission combustion for a range of liquid fuels. Next, we present important findings and results from several application examples by our research group.

4 Low-Emission Diesel Combustion

As first step, the combustion performance of a commercially available AB injector and an in-house-built FB injector is compared for fixed fuel and total airflow rates, while the split between the atomizing air and combustion air is varied. The injector performance is characterized by measuring the emissions of NO_X and CO. Details of the experimental setup are described by Panchasara et al. (2009a). Briefly, the combustion air enters the combustor through a swirler with swirl number of 1.5. The bulk axial inlet velocity of the primary air is 1.9–2.1 m/s, which resulted in the inlet Reynolds number varying from 5,960 to 6,750. The combustor is 8.0 cm ID and 46-cm-long quartz tube, and it is cooled by natural convection on the back side. A commercial AB atomizer (Delavan Siphon type 30609-2 SNA-0.20 nozzle) is used for baseline experiments. This commercial version creates a swirling flow of atomizing air to breakdown the fuel jet as it exits the orifice plate. The air swirler within the injector is replaced with a spacer tube to implement the FB concept presented in Fig. 2. In this latter configuration, air and fuel mix together before exiting through the orifice plate to form the spray downstream. One notable difference between the two injectors is the exit diameter of the injector; 0.4 mm for the AB injector versus 1.5 mm for the FB injector. Figure 3 illustrates the two injector designs and highlights the flow paths for fuel and atomizing air.

The flow rates of fuel and total air (combustion + atomizing) are constant, respectively, at 12 ml/min and 150 standard liters per minute (slpm). Combustion performance is strongly dependent upon the spray quality determined by the atomizing airflow rate. Thus, experiments were conducted by varying the percentage of the atomizing air (AA) flow rate in the range of 10–25 % of the total airflow rate, depending upon the fuel and injector type. Since the total air-to-fuel mass ratio is constant for all experiments, effects of atomizing airflow rate, injector

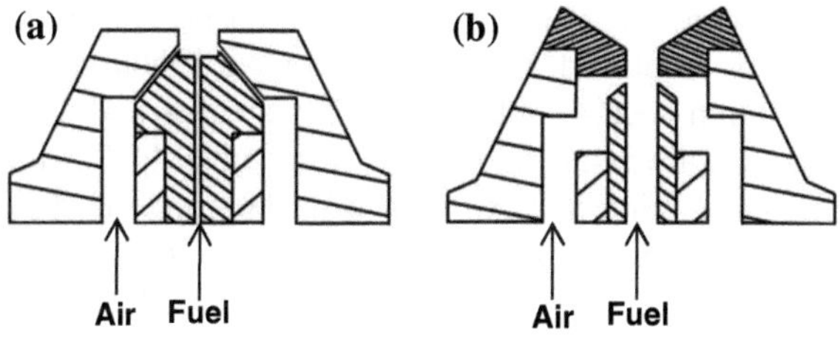

Fig. 3 Detailed view of fluid paths in: **a** air-blast and **b** flow-blurring injectors

type, and fuel on combustion performance can be delineated from the emissions measurements.

Direct photographs of flame were taken by a digital camera to obtain a qualitative understanding of the flame characteristics. These photographs are reproduced in black and white in Fig. 4 for both injector types using the diesel fuel. For each case, images are shown for atomizing air (AA) flow rates varying from 10 to 25 % of the total airflow rate. Images revealed a yellow–orange, sooty flame indicative of the fuel droplets burning in the diffusion mode for low AA rate, i.e., AA = 10 %. At higher AA flow rates, however, the flames showed a distinctive blue color typical of premixed combustion, suggesting that the fuel droplets prevaporize and premix with air before combustion takes place. A comparison of images for the two injectors revealed that the FB injector requires a smaller AA flow rate to produce the premixed blue flame. The transition from diffusion to premixed mode of combustion occurred between AA = 10 and 15 % for the FB injector and at AA = 20 % for the AB injector. The images in Fig. 4 make a clear distinction between the two injectors; the FB injector produces cleaner flames as compared to the AB injector. Figure 5 quantitatively supports these findings in terms of the emissions measurements at the combustor exit plane. CO and NO_X emissions decrease with increasing AA flow rate, and they are always 3–5 times lower for the FB injector as compared to the AB injector, evidently because of the smaller fuel droplets produced by the FB injector.

5 Spray Characterization

Figure 6 visually illustrates the differences in water sprays formed by AB and FB injectors for identical flow inlet condition with atomizing air-to-liquid ratio by mass or ALR = 3.5. These images represent a field of view (FOV) of 7 mm × 7 mm immediately downstream of the injector exit albeit with different spatial and temporal resolutions to highlight the key features; higher exposure time was necessary to acquire the image for the FB injector since it produced smaller droplets, which limits the scattered light for visual imaging. Figure 6 reveals that the AB injector produces the typical liquid core in the near field while the FB injector forms a fine spray, illustrating the significantly improved atomization capability of the latter.

The fineness of the FB injector spray offers the opportunity to employ advanced flow diagnostics techniques to understand the liquid breakup process in the near field. Thus, we have employed a high-speed camera (Photron FASTCAM SA5) mounted with 100-mm focal-length lens, a 2x extender and 130 mm length of extension tubes to acquire visual images for rather small FOV with dimensions of 2.3 mm × 1.4 mm at the injector exit. For acquisition rate of 100,000 frames per second (fps), the image renders spatial resolution of 7.16 μm per pixel, to distinguish individual though stationary and/or large droplets. Background illumination is provided by a very high-intensity white light source (Energetiq LDLS) produced by laser-driven plasma to minimize the blurring caused by the droplet motion

Fig. 4 Effect of atomizing air on visual images of diesel flames; air-blast injector (*top*); FB injector (*bottom*)

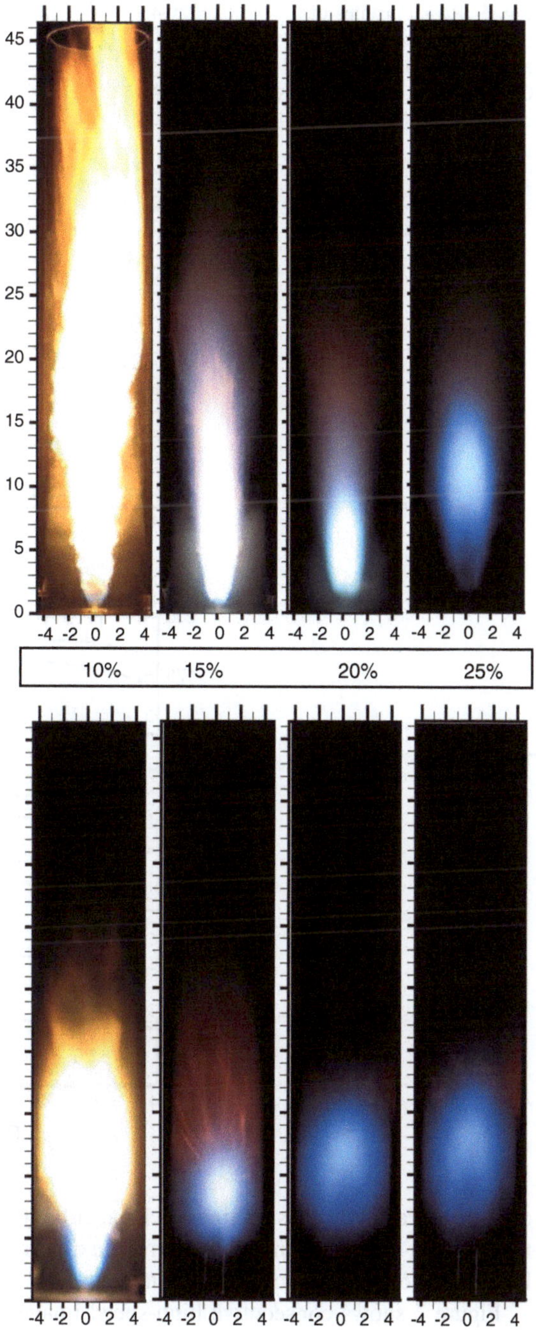

Fig. 5 Radial profiles of emissions at the combustor exit plane in diesel flames; CO concentration (*top*), NO_X concentration (*bottom*)

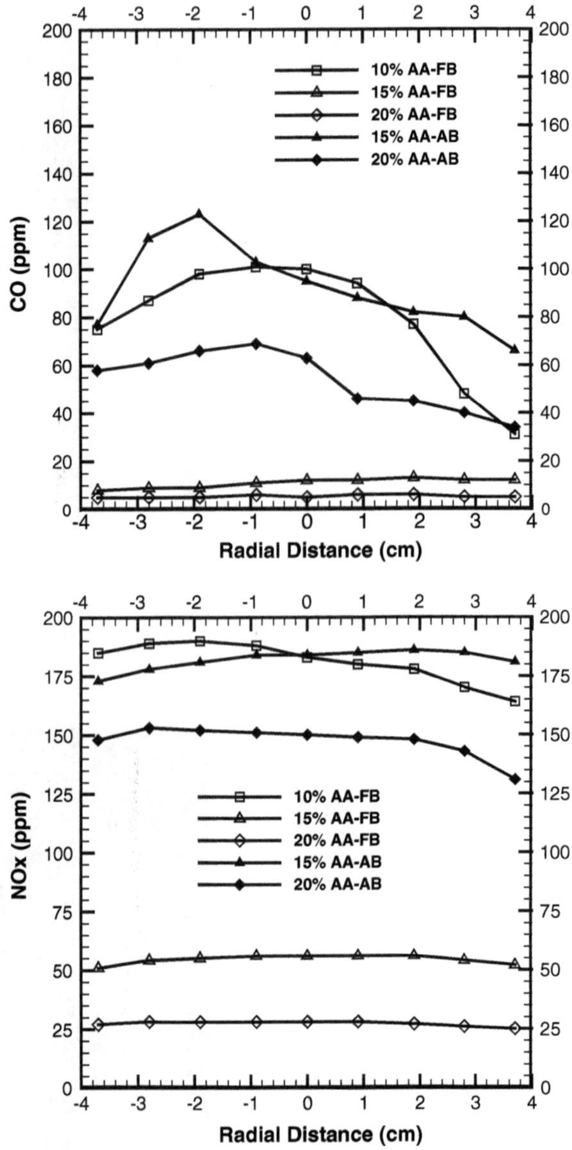

during the camera exposure time of 1 μs; still a droplet at velocity of 100 m/s will move about 100 μm during the camera exposure time. Thus, only slow moving, larger droplets can be accurately detected.

Figure 7 shows consecutive high-speed images of FB spray in the near field of the injector. In contrast to the low-speed image (30 fps at exposure time of 33 ms) of the FB spray shown in Fig. 6, the very small exposure time in the images of Fig. 7 makes it experimentally feasible to visualize the instantaneous spray in detail.

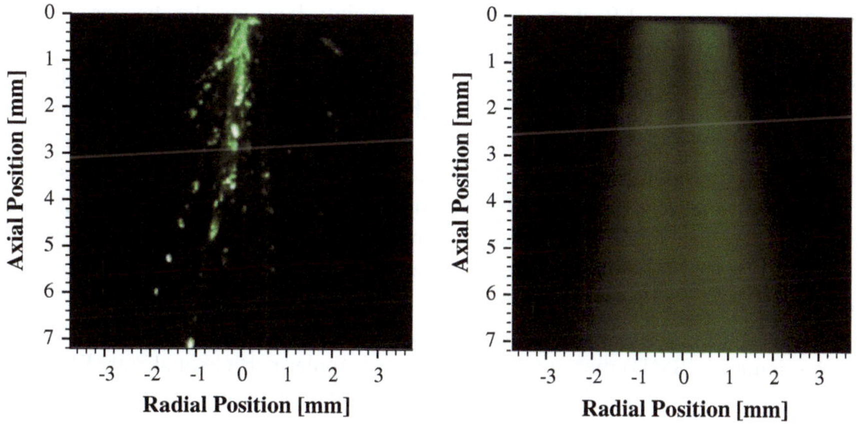

Fig. 6 Water spray at ALR = 3.5; AB injector with spatial resolution of 20 μm/pixel and exposure time of 1 ms (*left*); FB injector with spatial resolution of 33 μm/pixel and exposure time of 10 ms (*right*)

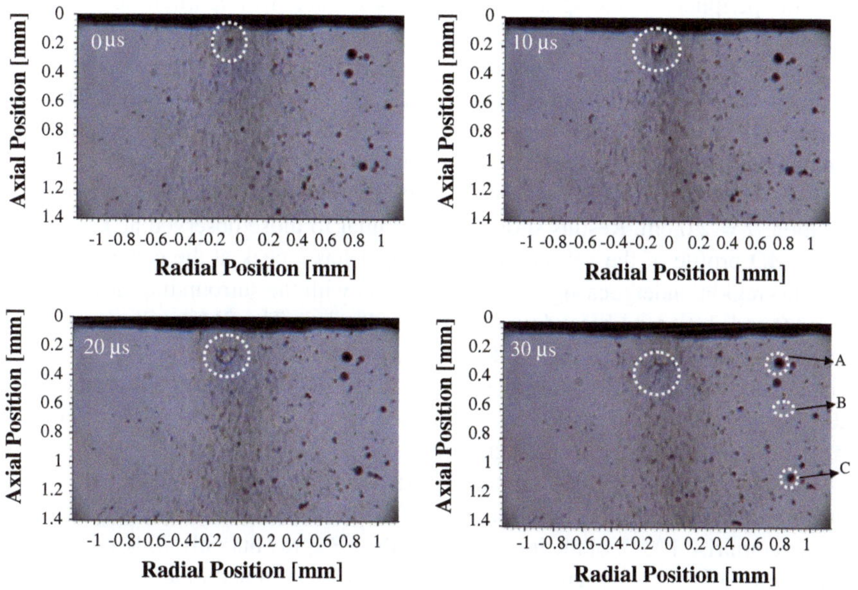

Fig. 7 Ultra-high-speed visualization of water spray in the FB injector near field for ALR = 2 at spatial resolution of 7.16 μm/pixel and exposure time of 1 μs

Figure 7 depicts an air bubble at the injector exit that expands, explodes, and shatters the surrounding liquid into droplets, visually illustrating the liquid breakup by FB atomization. Surprisingly, the majority of the liquid has already been atomized into fine droplets at the injector exit indicating that the liquid breakup by bubble explosion or FB atomization occurs mainly between the discharge orifice and injector exit. This process is hereby denoted as the primary breakup by bubble explosion or FB atomization. Most of the droplets cannot be spatially resolved in the images because of the blurring caused by the droplet motion in spite of the small camera exposure time. However, larger droplets on the injector periphery are spatially and temporarily resolved. In Fig. 7, the diameter of the droplet A is about 100 μm, droplet B is about 21 μm, and that of droplet C is about 72 μm. These results clearly demonstrate, for the first time, that the bubble explosion or primary breakup occurs mainly within the FB injector and the larger droplets appear only occasionally at the injector periphery.

A quantitative description of the spray quality is given by radial profiles of sauter mean diameter (SMD) measured by the phase Doppler particle analyzer (PDPA) at different axial locations in the water spray as shown in Fig. 8 for ALR = 2.5 (Simmons et al. 2009; Simmons and Agrawal 2011). Figure 8a shows the SMD profiles of the AB and FB atomizers at $Y = 1.0$ cm, the axial distance from the atomizer exit. For the FB atomizer, the peak SMD of 13 μm at the spray center decreases with increasing radius until a minimum of about 9 μm occurs at radial distance, $r = 4$ mm. Thereafter, the SMD increases to reach a value of about 12 μm near the periphery of the spray located at $r = 8$ mm. An important observation is that the FB atomizer produces droplets of 9–13 μm or within the narrow range of 4 μm. For the AB atomizer, the minimum SMD of 7 μm at the center increases gradually to about 20–25 μm at the periphery of the spray at $r = 12$ mm. Thus, the droplet SMD range is much wider (>15 μm) with the AB atomizer. Clearly, the FB atomizer offers unique features to distinguish it from the AB atomizer.

Figure 8b shows the SMD profiles for the two atomizers at a downstream location, $Y = 2.0$ cm. Results show trends similar to those observed at $Y = 1.0$ cm. The SMD profile of the FB atomizer no longer has a peak at the center indicating that this region undergoes significant interaction with the surrounding flow between $Y = 1.0$ and 2.0 cm to breakdown the larger droplets. The SMD still varies over a narrow range of 9 and 13 μm while the spray width increases to about $r = 10$ mm. The AB atomizer still demonstrates a wide range of SMDs, from around 6 to over 20 μm, with the smallest droplets contained in the center of the spray, and the largest droplets confined to the periphery.

Cumulative mass fraction versus droplet diameter distribution plot in Fig. 9a shows that toward the edge of the spray, $Y = 1.0$ cm, $r = 0.7$ cm, droplets smaller than 30 μm comprise 100 % of the liquid mass in the FB spray but only 70 % of the liquid mass in the AB spray, where the remaining mass is contained in larger (30–75 μm) droplets. Figure 9b shows that at the center of the spray at $Y = 3.0$ cm, a greater percentage of smaller droplets are produced by the AB atomizer. At $Y = 3.0$ cm and $r = 0.5$ cm, Fig. 9c shows that droplets smaller than 20 μm contain 95 % of the liquid mass in the FB spray as compared to 85 % of the liquid mass in the AB spray. These results demonstrate the superiority of the FB atomizer over AB atomizer, especially in the outer regions of the spray.

Next, Fig. 10a shows the radial profiles of SMD in non-reacting sprays of diesel and VO produced by the FB injector at the same operating conditions. The diesel

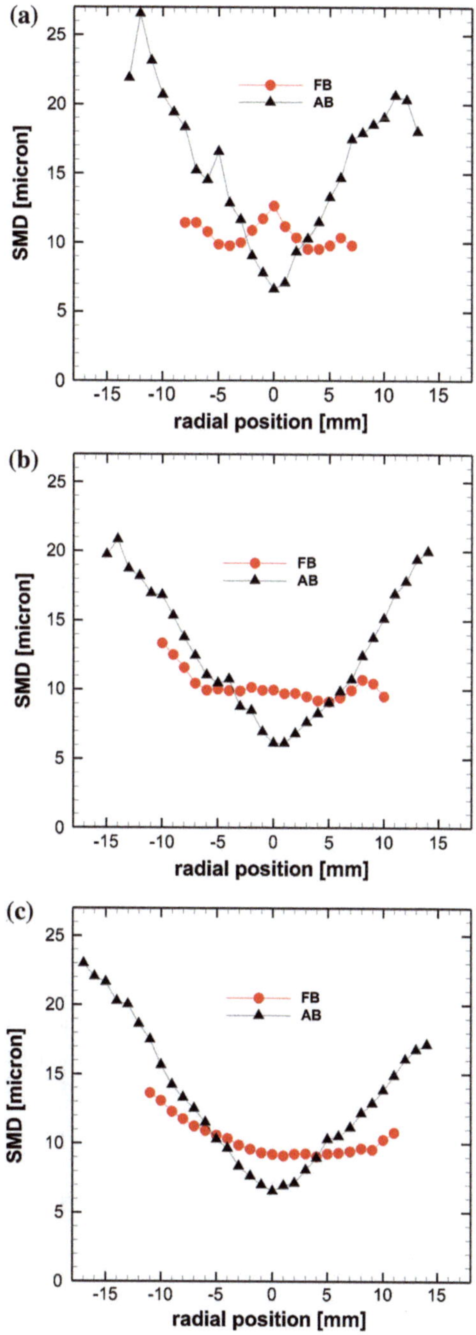

Fig. 8 SMD for FB and AB atomizers **a** $Y = 1.0$ cm; **b** $Y = 2.0$ cm; **c** $Y = 3.0$ cm

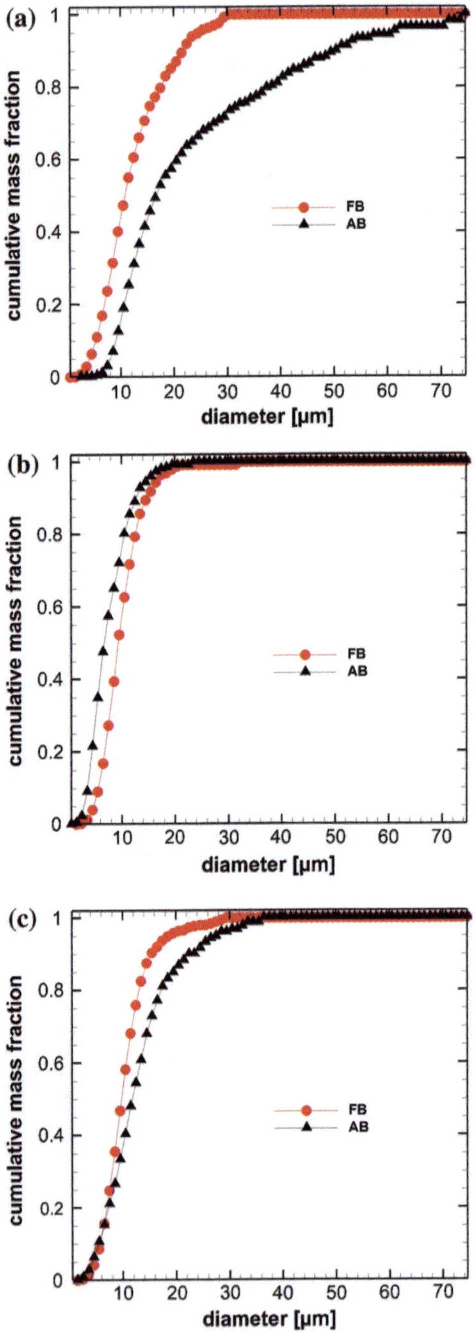

Fig. 9 Cumulative volume fraction distributions for FB and AB atomizers
a $Y = 1.0$ cm, $r = 0.7$ cm;
b $Y = 3.0$ cm, $r = 0.0$ cm;
c $Y = 3.0$ cm, 0.5 cm

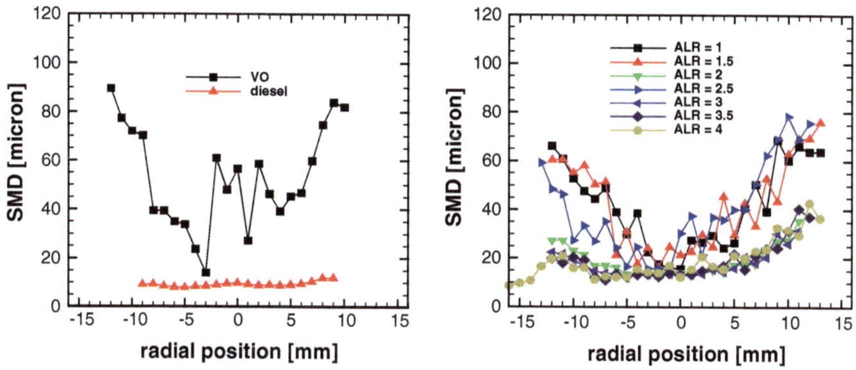

Fig. 10 Fuel impact on SMD at ALR = 2.5, $y = 2.0$ cm (*left*); ALR effect on SMD in VO spray at $y = 3.0$ cm (*right*)

spray contains very fine droplets, all around 10 μm. The droplets in the VO spray are much larger, up to around 90 μm at the edge of the spray. The VO spray also has a greater range of droplet sizes. The sizes of the droplets vary over an 80 μm range for the VO spray, while the droplets in the diesel spray only vary over a few μm. Figure 10b shows the effect of ALR on SMD in non-reacting VO spray at $Y = 3$ cm. The SMD is higher for ALR = 1, 1.5, and 2.5, although the differences between different ALRs are less pronounced. Results show that for ALR = 3.5 and higher, the FB injector can produce a very fine spray of straight VO with droplet diameters of about 20 μm. Note that in a reacting spray, the thermal feedback from the flame will help prevaporize the fuel, and thus, droplets with even smaller diameter distributions can be expected.

To illustrate this point, transverse profiles of the measured SMD in flames of glycerol co-combusted with methane are presented in Fig. 11. The percentage of the total heat release rate from glycerol and methane was varied, with some methane being injected through the injector together with the atomizing air. For example, G68 and G45 represent 68 and 45 % of the total heat release rate from glycerol, respectively, with the balance coming from methane. The term G45-29 signifies that 29 % of the total methane flow rate is supplied through the atomizer. All measurements were taken at approximately 10 cm from the FB injector exit as this was the most downstream point of optical access. The G68 flame produces the largest droplets even though there is significant overlap among all cases. The two G45 cases produce very similar results with the methane through the swirler case resulting in slightly smaller droplets. In all sprays, the SMD is generally around 25–40 μm. These values are smaller than the measured SMD in non-reacting VO sprays in Fig. 10, indicating that the present glycerol measurements have benefitted from the thermal feedback from the flame when making comparisons. Data in Fig. 11 were used with the corresponding velocity measurements to calculate a flow-weighted SMD for the entire spray. It resulted in flow-weighted SMDs of 40, 35, and 37 μm

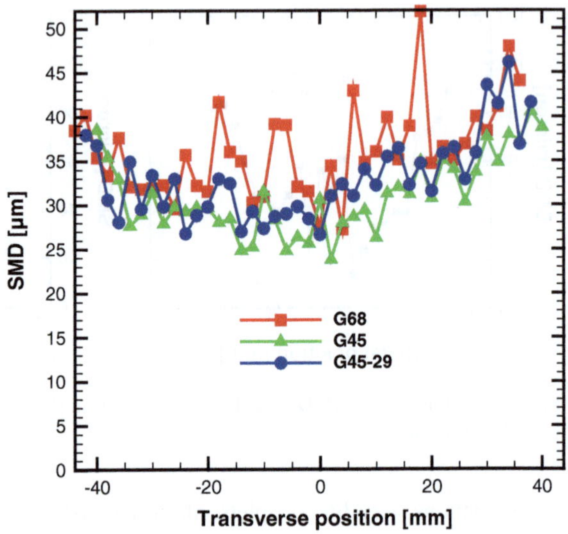

Fig. 11 SMD measured in different glycerol flames

for G68, G45, and G45-29, respectively. Clearly, the flow-weighted SMD for all three flames is similar and again highlights the effectiveness of the FB technique to atomize an extremely high-viscosity liquid for a range of operating conditions.

6 Biofuel Combustion

In this section, experimental results are presented for FB injector produced flames of biodiesel and straight VO as the two types of biofuels along with diesel (as baseline) and 70–30 diesel-VO blend. Experiments are conducted in an atmospheric pressure burner for fixed heat release rate and total airflow rate (Simmons et al. 2008). The total airflow rate is split between the atomizing air and primary combustion airflow rates. The air-to-fuel mass ratio through the injector is varied to document the injector fluid dynamics effects on emissions. Figures 12 and 13 show flame images taken for all fuels at atomizing airflow rates, AA = 10 and 15 %, respectively. In Fig. 12, for AA = 10 %, diesel, 70–30 diesel-VO blend, and biodiesel flames all have a blue premixed region overlapped with a yellow diffusion flame in the downstream region. The images revealed that the VO has a very distinct blue premixed flame. Figure 13, for AA = 15 %, showed that all fuels result in a blue premixed flame. The only visually conspicuous difference is that the straight VO flame is narrower and longer than the flames for the other fuels. At both atomizing airflow rates, all fuels exhibited stable flames with no particulate matter depositing on the combustor wall or escaping in the combustion products.

Figure 14 shows the radial profiles of CO and NO_X concentrations measured at the combustor exit plane for AA = 10 % and 15 %. Figure 14a shows that the CO emissions for AA = 10 % are similar for all fuels, around 10–15 ppm, with slightly

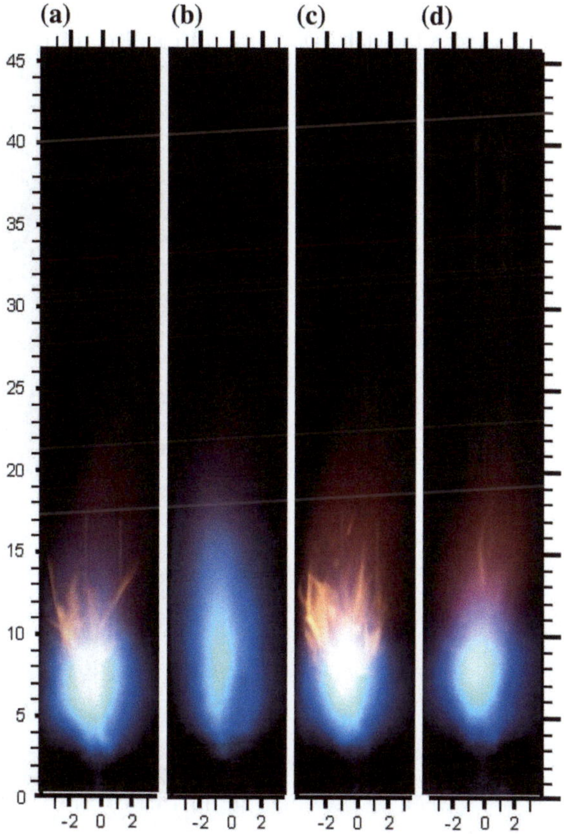

Fig. 12 Flame images at 10 % AA for **a** Diesel, **b** VO, **c** 70-30 diesel-VO, **d** Biodiesel

higher values for 70-30 diesel-VO blend and VO. Figure 14b shows that the NO_X emissions at AA = 10 % are highest for 70-30 diesel-VO blend, followed by diesel, biodiesel, and VO. The lower NO_X emissions for VO and biodiesel fuels are likely caused by their lower heating value in comparison with diesel. Note that the high heating value will result in slightly higher adiabatic flame temperature in case of diesel. Figure 14c shows the radial profiles of CO emissions for AA = 15 %. For all fuels tested, there is a significant decrease in CO emissions with increase in the atomizing airflow rate from AA = 10 to 15 %. For AA = 15 %, 70-30 diesel-VO blend shows the highest CO emissions, with profiles for other fuels somewhat overlapping with each other. No fuel showed CO emissions higher than 10 ppm for the higher atomizing airflow rate. Figure 14d shows the radial profiles of NO_X emissions for AA = 15 %. Results show a reduction by a factor of up to 3, as compared to the NO_X emissions for AA = 10 %. The diesel and VO produced the highest NO_X emissions, while biodiesel NO_X emissions were the lowest. However, all NO_X values were very low; between 15 and 20 ppm (Simmons and Agrawal 2012).

Results show low CO and NO_X emissions for all fuels at different atomizing airflow rates. CO and NO_X emissions decreased with increasing atomizing airflow

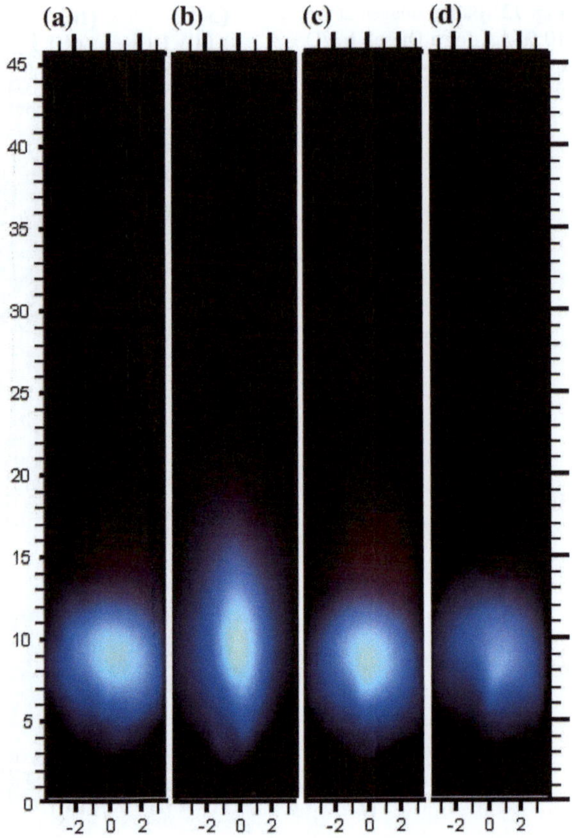

Fig. 13 Flame images at 10 % AA for **a** Diesel, **b** VO, **c** 70-30 diesel-VO, **d** Biodiesel

rate. Straight VO operated well under several conditions, and we were able to combust it cleanly without preheating or using other methods to modify the fuel properties and/or atomization characteristics.

7 Glycerol Combustion

FB injector was used to combust straight glycerol having extremely high kinematic viscosity as shown in Table 1. Glycerol also has a high ignition temperature. Thus, a glycerol flame can easily extinguish if sufficient thermal feedback from the flame is not available to preheat the incoming fuel stream (Metzger 2007). One potentially dangerous aspect of glycerol combustion is the production of acrolein. Acrolein is an extremely toxic compound at low levels (approximately 2 ppm) and is produced during decomposition of glycerol (Metzger 2007). Thus, the combustion system must oxidize decomposed species and ensure complete combustion. Preliminary experiments revealed a rather weak glycerol flame for combustion air at ambient

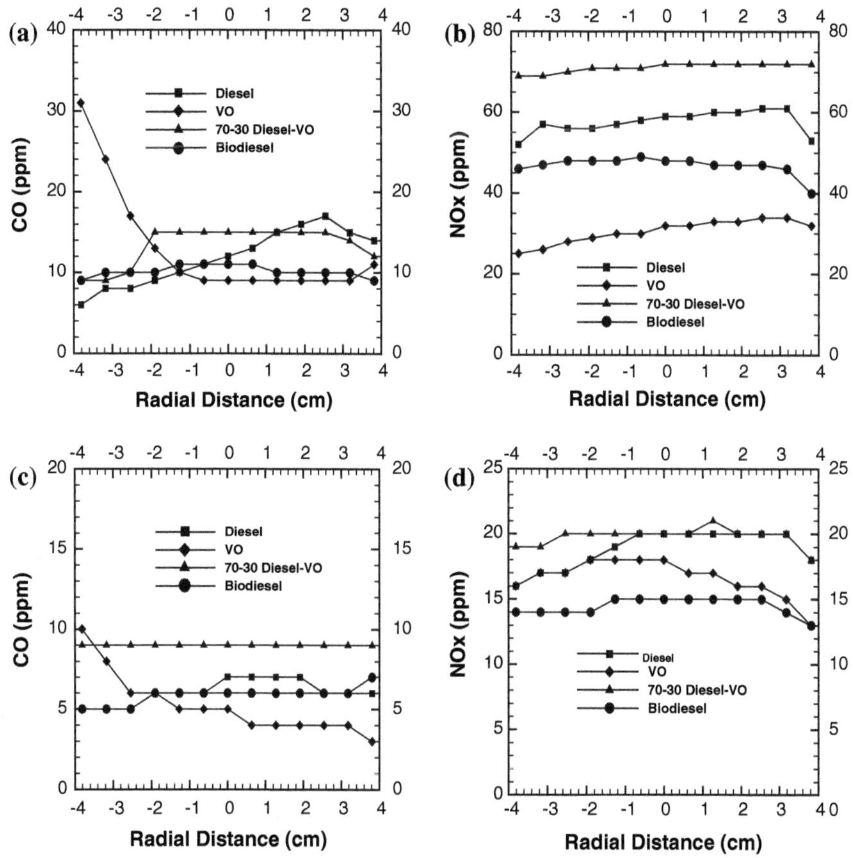

Fig. 14 Radial profiles of CO and NO_X concentrations (**a**, **b**) 10 % AA, (**c**, **d**) 15 % AA

temperature, resulting in poor thermal feedback from the flame to prevaporize the fuel droplets. Several options were considered to increase the fluid temperature in the near field of the spray: exhaust gas recirculation, preheated combustion air, and vitiated combustion air. However, a simpler solution was to combine glycerol combustion with methane combustion. The methane flame around the glycerol spray provided thermal feedback to prevaporize glycerol droplets and to help preheat the resulting fuel–air mixture to the required ignition temperature. The methane combustion also increased the overall reaction rates and helped ensure sufficient residence time to oxidize the decomposed species produced during glycerol combustion (Simmons et al. 2010).

In this section, the range of stable operating conditions and resulting emissions of glycerol–methane flames in a swirl-stabilized combustor operated at atmospheric pressure are reported. These results have subsequently been used to refine the burner concept to achieve even greater fuel flexibility while minimizing the environmental impact (Jiang and Agrawal 2014; Jiang et al. 2014). First, the heat release rate (HRR) was kept constant at approximately 7 kW, and the total air (combustion air plus atomizing air) flow rate (Q) was kept constant at 150 slpm pertaining to pure glycerol combustion at overall equivalence ratio of 0.75. At these conditions, 50/50 glycerol and methane fuel blend (meaning 50 % heat release from glycerol and 50 % from methane) was used, while the AA flow rate through the injector was varied to change the air-to-liquid mass ratio (ALR). Flame images and emissions data were acquired to determine the optimum ALR for atomization. This procedure was repeated for different HRRs to find the optimum ALR for each case. Finally, for a fixed heat release rate of 7 kW, the fuel blend was varied between 0 % glycerol (100 % methane) and 100 % glycerol (0 % methane).

7.1 Effect of Air-to-Liquid Mass Ratio

The visual flame images for 50/50 glycerol-methane fuel, HRR = 7.0 kW, and Q = 150 slpm showed that the flame becomes shorter and less luminous as the ALR is increased. At smaller ALRs, the flame exhibited two distinct regimes; a blue region in the near field and downstream orange region. The blue region pertains to combustion of methane and evaporated glycerol from smaller droplets. The larger glycerol droplets burnt mostly in diffusion mode in the downstream yellow region. With increasing ALR, these two distinct regions merged to form a fairly homogeneous reaction zone. For higher ALR, the FB injector is expected to produce smaller droplets that evaporate quickly to form a combustible fuel–air mixture closer to the injector exit. Thus, the resulting flame becomes shorter, less luminous, and more premixed. If the ALR was too high, the flame became unstable and eventually blew off. If the ALR was too low, unburned reactants were released in the products indicating incomplete combustion resulting from poor atomization. Based on these observations, an ALR of 2.6 was determined to be the best choice.

Figure 15 shows the effect of ALR on CO and NO_X emissions for test conditions discussed above. The emissions measurements were taken at the center point of the combustor exit plane. Figure 15 shows that the CO emissions decrease as the ALR is decreased. Evidently, the large concentrations of CO emissions produced near the wall of the quartz combustor were not captured by the single-point measurements presented in Fig. 15. The NO_X emissions were low, within 2 ppm, for all values of ALR.

7.2 Effect of Heat Release Rate

Heat release rate is an important parameter affecting the fuel and airflow rates, and thus, the flow mixing processes inside the combustor. Visual images revealed that at the lowest flow rates (or HRR), the fuels do not burn as a homogenous flame. At higher heat release rates, the flames appeared completely homogenous.

Fig. 15 Effect of ALR on CO and NO_X emissions for 50/50 mixture, HRR = 7 kW

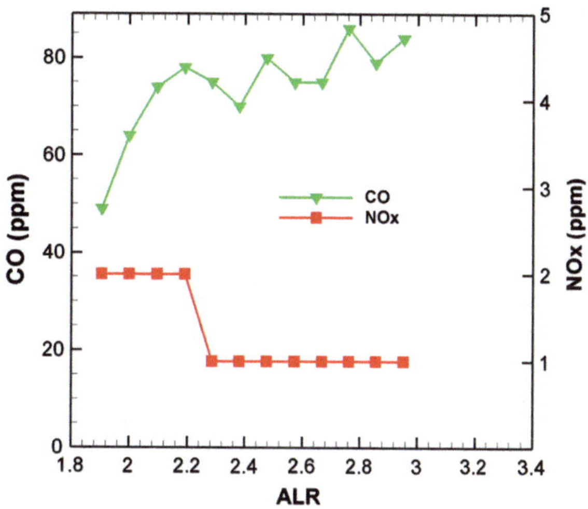

Fig. 16 Effect of total heat release on CO emissions for 50/50 mixture

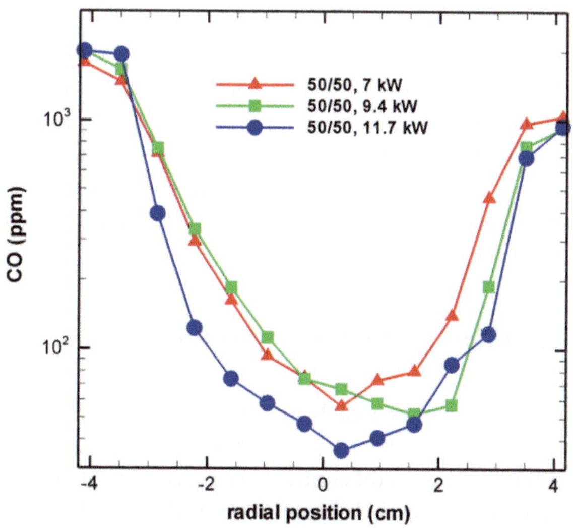

With increasing HRR, the resulting flame images became increasingly brighter, presumably because of the higher concentrations of radicals in the reaction zone. Next, Fig. 16 shows radial profiles of CO emissions at the combustor exit plane for the three heat release rates. Results show high CO concentrations of around 3,000 ppm near the combustor wall. Two factors account for this observation: (1) quenching of reactions because of heat loss through the wall and (2) larger droplets migrating and burning in diffusion mode near the wall. The CO concentrations decrease to 40–60 ppm toward the center of the combustor.

Although HRR = 11.7 kW case shows the lowest CO emissions over part of the plot, all cases give similar CO values overall. Figure 17 shows the NO_X emissions

Fig. 17 Effect of total heat release on CO emissions for 50/50 mixture

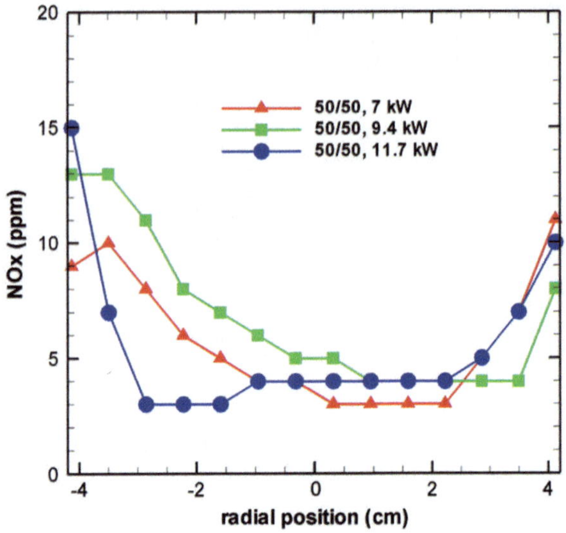

for the same conditions as Fig. 16. NO_X emissions peaked to between 8 and 15 ppm near the wall for all cases. Larger droplets migrating to the combustor periphery and burning in diffusion mode are responsible for this result. The NO_X concentrations were within 5 ppm at the center of the combustor for all cases. The heat release rate did not seem to affect the NO_X emissions for the cases studied. Note that measurements at the center represent values in a large combustor.

7.3 Effect of Fuel Composition

Figure 18 shows radial profiles of CO emissions for 50/50 and 75/25 glycerol–methane fuel compositions. The 0/100 and 25/75 cases were not of interest because of their low glycerol content. Data were not taken for the 100/0 case because of the unburned glycerol in the products. CO concentrations for the 50/50 case are around 3,000 ppm near the wall and 50 ppm at the center of the combustor. In comparison, the 75/25 case results in CO concentrations of around 5,000 ppm near the wall and 300 ppm at the center of the combustor.

The lower methane content produces leaner mixtures with slower reaction rates, and hence, poor thermal input to evaporate glycerol droplets, which leads to incomplete combustion. Figure 19 shows the NO_X emissions for the same conditions as Fig. 18. The NO_X concentrations were generally higher for the 75/25 case. The peak NO_X emissions (near the wall) for the 75/25 case were around 20 ppm while those for the 50/50 case were about 10 ppm. The smaller methane content for the 75/25 case resulted in longer evaporation time for the glycerol spray, which led to more droplets burning in diffusion mode to produce higher flame temperatures or thermal NO_X.

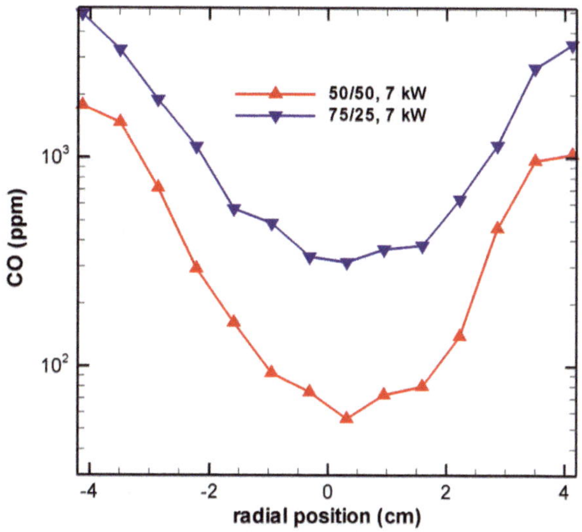

Fig. 18 Effect of fuel composition on CO emissions in glycerol flames

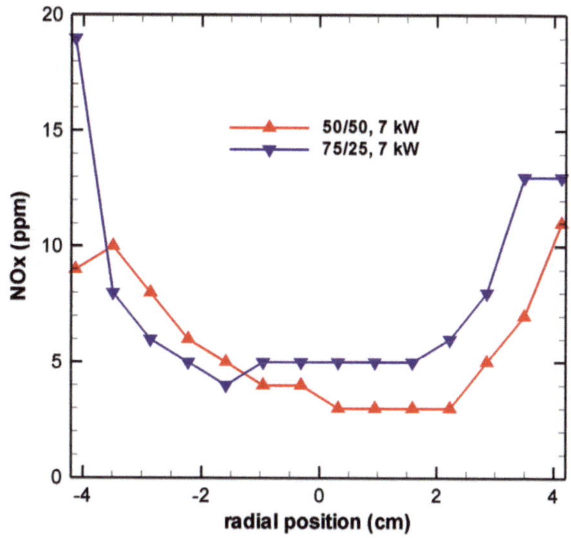

Fig. 19 Effect of fuel composition on NO_X emissions in glycerol flames

OH* chemiluminescence images were acquired to reveal that the reaction zone lengthens as the glycerol content in the flame is increased. For 75/25 case, the reaction zone encompassed the entire length of the combustor, demonstrating the difficulty of providing sufficient thermal feedback to vaporize glycerin droplets in high glycerol content flames. For 100 % glycerol case, the combustion process exhibited greater instabilities with occasional flame blow-off resulting from inadequate thermal feedback. Combustion is expected to improve by increasing the swirl number of the incoming combustion air, thereby increasing heat recirculation from the flame.

8 Concluding Remarks

We have investigated low-emission, fuel-flexible combustion of liquid fuels using a novel flow-blurring (FB) atomization technique. FB injector provides a simple atomization method that can reduce CO and NO_X emissions by a factor of 3–5 when compared to an AB injector. FB atomization method is inherently different than AB process in that atomization mechanism is based on expansion of gas in two-phase flow instead of shear between the fuel and air. FB atomizer produces water sprays with smaller droplets and a narrower range of droplet sizes as compared to AB atomizer. FB atomizer is capable of cleanly atomizing and combusting straight vegetable oil (VO) with no fuel processing or preheating. FB atomizer operating with VO produces lower emissions than AB atomizer operating with biodiesel, meaning that transesterification of VO to produce biodiesel is not necessary to achieve clean combustion. An insulated combustion environment is necessary to ensure proper thermal feedback in glycerol flames. FB combustor can be used to atomize and combust glycerol in an insulated environment without any fuel mixing or preheating. Sufficient residence time within hot environment is critical to complete oxidation of glycerol. Although the FB injector was effective for fuel atomization, the thermal feedback from the pure glycerol flame was inadequate to fully prevaporize the fuel and/or preheat the reactant mixture. Combining the glycerol combustion with methane combustion was shown to effectively solve these issues, resulting in wider operating limits while minimizing CO emissions. While co-firing glycerol with methane, the methane flow can be split between combustion and atomizing air lines to minimize emissions. Emissions were mostly formed near the combustor wall, where large droplet migrated and burned in diffusion mode. Subsequent work would delineate the coupling between thermal and flow aspects of the system to ensure stable combustion of many different fuels in the same system.

Acknowledgments Author would like to thank his former graduate students Pankaj Kolhe, Heena Panchasara, Benjamin Simmons, and Lulin Jiang for their contributions to this research. Financial support from the US Department of Energy, US Department of Education, and the University of Alabama facilitated this research.

References

Bolszo CD, McDonell VG (2009a) Evaluation of plain-jet air blast atomization and evaporation of alternative fuels in a small gas turbine engine application. Atomization Sprays 19(8):771–785

Bolszo CD, McDonell VG (2009b) Emissions optimization of a biodiesel fired gas turbine. Proc Combust Inst 32(2):2949–2956

Gañán-Calvo AM (2005) Enhanced liquid atomization: from flow-focusing to flow blurring. Appl Phys Lett 86:2141–2143

Graboski MS, McCormick RL (1998) Combustion of fat and vegetable oil derived fuels in diesel engines. Prog Energy Combust Sci 24(2):125–164

Jiang L, Agrawal AK (2014) Combustion of straight glycerol with/without methane using a fuel-flexible, low-emissions burner. Fuel 136:177–184

Jiang L, Agrawal AK, Taylor RP (2014) Clean combustion of different liquid fuels using a novel fuel injector. Exp Thermal Fluid Sci 57:275–284

Lefebvre AH (1989) Atomization and spray. Hemisphere Publishing

Lefebvre AH (1999) Gas turbine combustion, 2nd edn. Taylor and Francis Publications

Li TX, Zhu DL, Akafuah NK, Saito K, Law CK (2011) Synthesis, droplet combustion, and sooting characteristics of biodiesel produced from waste vegetable oils. Proc Combust Inst 33(2):2039–2046

Mansour A (2005) Gas Turbine fuel injection technology. *ASME Paper GT-2005-68173*

McDonell VG, Samuelsen S (1991) Gas and drop behavior in reacting and non-reacting airblast atomizer sprays. J Propul Power 7:684–691

Metzger B (2007) Glycerol combustion, M.S. Thesis, North Carolina State University, NC

Nakamura S, McDonell V, Samuelsen S (2006) The effect of liquid-fuel preparation on gas turbine emissions. ASME Paper GT-2006-90730

Pan KL, Li JW, Chen CP, Wang CH (2009) On droplet combustion of biodiesel fuel mixed with diesel/alkanes in microgravity condition. Combust Flame 156(10):1926–1936

Panchasara H, Sequera D, Schreiber W, Agrawal AK (2009a) Emission reductions in diesel and kerosene flames using a novel fuel injector. J Propul Power 25(4):984–986

Panchasara HV, Simmons BM, Agrawal AK (2009b) Combustion performance of biodiesel and diesel-vegetable oil blends in a simulated gas turbine burner. J Eng Gas Turbines Power 131 (3):1–11 031503

Park SH, Cha J, Lee CS (2011) Spray and engine performance characteristics of biodiesel and its blends with diesel and ethanol fuels. Combust Sci Technol 183(8):802–822

Porter L (2007) NASA's new aeronautics research program overview. Presented at the AIAA Aerospace Science Meeting, Reno, NV, Jan., 2007. http://www.aeronautics.nasa.gov/pdf/armd_overview_reno_4.pdf

Raghavan V, Rajesh S, Parag S, Avinash V (2009) Investigation of combustion characteristics of biodiesel and its blends. Combust Sci Technol 181(6):877–891

Razdan MK (1998) Fuel/Air preparation in the design of low emissions gas turbine combustion systems. Proceedings of fourteenth NATO RTO meeting on gas turbine combustion emissions and alternative fuels, Paper No. 34

Simmons BM, Panchasara HV, Agrawal AK (2008) Effect of fuel injection concept on combustion performance of liquid biofuels. *Session B3: Biofuels*, 2008 technical meeting of the central states section of the combustion Institute, Tuscaloosa, AL, 20–22 Apr 2008

Simmons B, Panchasara H, Agrawal AK (2009) A Comparison of air-blast and flow-blurring injectors using the Phase Doppler Particle Analysis technique. ASME Paper 2009-GT-60239

Simmons BM, Kolhe PS, Taylor RP, Agrawal AK (2010) Glycerol combustion using flow-blurring atomization. Proceedings of the 2010 central states meeting of the Combustion Institute, March, 2010

Simmons BM, Agrawal AK (2011) Spray characteristics of a flow-blurring atomizer. Atomization Sprays 20(9):821–825

Simmons B, Agrawal AK (2012) Flow blurring atomization for low-emission combustion of liquid biofuels. Combust Sci Technol 184:660–675

Song J, Alam M, Boehman AL (2007) Impact of alternative fuels on soot properties and DPF regeneration. Combust Sci Technol 179(9):1991–2037

Sovani SD, Sojka PE, Lefebvre AH (2001) Effervescent atomization. Prog Energy Combust Sci 27:483–521

Wang X, Huang Z, Kuti OA, Zhang W, Nishida K (2011) An experimental investigation on spray, ignition and combustion characteristics of biodiesels. Proc Combust Inst 33(2):2071–2077

Part V
Sustainable Energy Systems and Efficiency Improvements

In-Depth Performance Evaluation of RDF from Landfill Reclamation for Green Electricity Generation in a Downdraft Gasifier

Somrat Kerdsuwan and Krongkaew Laohalidanond

Abstract Landfill reclamation is one of the possibilities to reclaim land resources and recover energy hidden in garbage into green and renewable energy. This study aims to conduct a feasibility study of using waste reclaimed from old landfill, called mined Municipal Solid Waste (mined MSW), for power generation in a 50 kW downdraft gasification system and to optimize its operating conditions. Waste reclaimed from dumpsite in Phuket municipality, southern part of Thailand, consists of mostly plastic fraction and has a higher heating value of 28.68 MJ/kg. In this study, waste reclaimed is first solar dried and cut into small size before using as loose-RDF in 10 kg/h downdraft gasification process to study the gasification process. Three experiments with different air flow rates of 12, 18, and 21 Nm^3/h were carried out, and it was found that the heating value of producer gas from loose-RDF ranged from 0.89 to 2.71 MJ/Nm^3 which was considered to be very low for further use in diesel engine for power generation. Hence, the densification of RDF was performed in order to improve the RDF properties. After densification process, three experiments of dense-RDF gasification with different air flow rates of 18, 21, and 24 Nm^3/h were repeated again. The results indicated that by using dense-RDF, the heating value of producer gas increased to 2.46–2.99 MJ/Nm^3 and the cold gas efficiency reached 43.26–56.22 %. At the air flow rate of 21 Nm^3/h, the producer gas has the maximum heating value and maximum cold gas efficiency. Although dense-RDF can produce the higher heating value of producer gas and can achieve the higher cold gas efficiency, it consumed more energy in preparation process which was accounted for 0.12 kWh per dense-RDF 1 kg. For the experiments in a 50 kg/h prototype scale downdraft gasifier, dense-RDF was gasified with three different air flow rates of 73, 85, and 101 Nm^3/h. It was found that the maximum heating value of producer gas of 3.12 MJ/Nm^3 was achieved at the air flow rate of 101 Nm^3/h. After passing through gas-cleaning process, the producer

S. Kerdsuwan (✉) · K. Laohalidanond
The Waste Incineration Research Center, Department of Mechanical and Aerospace Engineering, Faculty of Engineering, King Mongkut's University of Technology North Bangkok, Wongsawang, Bangsue, Bangkok 10800, Thailand
e-mail: srk@kmutnb.ac.th

A.K. Agarwal et al. (eds.), *Novel Combustion Concepts for Sustainable Energy Development*, DOI 10.1007/978-81-322-2211-8_21

gas can be used to fuel a diesel engine in continuous mode for power production. The electrical energy and overall efficiency as well as the quality of lube oil and wear of engine part have been reported.

Keywords Landfill reclamation · Mined municipal solid waste · Power generation · Downdraft gasification · RDF · Producer gas · Heating value

1 Introduction

Municipal solid waste (MSW) is non-hazardous substance generated from routine activities of human life. Sources of MSW can be household, industrial, commercial and institutional sectors, markets, yards, and streets (Schübeler et al. 1996). The increase in world population and the economic boom have led to the increase in the amount of MSW. The World Bank predicted that the global MSW amount will increase from 1.3 billion ton in 2012 to 2.2 billion ton in 2025 (Hoornweg and Bhada-Tata 2012). The substantial amount of MSW requires the effective MSW management which includes collection, storage, and disposal. Focusing on the MSW disposal, most of developed countries or high income countries dispose their MSW by sanitary landfill or thermal treatment, whereas controlled dumping is used for MSW disposal in middle income. Low income countries or developing countries dispose their MSW by open dump with no measurement for air and water pollution control, especially in local administrative area (Hoornweg and Bhada-Tata 2012; Henry et al. 2006; Manaf et al. 2009). The sanitary landfilling, controlled dumping, and open dumping need a large authorized land area is essential, and the increasing amount of MSW will results the lack in land area for MSW disposal. Aside from the problem of the lack in land area, these MSW disposal methods cannot recover the useful energy hidden in MSW. As MSW contains high portion of combustible materials, for example, organic waste ranges from 28 to 64 wt%, paper and plastic waste contributes to 5–31 and 8–12 wt%, respectively, as shown in Table 1 for the MSW composition in different level of country's incomes (Hoornweg and Bhada-Tata 2012), these combustible materials have a high energy content which can further be converted into the green and clean energy.

The landfill reclamation or landfill mining can be one possibility not only to provide more land for MSW disposal but also to recover energy hidden in MSW into other forms of energy. During landfill, the organic material in MSW is decomposed into soil-like material which is the main product of landfill reclamation, and the remaining are plastic, textile, wood, and paper with some amounts of non-combustible materials, as shown in Table 2 for different size distributions of mined MSW (Prechthai et al. 2008).

After the separation of soil and non-combustible material, the heating value of mined MSW can increase to 28–37 MJ/kg, resulting from the high content of plastic waste (Prechthai et al. 2008). This heating value of mined MSW is much higher

Table 1 MSW composition in different level of country's incomes

Physical composition (wt%)	Country's income levels			
	Low income	Lower middle income	Upper middle income	High income
Organic waste	64	59	54	28
Paper	5	9	14	31
Plastic	8	12	11	11
Glass	3	3	5	7
Metal	3	2	3	6
Others	17	15	13	17

Table 2 Physical composition of mined MSW for different size distribution

Physical composition (wt%)	Size distribution	
	25–50 mm	>50 mm
Soil-like material	47.3	14.0
Plastic	14.5	34.3
Textile	3	10.5
Wood	4.5	10.7
Paper	2.8	4.3
Rubber	0.8	5.8
Foam	0.8	1.2
Ceramic	9.2	3.0
Glass	10.3	7.5
Metal	5.3	8.3
Battery	1.4	0.4

than the acceptable heating value of MSW for thermal treatment technology (at least 7 MJ/kg) (World Bank 1999). Gasification technology coupled with gas engine is one of the solutions to recover mined MSW to green and clean energy compared to conventional direct combustion integrated with steam turbine because it has high electricity production efficiency (30–40 % for gasification compared to 15–25 % for direct combustion (BioenergyNet 2014). Since the mined MSW is non-homogeneous with high size and composition variation, it is not suitable to be directly used as feedstock for gasification technology, the mined MSW has to be prepared into the homogenous form. This study aims to conduct the feasibility study of using reclaimed waste from landfill for power generation in a 50 kW downdraft gasification system and to optimize the operating condition for power generation in continuous mode. The downdraft gasification technology is selected in this study because it has the following advantages (Obernberger and Thek 2008; Belgiorno 2003; Knoef 2005; WIRC 2006):

- Flexible feedstock size (20–100 mm) and moisture content in feedstock (<20 %).
- Provides clean producer gas with low tar content which can meet the engine requirements.
- Suitable for decentralized small scale waste to energy power plant.
- Low investment as well as operation and maintenance cost.
- Ease to operate.

2 Materials and Methods

2.1 Materials

Five-year-old mined MSW reclaimed from the landfill site in Phuket waste management center, southern city of Thailand, as shown in Fig. 1, was used in this study. After mining, the MSW was solar dried and its physical composition was investigated by quartering method. Table 3 shows the physical composition of mined MSW compared to as received MSW at Phuket waste management center.

Fig. 1 Mined MSW from the landfill site in Phuket waste management center

Table 3 Physical composition of mined MSW and as received MSW

Physical composition	Mined MSW	As received MSW (WIRC 2010)
Soil-like material	43.50	0
Plastic	39.98	18.43
Textile/rubber	4.75	15.79
Paper	0.44	5.26
Glass	10.05	3.36
Metal	0.43	0.01

Table 4 Fuel characteristics of mine MSW

	Mined MSW	As received MSW (WIRC 2010)	ASTM
Proximate analysis (wt%)			
Moisture content[a]	1.23	46.34	D 3173-87
Volatile matter[b]	74.82	41.62	D 3175-89a
Fixed carbon[b]	6.95	0.82	D3172-89
Ash[b]	18.23	11.22	D 317489-
Ultimate analysis[b] (wt%)			
C	53.60	46.10	D 3178-89
H	6.03	6.38	D 3178-89
N	Not detected	1.10	D 3179-89
S	0.16	0.12	D 4239-85c
Other properties			
Bulk density (kg/m^3)[a]	90	128.91	
Lower heating value (MJ/kg)[b]	25.82	8.03	D 3286-91

[a] As received basis
[b] Dry basis

From Table 3, mined MSW contains high quantity of soil (43.50 wt%) and plastic waste (39.98 wt%). Other compositions are paper, textile/rubber, and organic waste, as well as non-combustible materials.

After separation of non-combustible material, a mined MSW sample was used to determine its moisture content, proximate and ultimate analysis, and heating value according to the ASTM standard. Table 4 shows the fuel characteristics of mined MSW in comparison with as received MSW. It can be seen that the moisture content of mined MSW is much lower than that of the as received waste, whereas volatile matter and ash of mined MSW are higher than those of as received waste since the decomposition of moist kitchen waste into soil-like material and high content of plastic waste in mined MSW, consequently, higher heating value.

2.2 Method

2.2.1 Feedstock Preparation

Feedstock used in this study is Refuse Derived Fuel (RDF) from mined MSW. After solar drying and separation process, mined MSW was cut into the small size with a width of 5 cm and a length of 15 cm by the 30 hp-shredder. The small size mined MSW was thereafter heat extruded into the cylinder form with a diameter of 4 cm by a 480 kg/day extruder. Finally, extruded mined MSW was cut to the length of 5–7 cm which is called dense-RDF, as shown in Fig. 2.

Fig. 2 **a** Shredded mined MSW, **b** Extruded mined MSW, and **c** RDF

2.2.2 Experimental Setup

In the first step of this study, the RDF gasification experiments were carried out in a 10 kg/h laboratory scale downdraft gasifier in order to study characteristic of producer gas formation. A 10 kg/h laboratory scale downdraft gasifier system consists of a downdraft gasifier, air blower, air preheater, and gas-cleaning unit, as shown in Fig. 3.

After studying the characteristic of producer gas formation behavior, the accumulated 200 h of experiments on a prototype scale downdraft gasification system was conducted for the feasibility of up-scaling. The gasification system consists of a 50 kg/h batch type downdraft gasifier, a shell and tube heat exchanger, a gas-cleaning system, and a internal combustion engine-generator, as presented in Fig. 4. The downdraft gasifier is separated in three zones along the height of gasifier; the fuel hopper, the reaction zone, and the ash discharging zone. A shell and tube heat exchanger was applied in order to preheat the air supplied into the reaction zone for the gasification process. The gas-cleaning system consisted of a cyclone, a venturi scrubber, and dust removal unit. The internal combustion engine-generator was a 4 stroke diesel engine with 4 cylinders and could produce 50 kW of electricity, 380/400 V, and 50/60 Hz.

2.2.3 Experimental Procedure

In order to study the characteristic of producer gas formation in a laboratory scale downdraft gasification, 2 kg of lignite coal was first introduced into gasifier for ignition purpose. After some part of coal was ignited and the combustion process occurred, approximately 8–10 kg of either shredded-RDF or dense-RDF was fed into gasifier and then air was supplied into reaction zone of downdraft gasifier as gasification agent at constant flow rate of 12, 18, and 21 Nm^3/h for shredded-RDF and 18, 21, and 24 Nm^3/h for dense-RDF. RDF was ignited and the gasification process started. During the gasification process, the samples of producer gas were taken for producer gas analysis.

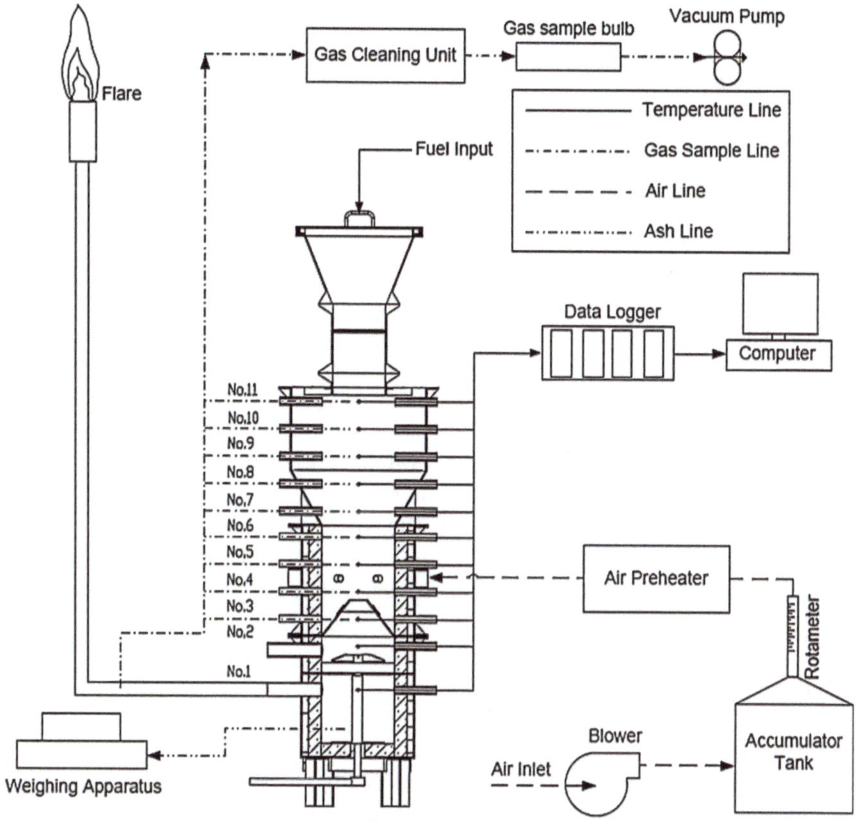

Fig. 3 A 10 kg/h laboratory scale downdraft gasification system

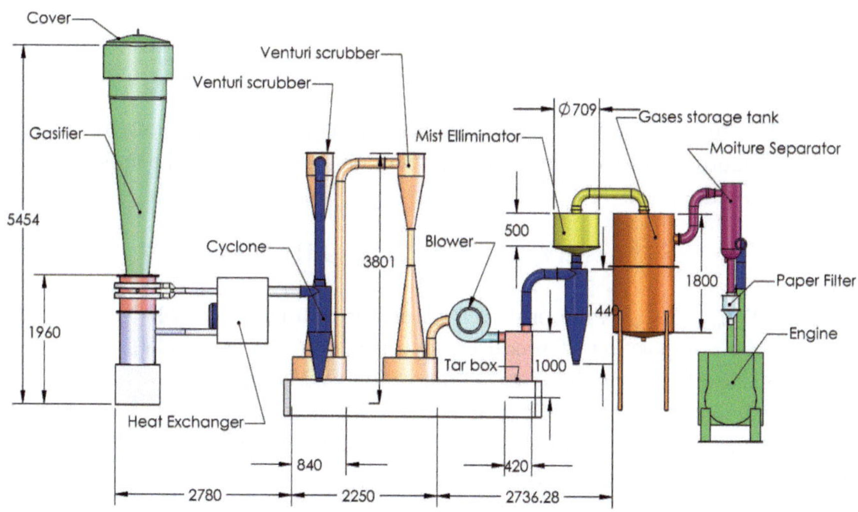

Fig. 4 Experimental setup for a downdraft gasification system

For the up-scaling test, 30 kg of charcoal was fed into the bottom of gasifier and ignited in order to preheat the gasifier, and 300 kg of dense-RDF was introduced into the downdraft gasifier per batch. To start the gasification process, the air flow rate of 75 Nm^3/h was supplied into the gasifier. During the gasification process, the samples of producer gas were taken for analysis. To optimize the operating condition, the experiment was repeated with the air flow rate of 85 and 101 Nm^3/h.

To study the feasibility of long-term test run for 200 accumulated operating hour, 60–100 kg of charcoal was fed into the bottom of gasifier and ignited in order to preheat the gasifier; whereas wood chip was also supplied together with dense-RDF at the different weight ratio of dense-RDF to wood chip of 50:50, 75:25, and 100:0. After all feedstock were put into the gasifier, the ignition process was conducted and then air was supplied into the gasifier at the optimum condition to promote the gasification process. During the steady state experiment, the producer gas was taken at the exit of gasifier for the investigation of producer gas composition and heating value. The system performance for power production was also evaluated at the same time. The gasification process was terminated when it was observed that no producer gas generated. The whole experiment was repeated again until the operation test run reached 200 h. In the accumulated 200 h test run, the dual fuel engine-generator test using diesel and producer gas was performed.

2.2.4 Analysis

The composition of producer gas in terms of CO, H_2, CO_2, CH_4, and N_2 was analyzed by gas chromatography (GC) according to the ASTM standard. The producer gas velocity is measured by the gas flow meter, and the producer gas yield at standard condition can be calculated by Eq. 1, while the heating value of producer gas was calculated by Eq. 2.

$$Q_{\text{Producer gas}} = \frac{Q_1 \times T_2}{T_1}, \tag{1}$$

where Q_1 is the volume flow rate of producer gas at the exit condition (m^3/h), T_1 is the temperature of producer gas at the gasifier exit (°C), and T_2 is the temperature at standard condition (25 °C).

$$\text{LHV}_{\text{Producer gas}} = \Sigma(X_i \times \text{LHV}_i) \tag{2}$$

where $\text{LHV}_{\text{producer gas}}$ is the producer gas lower heating value (MJ/Nm^3), X_i represents each composition of producer gas yield (Nm^3/h), and LHV_i is the lower heating value of each producer gas composition (MJ/Nm^3).

The efficiency of the gasification process was expressed in term of cold gas efficiency defined as the ratio of energy input into the gasifier to the energy output from the gasifier, as given in Eq. 3.

$$\eta_{\text{gasifier}} = \frac{\text{LHV}_{\text{producer gas}} \times Q_{\text{producer gas}}}{\text{LHV}_{\text{fuel}} \times m_{\text{fuel}}} \times 100\ \% \quad (3)$$

where η_{gasifier} is the cold gas efficiency. LHV_{fuel} and m_{fuel} are the lower heating value (MJ/kg) and weight (kg) of all fuel used in each experiment, respectively.

Since the engine-generator for power production was connected for the feasibility study of 200 h test run, the overall efficiency for power production should be consider. In this study, the overall efficiency for power production was determined by Eq. 4.

$$\eta_{\text{overall}} = \frac{P}{\text{LHV}_{\text{fuel}} \times m_{\text{fuel}}} \times 100\ \% \quad (4)$$

where η_{overall} is the overall efficiency for power production. P is the electrical power output from the engine-generator. In this study, the power production was kept constant at 24 kW.

Since the engine operated in the dual fuel mode, the diesel replacement rate (DR) was taken into account. Equation 5 expressed the diesel replacement.

$$\text{DR} = \frac{Q_{\text{d,o}} - Q_{\text{d,d}}}{Q_{\text{d,o}}} \times 100\ \% \quad (5)$$

where $Q_{\text{d,o}}$ and $Q_{\text{d,d}}$ are the volume flow rate of diesel consumption in case of using pure diesel and in case of using dual fuel (l/h), respectively.

The solid fuel consumption rate per a unit of power production (SCR in kg/kWh) is the amount of solid fuel consumed for 1 kWh power production and it was calculated by Eq. 6.

$$\text{SCR} = \frac{m_{\text{fuel}}}{P \times \text{time}} \quad (6)$$

where time is the experimental time for each experiment.

3 Results and Discussion

3.1 Characteristic of Producer Gas in a Laboratory Scale Downdraft Gasifier

From the study of the characteristic of producer formation in a laboratory scale, it can be found that shredded-RDF and dense-RDF has a potential to produce the high heating value producer gas. Table 5 shows the result of the laboratory test.

From Table 5, the producer gas composition (at the stable condition as average value) depended on the air flow rate supplied into the gasifier, consequently, lower heating value of producer gas. For the test with shredded-RDF, the lower heating

Table 5 Result of the laboratory scale test

Air flow rate (Nm^3/h)	Producer gas composition (vol%)				Lower heating value (MJ/Nm^3)
	CO	CH_4	H_2	CO_2	
Shredded-RDF					
12	2.92	0.87	2.04	6.22	0.89
18	4.39	2.17	3.71	8.05	1.73
24	5.38	4.61	3.51	7.30	2.71
Dense-RDF					
18	7.62	1.69	8.23	9.78	2.46
21	6.64	3.91	6.79	10.16	2.99

value varied from 0.89 to 2.71 MJ/Nm^3, which was considered to be very low for further used in the dual fuel engine-generator. The improvement of fuel characteristics was carried out by the densification step, as describe in Sect. 2.2.1. After fuel improvement, it can be observed that the dense-RDF provided the higher heating value of producer gas compared to shredded-RDF. The lower heating value of producer gas from dense-RDF ranged from 2.46 to 2.99 MJ/Nm^3 and reached the maximum value of 2.99 MJ/Nm^3 at the air flow rate of 21 Nm^3/h. Although the test was conducted at the air flow rate of 24 Nm^3/h, the gasification process occurred unstably. Therefore, the analysis of producer gas at this flow rate was not considered.

3.2 Up-Scaling of a Prototype Scale Downdraft Gasifier

The test laboratory indicated that dense-RDF provided the higher heating value of producer gas than shredded-RDF; hence, only dense-RDF was used as feedstock in the up-scaling of a prototype scale downdraft gasifier. The experimental results for a prototype scale downdraft gasifier are shown in Table 6.

From Table 6, the producer gas from the prototype scale downdraft gasifier had the higher heating value compared to the test in the laboratory scale downdraft

Table 6 Result of the prototype scale test

Air flow rate (Nm^3/h)	Producer gas composition (vol%)				Lower heating value (MJ/Nm^3)
	CO	CH_4	H_2	CO_2	
Dense-RDF					
73	9.95	0.49	6.88	9.84	2.13
85	14.99	0.47	10.40	7.31	3.12
101	14.44	1.20	6.41	8.51	2.86

gasifier. The maximum lower heating value of producer gas of 3.12 MJ/Nm^3 was achieved at the air flow rate of 85 Nm^3/h.

3.3 Accumulated 200 h Operation in a Prototype Scale Gasifier

The preliminary test in a prototype scale gasifier (Sect. 3.2) indicates that the test in a prototype scale can obtain the higher heating value of producer gas than the test in laboratory scale. To study the effect of operating condition on power production in the long-term run, the accumulated 200 h test was performed. In this test run, the composition of producer gas, its lower heating value, and the cold gas efficiency were considered as the gasification performance; at the same time, the engine-generator performance, for example diesel replacement rate, fuel consumption rate for a unit of power production and the overall power production efficiency were taken into consideration. Table 7 presents the operating conditions for each run.

The different weight ratio of dense-RDF to wood chip of 50:50 was applied to test no. 1 to 10, 75:25 for test no. 11 to 17 and 100:0 for test no. 18. After performing the test no. 1 and 2, it was found that 60 kg of charcoal cannot provide sufficiently high temperature for the combustion process in the combustion zone of

Table 7 Operating condition for the accumulated 200 h test run

Test no.	Operating hour (h)	Dense-RDF (kg)	Wood chip (kg)	Charcoal (kg)	Air flow rate (Nm^3/h)
1	10	100	100	60	73.48
2	10	116	116	60	74.89
3	10	75	75	70	74.89
4	10	75	75	70	101.74
5	10	75	75	70	90.43
6	7	75	75	70	84.78
7	7	75	75	70	62.17
8	8	75	75	70	96.08
9	6	75	75	70	96.08
10	9	75	75	70	96.08
11	9	112.5	37.5	70	73.48
12	12.5	150	50	70	73.48
13	10	114	38	70	73.48
14	14	183.75	61.25	70	73.48
15	10	135	45	70	73.48
16	23	285	95	70	73.48
17	15	187.5	62.5	70	73.48
18	20	284.5	0	100	73.48

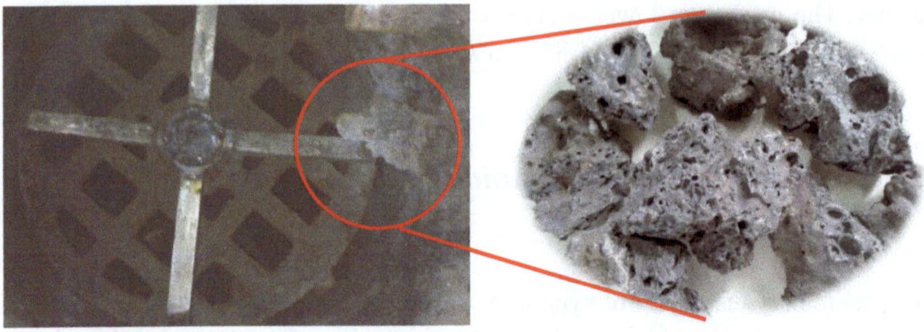

Fig. 5 Melted plastic agglomeration in the low temperature zone of gasifier

gasifier. This led to the plastic melting and a large amount of melted plastic solidified and agglomerated at the low temperature zone at the bottom of gasifier, as shown in Fig. 5. Therefore, the amount of charcoal was increased to 70 kg for test no. 3 to 17 and 100 kg for test no. 18 as the increasing amount of dense-RDF to 284.5 kg.

By increasing the amount of charcoal supplied in the gasifier, the high temperature can be promoted and the temperature in the combustion zone was sufficiently high in order to provide the heat transfer throughout the gasifier for the continuous gasification process in the test no. 3 to 17. However, the problem of melted plastic agglomeration occurred again in test no. 18 which used 100 % dense-RDF as feedstock, as a result, the gasification process was interrupted. It can be concluded that 100 % dense-RDF cannot be used as single fuel in the gasification process. This can be explained by the high amount of volatile matter in fuel. It has to be mixed with wood chip or fuel with high amount of fixed carbon.

Considering the producer gas composition, CO and H_2 was the dominant composition in producer gas with the value ranging from 13.40 to 20.59 vol% and 7.81 to 14.49 vol%, respectively. The lower heating value of 3.12–3.99 MJ/Nm^3 was obtained. Table 8 shows the performance of gasification process for accumulated 200 h test run in terms of producer gas composition, producer gas lower heating value, and cold gas efficiency.

With regard to the performance of engine-generator for power production, the producer gas can replace 64–80 % of diesel and the overall efficiency for power production via dense-RDF gasification process varied from 9.74 to 18.55 % at the constant power production 24 kW. The rate of solid fuel consumption per a unit of power generation ranged between 0.80 and 1.50 kg/kWh. The performance of engine-generator for power production is listed in Table 9.

Although there was gas-cleaning process, the paper filter was placed to trap the particulates in the producer gas before the producer gas entering the engine. The amount of particulate in the clean producer gas ranged from 14 to 56 mg/Nm^3, which is lower than the engine requirement (less than 50 mg/Nm^3; Bandi 2003). Even some soot and particulates were removed from producer gas, during the

Table 8 Performance of gasification process

Test no.	Air flow rate (Nm^3/h)	Producer gas yield (Nm^3/h)	Producer gas composition (vol%)				Lower heating value (MJ/Nm^3)	Cold gas efficiency (%)
			CO	CH_4	H_2	CO_2		
1	73.48	123.99	20.51	0.26	9.27	5.86	3.55	62.86
2	74.89	123.99	18.39	0.89	13.1	8.21	3.99	62.91
3	74.89	139.89	19.86	0.42	10.63	6.96	3.69	87.13
4	101.74	211.42	20.59	0.37	8.92	7.61	3.56	93.95
5	90.43	206.65	20.55	0	7.81	5.93	3.29	84.87
6	84.78	211.42	15.97	0.77	10.84	10.83	3.4	63.68
7	62.17	127.17	15.04	1.49	10.36	9.53	3.49	65.73
8	96.08	214.60	14.97	0.2	11.58	11.08	3.16	68.34
9	96.08	–	15.95	0.93	9.64	8.38	3.31	–
10	96.08	–	17.44	1.03	11.9	10.6	3.78	–
11	73.48	127.17	15.08	1.44	14.22	11.43	3.93	76.27
12	73.48	127.17	14.22	0.41	11.37	7.91	3.12	67.58
13	73.48	127.17	16.00	0.63	14.49	8.05	3.77	80.19
14	73.48	130.35	14.01	0.82	11.08	9.23	3.21	68.19
15	73.48	155.78	14.59	0.81	11.40	7.71	3.32	76.82
16	73.48	143.07	13.40	0.81	12.28	8.25	3.28	86.56
17	73.48	149.42	15.47	1.08	8.20	6.52	3.14	80.41
18	73.48	127.17	13.46	1.14	10.99	7.03	3.26	80.08
Max.	101.74	214.60	20.59	1.49	14.49	11.43	3.99	93.95
Min.	62.17	123.99	13.40	0.00	7.81	5.86	3.12	62.86

Remark The producer gas yield for test no. 9 and 10 was not measured

engine operation a few amount of particulates, soot and tars entrained in producer gas were agglomerated and accumulated on the surface of engine compartments, for example air cooler, as shown in Fig. 6. However, it did not play an important role in the engine performance.

After accumulated 200 operating hours test, the viscosity of lubricating oil was slightly reduced from 80.80 to 76.06 mm^2/s at 40 °C. The decrease in viscosity can result from the dilution of fuel oil or the cracking of lubrication oil due to the higher temperature operation (Udonne 2011). The substantial decrease in viscosity can lead to a decrease in film thickness on the surfaces of engine parts and, consequently, engine's wear. However, the small decrease in viscosity from this study did not affect the engine performance.

Table 9 Performance of engine-generator for power production

Test no.	Air flow rate (Nm^3/h)	Solid fuel consumption rate per a unit power (kg/kWh)	Diesel replacement rate (%)	Overall efficiency (%)
1	73.48	1.08	70.00	13.72
2	74.89	1.22	68.00	12.18
3	74.89	0.92	73.00	16.24
4	101.74	0.92	68.00	16.24
5	90.43	0.92	67.00	16.24
6	84.78	1.29	65.70	11.37
7	62.17	1.29	64.30	11.37
8	96.08	1.14	72.50	12.99
9	96.08	1.50	–	9.74
10	96.08	1.01	80.00	14.62
11	73.48	1.01	78.90	13.81
12	73.48	0.91	73.60	15.47
13	73.48	0.93	71.00	15.20
14	73.48	0.95	75.00	14.77
15	73.48	1.04	67.00	13.42
16	73.48	0.84	78.30	17.18
17	73.48	0.90	77.30	15.57
18	73.48	0.80	65.00	18.55
Max.	101.74	1.50	80.00	18.55
Min.	62.17	0.80	64.30	9.74

Fig. 6 Trap of soot at air cooler of the engine

4 Conclusion

Even 3Rs concept (Reduce, Reuse, and Recycle) is necessary for sustainable waste management; however, the 4th R of Energy Recovery from waste must be taken into consideration before final disposal by landfilling. Energy recovery from waste is considered as green, clean, and renewable. Energy recovery from mined landfill waste creates benefit in term of land reclamation and reducing of fossil fuel used.

Gasification is considered as promising alternative to recover energy from mined landfill waste. However, the process of prepreparation of mined landfill waste in the form of densed-RDF is necessary to use in a downdraft gasifier. Since densed-RDF from mined landfill waste has a high fraction of volatile matters and plastic with low melting temperature, it is needed to be mixed with biomass in order to increase the fraction of fixed carbon and help easing of operation. The performance test of 200 h operated downdraft gasification system for power production from mined landfill waste shows the potential of green and clean energy recovery from waste for the sustainable development of the society.

Acknowledgment The authors would like to express their grateful to The National Metal and Materials Technology Center (MTEC) of The National Science and Technology Development Agency (NSTDA) for financial support of this project. Also, the Waste Incineration Research Center (WIRC), Department of Mechanical and Aerospace Engineering, Faculty of Engineering and Science and Technology Research Center (STRI) of King Mongkut's University of Technology North Bangkok for the facilities support.

References

Bandi A (2003) Gasreinigungsverfahren, Fachtagung "Regenerative Kraftstoffe" Entwicklungstrend, Forschung- und Entwicklungsansaetze, Perspektiven, FVS-FAchtagung, Stuttgart

Belgiorno V, De Feo G, Della Rocca C, Napoli RMA (2003) Energy from gasification of solid wastes. Waste Manag 23:1–15

BioenergyNet (2014) Comparing gasification with other biomass thermal conversion alternatives. http://www.bioenergynet.com/articles/technology/innovation/1320-comparing-gasification-with-other-biomass-thermal-conversion-alternatives. Access 25 July 2014

Henry AK, Yingsheng Z, Jun D (2006) Municipal solid waste management challenges in developing countries—Henyan case study. Waste Manag 26:92–100

Hoornweg D, Bhada-Tata P (2012) World Bank 2012, WHAT A WASTE—a global review of solid waste management, World Bank

Knoef HAM (2005) Practical aspects of biomass gasification, handbook biomass gasification, BTG Biomass Technology Group BV

Manf LA, Abu Samah MA, Mohd Zukki NI (2009) Municipal solid waste management in Malaysia: practices and challenges. Waste Manag 29:2902–2906

Obernberger I, Thek G (2008) Combustion and gasification of solid biomass for heat and power production in Europe state-of-the-art and relevant future development. Proceedings of 8th European conference on industrial furnaces and boilers, April, 2008, Vilamoura, Portugal

Prechthai T, Padmasri M, Visvanathan C (2008) Quality assessment of mined MSW from an open dumpsite for recycling potential. Resour Conserv Recycl 53:70–78

Schübeler P, Wehrle K, Christen J, 1996, Conceptual Framework for Municipal Solid Waste Management in Low-Income Countries, World Bank

Udonne JD (2011) A comparative study of recycling of used lubrication oils using distillation, acid and activated charcoal with clay methods. Petroleum Gas Eng 2:12–19

WIRC (2006) The technological evaluation of energy production from municipal solid waste. The Waste Incineration Research Center, Report. JGSEE BM-T-Ws-019

WIRC (2010) Feasibility Study and Detail Design of the Phuket Integrated Waste Management Center, Report

World Bank (1999) Decision Makers' Guide to Municipal Solid Waste Incineration, World Bank

Ceramics for Sustainable Energy Technologies with a Focus on Polymer-Derived Ceramics

Thomas Konegger, Jessica Torrey, Octavio Flores, Tobias Fey, Bruno Ceron-Nicolat, Günter Motz, Franziska Scheffler, Michael Scheffler, Peter Greil and Rajendra K. Bordia

Abstract Due to their high hardness, high temperature stability, and high chemical stability, ceramic materials have significant uses and potential in existing and emerging sustainable technologies. In this paper, we provide a thorough overview of ceramics in a variety of sustainable applications. This is followed by a detailed discussion of an emerging process to make ceramics called polymer (or precursor)-derived ceramics. It is shown that due to the versatility of this process in making a wide range of shapes—fibers, coatings, and porous ceramics, this is an attractive route to make ceramics that will be a critical element in the next generation of sustainable technologies.

T. Konegger · R.K. Bordia (✉)
Department of Materials Science and Engineering, Clemson University,
161 Sirrine Hall, Clemson, SC 29634, USA
e-mail: rbordia@clemson.edu

T. Konegger
Institute of Chemical Technologies and Analytics, Vienna University of Technology,
Getreidemarkt 9/164-CT, 1060 Vienna, Austria

J. Torrey
U.S. Department of the Interior, Bureau of Reclamation, Denver Federal Center
Box 25007, Denver, CO 80225, USA

O. Flores · G. Motz
Ceramic Materials Engineering, University of Bayreuth, 95447 Bayreuth, Germany

T. Fey · B. Ceron-Nicolat · P. Greil
Department Material Science, Glass and Ceramics, Friedrich-Alexander-Universität
Erlangen-Nürnberg (FAU), Martensstr. 5, 91058 Erlangen, Germany

F. Scheffler
Institute for Chemistry, Otto-von-Guericke-University, Magdeburg, Universitaetsplatz 2,
39106 Magdeburg, Germany

M. Scheffler
Institute of Materials and Joining Technology, Otto-von-Guericke-University,
Magdeburg, Universitaetsplatz 2, 39106 Magdeburg, Germany

A.K. Agarwal et al. (eds.), *Novel Combustion Concepts for Sustainable Energy Development*, DOI 10.1007/978-81-322-2211-8_22

Keywords Ceramics · Polymer-derived ceramics · Coatings · Sustainable technologies

1 Introduction

A broad range of scientists and policy makers have raised the alarm regarding the change in global climate change due to emission of greenhouse gases. The most widely recognized and thorough review has been conducted by the Intergovernmental Panel on Climate Change (IPCC). The most recent report from this multinational group was released in 2014 (IPCC 2014). While the earlier reports focused on the science of climate change and its impact (global temperature increase, decrease in ice cover, changes in crop pattern and productivity, severe climate events, etc.), as well as the primary causes of it (greenhouse gases such as CO_2 produced as a result of human activity), this report focuses on strategies to mitigate climate change. It is widely recognized that one of the major sources of greenhouse gases is the combustion of coal and hydrocarbons (gas- and petroleum-based liquid fuels). These are the primary fuels of the industrial society and are used for everything from electricity generation, transportation, industrial processes, and residential needs and comforts. The best available science summarized in the IPCC report mentioned above clearly presents a road map for the mitigation of climate change. The three pillars of this strategy are as follows: energy efficiency, alternate energy technologies that reduce or eliminate greenhouse gases, and strategies to sequester greenhouse gases.

In this paper, we show that ceramics already play and will play an increasingly important role in all the three important mitigation strategies. In particular, we focus on polymer-derived ceramics since this emerging technology is an attractive approach to make a wide range of ceramic forms and components of interest to one of the central grand challenges of our times—to reduce greenhouse gas emissions while increasing the quality of life for all human beings.

1.1 Ceramics for Sustainable Applications

Ceramics, due to their refractory nature, high hardness, and environmental inertness, are used in a broad spectrum of industries and energy conversion technologies. However, the urgency of the global climate change has led to much higher demands on the material systems. In many cases, conventional materials systems—including ceramics—have been found to be inadequate to meet these increased demands in performance and/or cost. The emerging need for novel ceramic materials for energy, environment, and sustainability have been summarized in many publications, including, for example, the road map for the development of advanced

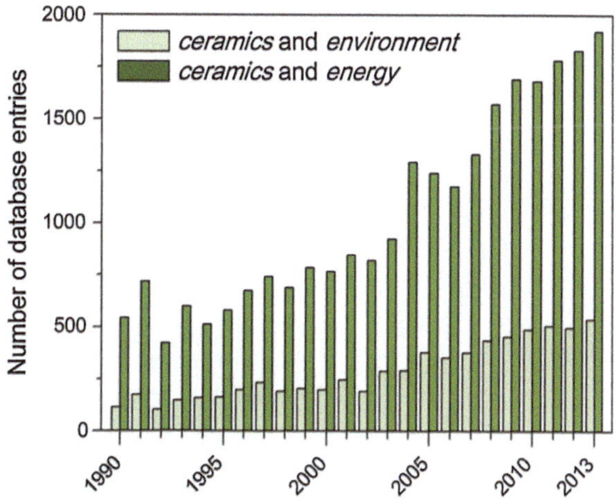

Fig. 1 Number of scientific database entries per publication year for the search queries "ceramics and environment" and "ceramics and energy" (*Source* Scopus scientific database, Elsevier B.V. Search conducted August 2014)

ceramics for the years 2010–2025 (Rödel et al. 2009). This attention has led to rise in research and development activity including an increasing number of scientific publications (Fig. 1). A large number of initiatives and funding programs by governmental and non-governmental agencies have been developed. Examples are the *Science, Engineering and Education for Sustainability* initiative of the US National Science Foundation with emphasis areas such as *Sustainable Chemistry, Engineering, and Materials* (Madsen and White 2014) and numerous initiatives on renewable energy and energy efficiency by the Chinese National Science Program. Another high-profile example of the importance of this area is the allocation of a total budget of € 8.8 billion on the challenges *Secure, Clean and Efficient Energy* as well as *Climate Action, Environment, Resource Efficiency and Raw Materials* as defined within the Horizon 2020 Framework for research and innovation of the European Commission for the time period of 2014–2020.

In addition to conventional uses of ceramics in energy- and environment-related areas, a large number of novel applications have emerged utilizing ceramic materials. This section gives an overview of both conventional and novel applications of ceramics in a wide variety of energy and environmental systems (Fig. 2), with the primary focus on developments within the last decade.

1.1.1 Emerging Energy Conversion Approaches

Ceramics have been a core class of materials in the energy sector, with a wide variety of applications in energy conversion, storage, distribution, and energy conservation and efficiency. In the field of emerging energy conversion, the use of advanced ceramics as components in fuel cells has been of major interest, both scientifically and economically. In addition, ceramic materials are also utilized or being developed to perform a number of approaches including thermoelectric

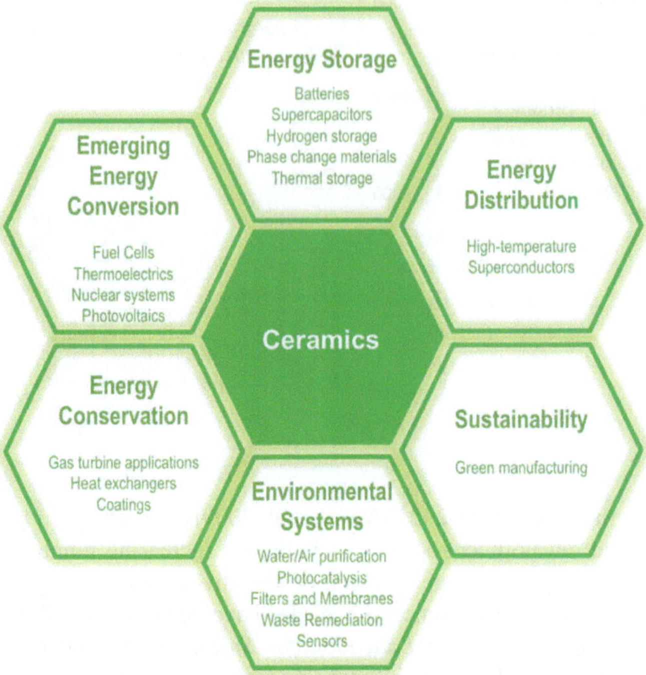

Fig. 2 Examples for applications of ceramics in energy, environmental, and sustainable technologies

systems, next-generation nuclear energy systems, photovoltaics, wind, and geothermal systems.

Solid oxide fuel cells (SOFC) have gained major economic and scientific interest as a result of their high energy conversion efficiency, their flexibility in terms of fuels used (natural gas, biogas, methane), and the wide range of possible applications, from combined heat and power plants producing several hundred kW down to small units suitable as household energy supplies (e.g., Menzler et al. 2010). In SOFC designs, oxygen ion-conducting ceramic compounds are used as solid electrolyte materials. The main materials currently used are partially cation-substituted zirconia (ZrO_2), ceria (CeO_2), $LaGaO_3$, as well as apatites such as $La_{10}Si_6O_{27}$. In addition to electrolytes, ceramics are also used in other parts of SOFC, including anodes, cathodes, as interconnects, and as sealants in the form of glass ceramics.

Thermoelectric (TE) materials and systems present a way to generate electrical power from heat by employing the Seebeck effect, thus using a temperature gradient to create an electromotive force (i.e., voltage). The absence of moving parts, a high power density, and the ability to convert waste heat to electricity have fueled an increasing interest in TE energy conversion systems. Although the primary TE materials are metallic or intermetallic, there is a growing interest in developing ceramic TE since they have the potential to operate at higher temperature and be stable under harsh environmental conditions, e.g., flue gases. In TE applications, both *n*-type and *p*-type materials are required. Several candidate ceramic *p*-type materials have been investigated including calcium cobaltite, $Ca_3Co_4O_9$. There are limited studies on *n*-type ceramics, one promising candidate being (Sr(Ba) Nb_2O_6) (Dandeneau et al. 2013). Current and potential applications for TE ceramics include the recovery of waste heat in industrial furnaces or incineration processes and the harnessing of electric energy from alternative renewable energy sources such as solar thermal or geothermal.

An increasing concern about greenhouse gas emissions and rising energy demands has resulted in revitalized interest in the development of next-generation *nuclear power* plants. Aside from the classic use of ceramics as nuclear fuels (UO_2, PuO_2, ThO_2, mixed oxides) or the application of boron carbide (B_4C) as nuclear radiation absorbent, e.g., in control rods, novel reactor designs have resulted in an increased demand for materials with specific requirement profiles in terms of mechanical, thermal, and radiation stability, many of which can be fulfilled by advanced ceramics. Due to their ability to resist high temperatures and its tolerance to neutron radiation, silicon carbide (SiC) and SiC-based composites have been intensively investigated for both advanced fission and fusion energy applications, e.g., as core or in-vessel components in advanced fission reactors or as structural components or flow-channel inserts in fusion reactors. In these types of harsh applications, SiC/SiC composites made from chemically vapor-infiltrated (CVI), high-purity crystalline SiC fibers have shown superior performance compared to conventional systems (Katoh et al. 2003).

1.1.2 Energy Storage

The emerging use of intermittent renewable energy sources such as wind or solar energy results in the need for high-efficiency and capacity energy storage systems. This challenge is even more prevalent in the transportation sector, where efforts of reducing CO_2 emissions have led to the anticipated replacement of vehicles with internal combustion engines by "zero-emission" vehicles such as electric vehicles. Electrochemical systems, primarily batteries and supercapacitors, provide a way to effectively store and release electric energy on demand and are therefore of major economic and technological importance.

Owing to their high specific energy, efficiency, and longevity, Li-ion *batteries* are currently found in a wide variety of mobile and stationary applications, ranging from mobile communication devices and consumer electronics to electric vehicles. A variety of ceramics are used as solid electrolytes in Li-ion batteries. A promising example for a lithium-ion-conducting oxide is the perovskite (La, Li)TiO_3. Other ceramics exhibiting high lithium-ion conductivity are Li_9SiAlO_8 or garnet-structure-related compounds such as $Li_5La_3Ta_2O_{12}$ (Fergus 2010). Aside from using ceramics as electrolytes, nanoscaled Al_2O_3, SiO_2, or ZrO_2 (or respective mixtures) has been used as composite separators in liquid electrolyte Li-ion batteries, increasing the safety of battery cells by providing high-dimensional stability at high temperatures, thus preventing accidental electrode contact.

Even though rapid progress has been made in terms of storage capacity in recent years, the large-scale storage of electrical energy from intermittent energy sources is still a big challenge, primarily due to high costs. The most important alternatives are the storage of produced energy by chemical (hydrogen storage) or thermal means. The "hydrogen economy" has been advocated as an alternative to the current fossil-fuel-based economy and was declared as a dedicated aim for future developments by many governments. In this proposed system, the safe and economic *storage of hydrogen* is of significant importance. Several ceramic materials have been considered for this function, primarily zeolites (Morris and Wheatley 2008).

An alternative way to overcome the gap between the power generation and demand is the use of *thermal energy storage*, where excess thermal energy is used to warm up or melt a material, thus storing it in the form of sensible or latent heat. When demand rises, the stored energy can be released by cooling of the storage material. A classic example for a sensible heat storage (SHS) material is ceramic bricks, or, more recently, porous ceramics with designed pore structure and thermal properties. In latent heat storage (LHS), phase-change materials with specific properties (including large specific heat of phase change, high thermal conductivity, small volume change, and relatively low cost) are used to absorb and release heat during a phase-change process. Advantages of this technique include the high energy storage density as well as the small temperature range required during operation. Porous ceramics have been proposed as containers for storing phase-change materials.

1.1.3 Energy Distribution

In order to increase the efficiency of energy grids, a substitution of metallic conductors by superconductive materials has been proposed, offering higher capacities and avoiding resistive losses. Since the discovery of ceramic-based high-temperature superconductors (HTS) in the form of cuprates such as yttrium barium copper oxide (YBCO) in the 1980s, extensive research has been conducted in developing materials with even higher transition temperatures, with the record currently being held by Hg-containing cuprates at over 130 K. While not yet quite ready to be implemented on a large scale, small-scale field tests of next-generation ceramic HTS systems are currently underway around the globe (Haught et al. 2007).

1.1.4 Sustainable Manufacturing Systems

Finally, ceramics are not only utilized as novel materials in sustainable technologies as described above, but the emergence of sustainable technologies also has implications on the manufacturing of ceramics themselves. The adoption of more sustainable concepts for production processes, sometimes referred to as *green manufacturing*, allows not only for the compliance with increasingly stringent legislative regulations in the fields of energy consumption and emission control, but also for a higher efficiency and, subsequently, a higher cost-effectiveness. Specifically, there is significant interest in developing refractories with higher thermal insulation capacity. Another area of focus is the development of processes that can reduce the energy requirements and greenhouse gas emission in the production of commodity building ceramics such as bricks, cement, and glass. Most of this work is being conducted in house by manufacturers, and there is little on this topic in the open literature.

1.1.5 Environmental Systems

Environmental systems include technologies for the reduction and/or purification of solid, liquid, and gaseous emissions, the remediation of waste, as well as the monitoring of potentially harmful or undesired compounds in natural and industrial environments. Ceramic materials play an important role in many of these applications. Specific advantages of ceramics for *filtration and purification* stem from the combination of high temperature and corrosion stability with high mechanical strength, leading to typical uses in the filtration of particulates from diesel engine exhaust gases, or uses in industrial processes where hot corrosive gases are present (Scheffler and Colombo 2005). Since porous structures are generally required for these applications, a wide variety of methods have been developed for the generation of processing of porous ceramics, including replicate techniques, direct foaming, or the use of sacrificial porogens.

Porous ceramics, e.g., SiC, have also been used in heterogeneous *catalysis* applications, primarily as catalyst supports. A combination of both filtration and catalysis functions exists in *multifunctional reactors*, promising a reduction in financial and instrumental expenditure. Ceramic devices for the filtration of particulates from diesel engine exhaust gases with subsequent soot–combustion functionalities are an economically important example for this concept (Scheffler and Colombo 2005). Several authors of this paper have been engaged in research on porous ceramics, and their work is summarized in Sect. 2.3.

Ceramic membranes are being increasingly considered for gas separation applications (Li 2007). Examples for the use of ceramic gas separation membranes are the extraction of energy carriers such as H_2 or CH_4 from product gas streams in coal gasification, steam reforming, or biogas production, or the removal of H_2O and H_2S impurities from hydrocarbon streams. Gas separation in ceramic membranes can take place either by selective ion conduction using mixed ion-conducting ceramics such as perovskite-type compounds for oxygen or hydrogen separation, or by utilizing microporous structures for size exclusion phenomena, e.g., by using zeolites or microporous silica. Closely related are advances in CO_2 capture and storage (CCS) technology, allowing for the sequestration of CO_2 from flue gas of combustion processes, with an aim to effectively inhibit its emission into the atmosphere. In this application field, the use of ceramic membranes—again primarily based on zeolites or silica—poses a promising alternative to conventional chemical absorption techniques (D'Alessandro et al. 2010).

In the application of porous ceramics for liquid media filtration, e.g., for *water purification*, chemical inertness and favorable anti-fouling properties are major advantages. Filters prepared from Al_2O_3, SiC, and TiO_2 present economical solutions for conventional filtration as well as for micro- and ultra-filtration. A modification of these inorganic filters with metals or organic encapsulating compounds permits the removal of a wide variety of pollutants, including heavy metals, radioactive compounds, or complex organic compounds. Recent developments in the field of photocatalysis, primarily involving nanoscaled TiO_2, have also shown great potential for the removal of organic contaminants present at low levels (Yu et al. 2007).

A specific environmental application for ceramics and glass ceramics is *hazardous waste remediation* in the form of immobilization of both conventional and radioactive wastes. Immobilization involves the chemical incorporation of waste products into a suitable matrix, e.g., glass, in order to prevent further release. While the vitrification of nuclear wastes in glass is suitable in most cases, some types of high-level waste, especially refractory and volatile wastes, require more advanced techniques involving the use of crystalline ceramic (monazite, zirconia, zircon, among others) or glass ceramic immobilization materials. The use of geopolymers, a family of compounds in the alumino-silicate system, has been proposed as an economic and sustainable alternative for the immobilization of hazardous, toxic, or nuclear wastes (Komnitsas and Zaharaki 2007).

Advanced ceramic-based sensor systems allow not only for the monitoring of hazardous species or pollutants, but are also beneficial for energy efficiency and emission control in industrial processes. Progress in the technological field of advanced ceramic sensors aims toward the development of low-cost sensors with quick response times, high sensitivity, and high reliability, even in harsh environments, permitting the detection of a wide variety of gases such as O_2, CO, CO_2, NO_x, or hydrocarbons. Next to well-established sensor systems such as YSZ-based potentiometric oxygen sensors for automotive applications, examples for ceramic sensors include semiconducting CO sensors made of SnO_2, electrochemical CO_2 sensors containing Li_3PO_4 as electrolyte, or electrochemical NO_x sensors, also based on YSZ (Akbar et al. 2006).

1.1.6 Energy Conservation

The efficient use of both conventional and renewable energy carriers is a primary factor contributing to current energy conservation efforts. One way to increase the efficiency in combustion processes is the operation at higher process temperatures. This, in turn, results in increased demands on the thermal and structural properties of the materials used in these processes.

Ceramic *thermal barrier coatings* (TBC) are used in a variety of applications, including gas turbines, in order to protect underlying metallic or composite structures from thermal stress and ensure operation at higher temperatures (and hence higher efficiency) that would be possible without the coatings. TBC in turbine applications generally consist of a porous ceramic top layer, responsible for the thermal insulation, and a metal alloy bond coat layer providing for a good interface to the substrate (generally Ni-based superalloys). The current material of choice for the TBC top-coat layer is yttria-stabilized zirconia in its metastable tetragonal modification. The criterion to withstand high temperatures in oxidizing environments for long periods of time has led to a focus on refractory oxides in search of novel and alternative TBC materials. Proposed candidates include fluorite-structured oxides (HfO_2, CeO_2, ThO_2), zirconate pyrochlores such as codoped $La_2Zr_2O_7$, garnets ($Y_3Al_xFe_{5-x}O_{12}$), or monazite, $LaPO_4$ (Clarke and Levi 2003).

Closely related to TBC are *environmental barrier coatings* (EBC), which protect an underlying substrate from aggressive combustion atmospheres. Furthermore, the coating acts as a diffusion barrier, inhibiting the interaction of potentially critical gas-phase components with the base material. Several authors of this paper have been engaged in the development of environmental barrier coatings, and their contributions are summarized in Sect. 2.2.

In thermal processes, a further increase in process efficiency can be achieved by the recovery of thermal energy released during the process by using *heat exchangers*. Heat exchange is also relevant for the utilization of thermal energy sources such as solar thermal or geothermal energy. Ceramic materials have found niches especially in high-temperature applications due to their resistance against thermal loads and corrosive environments. While a number of ceramic materials (Al_2O_3, ZrO_2, Al_2TiO_5, AlN, Si_3N_4) and ceramic matrix composites (C/C-SiC, SiC/Al_2O_3) have been used in heat exchange applications, SiC can generally be considered as one of the most promising ceramic materials for this application.

In summary, while the unique thermal and mechanical properties of ceramic materials define their application profile in energy and environmental applications, their shortcomings also have to be considered during design and production of components and systems. Owing to the inherent high temperature stability and mechanical resistance, the forming of parts remains a major challenge in many of the applications described above. This is particularly true for silicon-based ceramic materials. As a result, in the recent past, there has been significant research in the area of the precursor (or polymer)-derived ceramic route as an alternative to conventional ceramic processing methods. This approach, as discussed below, has specific advantages and is the focus of the rest of this paper.

1.2 Precursor or Polymer-Derived Ceramics

Preceramic polymers are organometallics that lose their organic components upon pyrolysis to form an amorphous or nanostructured ceramic. There are many types of preceramic polymers now available on the market; precursors to carbides, nitrides, and oxides are used to form such ceramics as SiC, B_4C, TiC, Si_3N_4, BN, AlN,

Al_2O_3, and SiO_2. The Si-based precursor family is the most investigated; however, research has also been done on aluminum-, titanium-, zirconium-, and boron-based systems (Greil 1995).

Preceramic polymers have certain polymeric characteristics, leading to many advantages with respect to processing that are unavailable to processing of ceramics using traditional techniques (vapor phase or powder processing). For example, by altering the main chain chemistry or that of the side groups, coupled with the introduction of a reactive atmosphere during pyrolysis, the final ceramic composition, porosity, pore structure, and grain morphology can be tailored (Colombo et al. 2010). Precursors can be made in liquid form, which allows ease of processing by slurry methods. This also allows for control of the viscosity of the precursor which can be useful in processes such as coating. As liquids, precursors are easily purified; this leads to a reduction in the impurity content of the ceramic product.

Another significant advantage of preceramic polymers is that they have relatively low pyrolysis temperatures compared with traditional ceramic processing methods, often forming dense, amorphous ceramics at temperatures as low as 600–800 °C (Torrey and Bordia 2007). While traditional ceramic processing methods such as powder compaction can be used, there are also many polymer processing techniques that can be employed when forming polymer-derived ceramics. These include fiber spinning, pressure casting, dip and spin coating, painting, spraying, printing, injection molding, UV curing, inkjet printing and biotemplating (Yajima et al. 1975; Torrey and Bordia 2008a, b; Walter et al. 1996; Scheffler et al. 2005b).

The main disadvantage of preceramic polymer technology is the large density change that occurs when the polymer precursor, typically a density of 1 g cm^{-3}, is converted to the ceramic product, often >2 g cm^{-3}. Densification leads to volumetric shrinkage on the order of 50–75 %, which makes production of net-shape components difficult. This is often accompanied by pore formation during cross-linking and pyrolysis due to the off-gassing of organic side groups. This disadvantage has been largely solved by the use of passive or active filler materials and by careful heat treatment to minimize the shrinkage on pyrolysis. As described by Greil (1995), these preceramic polymers can be filled with either inactive or active fillers. In a process known as AFCOP, active fillers such as Si, Al, Ti or intermetallics react either with the residual carbon during pyrolysis or with the gases used during processing. Typically, fillers are used such that the specific volume of the product phase is greater than that of the filler. This expansion partially compensates the shrinkage of the polymer leading to significantly lower, and in some cases, zero net shrinkage. The final phases in these materials depend on the pyrolysis products of the precursors and the reacted fillers.

Due to the advantages mentioned above, the polymer precursor approach is now used to make ceramics in many forms including fibers, coatings, porous ceramics and ceramic foams, high-temperature MEMS devices, and bulk composites, to name a few. Arguably, the most commercially successful application of preceramic polymers to date has been their use in the fabrication of fibers and nanorods (Motz et al. 2000; Scheffler et al. 2004). The process for mass production of SiC fibers from poly(carbosilane) was first published in 1975 (Yajima et al. 1975), and the fibers have since been commercialized by the Nippon Carbon Company under the trade name Nicalon®. This topic is elaborated further in Sect. 2.1.

Liquid preceramic polymers are ideal for coating via dipping, spinning, painting, or spraying. Both filled and unfilled coatings have been used for applications ranging from environmental protection to wear resistance (Torrey et al. 2006; Torrey and Bordia 2008a, b; Wang et al. 2014; Günthner et al. 2011; Wang et al. 2011). Joining and sealing of ceramic materials has also been achieved using preceramic polymers (Colombo et al. 2000; Lewinsohn et al. 2005). This topic is discussed below in Sect. 2.2.

An abundance of research has been published recently regarding preceramic polymer foams as potential materials for ceramic filters, catalytic supports, thermal insulation, and lightweight structural reinforcements (Scheffler et al. 2005a, Scheffler and Colombo 2005; Zeschky et al. 2003). Because of their natural foaming ability upon cross-linking, many preceramic polymers are suitable for direct foaming yielding macroporous components. Other polymers can be used with sacrificial templates to create microporous foams. Preceramic polymers also offer the potential for production of lightweight metal-ceramic composites. This topic is further elaborated in Sect. 2.3.

Precursors can be substituted for traditional organic binders during processing, and instead of burning off and creating porosity, preceramic polymer binders will instead convert to a ceramic. This aids in the creation of higher-density parts as well as increases the purity of the final ceramic by reducing the amount of organic contaminants in the green bodies. Preceramic polymers are quite attractive for their potential as matrix materials in ceramic matrix composites (CMCs). They offer the ease of polymeric infiltration with the high temperature stability and chemical resistance of ceramics (Seifert et al. 2014; Konegger et al. 2013a, b).

In addition to the processes mentioned above, several other polymer processing approaches such as injection molding or extrusion can be used. Several innovations have been proposed in this area to further improve the processing and microstructures. One example is a supercritical CO_2 blowing process with continuous extrusion of preceramic polymers, yielding rectangular, round, or hollow profiles with a gradient porous distribution (Wolff et al. 2012).

2 Precursor or Polymer-Derived Ceramics for Sustainable Applications

Here, we present illustrative examples of the development of polymer-derived ceramics for sustainable applications in three forms—fibers, coatings, and porous ceramics. We primarily highlight the research conducted by our research groups. We emphasize that this is not a comprehensive review of this approach to make ceramics.

2.1 Ceramic Fibers

Ceramic fibers are high-performance fibers, which combine high thermal and creep resistance with structural stability. Their high tensile strength >2 GPa and stiffness >200 GPa, together with the capacity to withstand temperatures >800 °C, make them attractive materials to be used as reinforcement in metal (MMCs) and CMCs. Ceramic fibers can be classified into two groups: oxide and non-oxide fibers. Oxide ceramic fibers are mainly Al_2O_3 (e.g., Nextel 610) and Al_2O_3/SiO_2 (e.g., Nextel 720)-based fibers. These types of fibers are very stable against oxidation at high temperatures. In spite of their high oxidation resistance, the creep resistance of oxide fibers is limited to temperatures up to 1,100 °C due to the formation of larger grains and, consequently, fiber failure. Non-oxide ceramic fibers are mostly based on SiC(O) and SiCN(O) systems, but studies with SiBCN and BN are frequently found in the literature. In contrast to oxide fibers, non-oxide fibers have higher creep resistance at temperatures up to ~1,450 °C and their oxidation behavior depends on the fiber processing conditions and the fiber microstructure.

2.1.1 Introduction to Processing, Structure, and Properties of Ceramic Fibers

Ceramic fibers, which are currently commercially available, typically are processed either using the solgel process and chemical vapor deposition (CVD), or using preceramic polymer processing.

In preceramic polymer processing, organometallic polymers are used as precursors for fiber spinning. The use of organometallic polymers, more specifically organosilicon ones, dates back several decades when Fritz et al. (1965) and Yajima et al. (1975) reported the first SiC materials obtained from polycarbosilane precursors. These findings marked the beginning of extensive research on silicon-based inorganic polymers, leading to the development of ceramic Si_3N_4 and SiC fibers. A recent review of the ceramic fibers from preceramic precursors provides a good overview of this approach and the original references for the work summarized below (Flores et al. 2014).

A high molecular weight and adequate viscoelastic properties of the melt are necessary properties of the organosilicon polymer to enable a stable spinning process and stretching of the fibers without breaking of the filament. In a melt spinning process, uncured green fibers are processed and subsequently cured chemically, thermally, or by electron beam irradiation. This curing step is necessary to avoid the remelting of fibers during the last stage, which is the conversion of the polymer to a ceramic fiber by pyrolysis. Figure 3 shows the manufacturing process of ceramic fibers employing the preceramic polymer route, as well as important parameters for a controllable processing of ceramic fibers.

Polymer-derived ceramic fibers show a strong correlation between the microstructure, mechanical, and thermal properties, which in turn depends on the

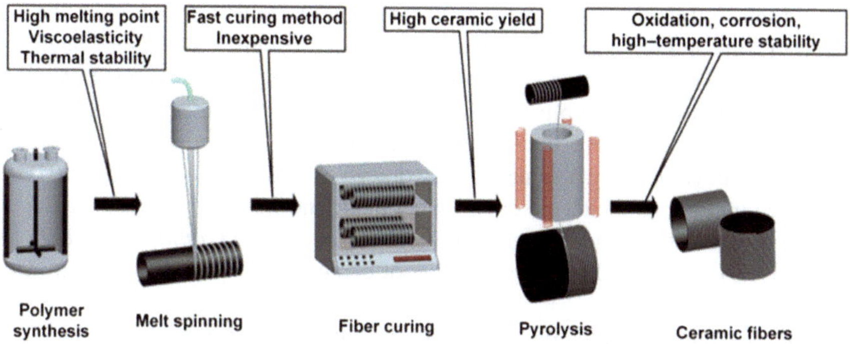

Fig. 3 Manufacturing steps for ceramic fibers employing the preceramic polymer route. Reprinted with permission from Flores et al. 2014. Copyright 2014, John Wiley and Sons

production conditions. An important requirement key for the production of a polymer-derived ceramic fiber is a homogeneous and defect-free microstructure, which allows high tensile strength. Their creep resistance and thermal resistance are primarily determined by the stoichiometry and, specifically, the oxygen content. Generally, polymer-derived ceramic fibers with lower oxygen content have better high temperature stability and creep resistance.

2.1.2 Polymer-Derived SiC Fibers

Over the course of the last 50 years, research in the field of polymer-derived ceramic fibers has led to the development of three generations of high-performance ceramic SiC fibers with diameters <30 μm based on organometallic polymers. Although the base of these fibers is SiC, typical properties of crystalline SiC, such as Young's modulus of about 400 GPa and oxidation resistance up to 1,600 °C, were not realized in the first generation of commercialized ceramic SiC fibers. Continued research and development have led to the commercialization of two more generations of SiC fibers, with the most advanced third-generation fibers showing thermal properties approaching those of crystalline SiC.

In the *first generation,* the SiC fibers achieved sufficient tensile strengths (>2 GPa) to reinforce matrices, but both Young's modulus (~200 GPa) and thermal stability were significantly lower than in crystalline SiC ceramics due to the formation of an amorphous Si–O–C phase in the fiber at the typical pyrolysis temperature of about 1,200 °C. The first company to introduce continuous polymer-derived ceramic SiC fibers in the market was Nippon Carbon Corporation Ltd, Japan. Based on the work done by Yajima in the development of organosilicon polymers, Nippon launched the trade name Nicalon, which is the family name of three grades (Nicalon, Hi-Nicalon, and Hi-Nicalon Type S) of SiC fibers. *First-generation* ***Nicalon***™ fibers are produced from melt spinnable polycarbosilane (PCS). Polymeric fibers are spun from the polycarbosilane (PCS) melt at 350 °C and subsequently cross-linked in air at 190 °C. After pyrolysis of the cured fibers at 1,200 °C in an inert gas atmosphere (Ar or N_2), ceramic SiC fibers having a diameter of ~14 μm and an oxygen content of ~12 wt% are obtained. Another line of SiC fibers were commercialized by Ube Industries under the trade name Tyranno SiC fibers. These fibers are derived from a novel polytitanocarbosilane (PTC), also developed by Yajima and coworkers. Similar to Nicalon fibers, PTC is processed into fibers, which are cross-linked in air and further pyrolyzed leading to ceramic fibers with oxygen contents up to about 20 wt%. ***Tyranno S*** and ***LOX-M*** (Si–Ti–C–O) represent the *first generation* of ceramic SiC fibers produced by Ube Industries.

With the understanding of the oxygen influence, amorphous state, and C/Si relation in the microstructure of these fibers, the *second generation* of SiC fibers with reduced oxygen but high amounts of carbon content was produced. A typical way used to avoid the oxygen is curing the polymeric fibers with γ-ray or electron beam irradiation in inert atmosphere. The almost complete removal of oxygen in the second generation of SiC fibers resulted in a microstructure with almost no amorphous phase. Larger SiC grains are formed at a higher pyrolysis temperature, and a Young's modulus up to 270 GPa was achieved. However, the excess free carbon in the microstructure of the fibers limited the growth of the SiC grains. These fibers are processed at about 1,300 °C. ***Hi-Nicalon***™ fibers represent the *second generation* of SiC fibers in the Nicalon family, characterized by a reduced oxygen content of <1 wt%. These fibers are produced by using the same manufacturing process as the classic Nicalon fibers, except for the curing method. Fibers are cured by using electron beam irradiation doses between 10 and 15 MGy in an inert atmosphere to avoid oxygen contamination. *Second-generation* ***Tyranno ZMI*** ceramic fibers based on a Si–Zr–C–O structure are produced by Ube. These fibers have reduced oxygen content (9 wt%) and improved strength retention at high temperatures compared to the titanium-containing Tyranno S fibers. They are produced from a modified preceramic polymer based on a Si–Zr–C–O structure (instead Si–Ti–C–O), but the polymer fiber is still cross-linked in air.

CMCs reinforced with the first and second generations of SiC fibers did not meet the high-temperature requirements of advanced applications such as high-efficiency turbines. Therefore, further developments led to the production of *third generation* of SiC fibers based on highly crystalline SiC materials. The strength of these fibers is >2 GPa and the modulus >300 GPa, and the fibers are thermally stable up to temperatures of 1,400 °C. *Third-generation* ***Hi-Nicalon***™ **Type S** fibers, with the highest modulus of about 420 GPa and improved thermal properties in comparison with the other fibers from the Nicalon family from Nippon Carbon, are manufactured taking advantage of the curing method via electron beam irradiation used for the processing of low-oxygen-content SiC fibers. The enhanced properties of these *third-generation* SiC fibers were achieved by controlling the pyrolysis using hydrogen as process gas to reduce the carbon content of the fibers from C/Si = 1.39 (second generation) to near stoichiometric ratios (C/Si = 1.05) and further sintering at higher temperatures to achieve polycrystalline ceramic fibers with SiC grain sizes of 50–100 nm. As a result, Hi-Nicalon Type S fibers exhibit an improved creep and oxidation resistance up to 1,400 °C. Ube Industries also launched the *third-generation* ***Tyranno SA3*** SiC fiber. These fibers are obtained from polyaluminocarbosilane by curing in air and further by pyrolysis up to 1,300 °C under inert gas. The ceramic Si–Al–C–O fibers with excess of carbon and oxygen content of about 12 wt% are converted into sintered SiC by decomposition of oxide phases and outgassing of CO at temperatures from 1,500 to 1,700 °C and sintering at temperatures >1,700 °C. Tyranno SA3 fibers show considerable improvement in creep resistance compared to the earlier generations Tyranno S, LOX-M, and ZMI.

The company ATK COI Ceramics, USA, developed *third-generation* SiC fibers based on the approach adopted by Ube Industries to make the Tyranno LOX-M. ***Sylramic*** and ***Sylramic-iBN*** fibers are produced from PTC polymer fibers. PTC fibers are cured by oxidation and doped with boron. The cured doped fibers are pyrolyzed at around 1,400 °C to eliminate the excess of carbon and oxygen as volatile species and sintered at higher temperatures to form a near stoichiometric SiC fiber with smaller grains of TiB_2 and B_4C. Sylramic-iBN fibers are considered to have reduced creep and increased oxidation resistance in comparison with the other commercial SiC fibers. These fibers are the result of a further thermal treatment of Sylramic fibers in a nitrogen-containing gas to allow the diffusion of boron from the bulk to the fiber surface, where it reacts with nitrogen to form an in situ BN coating needed for fiber-reinforced CMCs.

2.1.3 Polymer-Derived SiCN Fibers

In 1972, Verbeek synthesized a meltable carbosilazane resin for the processing of ceramic fibers in the SiCN systems. Since then, many studies concerning polyorganosilazanes (SiCN(O)-based polymers) and metal-modified polyorganosilazanes (e.g., SiBCN-based polymers) were published and attempts done to commercialize ceramic SiCN fibers. Polyorganosilazanes have higher thermal stability in comparison with polycarbosilanes due to the formation of a more stable amorphous SiCNO phase. In this amorphous phase, the nitrogen delays the SiC crystal nucleation, shifting the decomposition of ceramic SiCNO fibers—in comparison with oxygen-rich SiC fibers—to higher temperatures.

At the institute of Ceramic Materials Engineering, University of Bayreuth, Germany, a method was developed to synthesize a solid and meltable polyorganosilazane, named ABSE, by ammonolysis of 1,2-bis(dichloromethylsilyl)ethane on a pilot plant scale. This polymer has a thermal stability up to 220 °C and good moisture resistance. Only at higher temperatures, the material starts to cross-link, which takes place via condensation releasing ammonia, and changes the polymer to an unmeltable resin (Motz et al. 2001; Hacker et al. 2001a). In cooperation with the Institute of Textile Chemistry and Chemical Fibers (ITCF) (Denkendorf, Germany), further improvements concerning the rheological properties of ABSE were successful to achieve a polymer melt with viscoelastic properties necessary for a stable melt spinning process of flexible green fibers. In contrast to Nicalon fiber types, very low doses of $\sim$300 kGy are necessary to cure ABSE-derived green fibers with electron beam irradiation in inert atmosphere. The use of lower electron beam doses reduces significantly the cost of the processing of ceramic SiCN fibers. Ceramic SiCN fibers from cured ABSE fibers are processed by continuous pyrolysis at a low temperature of 1,100 °C. The SiCN fibers have adjustable diameters from 15 up to 120 μm, a tensile strength up to 1.2 GPa and an oxidation stability comparable to that of the Hi-Nicalon fibers (Hacker et al. 2001b; Motz et al. 2002a, b; Motz 2006; Kokott and Motz 2007). The fracture surfaces of a Hi-Nicalon ceramic SiC fiber and an ABSE-derived ceramic SiCN fiber are shown in Fig. 4.

The rheological properties of ABSE and the mechanical stability of the green fibers were further improved by adding dispersed multiwalled carbon nanotubes (MWCNTs) to the polysilazane matrix. The addition of just 0.5 wt% carbon nanotubes improves the viscoelasticity of the polymer melts, allowing the processing of reinforced ceramic C/SiCN fibers via melt spinning and curing via electron beam (300 kGy). The ceramic C/SiCN fibers pyrolyzed at 1,100 °C can reach tensile strengths about 50 % higher than the ABSE-derived ceramic SiCN fibers without carbon nanotubes (Kokott et al. 2008).

In a cooperation between the University of Bayreuth and the company Clariant GmbH, a new method was developed to modify two types of commercially available liquid oligosilazanes, ML33 and HTT1800 (Fig. 5 shows their chemical structures), into solid meltable polysilazanes by a selective cross-linking via dehydrocoupling of N–H and Si–H groups with tetra-n-butylammonium fluoride (TBAF) as a catalyst. The termination of the reaction with calcium borohydride as inhibitor allows the

Fig. 4 Fracture surfaces of non-oxide ceramic fibers: **a** Hi-Nicalon and **b** ABSE-derived SiCN fiber. Reprinted with permission from Flores et al. 2014. Copyright 2014, John Wiley and Sons

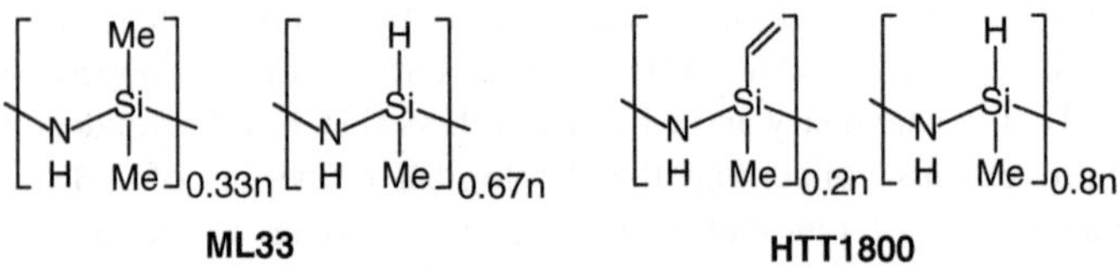

Fig. 5 Chemical structures of liquid oligosilazanes ML33 and HTT1800. Reprinted with permission from Flores et al. 2014. Copyright 2014, John Wiley and Sons

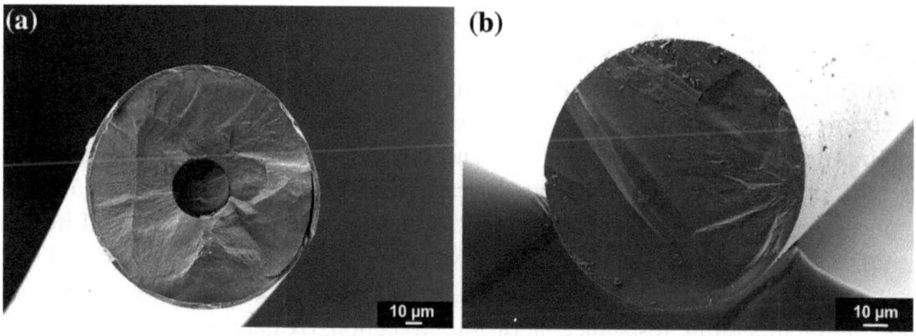

Fig. 6 Fracture surface of monofilament (large diameter) non-oxide ceramic fibers: **a** SCS-6 SiC fiber and **b** ML33S-derived SiCN fiber. Reprinted with permission from Flores et al. 2014. Copyright 2014, John Wiley and Sons

processing of meltable solid polysilazanes (ML33S and HTTS) with tailored chemical and thermal properties, to fulfill the requirements for the melt spinning of mechanically stable and homogeneous polymeric fibers (Flores et al. 2013, 2014).

These polymers have a defined glass transition temperature and viscoelastic melt behavior, which enables the continuous melt spinning of mechanically stable green fibers. Green fibers from ML33S can be cured with electron beam doses of 600 kGy, while HTTS green fibers are infusible after curing with a very low dose of 200 kGy due to reactive Si-vinyl groups. After pyrolysis at 1,100 °C, continuous amorphous ceramic SiCN fibers from both ML33S and HTTS were successfully synthesized. ML33S- and HTTS-based ceramic SiCN monofilaments with diameter >75 µm were recently developed at the institute of Ceramic Materials Engineering, University of Bayreuth (see Fig. 6).

The ceramic SiCN monofilaments are a partial result of a cooperation between the University of Bayreuth, the Technical University of Chemnitz, and the German Aerospace Center (DLR) in Germany. This project aims at the development of metal matrix composites reinforced with ceramic SiCN monofilaments on the basis of cost-effective technologies, such as precursor technology for the processing of the fiber and infiltration of the matrix with metals instead of hot isostatic processing for the consolidation of the composite.

2.2 Ceramic Coatings

With the help of coatings, surfaces of materials can be adapted to a specific application. The coatings can protect surfaces and extend the service life of the support materials. Surface modifications can be used to tailor multiple types of properties, including mechanical, electrical, optical, magnetic, chemical, and thermal properties. Currently, coatings are of immense interest in the context of careful handling of resources and energy saving. For example, the economic loss by wear, oxidation, and corrosion of metals are estimated at several billions $ per year. New developments in the industry as well as political regulations demand coatings with excellent chemical and thermal resistance for applications in power plants, in electronics, in energy conversion or storage, in the chemical industry, or in the automotive sector. Varnishes, paints, and enamel layers have traditionally been used as protective layers. They are characterized by simple and well-established application techniques. However, organic varnishes and paints are limited in temperature stability as well as resistance to wear and solvents, whereas enamel layers are very sensitive to mechanical deformations. Ceramic coating systems can provide high-temperature resistance, abrasion protection, oxidation protection, thermal insulation, as well as functional properties. Non-oxide and oxide ceramic layers or even amorphous and diamond-like carbon coatings are now applied mainly with CVD or PVD processes to improve wear and hardness of the surface. Drawbacks of these well-established techniques are the high cost due to specialized equipment (e.g., plasma or high vacuum technology), the limited adhesion of the layers, as well as the formation of cracks, especially under thermal stress. In addition, it is still difficult to coat complex-shaped components.

In technical applications, surface protection with hard coatings such as TiC, SiC, or Si_3N_4 can extend the lifetime of moving parts by minimizing abrasion and wear. Thermal barrier coatings (TBC), for example, yttria-stabilized zirconia (YSZ), TiB_2, or WC, perform the important function of insulating components, such as gas turbines and aeroengine parts, operating at elevated temperature (Clarke and Levi 2003). However, high costs, scale-up problems, and the limited geometry of the substrates are crucial disadvantages of conventional techniques used to apply these non-oxide or oxide ceramic layers, including physical or chemical vapor deposition or plasma spraying.

As an alternative approach, preceramic polymers can be utilized for coating processes and converted into cross-linked polymer coatings (at temperatures below 300 °C) or into polymer-derived ceramic (PDC) coatings (at temperatures higher than 500 °C). They often possess a composite structure, due to more than one phase formed during thermal conversion or after reaction with substrate of filler material. Figure 7 shows an overview of the polymer-derived ceramic process in coating applications. The following advantages have been identified for polymer-derived ceramic coatings over the last two decades of research:

- Use of inexpensive coating techniques such as spray coating, spin coating, or dip coating,
- Application as solid (dissolved in a liquid) or liquid polymeric precursors,
- Chemical resistance and thermal stability up to 500 °C in the polymer state,
- High ceramic yield of the precursors after thermal conversion (pyrolysis),
- Coating of complex-shaped parts or large-volume parts,
- Control of the polymer-to-ceramic conversion and resulting properties by temperature, as well as
- Easy incorporation of functional properties.

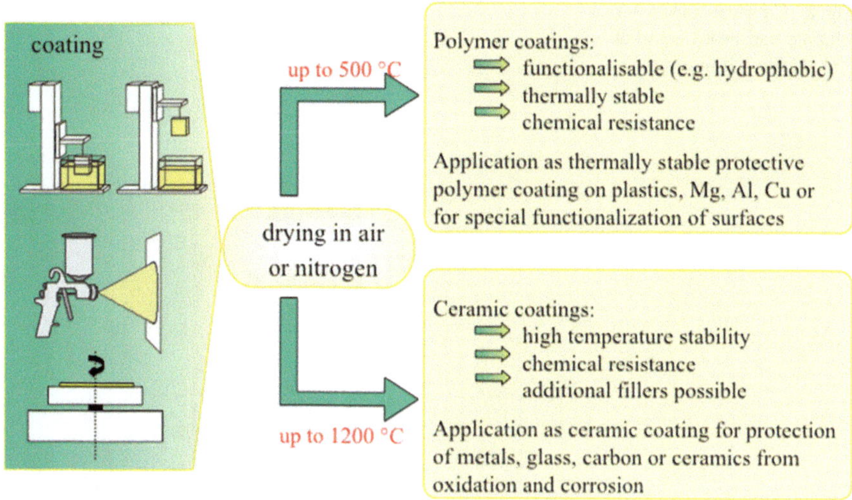

Fig. 7 Typical coating techniques involving preceramic polymers, as well as classification and applications of the resulting coatings

The precursor technique thus exhibits a relatively low cost and easy approach to produce polymeric and ceramic coatings. The coatings are mainly obtained from silicon-based precursors such as polysiloxanes (Torrey and Bordia 2008b), or polysilazanes (Günthner et al. 2009a). In order to use preceramic polymers for coating processes, either liquid state or solubility in organic solvents is necessary. A tailorable rheology is a precondition for applying the suitable coating technology to get the optimal coating performance.

2.2.1 Polymer Coatings Based on Polysilazanes (T < 500 °C)

In contrast to the limited thermal stability of coating systems based on organic polymers, preceramic polymers have a higher temperature stability after they have been cross-linked. In the case of the moisture-sensitive silazanes, this reactivity leads to an additional cross-linking and is very advantageous for the adhesion of the coating with the oxidic surface of the substrate due to reaction of functional groups of the silazane precursor with −OH groups of, e.g., metals. Figure 8 shows a copper sheet dip-coated with a silazane in the lower region and heat treated in air at 500 °C. Though the coating thickness is only 1 μm, the coating protects copper from oxidation up to 500 °C. In contrast, the uncoated region shows the typical oxidation of copper into black copper oxides (Günthner et al. 2007).

Polysilazane can contain various functional reactive groups not only suitable for cross-linking reactions. Functionalities of Si–H, Si-vinyl, or N–H groups are predestinated for further modifications due to their reactivity. Reactions with molecules containing very polar groups such as carboxylates can lead to the modification of

Fig. 8 Cu sheet coated with a silazane and heat treated at 500 °C for 1 h in air

the surface regarding hydrophilicity. The contact angle of such a coated glass substrate to water is less than 10° (Fig. 9). This anti-fog coating offers interesting potential in a broad range of applications, from mirrors in bathrooms to automotive windshields. Recently, it has been shown that the surface energy of silazane-based coatings can be systematically controlled by controlled heat treatment (Wang et al. 2012, 2013).

The high reactivity of silazane-based coatings with moisture allows for the transformation into glasslike SiO_x coatings at very low temperatures of up to 200 °C (Günthner et al. 2012; Wang et al. 2013). This behavior leads not only to an increase in hardness and chemical stability, but also to special optical properties. Recently, a coating system on the basis of silazanes with a high emissivity for flexible thin-film solar cells consisting of Cu(In,Ga)Se_2 (CIGSe) on polyimide was

Fig. 9 Superhydrophilicity and resulting anti-fog coating on the basis of silazanes on glass substrates

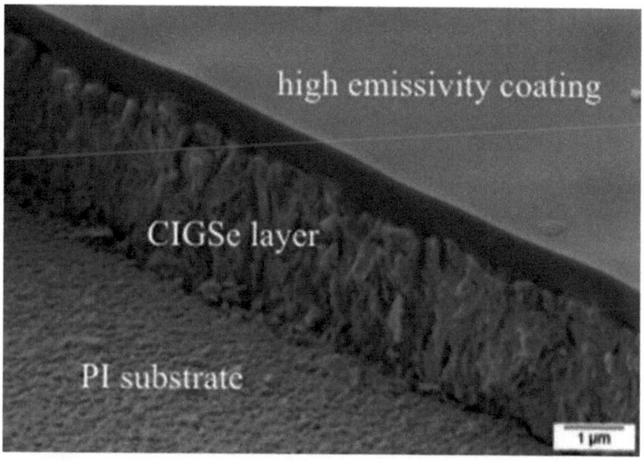

Fig. 10 High emissivity coating on the basis of a silazane on a flexible CIGSe solar cell

developed (Günthner et al. 2014). Due to their superior radiation hardness and high specific power to weight ratio, CIGSe cells have received much attention for power generation in space applications, e.g., as electrical power supply for satellites. A significant drawback of CIGSe solar cells is their low emissivity ($\varepsilon < 0.2$) because of the high infrared reflectance of the cell. This leads to an increase in the cell temperature of more than 100 °C during operation and a decrease in the efficiency. Additional optical coatings with a high emissivity like SiO_2 or Al_2O_3 reduce the increase of temperature remarkably. Therefore, a solution of a polysilazane was used to apply a coating with a thickness of 1–2 µm on the CIGSe solar cell (Fig. 10). After curing in air at 200 °C, the silazane coating converts into a flexible SiOC layer with a high emissivity of 0.8, thus reducing the operation temperature by more than 40 °C and subsequently leading to a significant increase of the efficiency under space conditions (Günthner et al. 2014).

2.2.2 Ceramic Coatings Based on Polysilazanes and Polysiloxanes (T > 600 °C)

The biggest drawback of the PDC technology is the shrinkage of the precursor during pyrolysis due to the mass loss of 10–30 wt% but especially due to the increase in density from 1.1 g cm^{-3} in the polymer state up to 2.4 g cm^{-3} after annealing at up to 1,000 °C. These effects result in shrinkage of the coating thickness up to 80 %, leading to a critical coating thickness of 1–2 µm. Despite the limited coating thickness of only 1 µm, a silazane-based layer protects stainless steel from oxidation up to an annealing temperature of 1,000 °C (Fig. 11) (Günthner et al. 2009b).

While the described thin coatings are very effective for protection of metals from oxidation, the low thickness limits use in an abrasive environment typical for heat exchangers or incinerators. Furthermore, a thicker coating also facilitates a better protection of the metal surface from aggressive environments, especially at higher temperatures. A simple strategy to increase the coating thickness is to use ceramic

Fig. 11 Polysilazane-based coating on stainless steel 1.4301 annealed at 1,000 °C in air

particles as passive fillers (Günthner et al. 2007; Kraus et al. 2009). In this case, the precursor acts as a binder between the particles and the ceramic fillers reduce the shrinkage of the coating. The used slurry for coating applications consists typically of a precursor, a solvent, ceramic particles (e.g., ZrO_2, Al_2O_3, SiC, Si_3N_4, B_4C, BN), and additives to homogenize and to stabilize the suspension. Figure 12 shows

Fig. 12 Steels 1.4301 (*left*) and ST37 (*right*) coated with silazane/h-BN and annealed at 600 °C in air

Fig. 13 Steel 1.4301 coated with a silazane/h-BN, annealed at 800 °C (for 1 h at 800 °C) in an aluminum melt

examples for two types of steel coated with a mixture of hexagonal BN (h-BN) and silazane as a binder. After annealing at 600 °C, the crack-free coating has a thickness of about 20 μm.

This type of coating not only leads to an improved protection from oxidation and abrasion, but also the use of special fillers also allows for the generation of functional properties. The ceramic filler h-BN is well known as releasing agent to enable the use of steel in contact with aggressive compounds such as glass or aluminum melts. Figure 13 demonstrates that the silazane/h-BN layer protects the steel substrate from the aluminum melt and acts as releasing agent. It is very easy to remove the aluminum from the substrate surface.

Several environmental barrier coating systems based on SiOC precursors have been developed to protect metals from oxidation and corrosion. In order to make thick coatings, the polymeric systems have been filled with reactive fillers that expand during the polymer-to-ceramic conversion (Torrey et al. 2006; Torrey and Bordia 2007; Wang et al. 2011, 2014). These coatings have been shown to demonstrably improve the oxidation resistance of both Fe-based and Ni-based superalloys. They have been developed for a broad range of energy-related applications, including protection of metallic systems in coal-fired power plants and in waste incineration plants. For this application, a hard and chemically very resistant, dense coating with a thickness of >50 μm is required, which was achieved by development of a composite coating on the basis of a silazane as a binder, ZrO_2 as hard and passive ceramic filler with a CTE similar to the used steel (13CrMo4), as well as special glass fillers as sealing agent (Günthner et al. 2011). Since the thermal stability of the steel is limited to 800 °C, it was necessary to pyrolyze the coating only at 700 °C. At this temperature, the ceramization of the precursor is almost finished and the glass powders start to melt, leading to a relaxation of the stresses caused by the shrinkage of the precursor. Additionally, a very thin silazane layer was applied on the metal surface as a protection of the mild steel from oxidation during the densification of the thick composite layer to improve the adhesion of the whole coating system. As a result, it was possible to manufacture a ceramic composite coating with a thickness of more than 100 μm (Fig. 14). This coating was successfully tested in salt melts at 540 °C and with special abrasion

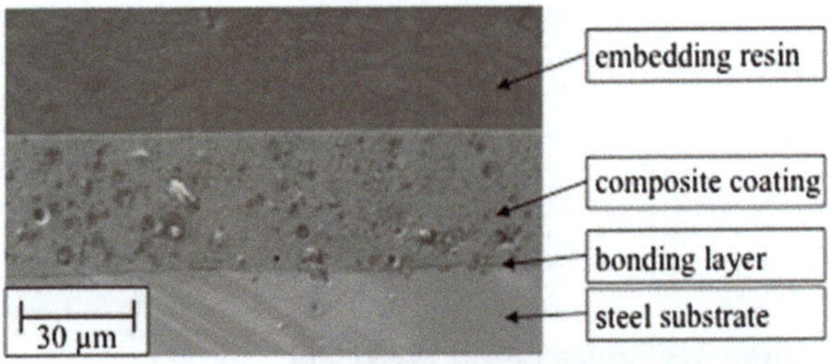

Fig. 14 Silazane/ZrO_2/glass composite coating system on mild steel 13CrMo4, pyrolyzed at 800 °C in air

tests at the same temperature to simulate its use in a waste incineration plant (Schütz et al. 2012). After thermal cycling, no cracks or spallation were detected, which can be attributed to the similar CTE values. The closed porosity was shown to have a positive influence on the adhesion of the coating during thermal cycling because it limits the Young's modulus of the layer.

2.2.3 Control of Surface Area by Demixing and Dewetting Transitions

The coating process of substrate materials itself can be used to control the coating structure. In a recent contribution (Stern et al. 2009), two preceramic polymers were mixed with SiC fillers and the rheological properties were adapted to carry out a coating process as described in Sect. 2.2.2. After drying, a netlike structure was observed, which remains intact even after pyrolysis in argon or in nitrogen. An example of the structure of such a coating is shown in Fig. 15.

A similar structure was also found in another study, where a different preceramic polymer system and Si_3N_4 fillers were used for the dip coating process (Kappa et al.

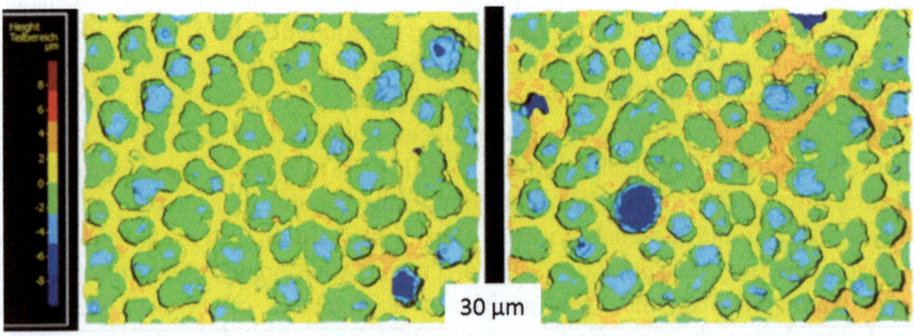

Fig. 15 Confocal microscope image of a PDC-coated alumina tape after cross-linking at 110 °C (*left*) and after pyrolysis at 1,100 °C in argon (*right*); height of the cell walls color-coded; scale: ±10 µm (courtesy of A. Laskowsky, Institute of Materials and Joining Technology, Otto-von-Guericke-University, Magdeburg)

2010). To get a deeper understanding of the structure forming process and parameters relevant for the control of the netlike structure, the solubility of the preceramic polymers in different organic solvents was investigated (Woiton et al. 2011). It was found that a poly(methyl silsesquioxane), PMS, and a poly(methyl phenyl silsesquioxane), PMPS, are immiscible, which results in a demixing process during evaporation of the methanol in filler-free coating systems. In filler-containing systems, the filler size and the type of filler play an important role for the structure forming process, and it was found that dewetting of the preceramic polymer from the filler particles is the major driving force for the structure formation. More details might be found in (Woiton 2011).

2.3 Porous and Cellular Ceramics

Preceramic polymers have been used for the manufacturing of porous and cellular ceramics, and some of these ceramics have been provided with an additional functionality. Independent of the material, porous ceramics can be obtained by one of three different general processing techniques:

- *Templating* uses an organic template, e.g., a polymeric foam, which is impregnated with a ceramic slurry. A thermal treatment decomposes the organic template, and sintering of the ceramic particles in the shape of the template occurs. One main disadvantage is hollow struts of the ceramic structure. This process predominantly yields open-cell structures with a high pore connectivity. Slurries of preceramic polymers with active or inert fillers are widely used for the impregnation of polymer preforms to form ceramic foams (Adam et al. 2014; Krishnan et al. 2014; Ceron-Nicolat et al. 2010; de Sousa et al. 2008).
- *Sacrificial particulate fillers* such as polymeric spheres are added to a ceramic slurry or to a powder bed. After shaping and forming, the sacrificial templates are thermally decomposed to form volatile gases; the initial pore shape less some shrinkage remains in the final ceramic. This leads to predominantly closed-cell structures, with cell windows mostly developing at the contact area of two sacrificial template particles.
- *Direct foaming* is an in situ foaming method which involves the generation of gaseous products either by reaction, by the use of foaming agents in a ceramic slurry, or by introduction of an external gas in a blowing process. The stabilization of the foamed structures is the most critical part of the process. Due to the foaming process (pore forming and release of gas), the formation of a gradient structure is often the case. These gradient structures, if properly controlled, are of interest for filtering applications or for thermal insulation (Wolff et al. 2012; Ceron-Nicolat et al. 2012).

These three techniques are suitable for the generation of a wide variety of porous ceramics for different applications and can be used and combined to design the microstructure (viz., porosity, permeability, and interconnectivity) and, subsequently, the resulting material properties (e.g., mechanical, thermal, catalytic properties). In addition, more functionality can be imparted to the cellular ceramics by impregnation processes.

2.3.1 Polymer-Derived Ceramic Foams

A specific approach of foam formation with preceramic polymers makes use of the intrinsic hydroxy and alkoxy groups of silicone materials for direct foaming. A common material of choice is a poly(methyl phenyl silsesquioxane) (commercial name Silres H44, supplied by Wacker Silicones, Burghausen, Germany) having the general formula $[Ph_{0.62}Me_{0.31}R_{0.07}SiO_{1.5}]_n$ with $n > 20$, Ph = phenyl groups, Me = methyl groups, and R = $-OC_2H_5$ and −OH. The solid powder has a glass transition temperature of ~43°C. The average molecular weight is 2,100 g mol^{-1}, the density is 1.1 g cm^{-3}, and the amount of cross-linking-active hydroxy- (−Si–OH) and ethoxy- ($-Si-OC_2H_5$) groups is 7 mol%, related to a formula unit as shown above. Heating of the H44 preceramic polymer results in condensation reactions and siloxane bond formation at the atomic level (see reaction scheme in Eqs. (1a) and (1b)):

$$\equiv Si-OH + HO-Si \equiv \rightarrow \equiv Si-O-Si \equiv + H_2O \uparrow \quad (1a)$$

$$\equiv Si-OC_2H_5 + HO-Si \equiv \rightarrow \equiv Si-O-Si \equiv + C_2H_5OH \uparrow \quad (1b)$$

Under the right conditions, foam formation occurs on the macroscopic level. This is because water and/or ethanol formed in Eqs. (1a) and (1b) act as blowing agents, and the foam is prevented from collapsing due to the siloxane bond formation and the resulting increase in size of the polymeric molecules and its viscosity. Reaction 1 also results in the formation of a cross-linked thermoset material, resulting in a product that is neither meltable nor dissoluble in solvents. For the tailoring of the material properties, the starting material can be loaded with particulate fillers such as silicon carbide (SiC), alumina (Al_2O_3), silica (SiO_2), or silicon (Si). Subsequent pyrolysis at temperatures between 800 and 1,500 °C results in a ceramic composite material. The properties of those ceramics are controllable by the temperature, the pyrolysis atmosphere, and the pyrolysis time and by the composition and the polymer/filler ratio of the starting material (Gambaryan-Roisman et al. 2000; Zeschky et al. 2003, 2005).

The material properties of PDC foams manufactured as described above can be modified either by the selection and amount of additional filler materials to the starting material or by a post-manufacturing process. Examples for the direct modification are the addition of pore formers leading to a pronounced strut porosity (Reschke et al. 2011), or the addition of specific agents to influence the properties of the foam, such as the addition of copper (I) oxide (Cu_2O) which results in an increase of the electric conductivity by several orders of magnitude even at low Cu_2O concentrations (Colombo et al. 2001).

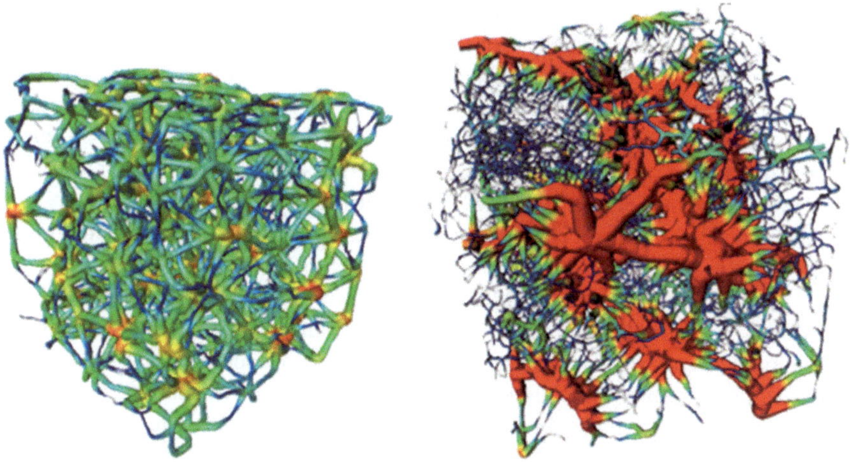

Fig. 16 *Left* Pore network of replica foam and *right* combined replica and direct foaming foam

In a combined approach, a macrocellular foam was manufactured by a reticulation process from preceramic polymers having a cell size of about 4 mm, and the cells were filled with a direct-foamable preceramic polymer slurry. In the subsequent foaming process, the cell size of the "filled" foam was retained and a secondary pore system was formed with significant smaller cells of about 0.5–1 mm in diameter (Ceron-Nicolat et al. 2010). Those hierarchical structures are of increasing interest for hosting active materials including applications for heat storage or heterogeneous catalysis. Figure 16 shows the connectivity and the structure of foams from microtomography investigations for the macrocellular foam (left) and the hierarchical foam (right).

In another approach, cylindrical silicon oxycarbide foams with a radial gradient in pore size were manufactured by an extrusion process of a preceramic polymer (Ceron-Nicolat et al. 2012). The inner material possessed a cell size of ~200 μm, and the cells close to the surface had a cell size of ~20 μm. Those materials are candidates for high-temperature stable lightweight materials and systems.

2.3.2 Zeolite-Coated PDC Foams

Zeolites and zeolite-like structures are of special interest for a broad range of applications. Due to the combination of pore geometry and small pore diameter, the regular three-dimensional pore system provides a large internal surface area (hundreds of $m^2\ g^{-1}$). In conjunction with the chemical composition, the surface properties of the particular zeolite type can be tailored for a large number of separation, sorption, and catalytic processes. As a result, zeolites are widely used in a variety of energy-related applications including petrochemical industry, for exhaust gas and waste water cleaning and also for thermochemical energy storage.

For many of these technical applications, the zeolite particles (typical size of a few μm) have to be partially consolidated or attached to substrates. Usually, spheres, granules, rods, or tubes with a diameter of millimeters are used in different types of packed-bed reactors. However, there is still need to optimize these catalytic systems. For example, for fast reactions, like the partial oxidation of benzene to phenol, the use of thin catalytically active layers of small zeolite crystals would be preferable (Hiemer et al. 2004). A promising approach to comply with both requirements is the use of cellular monoliths such as ceramic foams coated with the desired zeolite type.

In principle, there are three different methods to obtain zeolite-covered cellular monoliths. The simplest one is dip coating the foam in a slurry that contains the zeolite particles, a dispersant/solvent, and a binder. However, by using this method, only foam monoliths with a large cell size are coatable, and the homogeneity of the layer thickness inside the cells is not very high. A second approach is the synthesis of a desired zeolite structure directly on the foam surface. The hydrothermal crystallization of a zeolite-forming reaction mixture in the presence of PDC foams can be used even for support foams with relatively small cell sizes and leads to a zeolite layer of very homogenous thickness (Zampieri et al. 2004). The layer thickness can be varied by synthesis parameters such as synthesis time or the ratio of support/reaction mixture. However, in this case, the zeolite growth reduces the cell diameter and can cause a blocking of the cells.

A more sophisticated technique is the partial transformation of the three-dimensional cellular structure. In case of filler-containing polymer-derived foams, the type of filler can be chosen with respect to the requirements for the zeolite synthesis. Silicon powder can be used to obtain Si-rich zeolites such as MFI-type zeolite (which is an important catalyst for isomerization reactions, for the decomposition of nitric oxides, and also for partial oxidation reactions) after a hydrothermal treatment in the presence of a structure-directing agent (SDA) (Scheffler and Scheffler 2006, 2008). By ^{29}Si-MAS NMR spectroscopic investigations, it could be shown that the silicon filler in the starting mixture is transformed into the zeolite coating during the hydrothermal treatment (Scheffler et al. 2005a). Al-rich zeolites, which are important components in hydrocracking catalysts as well as in separation processes such as pressure or temperature swing air separation, might be obtained by the partial transformation of PDC foams filled with Al powder as shown in Fig. 17. The properties of the zeolite layer such as crystal morphology can be influenced by varying the reaction parameters. The special feature of these materials is the very good interdigitation between the support structure of the foam and the zeolite layer formed by the partial transformation of the filler component, even if more than 50 % of the foam is transformed. Al-rich zeolites and their composites are also good candidates for thermochemical energy storage via the reversible sorption of water vapor.

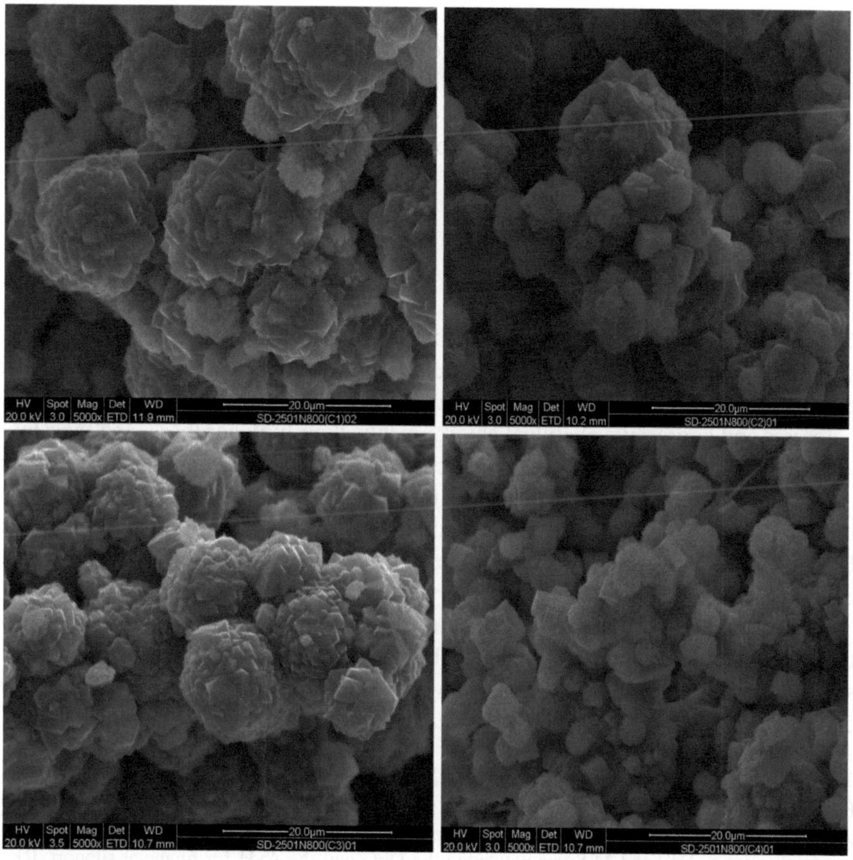

Fig. 17 SEM micrographs of in situ hydrothermally synthesized zeolite X obtained by the partial transformation of an Al-containing (13 %) PDC foam with varied Si/Al ratio (*top left* 6.25; *top right* 5.56; *bottom left* 5; *bottom right* 4)

2.3.3 Glass Foams for Environmental Catalysis

Glass foams with a high optical transmittance were prepared by a direct foaming approach with a poly(methyl phenyl silsesquioxane) (PMPS; Wacker Silres H44), particulate borosilicate glass fillers, and sodium borate (Ohl et al. 2011). In this case, PMPS was used as a binder, as a silica source for direct glass formation, and as the foaming agent. In a follow-up paper, a similar approach was applied to load glass foams with a photocatalyst (Bhattacharjee et al. 2011). A preceramic polymer (H44) was loaded with Duran glass particles and TiO_2 (anatase) nanoparticles, directly foamed, and thermally consolidated. Fourier-transformed infrared spectroscopic (FTIR) investigations showed that the anatase nanoparticles were chemically bound to the glass foam matrix indicated by Si–O–Ti bonds, thus preventing the transformation of the nanoparticles into the rutile modification. It was shown that this TiO_2-loaded glass foam performed better than TiO_2 nanoparticles in mineralizing methylene blue (3,7-bis(dimethylamino) phenazathionium chloride) which was used as a model substance to demonstrate the high catalytic performance as an environmental catalyst to clean contaminated water.

3 Summary, Conclusions, and Perspective

In this contribution, we have shown that ceramics are used in a wide range of sustainable technologies and are poised to play an ever-increasing role in these emerging technologies. This is in part due to their ability to withstand high temperatures and aggressive chemical environments, their high hardness, and their high wear resistance. However, a continuing challenge in the increased use of ceramics is the challenge in manufacturing of ceramics. The process that we discuss in detail in this paper, polymer-derived ceramics, addresses this central challenge of ceramics. Over the years, a variety of techniques have been developed to make ceramic fibers, coatings, and porous ceramics. An overview of these approaches is provided.

Due to the ability to make a wide variety of desirable forms, the ability to control the composition and purity, and the ability to make unusual micro- and nanostructures, ensures that this approach will provide differentiating solutions to the challenges offered by the emerging sustainable technologies.

Acknowledgments T.K. gratefully acknowledges funding through an Erwin Schrödinger Fellowship by the Austrian Science Fund (FWF), grant J3422. T.F. and P.G. gratefully acknowledge the Cluster of Excellence "Engineering of Advanced Materials" funded by DFG for financial support. F.S. and M.S. acknowledge the DFG and the AvH for financial support. RKB acknowledges support from the US Department of Energy Office of Nuclear Energy (DoE-NEUP) Grant through Idaho National Laboratory (Grant # 10-918) and US Air Force Office of Scientific Research (Grant FA9550-09-1-0633).

References

Adam M, Kocanis S, Fey T, Wilhelm M, Grathwohl G (2014) Hierarchically ordered foams derived from polysiloxanes with catalytically active coatings. J Eur Ceram Soc 34:1715–1725

Akbar S, Dutta P, Lee C (2006) High-temperature ceramic gas sensors: a review. Int J Appl Ceram Technol 3:302–311

Bhattacharjee S, Das PR, Ohl C, Wilker V, Kappa M, Scheffler F, Scheffler M (2011) Novel-type inorganic foams from preceramic polymers with embedded titania nanoparticles for photocatalytic applications. Adv Eng Mater 11:996–1001

Ceron-Nicolat B, Wolff F, Dakkouri-Baldauf A, Fey T, Münstedt H, Greil P (2012) Graded cellular ceramics from continuous foam extrusion. Adv Eng Mater 14:1097–1103

Ceron-Nicolat B, Fey T, Greil P (2010) Processing of ceramic foams with hierarchical cell structure. Adv Eng Mater 12:884–892

Clarke DR, Levi CG (2003) Materials design for the next generation thermal barrier coatings. Annu Rev Mater Res 33:383–417

Colombo P, Gambaryan-Roisman T, Scheffler M, Buhler P, Greil P (2001) Conductive ceramic foams from preceramic polymers. J Am Ceram Soc 84:2265–2268

Colombo P, Mera G, Riedel R, Soraru GD (2010) Polymer-derived ceramics: 40 years of research and innovation in advanced ceramics. J Am Ceram Soc 93:1805–1837

Colombo P, Riccardi B, Donato A, Scarinci G (2000) Joining of SiC/SiC ceramic matrix composites for fusion reactor blanket applications. J Nucl Mater 278:127–135

D'Alessandro DM, Smit B, Long JR (2010) Carbon dioxide capture: prospects for new materials. Angew Chem Int Ed 49:6058–6082

Dandeneau CS, Bodick TW, Bordia RK, Ohuchi FS (2013) Thermoelectric properties of reduced polycrystalline $Sr_{0.5}Ba_{0.5}Nb_2O_6$ fabricated via solution combustion synthesis. J Am Ceram Soc 96:1–8

de Sousa E, Rambo CR, Hotza D, Novaes de Oliveira AP, Fey T, Greil P (2008) Microstructure and properties of LZSA glass-ceramic foams. Mater Sci Eng A 476:89–97

Fergus JW (2010) Ceramic and polymeric solid electrolytes for lithium-ion batteries. J Power Sources 195:4554–4569

Flores O, Bordia RK, Nestler D, Krenkel W, Motz G (2014) Ceramic fibers based on SiC and SiCN systems: current research, development and commercial status. Adv Eng Mater 16:621–636

Flores O, Schmalz T, Krenkel W, Heymann L, Motz G (2013) Selective cross-linking of oligosilazanes to tailored meltable polysilazanes for the processing of ceramic SiCN fibres. J Mater Chem A 1:15406–15415

Fritz G, Grobe J, Kummer D (1965) Carbosilanes. In: Emeléus HJ, Sharpe AG (eds) Advances in inorganic chemistry and radiochemistry. Academic Press, New York, pp 349–418

Gambaryan-Roisman T, Scheffler M, Buhler P, Greil P (2000) Processing of ceramic foam by pyrolysis of filler containing phenylmethyl polysiloxane. Ceram Trans 108:121–130

Greil P (1995) Active-filler-controlled pyrolysis of preceramic polymers. J Am Ceram Soc 78:835–848

Günthner M, Albrecht Y, Motz G (2007) Polymeric and ceramic-like coatings on the basis of SIN (C) precursors for protection of metals against corrosion and oxidation. Ceram Eng Sci Proc 27:277–284

Günthner M, Kraus T, Dierdorf A, Decker D, Krenkel W, Motz G (2009a) Advanced coatings on the basis of Si(C)N precursors for protection of steel against oxidation. J Eur Ceram Soc 29:2061–2068

Günthner M, Kraus T, Dierdorf A, Decker D, Krenkel W, Motz G (2009b) Particle filled PHPS silazane based coatings on steel. Int J Appl Ceram Technol 6:373–380

Günthner M, Pscherer M, Kaufmann C, Motz G (2014) High emissivity coatings based on polysilazanes for flexible Cu(In, Ga)Se2 thin-film solar cells. Solar Energy Mater Solar Cells (SOLMAT) 123:97–103

Günthner M, Schütz A, Glatzel U, Wang K, Bordia RK, Greißl O, Krenkel W, Motz G (2011) High performance environmental barrier coatings, part I: passive filler loaded SiCN system for steel. J Eur Ceram Soc 31:3003–3010

Günthner M, Wang K, Bordia RK, Motz G (2012) Conversion behaviour and resulting mechanical properties of polysilazane-based coatings. J Eur Ceram Soc 32:1883–1892

Hacker J, Motz G, Ziegler G (2001a) Novel ceramic SiCN-fibres from the polycarbosilazane ABSE. In: Krenkel W, Naslain R, Schneider H (eds) High temperature ceramic matrix composites. Wiley-VCH Verlag, Weinheim, pp 52–55

Hacker J, Motz G, Ziegler G (2001b) Processing of SiCN-fibres prepared from polycarbosilazanes. In: Heinrich JG, Aldinger F (eds) Ceramic materials and components for engines. Wiley-VCH Verlag, Weinheim, pp 653–656

Haught D, Daley J, Bakke P, Marchionini B (2007) Overview of the U.S. Department of Energy (DOE) high-temperature superconductivity program for large-scale applications. Int J Appl Ceram Technol 4:197–202

Hiemer U, Klemm E, Scheffler F, Selvam T, Schwieger W, Emig G (2004) Microreaction engineering studies of the hydroxylation of benzene with nitrous oxide. Chem Eng J 101:17–22
IPCC (2014) Report Mitigation 2014: Summary for Policymakers. Retrieved 20 Aug 2014 from http://report.mitigation2014.org/spm/ipcc_wg3_ar5_summary-for-policymakers_approved.pdf
Kappa M, Kebianyor A, Scheffler M (2010) A two-component preceramic polymer system for structured coatings on metals. Thin Solid Films 519:301–305
Katoh Y, Kohyama A, Hinoki T, Snead LL (2003) Progress in SiC-based ceramic composites for fusion applications. Fusion Sci Technol 44:155–162
Kokott S, Motz G (2007) Cross-linking via electron beam treatment of a tailored polysilazane (ABSE) for processing of ceramic SiCN-fibres. Soft Mater 4:165–174
Kokott S, Heymann L, Motz G (2008) Rheology and processability of multi-walled carbon nanotubes—ABSE polycarbosilazane composites. J Eur Ceram Soc 28:1015–1021
Komnitsas K, Zaharaki D (2007) Geopolymerisation: a review and prospects for the minerals industry. Miner Eng 20:1261–1277
Konegger T, Liersch A, Gierl C, Scheffler M (2013a) Bulk ceramic composites derived from a preceramic polysilazane with alumina and zirconia fillers. Adv Eng Mater 15:394–406
Konegger T, Schneider P, Bauer V, Amsüss A, Liersch A (2013b) Structure and performance of polymer-derived bulk ceramics determined by method of filler incorporation. In: IOP conference series: material science and engineering, vol 47, p 012054
Kraus T, Günthner M, Krenkel W, Motz G (2009) cBN-particle-filled SiCN-precursor coatings. Adv Appl Ceram 108:476–482
Krishnan S, Fey T, Greil P (2014) Siliconboronoxycarbide (SiBOC) foam from methyl borosiloxane. Ceram Trans 243:47–60
Lewinsohn CA, Rao S, Bordia RK (2005) Effect of active fillers on ceramic joints derived from preceramic polymers. Ceram Eng Sci Proc 26:407–415
Li K (2007) Ceramic membranes for separation and reaction. Wiley, Chichester
Madsen LD, White AA (2014) Investments in ceramic science, engineering and education for sustainability by the U.S.A. National Science Foundation. J Electroceram 32:60–65
Menzler NH, Tietz F, Uhlenbruck S, Buchkremer HP, Stöver D (2010) Materials and manufacturing technologies for solid oxide fuel cells. J Mater Sci 45:3109–3135
Morris RE, Wheatley PS (2008) Gas storage in nanoporous materials. Angew Chem Int Ed 47:4966–4981
Motz G (2006) Synthesis of SiCN-precursors for fibres and matrices. Adv Sci Technol 50:24–30
Motz G, Hacker J, Ziegler G (2000) Special modified silazanes for coatings fibers and CMC's. Adv Ceram Mater Struct B 21:307–314
Motz G, Hacker J, Ziegler G (2001) Design of SiCN-precursors for various applications. In: Heinrich JG, Aldinger F (eds) Ceramic materials components for engines. Wiley-VCH Verlag, Weinheim, pp 581–585
Motz G, Hacker J, Ziegler G, Clauss B, Schawaller D (2002a) Low-cost ceramic SiCN Fibres by an optimized polycarbosilazane and continuous processing. In: Vincenzini P, Badini C (eds) Advanced inorganic structural fiber composites IV. Techna, Faenza, pp 47–54
Motz G, Hacker J, Ziegler G, Schawaller D, Clauss B (2002b) New SiCN fibres from the ABSE polycarbosilazane. Ceram Eng Sci Proc 23:255–260
Ohl C, Kappa M, Wilker V, Bhattacharjee S, Scheffler F, Scheffler M (2011) Novel open-cellular glass foams with high optical transmittance. J Am Ceram Soc 94:436–441
Reschke V, Laskowsky A, Kappa M, Wang K, Bordia RK, Scheffler M (2011) Polymer derived ceramic foams with additional strut porosity. Epitőanyag 63:57–60
Rödel J, Kounga ABN, Weissenberger-Eibl M, Koch D, Bierwisch A, Rossner W, Hoffmann MJ, Danzer R, Schneider G (2009) Development of a roadmap for advanced ceramics: 2010–2025. J Eur Ceram Soc 29:1549–1560
Scheffler F, Zampieri A, Schwieger W, Zeschky J, Scheffler M, Greil P (2005a) Zeolite-covered polymer-derived ceramic foams: novel hierarchical pore systems for sorption and catalysis. Adv Appl Ceram 104:43–48

Scheffler M, Bordia R, Travitzky N, Greil P (2005b) Development of a rapid crosslinking preceramic polymer system. J Eur Ceram Soc 25:175–180
Scheffler M, Colombo P (eds) (2005) Cellular ceramics. Wiley-VCH Verlag, Weinheim
Scheffler M, Greil P, Berger A, Pippel E, Woltersdorf J (2004) Nickel-catalyzed in situ formation of carbon nanotubes and turbostratic carbon in polymer-derived ceramics. Mater Chem Phys 84:131–139
Scheffler M, Scheffler F (2006) Zeolite coatings on porous monoliths. Adv Sci Technol 45:1260–1267
Scheffler M, Scheffler F (2008) Processing of ceramic foams and their surface modification with zeolites. Ceram Forum Int 85:E45–E50
Schütz A, Günthner M, Motz G, Greißl O, Glatzel U (2012) Characterisation of novel precursor derived ceramic coatings with glass filler particles on steel substrates. Surf Coat Technol 207:319–327
Seifert M, Travitzky N, Krenkel W, Motz G (2014) Multiphase ceramic composites derived by reaction of Nb and SiCN precursors. J Eur Ceram Soc 34:1913–1921
Stern E, Heyder M, Scheffler F (2009) Micropatterned ceramic surfaces by coating with filled preceramic polymers. J Am Ceram Soc 92:2438–2440
Torrey JD, Bordia RK (2007) Phase and microstructural evolution in polymer-derived composite systems and coatings. J Mater Res 22:1959–1966
Torrey JD, Bordia RK (2008a) Mechanical properties of polymer-derived ceramic composite coatings on steel. J Eur Ceram Soc 28:253–257
Torrey JD, Bordia RK (2008b) Processing of polymer-derived ceramic composite coatings on steel. J Am Ceram Soc 91:41–45
Torrey JD, Bordia RK, Henager CH, Blum Y, Shin Y, Samuels WD (2006) Composite polymer derived ceramic system for oxidizing environments. J Mater Sci 41:4617–4622
Walter S, Suttor D, Erny T, Hahn B, Greil P (1996) Injection molding of polysiloxane/filler mixtures for oxycarbide ceramic composites. J Eur Ceram Soc 16:387–393
Wang K, Günthner M, Motz G, Bordia RK (2011) High performance environmental barrier coatings, part II: active filler loaded SiOC system for superalloys. J Eur Ceram Soc 31:3011–3020
Wang K, Günthner M, Motz G, Flinn, Bordia RK (2013) Control of surface energy of silicon oxynitride films. Langmuir 29:2889–2896
Wang K, Unger J, Torrey JD, Flinn BD, Bordia RK (2014) Corrosion resistant polymer derived ceramic composite environmental barrier coatings. J Eur Ceram Soc 34:3597–3606
Wang K, Zheng X, Ohuchi FS, Bordia RK (2012) The conversion of perhydropolysilazane into SiON films characterized by X-Ray photoelectron spectroscopy. J Am Ceram Soc 95:3722–3725
Woiton M, Heyder M, Laskowsky A, Stern E, Scheffler M, Brabec CJ (2011) Self-assembled microstructured polymeric and ceramic surfaces. J Eur Ceram Soc 31:1803–1810
Wolff F, Ceron Nicolat B, Fey T, Greil P, Münstedt H (2012) Extrusion foaming of a preceramic silicone resin with a variety of profiles and morphologies. Adv Eng Mater 14:1110–1115
Yajima S, Hayashi J, Omori M (1975) Continuous silicon carbide fibre of high tensile strength. Chem Lett 4:931–934
Yu JG, Su YR, Cheng B (2007) Template-free fabrication and enhanced photocatalytic activity of hierarchical macro-/mesoporous titania. Adv Funct Mater 17:1984–1990
Zampieri A, Colombo P, Mabande GTP, Selvam T, Schwieger W, Scheffler F (2004) Zeolite coatings on microcellular ceramic foams: a novel route to microreactor and microseparator devices. Adv Mater 16:819–823
Zeschky J, Goetz-Neunhoeffer F, Neubauer J, Jason Lo SH, Kummer B, Scheffler M, Greil P (2003) Preceramic polymer derived cellular ceramics. Compos Sci Technol 63:2361–2370
Zeschky J, Höfner T, Arnold C, Weißmann R, Bahloul-Hourlier D, Scheffler M, Greil P (2005) Polysilsesquioxane derived ceramic foams with gradient porosity. Acta Mater 53:927–937

Solid-Fuel Rocket Motor Efficiency Improvement Scheme

Ryoichi S. Amano, Yi-Hsin Yen and Michael L. Hamman

Abstract Aluminum-based propellants are commonly used in solid rocket motors (SRM) due to their high energy densities. However, the alumina (Al_2O_3) particles produced during aluminum propellant combustion present performance issues. These particles flow along the combustion chamber to the nozzle in liquid form causing chemical and mechanical erosive damage. This erosive behavior should be avoided in an SRM because it decreases the ballistic performance. Since particle size and trajectory are believed to influence the impingement and accumulation of alumina droplets, which then affects erosive behavior, it is necessary to accurately predict both the particle size and trajectory. For design purposes, accurate prediction must allow for numerical simulation of particle size and trajectory for economic purposes. Recent work in particle size and trajectory using real time radiography (RTR) and numerical simulation demonstrated predictive capabilities for low solid-to-gas. Another study presented image processing methods to effectively process RTR images for larger particle sizes. Since the cost of experimental testing in is very high, due to high temperature and pressure, research in SRM field is more focused on numerical simulation. However, before simulation result could be used in SRM research CFD model validation is necessary. To provide validation for CFD modelling, a water-air two phase strait channel flow with controlled low temperature and pressure is used. In this chapter, two major parts will be covered, which include the comparison between water-air strait channel experiment and CFD results, and a quantification method for both experimental and CFD results is presented.

Keywords Ceramic fibers · Polymer-derived ceramics · Hazardous waste remediation · Advanced ceramic-based sensor systems · SiC fibers · SiCN fibers

---

R.S. Amano (✉) · Y.-H. Yen · M.L. Hamman
Department of Mechanical Engineering, University of Wisconsin-Milwaukee, Milwaukee, WI 53211, USA
e-mail: amano@uwm.edu

A.K. Agarwal et al. (eds.), *Novel Combustion Concepts for Sustainable Energy Development*, DOI 10.1007/978-81-322-2211-8_23

Nomenclature

Variables

$A_{c,a,w}$	Area; channel, air, water
a	Face area vector
B	Breakup ratio
$C_{\varepsilon 1}$	Turbulence production coefficient
$C_{\varepsilon 2}$	Dissipate coefficient
$C_{\varepsilon 3}$	Buoyancy coefficient
D_h	Hydraulic diameter ≡ $4A_c/Y$
D_ω	SST k-ω cross-derivative term
F_1	Blending function
f_β	Vortex-stretching modification coefficient
f_{β^*}	Dissipate vortex-stretching modification coefficient
f_c	Curvature correction factor
G_k	Turbulence production
G_{nl}	Nonlinear turbulent production
G_b	Turbulent production due to buoyancy
G_ω	Production of specific dissipation rate
g	Gravity
h	Height of liquid body
K	Temperature in Kelvin
k	Turbulent kinetic energy
k_0	Ambient turbulent kinetic energy
m	Mass
P	Pressure
Q	Volumetric flow rate
p'	Turbulent pressure fluctuation
Re	Reynolds number $\equiv \rho v D_h/\mu$
S	Strain rate parameter
$\boldsymbol{S}$	Strain rate tensor
T	Temperature
t	Time
u	Time-averaged mean velocity
u'	Turbulent fluctuating velocity component in Reynolds Stress Model
v, V	General velocity expression
ν	Kinematic viscosity
$\boldsymbol{V}$	Volume integral vector
v_g	Grid velocity
We	Weber number $\equiv \rho v^2 l/\sigma$
x	Mass flow-rate fraction
Y	Channel width

Symbol

β	Thermal expansion coefficient
β^*	SST k-ω thermal expansion coefficient
δ_{ij}	Kronecker delta $\equiv \begin{cases} 0 & \text{if } i \neq j \\ 1 & \text{if } i = j \end{cases}$
ε	Turbulent dissipation rate
ε_0	Ambient turbulent dissipation rate
γ_{eff}	Effective intermittency
γ_M	Dilation dissipation
γ_y	γap correction term
γ'	Conditional statement of effective intermittency
σ	Surface tension
σ_k, σ_ε, σ_ω	Turbulent Schmidt number
θ	Temperature in Reynolds stress model
μ	Dynamic viscosity
μ_t	Turbulent viscosity
ρ	Density of fluid
ω	Turbulence specific dissipation
ω_0	Ambient turbulence specific dissipation

Subscripts

a	Air
b	Two-phase breakup water outlet flow
e	Exit
i	Inlet
i, *j*, *k*	Coordinate direction/tenser index
m	Mean parameter
w	Water

Superscripts

T	Transpose

1 Introduction

Aluminum-based propellants are commonly used in solid rocket motors (SRM) (Xiao and Amano 2006) due to their high energy densities (Rapp 1968). However, the alumina (Al_2O_3) particles produced during aluminum propellant combustion present performance issues. These particles flow along the combustion chamber to the nozzle in liquid form causing chemical and mechanical erosive damage. This erosive behavior should be avoided in an SRM because it decreases the ballistic performance (Thankre and Yang 2009). Since particle size and trajectory are believed to influence the impingement and accumulation of alumina droplets, which then affects erosive behavior, it is necessary to accurately predict both the particle size and trajectory. For design purposes, accurate prediction must allow for numerical simulation of particle size and trajectory for economic purposes. Recent work in particle size and trajectory using real time radiography (RTR) and numerical simulation demonstrated predictive capabilities for low solid-to-gas ratios (Xiao et al. 2003). Another study presented image processing methods to effectively process RTR images for larger particle sizes (Xiao et al. 2005).

Since erosion primarily occurs in the nozzle throat vicinity, the alteration of the nozzle throat geometry could increase SRM thrust efficiency. This improvement is expected through mitigating impingement, accumulation, and flow of alumina in the nozzle. A throat-area increase greater than 5 % in most SRMs is considered excessive (Wong 1968). The working fluid velocity contour in SRM is affected by the multiphase flow in the combustion chamber; therefore, the performance is affected by the nonequilibrium behavior of exhaust gas limits. The motor combustion chamber and the nozzle geometry are ideal for molten alumina agglomerate formation, resulting in two-phase flow losses (Borass 1984; Salita 1992; Hess et al. 1992). Agglomerates lower the propulsive efficiency of the exhaust flow since the agglomerates cause additional drag force on the flow by not expanding in the nozzle (Xiao et al. 2003, 2005). Particular alumina particle-gas mixture ratio and particle size are known to reduce the specific impulse, with reductions as high as 6 % (Bandera et al. 2011; Holtzmann 1964). The severe operating environment of SRMs combustion chambers has temperatures ranging from 3,000 to 3,500 K with pressures of 2.0×10^7 Pa or greater. Given the alumina melting temperature of 2,327 K at atmospheric pressure and the evaporation temperature about 3,200 K; alumina exists in liquid form with possible evaporation. Erosive damage can be caused by the re-entrainment of the alumina film near the nozzle throat and impingement on the diverging section of the rocket nozzle. Therefore, it is necessary to obtain a better understanding of this phenomenon for SRM design.

The breakup mechanism of liquid alumina in SRMs is complicated. The attachment and accumulation of liquid alumina on and to the wall of the de Laval nozzle (Salita 1995) produces unstable shear-driven interactions with the surrounding air. Shear instability causes the development of a liquid film wave which leads to the onset of breakup on the wave crest. The breakup level grows with the increasing magnitude of surrounding gas velocity; as a result, the observance of liquid alumina droplet size reduction is seen. This results in decreased resistance to droplet discharge by the gas flow rather than the adherence of the liquid alumina to the nozzle wall.

A method of reducing the negative effects of aluminum-based propellants through geometric modification was employed. This method seeks to induce alumina breakup through SRM nozzle geometry change, a ramp is placed into the combustion flow to encourage adequate mixing to prevent nozzle damage. The ramp is used to induce turbulent flow, enhancing breakup and two-phase mixing. The alumina particles are expected to be largely carried past the nozzle in the exhaust flow. Experimental and numerical investigation of geometric modification as given in Amano et al. (2014a, b) is discussed for SRM nozzle design.

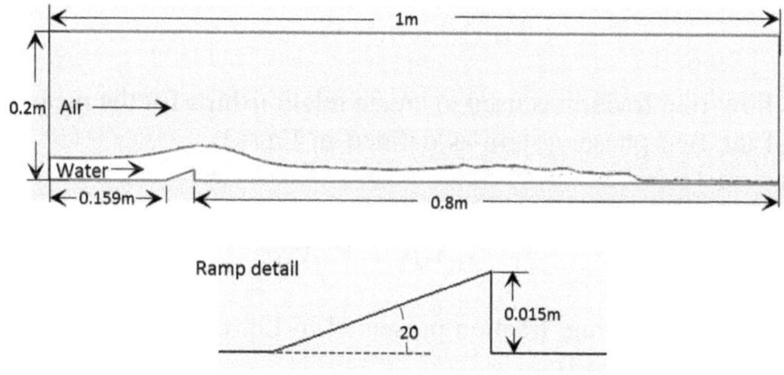

Fig. 1 Flow geometry

Since the cost of experimental testing is very high, due to high temperature and pressure, research in SRM field is more focused on numerical simulation. However, before simulation result could be used in SRM research, CFD model validation is necessary. To provide validation for CFD modeling, a water–air two-phase straight channel flow with controlled low temperature and pressure is used. In this chapter, two major parts will be covered, which include the comparison between water–air straight channel experiment and CFD results, and a quantification method for both experimental and CFD results is presented.

2 Dimension of Experiment and Simulation

The two-phase flow experiment and simulation were conducted using a rectangular channel for comparison basis and is shown in Fig. 1 with the dimensions included. The inlet velocity of water is maintained at 0.68 m/s with air inlet velocity ranging incrementally from 0 to 40 m/s. The experiment and simulation were performed at standard room conditions.

A rectangular, transparent acrylic channel was used to simulate the two-dimensional SRM combustion flow.

3 Physics Equations

The Reynolds' number is a dimensionless quantity used to predict flow patterns. Using the standard Reynolds for a rectangular duct defined as, $\mathrm{Re} = \rho V D_h / \mu$, as a basis for defining a Reynolds number for the two-phase rectangular duct flow. Where ρ, V, D_h, and μ are the fluid density, velocity, hydraulic diameter, and dynamic viscosity, respectively. Since it is advantageous to know the flow conditions that lead to breakup of the liquid flow, the velocity term is defined as the relative mean velocity as shown in Eq. (1).

$$V_{\mathrm{m}} = V_{\mathrm{m,air}} - V_{\mathrm{m,water}} \tag{1}$$

A mass flow-rate fraction is used to create relationships for the mean density and viscosity of the two-phase system as defined in Eq. (2).

$$x = \frac{V_{\mathrm{a}}A_{\mathrm{a}}\rho_{\mathrm{a}}}{V_{\mathrm{a}}A_{\mathrm{a}}\rho_{\mathrm{a}} + V_{\mathrm{w}}A_{\mathrm{w}}\rho_{\mathrm{w}}} \tag{2}$$

Using the mass flow-rate fraction presented in Eq. (2), the two-phase density is defined as shown in Eq. (3).

$$\rho_{\mathrm{m}} = \left(\frac{x}{\rho_{\mathrm{a}}} + \frac{1-x}{\rho_{\mathrm{w}}}\right)^{-1} \tag{3}$$

The two-phase viscosity is defined for the system in Eq. (4)

$$\mu_{\mathrm{m}} = \left(\frac{x}{\mu_{\mathrm{a}}} + \frac{1-x}{\mu_{\mathrm{w}}}\right)^{-1} \tag{4}$$

The specific Reynolds number equation for the system using Eqs. (1)–(4) is given in Eq. (5).

$$\mathrm{Re} = \frac{\rho_{\mathrm{m}} V_{\mathrm{m}} D_{\mathrm{h}}}{\mu_{\mathrm{m}}} \tag{5}$$

The specific Weber number is written using velocity and density definitions from Eqs. (1) and (3) and given as:

$$\mathrm{We} = \frac{\rho_{\mathrm{m}} V_{\mathrm{m}}^2 D_{\mathrm{h}}}{\sigma} \tag{6}$$

where σ is the surface tension of water with an air interface. Since it is desired to know the extent of breakup occurring over the experimental section, the mass conservation principle is applied to liquid water volume beneath the time-averaged boundary layer between the two-phase and liquid water flows. With the time-averaged condition, the conservation law is limited to the liquid water mass entering and leaving the control volume as given in Eq. (7) and (5)

$$m_{\mathrm{m,w,i}} = m_{\mathrm{m,w,b}} + m_{\mathrm{m,w,e}} \tag{7}$$

where m is the mass flow rate, and the indices m, w, i, b, and e denote; the mean, liquid water, inlet flow, water breakup in the two-phase outlet flow, and liquid exit flow, respectively. With constant density, Eq. (7) is reduced to volumetric balance as shown in Eq. (8). The constant density condition is valid unless the static pressure drops to or below the vapor pressure of water. In such cases, the mass

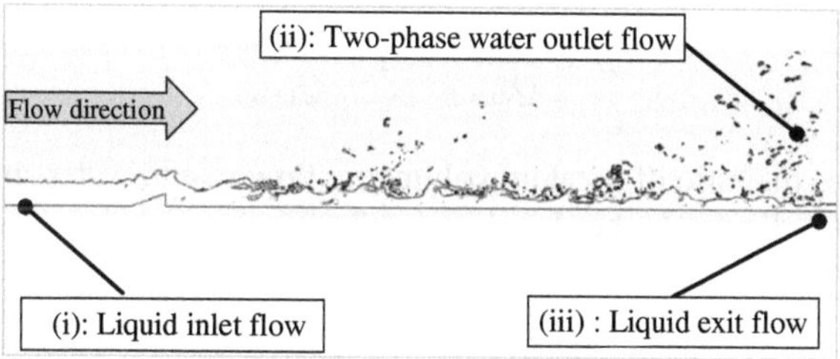

Fig. 2 Liquid water mass conservation illustration

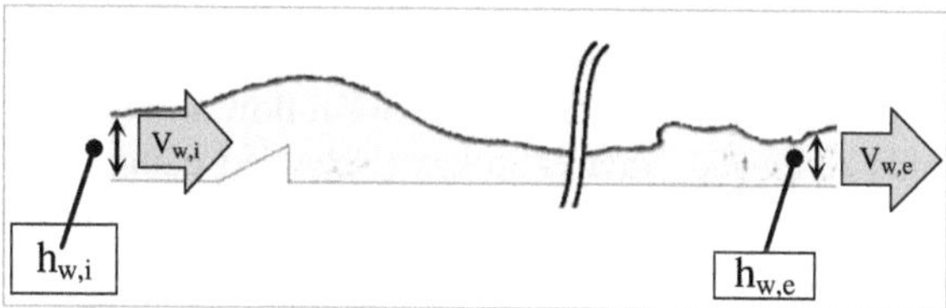

Fig. 3 Water level and velocity at inlet and exit

conservation in Eq. (7) must be employed. Figure 2 illustrates the mass conservation region in control volume for each term in Eq. (8).

$$\underset{\text{(i)}}{Q_{\mathrm{m,w,i}}} = \underset{\text{(ii)}}{Q_{\mathrm{m,w,b}}} + \underset{\text{(iii)}}{Q_{\mathrm{m,w,e}}} \tag{8}$$

where Q denotes volumetric flow rate, and the term (i) is measured directly from the flow meter as displayed in Fig. 3. The term (ii) is the volumetric flow rate of the sum of the dispersed liquid water particles in the two-phase flow exiting the channel. The term (iii) is the volumetric flow rate exiting the channel and is approximated by calculating the mean water exit velocity and boundary height. The extent of dispersed liquid water is a good indicator of breakup; however, it is not easy to measure dispersed liquid water or $Q_{\mathrm{m,w,b}}$ directly; therefore the measurement needs to be carried out by subtracting measured value of liquid inlet flow by exit flow. The volumetric flow rate of inlet and exit are defined in Eqs. (9) and (10).

$$Q_{\mathrm{m,w,i}} = h_{\mathrm{w,i}}\, \mathrm{v_{w,i}}\, \mathrm{Y} \tag{9}$$

$$Q_{\mathrm{m,w,e}} = h_{\mathrm{w,e}}\, \mathrm{v_{w,e}}\, \mathrm{Y} \tag{10}$$

where $h_{\mathrm{w,i}}$, $h_{\mathrm{w,e}}$, $\mathrm{v_{w,i}}$, and $\mathrm{v_{w,e}}$ are water level of channel and water velocity of inlet and exit, respectively. The Y in Eqs. (9) and (10) is the width of the flow channel. The illustration of the mass flow component in both equations is shown in Figure 3. Since $Q_{\mathrm{m,w,i}}$ and $Q_{\mathrm{m,w,e}}$ are known, $Q_{\mathrm{m,w,b}}$ can be calculated which permits the introduction of the breakup ratio B as expressed as:

$$B = \frac{Q_{\mathrm{m,w,b}}}{Q_{\mathrm{m,w,i}}} = 1 - \frac{Q_{\mathrm{m,w,e}}}{Q_{\mathrm{m,w,i}}} \tag{11}$$

Which describes the ratio of breakup volumetric flow rate to water supply flow rate, and a higher B indicates higher breakup level of water into the two-phase flow.

4 Numerical Analysis

4.1 Numerical Models

In this study, STAR-CCM+ was applied to perform flow simulations by solving the unsteady Reynolds-Averaged Navier–Stokes (URANS) equations. The system of equations was solved in an Eulerian multiphase state with an implicit unsteady solver applying the volume of fluid (VOF) approach which is a multiphase model simulating several fluid interfaces between various fluid phases (Hirt and Nichols 1981).

4.2 k-ε Model

The standard two-layer k-ε model combines the standard k-ε model with two-layer approach introduced by Rodi (1991). The k-ε model can be expressed as in Eqs. (12) and (13).

$$\frac{\mathrm{d}}{\mathrm{d}t}\int_V \rho k \,\mathrm{d}\boldsymbol{V} + \int_A \rho \mathrm{k}\left(v - v_g\right) \cdot \mathrm{d}\boldsymbol{a} = \int \left(\mu + \frac{\mu_t}{\sigma_k}\right) \nabla k \cdot \boldsymbol{da} + \int_V \left[G_k + G_{\mathrm{nl}} + G_b - \rho((\epsilon - \epsilon_0) + \gamma_M)\right] \mathrm{d}\boldsymbol{V} \tag{12}$$

$$\frac{\mathrm{d}}{\mathrm{d}t}\int_V \rho\epsilon \,\mathrm{d}\boldsymbol{V} + \int_A \rho\epsilon(v - v_{\mathrm{g}}) \cdot \mathrm{d}\boldsymbol{a} = \int \left(\mu + \frac{\mu_{\mathrm{t}}}{\sigma_\epsilon}\right) \nabla\epsilon \cdot \mathrm{d}\boldsymbol{a} + \int_V \frac{1}{T}\left[C_{\epsilon 1}(G_{\mathrm{k}} + G_{\mathrm{nl}} + C_{\epsilon 3} G_{\mathrm{b}}) - C_{\epsilon 2}\rho(\epsilon - \epsilon_0) + \rho\gamma_{\mathrm{y}}\right] \mathrm{d}\boldsymbol{V} \tag{13}$$

G_{nl} in nonlinear production term for nonlinear model, G_{b} is the production due to buoyancy and defined as in Eq. (14), G_{k} is the turbulent production and evaluated as in Eq. (15).

$$G_{\mathrm{b}} = \beta \frac{\mu_{\mathrm{t}}}{\sigma_{\mathrm{t}}} (\nabla T \cdot g) \tag{14}$$

$$G_{\mathrm{k}} = \mu_{\mathrm{t}} S^2 - \frac{2}{3} \rho \, \mathrm{k} \nabla \cdot v - \frac{2}{3} \mu_{\mathrm{t}} (\nabla \cdot v)^2 \tag{15}$$

where $\nabla \cdot v$ is the velocity divergence, S is the strain parameter, and $\mathbf{S}$ is the strain rate tensor given in Eqs. (16) and (17).

$$S = |\mathbf{S}| = \sqrt{2\mathbf{S} : \mathbf{S}^{\mathrm{T}}} = \sqrt{2\mathbf{S} : \mathbf{S}} \tag{16}$$

And

$$\mathbf{S} = \frac{1}{2} (\nabla \boldsymbol{v} + \nabla \boldsymbol{v}^{\mathrm{T}}) \tag{17}$$

Also, the product of two tensors is described in Eq. (18).

$$\mathbf{s} : \mathbf{s} = s_{ij} s_{ji} \tag{18}$$

4.3 SST k-ω Model

The shear–stress transport k-ω model was addressed by Menter (1994) and can be expressed as Eqs. (19) and (20):

$$\frac{d}{dt} \int_V \rho \, k \, d\boldsymbol{V} + \int_A \rho k (v - v_g) \cdot \mathrm{d}\boldsymbol{a} = \int_A (\mu + \sigma_k \mu_{\mathrm{t}}) \nabla k \cdot \mathrm{d}\boldsymbol{a} + \int_V (\gamma_{\mathrm{eff}} G_k - \gamma' \rho \beta^* f_{\beta^*} (\omega \mathrm{k} - \omega_0 k_0)) \cdot \mathrm{d}\boldsymbol{V} \tag{19}$$

$$\frac{\mathrm{d}}{\mathrm{d}t} \int_V \rho \omega \, \mathrm{d}\boldsymbol{V} + \int_A \rho \omega (v - v_g) \cdot \mathrm{d}\boldsymbol{a} = \int_A (\mu + \sigma_\omega \mu_{\mathrm{t}}) \nabla \omega \cdot \mathrm{d}\boldsymbol{a} + \int_V (G_\omega - \rho \beta \mathrm{f}_\beta (\omega^2 - \omega_0^2) + D_\omega) \cdot \mathrm{d}\boldsymbol{V} \tag{20}$$

where k_0 are the ambient turbulence values in source terms that counteract turbulence decay, γ_{eff} is the effective intermittency provided by the Gamma Theta Transition model, and $\gamma' = \min[\max(\gamma_{\mathrm{eff}}, 0.1), 1]$. The turbulent production G_k and G_ω can be written as follows in Eqs. (21) and (22).

$$G_{\mathrm{k}} = \mu_t f_c S^2 - \frac{2}{3}\rho k \nabla \cdot v - \frac{2}{3}\mu_t (\nabla \cdot v)^2 \tag{21}$$

$$G_\omega = \rho\gamma[(S^2 - \frac{2}{3}(\nabla \cdot v)^2) - \frac{2}{3}\omega \nabla \cdot v] \tag{22}$$

D_ω in Eq. (20) is a cross-derivative term defined in Eq. (23)

$$D_\omega = 2(1 - F_l)\rho\sigma_{\omega 2}\frac{1}{\omega}\nabla k \cdot \nabla\omega \tag{23}$$

4.4 Reynolds Stress Model

The Reynolds stresses model (RSM) was proposed by Launder et al. (2006) and is a seven-equation model solving six-component symmetric Reynolds stress tensor and turbulent dissipation equation. In two-dimensional computation, only four of Reynolds stress components need to be evaluated; therefore, the total number of equation is reduced to five. RSM can be expressed as in Eq. (24).

$$\underbrace{\frac{\partial}{\partial t}(\rho\overline{u_i'u_j'})}_{\text{(I)}} + \underbrace{\frac{\partial}{\partial x_k}(\rho u_k\overline{u_i'u_j'})}_{\text{(II)}} = \underbrace{-\frac{\partial}{\partial x_k}[\rho\overline{u_i'u_j'u_k'} + \overline{p'(\delta_{\mathrm{kj}}u_i' + \delta_{\mathrm{ik}}u_j')}]}_{\text{(III)}}$$
$$\underbrace{+\frac{\partial}{\partial x_k}[\mu\frac{\partial}{\partial x_k}(\overline{u_i'u_j'})]}_{\text{(IV)}} \quad \underbrace{-\rho(\overline{u_i'u_k'}\frac{\partial u_j}{\partial x_k} + \overline{u_j'u_k'}\frac{\partial u_i}{\partial x_k})}_{\text{(V)}}$$
$$\underbrace{-\rho\beta(g_i\overline{u_j'\theta} + g_j\overline{u_i'\theta})}_{\text{(VI)}} \quad \underbrace{+\overline{p'(\frac{\partial u_i'}{\partial x_j} + \frac{\partial u_j'}{\partial x_i})}}_{\text{(VII)}} \quad \underbrace{-2\mu\overline{\frac{\partial u_i'}{\partial x_k}\frac{\partial u_j'}{\partial x_k}}}_{\text{(VIII)}} \tag{24}$$

In Eq. (24), the term (I) is the local time derivative term with ρ as the density and u' as the turbulent fluctuating velocity component in RSM; the term (II) is the convection term; term (III) is the turbulent diffusion term; term (IV) is the molecular diffusion term with β as the viscosity of fluid; term (V) is the stress production term; term (VI) is the buoyancy term with β as the thermal expansion coefficient, g as the acceleration direction and θ as the temperature; term (VII) is the pressure strain term of turbulence pressure fluctuation; and term (VIII) is the dissipation term.

5 Simulation Setup

The simulation modeling setting is shown in Table 1. The initial and boundary condition are shown in Table 2.

Table 1 Simulation condition and setting

List of condition	Applied model		Value	Unit
CFD dimension	2			
Time derivative	Implicit unsteady			
Temporal discretization	1st-order			
Time step			5.0E−06	s
Inner iteration			2	Iteration
Multiphase model	Volume of fluid (VOF)			
Turbulent model	k-ε			
	SST k-ω			
	RSM			
Multiphase material	Air	Density	1.18	kg/m^3
		Dynamic viscosity	1.9E−05	Pa-s
	Water	Density	997.56	kg/m^3
		Dynamic viscosity	8.9E−04	Pa-s
Multiphase interaction	Water–air	Surface tension	7.4E−02	N/m

Table 2 Boundary and Initial condition

Boundary	Quantity	Value	Unit
Air inlet	Velocity	40	m/s
Water inlet	Velocity	0.67	m/s
Pressure outlet	Gauge pressure	0	Pa
Wall	Nonslip condition		
Initial condition	Velocity	0	m/s
	Gauge pressure	0	Pa
	Material in computational domain	Air: 1	
		Water: 0	

6 Quantification

6.1 Boundary Length Definition

The boundary length is an important indicator in water breakup studies; thus, the boundary length illustration is shown in Fig. 4. When there is no breakup, as shown in Fig. 4a, the two-phase boundary is controlled by the gravity and the boundary is a straight simple line. When breakup phenomenon occurs, as shown in Fig. 4b, both the wave boundary surface and droplets contribute to increase the length of the boundary resulting in longer boundary length. In this study, boundary length is used as quantification means of grid-independent study and simulation result.

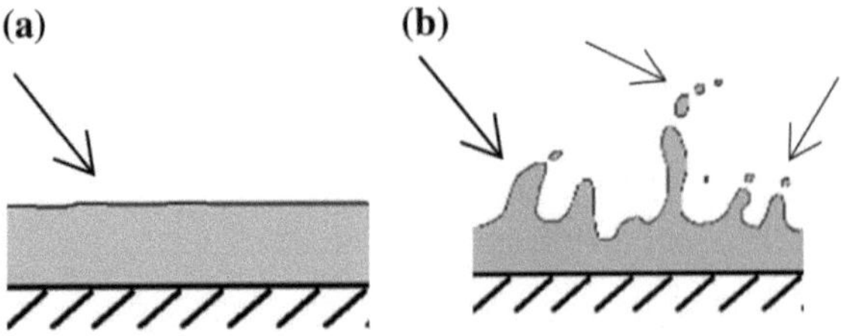

Fig. 4 Breakup level and boundary length. **a** No breakup. **b** Fractional breakup

6.2 Grid-Independent Study

Four mesh sizes with computational domain cells numbering 100,000, 300,000, 500,000, and 1,000,000, were used to test the grid resolution, the boundary length, and the breakup phenomenon. The simulations apply air and water as the two-phase materials. The reason to use water is that it is more likely to breakup than alumina due to its lower surface tension, density, and viscosity, which could help us to capture the fluid breakup details for all four different mesh size settings. The velocity settings of these grid-independent study simulations are 26 m/s and 5 m/s, for the air and water, respectively. The demonstrate of the boundary length for different domain cell numbers are shown in Fig. 5; the boundary length increases due to the better resolution of the mesh throughout the computational domain until the domain cell number reaches 500,000. However, from 500,000 to 1,000,000, the domain cell number increases 100 % compared with the boundary length increase of 2 %. As shown any domain cell number higher than 500,000 is sufficient to capture the breakup detail. For this reason, all the computations were made using 500,000 cells, which is considered more than enough mesh resolution to capture the smaller breakup of the liquid droplets.

Fig. 5 Boundary length for different mesh size setup

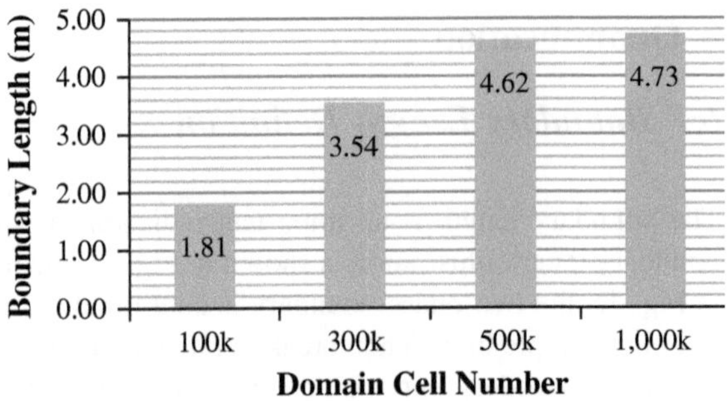

6.3 Time Averaging Boundary Length

Two-phase turbulent flows behave highly irregular and it is hard to capture a certain repeatable flow pattern. Figure 6 illustrates the simulation results with the same boundary condition and the numerical model setting over a period of different time step, the simulation condition detail as shown in Table 2. The breakup droplet size might be similar in their size; however, the similarity character is difficult to determine from those three figures even though the operating conditions are the same. Therefore, a method of figure superposition is introduced in this paper in order to visualize time-averaged contour through the results of all time steps.

In this demo (Table 3; Figs. 7, 8 and 9), three pictures with black and white pixels with pixel labels of 0 and 1 are shown in Fig. 7, where pixel value 0 represents pure black and pixel value 1 represents pure white color, the illustration of the grayscale map from corresponding pixel label value are shown in Table 3. The pixel superimposition is shown in Fig. 7. After superimposition of Fig. 7a–c,

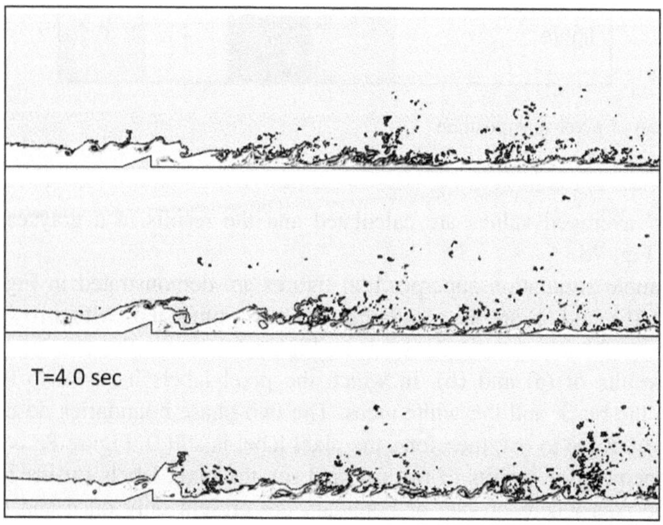

Fig. 6 VF of fluid boundary at different time step

Table 3 Grayscale color map

Pixel label value	Gray scale color map
0.00	
0.33	
0.50	
0.67	
1.00	

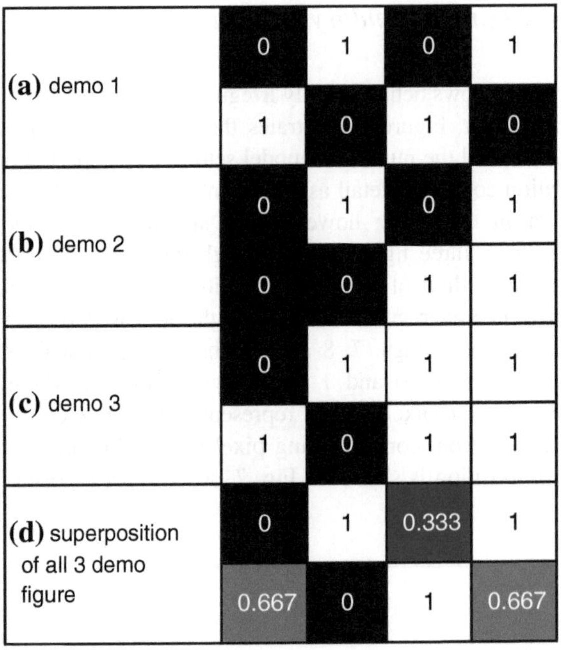

Fig. 7 Demo of pixel superposition

the pixels' averaged values are calculated and the results in a grayscale map is shown in Fig. 7d.

The sample simulation superposition figures are demonstrated in Fig. 8. From which, (a), (b), and (c) are the results from different time steps with arrows marking the location of interest with pixel labels. Figure 8d shows the two frames superposition results of (a) and (b), in which the pixel labels increase to 0.5 due to averaging the black and the white areas. The two-phase boundaries do not change too much from (a) to (b); therefore, the pixel label is still 0. Figure 8e is the three-frames superposition results of (a), (b), and (c), the pixel labels further increase to 0.67 at the area where droplet or boundary are present only once and those area become lighter color. As shown in these figures, at the boundary in the inlet region, the pixel label remains 0 for all the cases. Figure 8f shows the result of 260 time steps superposition, the grayscale figure shows clear two-phase boundary in inlet region shown in region (i), due to the two-phase boundary at the inlet entrance are still steady. When the breakup starts, the probability of two-phase flow existence at region (ii) will be an averaged value ranging between 0 and 1, therefore it becomes a grayscale region. The white area at the upper channel in region (iii) is where water droplets are less likely to be present. The superposition figure could be interpreted as a possibility map of the two-phase flow boundary; the darker the color higher the chance to have two-phase boundary in that area and vice versa. The software that performs the figure superposition is Picasa 3.9 (Picht).

Fig. 8 Superposition on actual simulation case

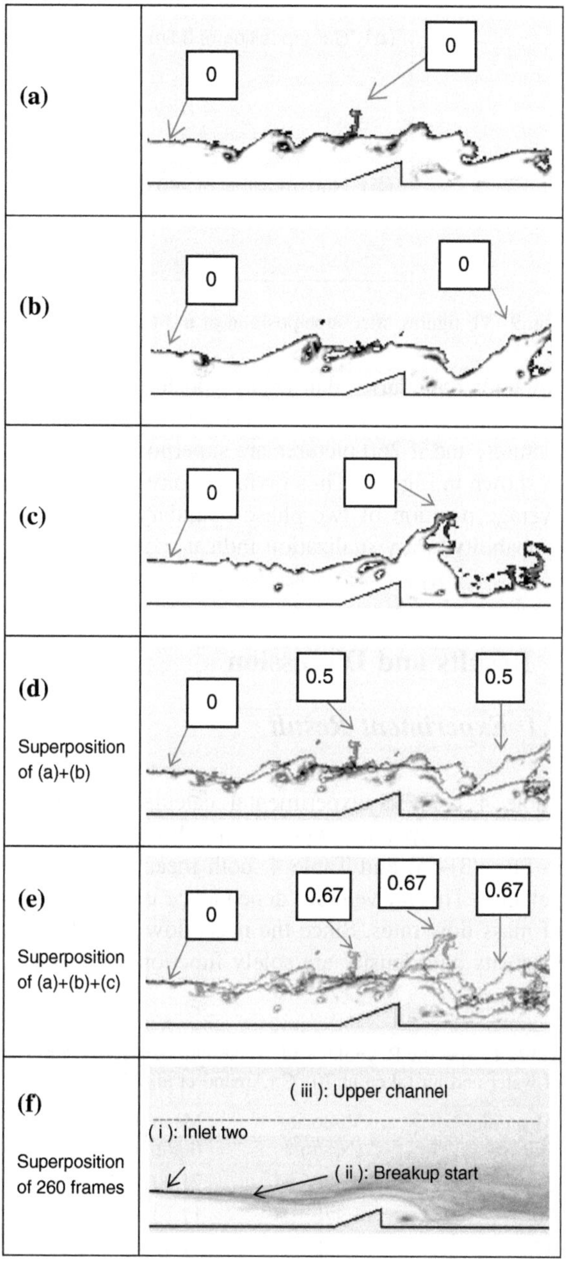

The three figures for different time frames shown in Fig. 6 are superposed with the results shown in Fig. 9a. In Fig. 9a the dark color is used for the position where boundaries appear at relatively the same position in the flow channel and lighter in

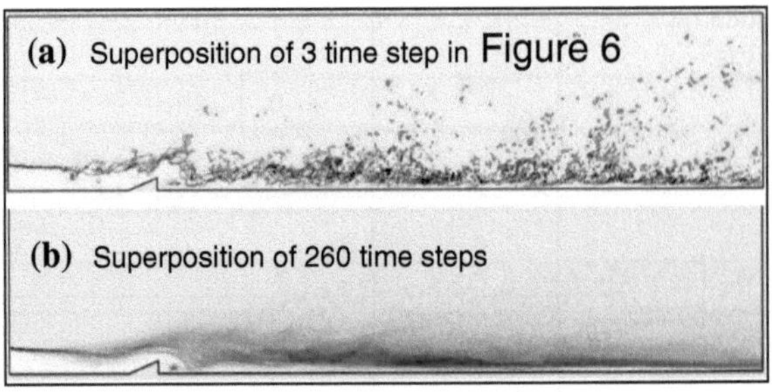

Fig. 9 VF figures after superposition of **a** 3 time step and **b** 260 time steps of RSM 40 m/s

color for boundaries that change position over three time frames. The probability distribution of the two-phase boundary movement by the superposition over three pictures, and if 260 pictures are superposed, the resulting image becomes smoother as shown in Fig. 9b. Thus giving a convenient comparison of the distribution of the average position of two-phase boundary. The time-averaged two-phase boundary probability is a visualization indicator of different flow characteristics.

7 Results and Discussion

7.1 Experiment Result

Table 4 presents experimental calculations for two-phase mean viscosity and density, and Specific Reynolds' and Weber numbers for varied air velocity. As seen in Eqs. (3)–(5) and Table 4, both mean viscosity and density vary with mean air velocity. This air velocity dependence is due to defining the mass fraction in terms of mass flow rates. Since the mass flow rate of water was held constant, the mean viscosity and density are solely functions of mean air velocity.

Table 4 Specific Reynolds and Weber numbers at 0.676 m/s mean water velocity, surface tension of water and air taken at 70 °C (Amano et al. 2014a, b)

Mean air velocity (m/s)	Viscosity (N s/m^2)	Mean density (kg/m^3)	Reynolds number (–)	Weber number (–)
5	2.47E−04	21.00	86	277
10	1.47E−04	11.20	165	686
15	1.08E−04	7.89	243	1,141
20	8.73E−05	6.23	320	1,639
25	7.42E−05	5.23	398	2,180
30	6.53E−05	4.56	476	2,764
35	5.89E−05	4.08	553	3,390
40	5.40E−05	3.72	631	4,058

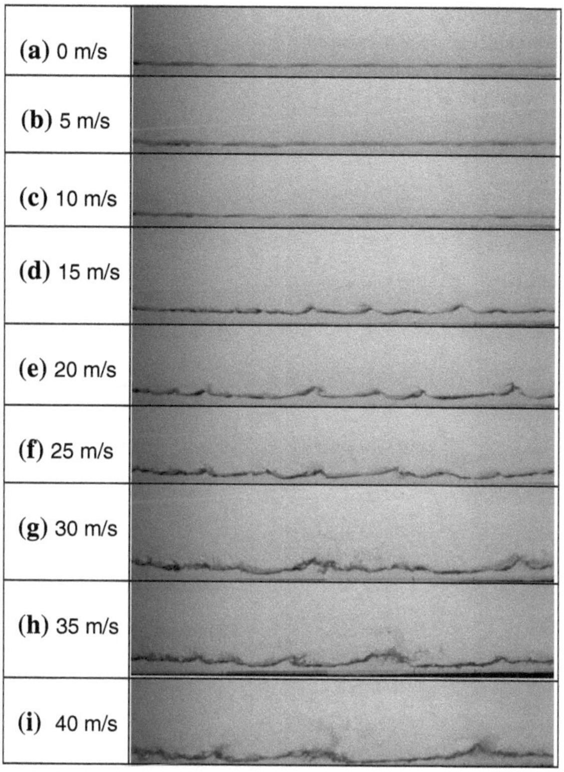

Fig. 10 Chamber with no wedge; varied velocity (Amano et al. 2014a, b)

Similarly, the Reynolds and Weber numbers vary as functions of the mean air velocity. Figure 10 presents the flow behavior observed with the absence of the wedge obstruction.

Figure 10 displays photographs representing the flow behavior observed with the ramp in the chamber obstructing the water flow. From Figs. 10 and 11, it is clear that water breakup begins to occur around the mean air velocity of 15 m/s. It is also notable that both the cases are similar in breakup magnitude differing only around the vicinity of the wedge, showing a clear enhancement of breakup behavior due to geometric modification. Figure 11d–i shows a distinguishable increase in breakup around the trailing edge of the wedge indicating that the wedge increases the breakup phenomena.

Knowing that the mean air velocity of 15 m/s with constant mean water velocity is the approximate condition for onset of the breakup, the condition for breakup can be thought of in terms of relative mean velocity between the two fluids. It is then convenient to use the Specific Reynolds and Weber numbers as a criterion for breakup. The Specific Reynolds and Weber numbers corresponding to breakup initiation are presented in Table 4.

Due to the focal length and the angle of view used to capture perspective image, distortion was observed in the photographs. The water stream boundary height was

Fig. 11 Breakup at various air velocities; constant mean water velocity (0.676 m/s) (Amano et al. 2014a, b)

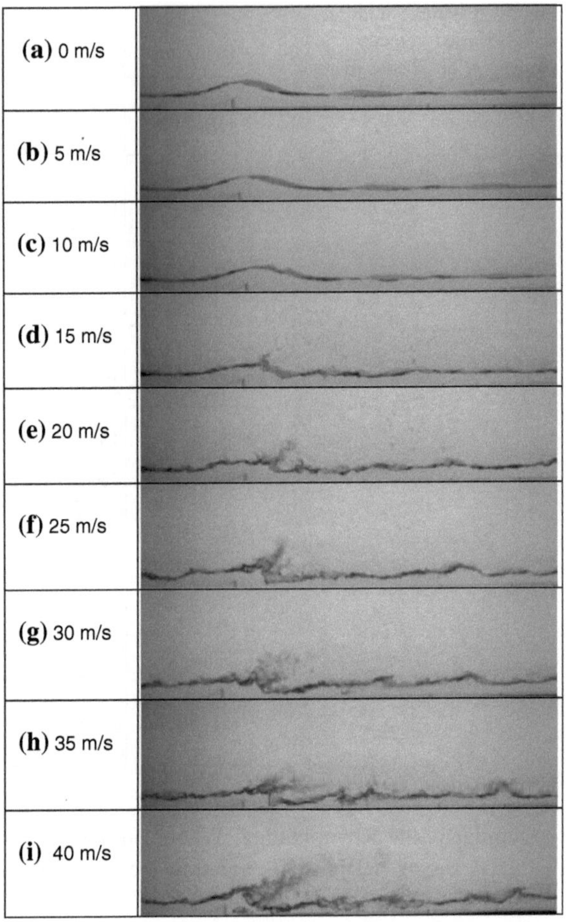

determined for the chamber inlet and outlet in terms of pixels after normalizing for perspective distortion. The normalization was based on the constant chamber height and center chamber height pixel reading.

The inlet water velocity was calculated using the volumetric flow rate, chamber depth, and measure approximate boundary height. The outlet water velocity, which is the mean water velocity beneath the boundary surface, was calculated using high-speed photography frames. The velocity was determined assuming mean velocity to be similar to the observed entrapped air bubble velocity. Figure 12 shows the first and eighth frame used for the mean outlet velocity for the 40 m/s trial. The outlet velocity is calculated by following the moving target for 0.01 m divided by the time of the recorded frames.

Similar treatment as seen in Fig. 12 was given for air velocities of 10, 15, 20, and 30 m/s. The water breakup ratios for these flow rates were calculated and are presented in Fig. 13. The breakup ratio in percent form is the mass flow rate of water when leaving the chamber in two-phase flow relative to the inlet water mass flow rate.

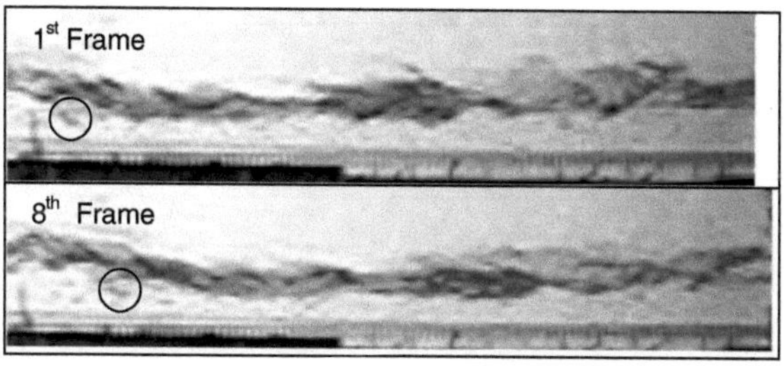

Fig. 12 Frames taken at 1,000 fps for exit velocity; mean air velocity 40 m/s (Amano et al. 2014a, b)

The increase in the breakup ratio with increased air velocity is displayed in Fig. 13. The breakup ratio of 3.0 % for air velocity of 10 m/s is not likely due to actual water breakup, but rather to error in calculation of boundary height and mean exit velocity. Figure 14 displays a distribution of experimental breakup and surface interface. The breakup phenomenon is again noticed in the simulation distributions of Fig. 16a, b, respectively.

In Fig. 14b–d liquid breakup is seen around the wedge location. Figure 14 is the distribution concentration displaying the concentration level of liquid water particles from the liquid and two-phase interface. This is graphically demonstrated in Fig. 15 with water droplet concentration range highest at the boundary interface depicted by the point along the arrow. The concentration drops with increase of distance from the interface. This is also seen in the concentration distribution slope where higher breakup is seen with a shallower slope from the two-phase boundary.

Figure 16 presents the time distributions of simulation for flow simulations with air velocities of 30 and 40 m/s.

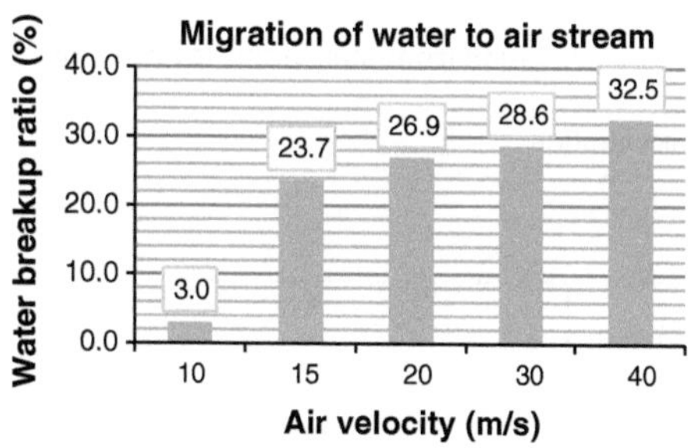

Fig. 13 Migration of liquid water leaving chamber in two-phase flow relative to the water inlet rate (Amano et al. 2014a, b)

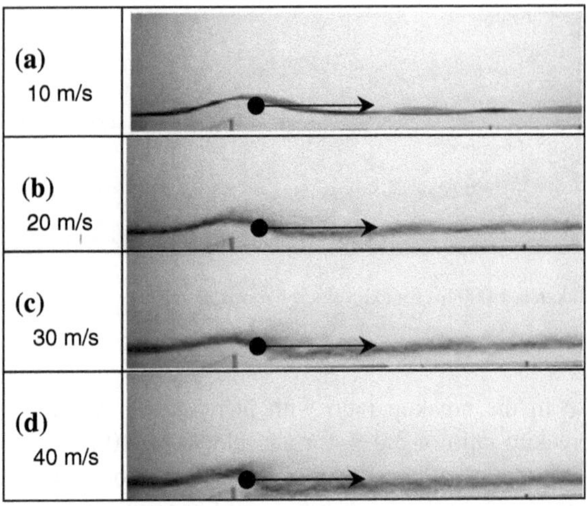

Fig. 14 Experimental time distribution (Amano et al. 2014a, b)

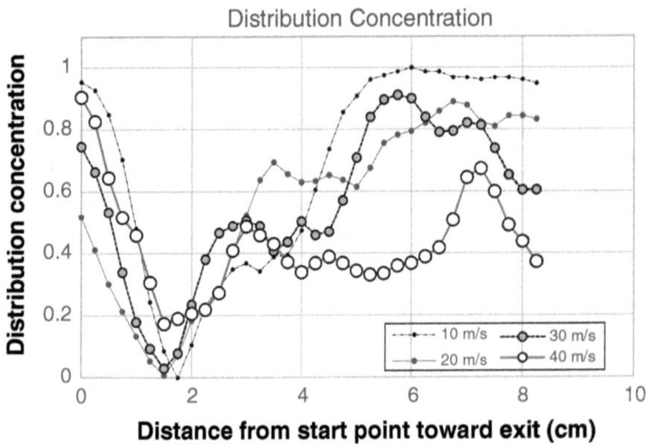

Fig. 15 Distribution concentration

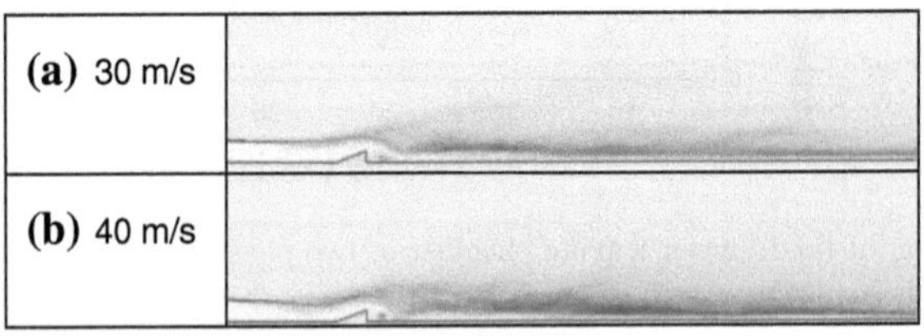

Fig. 16 Simulation time distribution (Amano et al. 2014a, b)

7.2 Simulation Result and Comparison

Figure 17 shows the VF of two-phase fluid of V_{air} = 40 m/s at physical time at 4 s after the both air and water start flowing from the left inlet of the channel domain by computing with three different turbulence models: k-ε, SST k-ω, and RSM. From the figure it can be seen that the RSM demonstrates the highest rate of a breakup behavior, SST k-ω has medium rate of breakups, and k-ε model gives the least amount of breakups among all the three turbulence models. It is difficult to tell which turbulence model best matches the experimental due to the nature of transient states; however, the experiment results will be clear after superposition treatment of the ramp region. Treatment of the superposition of the turbulent for the ramp region is given in the paragraph on the prehydraulic jump angle. Figure 18 summarizes the averaged boundary length for all three different turbulence models, with boundary length readings for k-ε, SST k-ω, and RSM of 11.85, 12.79, and 12.87 m, respectively. The RSM and SST k-ω show a similar boundary length value whereas the k-ε has the shortest boundary length which indicates that the k-ε model produces fewer breakups than the other two models. Standard deviations from the 260 time step of the respective models in the boundary length are 0.41, 0.40, and 0.61 m, as indicated by the error bars shown in Fig. 18. Comparing the RSM and SST k-ω despite having a similar boundary length, the standard deviation of the RSM is greater than the SST k-ω. This indicates that the two-phase flows of the RSM run more unsteady than that of the SST k-ω. The k-ε model has a similar standard deviation value with the SST k-ω

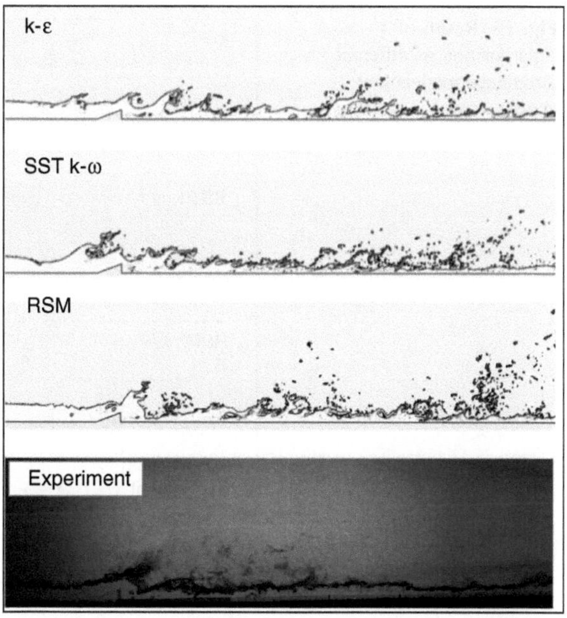

Fig. 17 Result of different turbulence model and experiment of V_{air} = 40 m/s at physical time T = 4.0 s (Amano et al. 2014a, b)

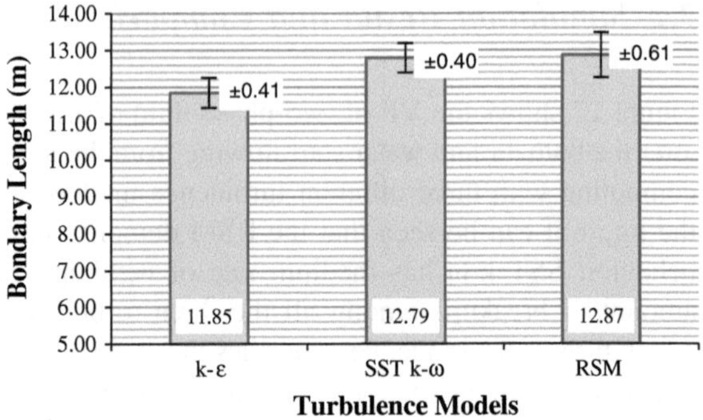

Fig. 18 Boundary length of three models with $V_{air} = 40$ m/s (Amano et al. 2014a, b)

Figure 19 shows the superposition of VF with the three turbulence model and result from experiment. The k-ε model shows the least thick breakup band at the region downstream from the ramp. The SST k-ω has a relatively uniformly distributed breakup band downstream of the ramp and this explains why the SST k-ω model produced a less standard deviation of boundary length due to lower unsteady breakup phenomenon. A wider and darker band downstream of the ramp is seen in the case of the RSM model, which could be interpreted as a more uniform breakup process occurring around region (i). The superposition of experimental images does not show the probability distribution clearly due to the nonuniform distributed light source; however, the profile of the two-phase boundary in the ramp vicinity is similar to RSM's profile.

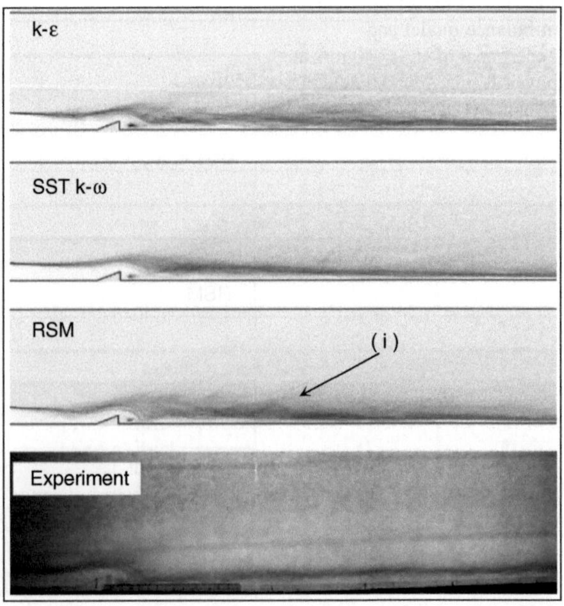

Fig. 19 Result of superposition of different turbulence models and experiments, $V_{air} = 40$ m/s (Amano et al. 2014a, b)

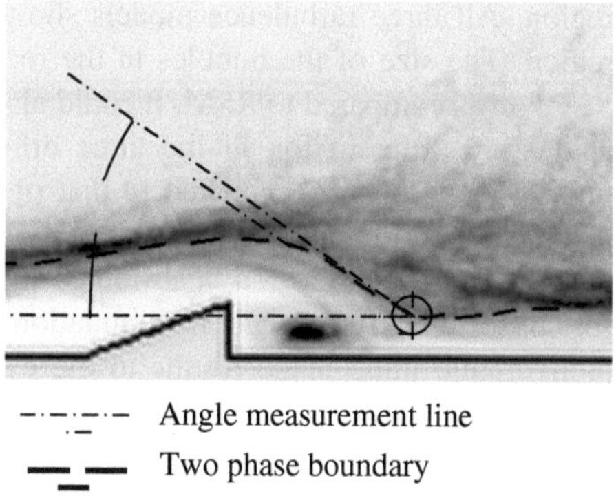

Fig. 20 Measurement of prehydraulic jump angle

Figure 20 shows the method to measure the prehydraulic jump angle near the ramp area. The angle measures start from the lowest point of two-phase boundary after the ramp, and measure the angle between the tangent line of two-phase boundary and the horizontal line. Figure 21 shows a close-up view of the ramp

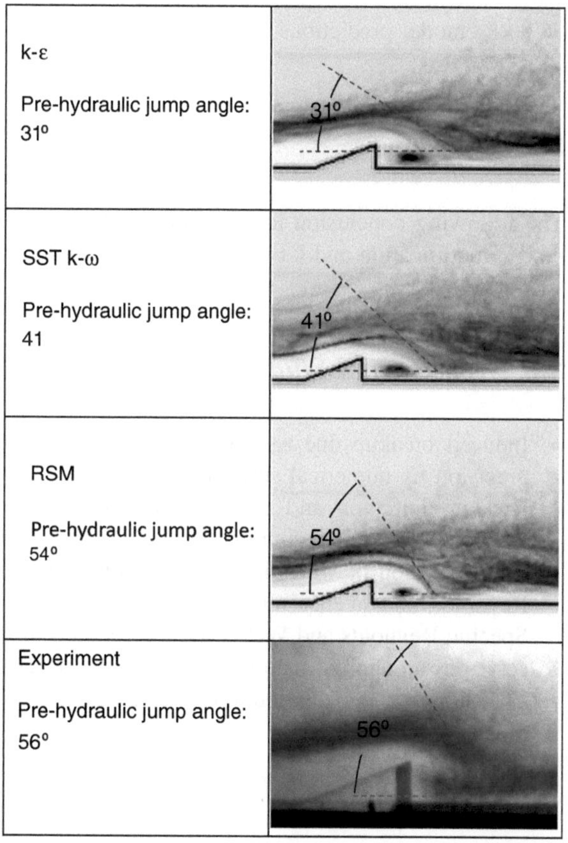

Fig. 21 Prehydraulic jump angles of all models at $V_{air} = 40$ m/s (Amano et al. 2014a, b)

region. All three turbulence models show separation bubbles in the recirculation region. The size of the bubbles in the recirculating regions are larger for k-ε and SST k-ω as compared to RSM. It could also be observed that the prehydraulic jump angles are different for all the three different models' cases. The RSM has the highest angle of 54° compared to that of the SST k-ω 41° and k-ε 31°, and it is significantly different with the computation by the RSM compared to the other two models. This phenomenon is attributed to the feature of the RSM, which accounts for anisotropic turbulence in simulation predictions. Comparing the numerical prehydraulic jump angle results to the experimental angle of 56°, it is clear that RSM in this sense gives a more accurate representation than the other models for complex flow prediction.

RSM is more accurate at solving complex flow fields than the k-ε and SST k-ω in 2D turbulence models since RSM solves transport equations for all components of the specific Reynolds stress tensors directly. Therefore, RSM can more accurately simulate rapid changes in strain rate or anisotropic effects due to swirling motion off the ramp's trailing edge. Since RSM can more accurately predict the irregular unsteady flow details, this explains the longer boundary length and higher standard deviation result from Fig. 18. RSM's prediction of the prehydraulic jump angle more accurately represents the experimental angle as compared to the k-ε and SST k-ω model predictions.

8 Conclusions

The following conclusion are presented below as preliminary to using geometric nozzle modification and CFD for improving SRM efficiency in SRM design

8.1 Experimental Study

- Induced breakup due to modified geometry is predictable to some degree of precision by numerical simulation.
- Breakup initiation and intensity can be represented by a geometric-specific Reynolds and Weber numbers.
- The migration of water into the air stream increases with increase in specific Reynolds and Weber numbers. Significant increase in breakup was seen at Specific Reynolds and Weber numbers values of 243 and 80, respectively. This serves as an approximate threshold for the onset of water breakup.
- The slope of the distribution concentration is shallow corresponding to the drastic breakup behavior.

8.2 Numerical Study

- The breakup length computed with the RSM is the highest compared to k-ε and SST k-ω model. The RSM also gives the highest standard deviation value among all the turbulence models due to the fact that the RSM can predict more faithfully the unsteady nature of the flow. The SST k-ω model shows the breakup length value similar to the RSM; however, the standard deviation value is lower and behaves more similar to the computations with the k-ε model.
- It could also be observed from the method of superposition that the RSM produces a broader breakup distribution band. The SST k-ω model shows a similar behavior but it results in a more uniformly distributed breakup pattern in the region downstream of the ramp. And for k-ε model, the breakup distribution band is much thinner compared with the other two models.
- Observing the prehydraulic jump angles, the experiment result shows the pre-hydraulic jump angles of 56°. RSM predicts significantly high value with 54° compared to the rest of SST k-ω and k-ε that are around 41° and 31°, respectively.
- Overall, RSM shows the capability of predicting irregular, transient state and complex flow in the two-phase flow simulation when comparing the results of k-ε and SST k-ω model.

References

Amano RS, Yi-Hsin Y, Hamman M (2014) Study of two phase flow breakup behavior of application on solid rocket motor nozzle. In: 4th joint US-European fluids engineering summer meeting (FEDSM2014), pp 21256, Chicago, IL

Amano RS, Yi-Hsin Y, Hamman M, Kristopher, Joshua (2014) Experimental investigation of liquid phase breakup in solid fuel rockets. In: 4th joint US-European fluids engineering summer meeting (FEDSM2014), pp 21224, Chicago, IL

Bandera, Maggi, Deluca (2011) Agglomeration of aluminized solid rocket propellants. In: AIAA/ASME/SAE/ASEE joint propulsion conference and exhibit, Denver, CO

Borass S (1984) Modeling slag deposition in the space shuttle solid rocket motor. J Spacecraft Rockets 27:47–54

Hess, Chen, Acosta, Brent, Fendell (1992) Effect of aluminized-grain design on slag accumulation. J Spacecraft Rocket 29:697–703

Hirt CW & Nichols BD (1981) "Volume of Fluid (VOF) method for the dynamics of free boundaries". J Comput Phys 39:201–225

Holtzmann (1964) Introduction—the nature of an advanced propellant. (Aerojet-General Corporation Von Karman Center.) Retrieved from www.web.anl.gov/PCS/acsfuel/preprint%20archive/Files/Volumes/Vol09-1.pdf

Launder, Reece, Rodi (2006) Progress in the development of a reynolds-stress turbulent closure. J Fluid Mech 63(3):537–566

Menter (1994) Two-equation eddy-viscosity turbulence modeling for engineering applications. AIAA J 32(8):1598–1605

Google (n.d.) *Picasa*. Retrieved from http://picasa.google.com/

Rapp (1968) High energy-density liquid rocket fuel performance. Sverdrup Technology Inc., NASA Lewis Research Center, Brook Park

Rodi (1991) Experience with two-layer models combining the k-ε model with a one-equation model near the wall. In: 29th aerospace sciences meeting, Reno, NV

Salita (1995) Deficiencies and requirements in modeling of slag generation in solid rocket motors. J Propul Power 11(1):10–23

Thankre, Yang (2009) Chemical erosion of refractory-metal nozzle inserts in solid-propellant rocket motors. Propul Power 25(1):40–50

Wong (1968) Solid roket nozzle design summary. In: 4th AIAA propulsion joint specialist conference, Cleveland, OH

Xiao Y, Amano RS (2006) Aluminized composite solid propellant particle path in combustion chamber of solid rocket motor. In: Advances in fluid mechanics. WIT Press, UK, pp 153–164

Xiao, Amano RS, Cai, Li (2003) Particle velocity on solid-propellant surface using X-ray real time radiography. AIAA J 41(9):1763–1770

Xiao, Amano RS, Cai, Li (2005) A new method to determine the velocities of particles on a solid propellant surface. ASME J Heat Transf 127:1057–1061

Index

A.K. Agarwal et al. (eds.), *Novel Combustion Concepts for Sustainable Energy Development*, DOI 10.1007/978-81-322-2211-8

MIX
Papier aus verantwortungsvollen Quellen
Paper from responsible sources
FSC® C105338
FSC
www.fsc.org

If you have any concerns about our products,
you can contact us on
ProductSafety@springernature.com

In case Publisher is established outside the EU,
the EU authorized representative is:
Springer Nature Customer Service Center GmbH
Europaplatz 3, 69115 Heidelberg, Germany

Printed by Libri Plureos GmbH
in Hamburg, Germany